Rings, Polynomials, and Modules

Marco Fontana • Sophie Frisch • Sarah Glaz
Francesca Tartarone • Paolo Zanardo
Editors

Rings, Polynomials, and Modules

Springer

Editors
Marco Fontana
Dipartimento di Matematica e Fisica
Università degli Studi Roma Tre
Rome, Italy

Sophie Frisch
Department of Analysis and Number Theory
Graz University of Technology
Graz, Austria

Sarah Glaz
Department of Mathematics
University of Connecticut
Storrs, CT, USA

Francesca Tartarone
Dipartimento di Matematica e Fisica
Università degli Studi Roma Tre
Rome, Italy

Paolo Zanardo
Dipartimento di Matematica
Università di Padova
Padova, Italy

ISBN 978-3-319-65872-8 ISBN 978-3-319-65874-2 (eBook)
https://doi.org/10.1007/978-3-319-65874-2

Library of Congress Control Number: 2017957649

Printed on acid-free paper

This Springer imprint is published by Springer Nature
The registered company is Springer International Publishing AG
The registered company address is: Gewerbestrasse 11, 6330 Cham, Switzerland

Preface

This volume is the outcome of two conferences: "Recent Advances in Commutative Ring and Module Theory" held in Bressanone/Brixen, Italy, June 13–17, 2016; and "Conference on Rings and Polynomials" held in Graz, Austria, July 3–8, 2016. The volume contains contributed as well as invited papers by the speakers at these conferences, and a small collection of invited papers by some of the leading experts in the area, carefully selected for the impact of their research on the major themes of the conferences.

The aim of the meetings was to present recent progress and new trends in the area of commutative algebra, with emphasis on commutative ring theory, module theory, and integer-valued polynomials along with connections to algebraic number theory, algebraic geometry, topology, and homological algebra. The wide range of topics is reflected in the table of contents of this volume.

The two conferences brought together over one hundred mathematicians from over 20 countries—renowned researchers as well as promising young newcomers—in a pleasant and peaceful atmosphere that engendered many fruitful collaborations.

In addition to the conference participants and authors of papers, a number of other people helped make these conferences and this volume of proceedings possible. Among those we count the organizing and scientific committees of both conferences. The organizing committee of the Bressanone conference consisted of Florida Girolami, Francesca Tartarone, and Paolo Zanardo, while the scientific committee included Valentina Barucci, Dikran Dikranjan, Brendan Goldsmith, Evan Houston, Bruce Olberding, Francesca Tartarone, and Paolo Zanardo. The organizing committee of the Graz conference consisted of Sophie Frisch, Carmelo Finocchiaro, and Roswitha Rissner, while the scientific committee included Karin Baur, Jean-Luc Chabert, Marco Fontana, Alfred Geroldinger, Sarah Glaz, and Irena Swanson. We wish to thank them all for their efforts, without which these conferences would not have taken place and this volume would not have seen the light of day. In addition, the Graz conference editors wish to thank the departmental secretary

Hermine Panzenböck for administrative support and many students for technical support. The Bressanone conference editors wish to extend special thanks to Marco Fontana, Stefania Gabelli, and Luigi Salce for useful suggestion.

We also thank the many organizations who sponsored these conferences and, most importantly, made it possible to provide support for graduate students and mathematicians not supported by their institutions. The Bressanone conference was sponsored by Istituto Nazionale di Alta Matematica (INdAM), the departments of mathematics of Università degli Studi di Padova and Sapienza Università di Roma, and the department of mathematics and physics of Università degli Studi Roma Tre. The Graz conference was sponsored by the Austrian Science Fund (FWF), the Austrian Mathematical Society, the province of Styria, and the faculty of mathematics and physics of Technische Universität Graz.

Last, but not least, we thank the editorial staff of Springer, in particular Elizabeth Loew, for their cooperation, hard work, and assistance with this volume.

Rome, Italy	Marco Fontana
Graz, Austria	Sophie Frisch
Storrs, CT, USA	Sarah Glaz
Rome, Italy	Francesca Tartarone
Padova, Italy	Paolo Zanardo
July 2017	

Contents

Reducing Fractions to Lowest Terms

Daniel D. Anderson and Erik Hasse

Abstract The purpose of this paper is to investigate putting or reducing a fraction to lowest terms in a general integral domain. We investigate the integral domains in which every fraction can be (uniquely) put in or reduced to lowest terms.

Keywords ACCP • Atomic domain • GCD domain • Lowest terms • gcd • Weak gcd

Subject Classifications Primary 13G05, Secondary 13A05, 13F15

We are all familiar with reducing fractions to lowest terms over the integers or polynomials over a field. The purpose of this paper is to study this in the context of general integral domains. We investigate when a fraction a/b can be put in lowest terms c/d (i.e., $a/b = c/d$ where c and d are relatively prime) or reduced to lowest terms $(\frac{a}{d})/(\frac{b}{d})$ (i.e., $(\frac{a}{d})/(\frac{b}{d})$ is in lowest terms for some common divisor d of a and b) and when the lowest terms representation for a/b is "unique". Of particular interest are the integral domains in which every fraction can be reduced to lowest terms.

Throughout D will be an integral domain with quotient field K. Let $a, b \in D^* := D - \{0\}$. We denote the gcd of a and b by $[a, b]$, if it exists. Of course, $[a, b]$ is only unique up to a unit factor. We write $[a, b] = 1$ ($[a, b] \neq 1$) if a and b are (not) relatively prime. A common divisor d of a and b is a *weak gcd for a and b* if $\left[\frac{a}{d}, \frac{b}{d}\right] = 1$. And D is a *(weak) GCD domain* if every pair $a, b \in D^*$ has a (weak) gcd. For a nonzero fractional ideal I of D, $I^{-1} := \{x \in K | xI \subseteq D\}$ and $I_v := (I^{-1})^{-1} = \cap\{Dx | x \in K \text{ with } I \subseteq Dx\}$. If $(a, b)_v = (d)$ (or equivalently $\text{lcm}(a, b) = \frac{ab}{d}$), then $[a, b] = d$, but not necessarily conversely (see Example 2). However, if D is a GCD domain with $[a, b] = d$, then $(a, b)_v = (d)$. We remark that the following three conditions are equivalent: (1) $\text{lcm}(a, b)$ exists, (2) $(a) \cap (b)$ is principal, and (3) $(a, b)_v$ is principal. And in this case $((a) \cap (b))\,(a, b)_v = (a)(b)$.

D.D. Anderson (✉) • E. Hasse
Department of Mathematics, The University of Iowa, Iowa City, IA, USA
e-mail: dan-anderson@uiowa.edu; erik-hasse@uiowa.edu

M. Fontana et al. (eds.), *Rings, Polynomials, and Modules*,
https://doi.org/10.1007/978-3-319-65874-2_1

If $(a, b) = (d)$, then $(a, b)_v = (d)$, but not conversely. A nonzero nonunit a of D is *irreducible* or an *atom* if $a = bc$ implies b or c is a unit and D is *atomic* if each nonzero nonunit of D is a finite product of atoms. An integral domain D satisfies the *ascending chain condition on principal ideals* (*ACCP*) if every ascending chain $(a_1) \subseteq (a_2) \subseteq \cdots$ of principal ideals of D stabilizes. And D is a *Bezout domain* if every finitely generated ideal, equivalently, every two-generated ideal of D, is principal. Thus a Bezout domain is a GCD domain in which the gcd for each pair a, b is a linear combination of a and b.

We begin with the following definitions.

Definition 1 Let D be an integral domain and let $a, b, c, d, e, f \in D^*$. We say that a/b *can be put in the form* c/d if $a/b = c/d$ and that a/b *can be reduced to the form* c/d if there is a common divisor e of a and b with $c = \frac{a}{e}$ and $d = \frac{b}{e}$. The fraction a/b is in (*strong, resp., absolute*) *lowest terms* if $[a, b] = 1$ ($(a, b)_v = D$, resp., $(a, b) = D$). Thus a/b can be put in lowest terms if $a/b = c/d$ where $[c, d] = 1$ and a/b can be reduced to lowest terms if $a/b = c/d$ where $c = \frac{a}{e}$ and $d = \frac{b}{e}$ for some common divisor e of a and b and $[c, d] = 1$. We will then sometimes say that c/d is a (reduced) lowest terms for a/b. Similar statements hold for strong and absolute lowest terms. The fraction a/b has *essentially unique* (*strong, resp., absolute, reduced*) *lowest terms* if there exists at least one c/d in (strong, resp., absolute, reduced) lowest terms with $a/b = c/d$ and if $a/b = e/f$ where e/f is in (strong, resp., absolute, reduced) lowest terms, then $e = uc$ and $f = ud$ for some unit u of D.

Remark 1 Let $a, b \in D$ with b nonzero. There is some ambiguity in the notation a/b as to whether a/b just denotes an element of K or the particular representation of that element. Indeed, an element $x \in K$ has many representations in the form a/b with $a/b = c/d \Leftrightarrow ad = bc$ ($a, b, c, d \in D$, b, d nonzero). However, when we write a/b we will usually mean the particular representation, even though we write $a/b = c/d$ to mean they are equal as an element of K, i.e., $ad = bc$.

Definition 2 The integral domain D is a *lowest terms* (*LT*) *domain* (resp., *reduced lowest terms* (*RLT*) *domain*) if each nonzero fraction a/b ($a, b \in D^*$) can be put in (resp., reduced to) lowest terms. And D is a *unique lowest terms* (*ULT*) *domain* if each nonzero fraction a/b ($a, b \in D^*$) has essentially unique lowest terms.

Remark 2 In an obvious way we could have defined the following integral domains: unique reduced lowest terms domain, strong (resp., absolute) lowest terms domain, unique strong (resp., absolute) lowest terms domain, reduced strong (resp., absolute) lowest terms domain, and unique reduced strong (resp., absolute) lowest terms domain. The reason we have not is because by Theorem 1 (5, resp., 6) they (resp., the last four) are all equivalent to the integral domain being a GCD domain (resp., Bezout).

Remark 3 So far we have only considered nonzero fractions a/b ($a, b \in D^*$). Suppose that $a = 0$ and consider $0/b$ where $b \in D^*$. Since $[0, b] = b$, $(0, b)_v = (b)$ and $(0, b) = (b)$, we see (with the obvious extension of the definitions in Definition 1) that $0/b$ can be reduced to $0/1$ and $0/b$ has essentially unique (strong resp., absolute) lowest terms $0/1$. Thus there is no loss in generality in only considering nonzero fractions.

We next determine when a fraction can be put in or reduced to (strong, absolute) lowest terms.

Theorem 1 *Let D be an integral domain and let $a, b, c, d, e, f \in D^*$.*

1. *a/b can be put in lowest terms if and only if there exists an $s \in D^*$ so that sa and sb have a weak gcd. If d is a weak gcd for sa and sb, then $a/b = \left(\frac{sa}{d}\right) / \left(\frac{sb}{d}\right)$ and the last fraction is in lowest terms. So D is an LT domain if and only if for each $a, b \in D^*$, there exists $s \in D^*$ so that sa and sb have a weak gcd.*
2. *a/b can be reduced to lowest terms if and only if a and b have a weak gcd. If a and b have a weak gcd d, then $a/b = \left(\frac{a}{d}\right) / \left(\frac{b}{d}\right)$ and the last fraction is in lowest terms. So D is an RLT domain if and only if D is a weak GCD domain.*
3. *The following are equivalent:*

   a. *$[a, b]$ exists, and*
   b. i. *If c is a common divisor of a and b, then $\left(\frac{a}{c}\right) / \left(\frac{b}{c}\right)$ can be reduced to lowest terms and*
      ii. *a/b has essentially unique reduced lowest terms.*
4. *a/b is in strong lowest terms if and only if $a/b = c/d$ implies there exists $e \in D^*$ with $c = ea$ and $d = eb$.*
5. a. *The following are equivalent:*

      i. *$(a, b)_v$ is principal (or equivalently, $(a) \cap (b)$ is principal or $\mathrm{lcm}(a, b)$ exists),*
      ii. *a/b can be reduced to strong lowest terms, and*
      iii. *a/b can be put in strong lowest terms.*

   If $(a, b)_v = (d)$, then $a/b = \left(\frac{a}{d}\right) / \left(\frac{b}{d}\right)$ where the last fraction is in strong lowest terms. Moreover, this strong lowest terms representation is unique in the following sense. If $a/b = e/f$ where $[e, f] = 1$, then $e = u\left(\frac{a}{d}\right)$ and $f = u\left(\frac{b}{d}\right)$ where u is a unit of D. Hence e/f is actually a strong lowest terms representation for a/b.

   b. *For the integral domain D, the following are equivalent:*

      i. *G is a GCD domain,*
      ii. *Every nonzero fraction of D can be reduced to strong lowest terms,*
      iii. *Every nonzero fraction of D can be put in strong lowest terms,*
      iv. *Every nonzero fraction of D has a essentially unique reduced lowest terms.*
      v. *D is an RLT domain and a ULT domain.*
6. a. *The following are equivalent:*

      i. *(a, b) is principal,*
      ii. *a/b can be reduced to absolute lowest terms, and*
      iii. *a/b can be put in absolute lowest terms.*

If $(a,b) = (d)$, *then* $a/b = (\frac{a}{d})/(\frac{b}{d})$ *where the last fraction is in absolute lowest terms. Moreover, this absolute lowest terms representation is unique in the following sense. If* $a/b = e/f$ *where* $[e,f] = 1$, *then* $e = u(\frac{a}{d})$ *and* $f = u(\frac{b}{d})$ *where* u *is a unit of D. Hence* e/f *is actually an absolute lowest terms representation for* a/b.

b. *For an integral domain D, the following are equivalent:*

   i. *D is a Bezout domain,*
   ii. *Every nonzero fraction of D can be reduced to absolute lowest terms,*
   iii. *Every nonzero fraction of D can be put in absolute lowest terms.*

Proof (1) Suppose there exists an $s \in D^*$ with sa and sb having weak gcd d. Then $\left[\frac{sa}{d}, \frac{sb}{d}\right] = 1$ and $a/b = sa/sb = \left(\frac{sa}{d}\right) / \left(\frac{sb}{d}\right)$. So a/b can be put in lowest terms. Conversely, suppose that a/b can be put in lowest terms c/d. Now $a/b = c/d$ implies $ad = bc$ and so $a|bc$. Thus a is a common divisor of ac and bc and $\left[\frac{ac}{a}, \frac{bc}{a}\right] = [c,d] = 1$, i.e., a is weak gcd of ac and bc. The last statement is now immediate.

(2) Note that d is a weak gcd for a and b if and only if d is a common divisor of a and b with $\left[\frac{a}{d}, \frac{b}{d}\right] = 1$. This just says that $a/b = \left(\frac{a}{d}\right) / \left(\frac{b}{d}\right)$ where the last fraction is in lowest terms. This proves the first statement and the second statement is now immediate.

(3) $(a) \Rightarrow (b)$ If $[a,b]$ exists and c is a common divisor of a and b, then $\left[\frac{a}{c}, \frac{b}{c}\right]$ exists and hence is the unique weak gcd for $\frac{a}{c}$ and $\frac{b}{c}$. Then apply (2).

$(b) \Rightarrow (a)$ Let $a,b \in D^*$. Since a/b can be reduced to lowest terms, by (2), a and b have a weak gcd d. We show that $[a,b] = d$. Certainly d is a common divisor of a and b. Suppose e is a common divisor of a and b. Then $\left(\frac{a}{e}\right) / \left(\frac{b}{e}\right)$ can be reduced to lowest terms, so again by (2) there is an $f \in D^*$ with $\left[\frac{a}{ef}, \frac{b}{ef}\right] = 1$. Now $\left(\frac{a}{d}\right) / \left(\frac{b}{d}\right) = a/b = \left(\frac{a}{ef}\right) / \left(\frac{b}{ef}\right)$ where the first and third fractions are in lowest terms. By uniqueness $\frac{a}{d} = u\frac{a}{ef}$ for some unit u. Hence $d = u^{-1}ef$, so $e|d$. Thus $[a,b] = d$. (We have shown that for $a,b \in D^*$, $[a,b]$ exists if and only if a and b have a unique (up to associates) weak gcd and for every $c|a,b$, $\frac{a}{c}$ and $\frac{b}{c}$ have a weak gcd.)

(4) $(\Rightarrow) a/b = c/d$ implies $ad = bc$, so $a|bc$. Since $(a,b)_v = D$, $a|c$ (see Remark 4). So $c = ea$ for some $e \in D^*$ and hence $d = \frac{bc}{a} = eb$. $(\Leftarrow)$ Suppose that $(a,b) \subseteq (\alpha/\beta)$ for $\alpha, \beta \in D^*$. Then $\beta(a,b) \subseteq (\alpha)$, so $a\beta = c\alpha$ and $b\beta = d\alpha$ for some $c,d \in D^*$. So $a/b = a\beta/b\beta = c\alpha/d\alpha = c/d$. Hence $c = ax$, and $d = bx$ for some $x \in D^*$. Then $a\beta = ax\alpha \Rightarrow \beta = x\alpha \Rightarrow \alpha/\beta = 1/x$. So $D \subseteq (\alpha/\beta)$. Hence $(a,b)_v = D$.

(5) (a) (i)$\Rightarrow$(ii) Suppose $(a,b)_v = (d)$. Then $d|a$ and $d|b$ and $\left(\frac{a}{d}, \frac{b}{d}\right)_v = \frac{1}{d}(d) = D$; so $a/b = \left(\frac{a}{d}\right) / \left(\frac{b}{d}\right)$ where the last fraction is in strong lowest terms. (ii)$\Rightarrow$(iii) Clear. (iii)$\Rightarrow$(i) Suppose $a/b = c/d$ where $(c,d)_v = D$. Now $ad = bc$ and $(c,d)_v = D$ implies $(a) = (ac,ad)_v = (ac,bc)_v = (a,b)_v c$; so $(a,b)_v = \left(\frac{a}{c}\right)$ is principal. This proves the equivalence of (i)–(iii) and the second statement. Next suppose $(a,b)_v = (d)$ and $a/b = e/f$ where $[e,f] = 1$. Now $e/f = a/b = \left(\frac{a}{d}\right) / \left(\frac{b}{d}\right)$

where $[e,f] = 1$ and $\left(\frac{a}{d}, \frac{b}{d}\right)_v = D$. By (4) $e = \frac{a}{d}g$ and $f = \frac{b}{d}g$ for some $g \in D^*$ So $1 = [e,f] = \left[\frac{a}{d}g, \frac{b}{d}g\right]$. Hence g must be a unit.

(b) The equivalence of (i)–(iii) and (i)⇒(iv),(v) follow from (a). (iv),(v)⇒(i) follows from (3).

(6) (a) This is similar to the proof of 5(a). Indeed, we can just delete the "subscript" v wherever it occurs.

(b) This follows from 6(a).

Remark 4 The proof of Theorem 1 (4) used the well-known fact that for $a, b, c \in D^*$ with $(a, b)_v = D$, then $a|bc \Rightarrow a|c$. For suppose $ar = bc$ for some $r \in D^*$. Then $(c) = c(a, b)_v = (ac, bc)_v = (ac, ar)_v \subseteq (a)$; so $a|c$. It is interesting to note that the converse is also true: If $a|bc \Rightarrow a|c$ for all $c \in D^*$, then $(a, b)_v = D$. As we will not need this result, the proof is left to the reader.

Thus it is not true in general that $a|bc$ with $[a, b] = 1$ implies $b|c$. The "proof" breaks down because $[a, b] = 1$ does not imply that $[ac, bc] = c$. If fact, for $a, b \in D^*$, $[ac, bc]$ exists for all $c \in D^*$ if and only if $(a, b)_v$ exists [1, Theorem 2.1]. It is easy to check that if $[ac, bc]$ exists, then $[a, b]$ exists and $[ac, bc] = [a, b]c$. If d is a (weak) gcd for a and b and $c|d$, then $\frac{d}{c}$ is a (weak) gcd for $\frac{a}{c}$ and $\frac{b}{c}$.

The above paragraphs explain why a/b having a strong lowest terms representation forces $(a, b)_v$ to be principal while a/b having a lowest terms representation does not force $[a, b]$ to exist.

Remark 5 An integral domain R is said to satisfy *Property D* if whenever $a, b, c \in R^*$ with $[a, b] = 1$ and $a|bc$, then $a|c$. Property D is equivalent to a number of other properties slightly weaker than being a GCD domain such as PSP2: $a, b \in R^*$ with $[a, b] = 1$ implies $(a, b)_v = R$ (also called Property λ in [5]). Property D implies that atoms are prime, so an atomic domain satisfying Property D is a UFD, and conversely. See [2] for a thorough investigation of these related properties. Via Theorem 1 (3)(a) the following are equivalent: (1) R satisfies PSP2, (2) if a fraction a/b $(a, b \in R^*)$ can be put in lowest terms, it can be put in strong lowest terms, and (3) any lowest terms representation of a fraction a/b $(a, b \in R^*)$ is actually a strong lowest terms representation.

Remark 6 R. Gilmer briefly considers fractions in (strong) lowest terms in [4, Exercise 5, p.183]. Let D be an integral domain and $a, b \in D^*$. There he defines a fraction a/b to be *irreducible* if $[a, b] = 1$ and to be in *canonical form* if $a/b = c/d$ for $c, d \in D^*$ implies there is a $x \in D^*$ with $c = ax$ and $d = bx$. The exercise asks to show that a/b is in canonical form if and only if $(a, b)_v = D$ which is our Theorem 1 (4) and that every fraction can be put in canonical form if and only if D is a GCD domain which is *(i)* ⇔ *(iii)* of our Theorem 1 (5)(b).

Remark 7 We can give a star-operation version of Theorem 1 (5,6). Recall that a *star-operation* $\star$ on D is a closure operation $\star$ on the set $F(D)$ of nonzero fractional ideals of D that satisfies $(aA)^\star = aA^\star$ and $(a)^\star = (a)$ for $a \in K^*$ and $A \in F(D)$. Examples of star-operations include the *v-operation* $A \to A_v$ and the *d-operation*

$A \to A_d = A$. For an introduction to star-operations, see [4, Section 32]. For $a, b \in D^*$, we say that a/b is in *$\star$-lowest terms* if $(a, b)^\star = D$. Then a/b can be put in (equivalently, reduced to) $\star$-lowest terms if and only if $(a, b)^\star$ is principal. Thus every fraction a/b can be put in (equivalently, reduced to) $\star$-lowest terms if and only if every nonzero doubly generated (equivalently, finitely generated) ideal A has $A^\star$ principal. Here Theorem 1 (5, resp., 6) is just the case where $\star = v$ (resp., d).

We next show the ubiquity of RLT domains.

Theorem 2 *Let D be an integral domain. If D is a GCD domain or satisfies ACCP, then D is a weak GCD domain and hence is an RLT domain.*

Proof The case where D is a GCD domain is immediate, so assume that D satisfies ACCP. Suppose there exists $a_0, b_0 \in D^*$, so that a_0, b_0 do not have a weak gcd. Then the set $S = \{(a) | a \in D^*,$ there exists a $b \in D^*$ so that a, b do not have a weak gcd$\}$ is a nonempty set of proper principal ideals. Let (a) be a maximal element of S. So there exists a $b \in D^*$ so that a, b do not have a weak gcd. In particular, $[a, b] \neq 1$. So there is a nonunit $e \in D^*$ with $e|a$ and $e|b$. But then $\frac{a}{e} \in D^*$ with $\left(\frac{a}{e}\right) \supsetneq (a)$. So either $\frac{a}{e}$ is a unit or $\left(\frac{a}{e}\right)$ is a proper principal ideal of D. Thus $\frac{a}{e}$ and $\frac{b}{e}$ have a weak gcd d, so $\left[\frac{a}{ed}, \frac{b}{ed}\right] = 1$. So ed is a weak gcd for a and b, a contradiction.

Corollary 1 *An integral domain D is a UFD if and only if D satisfies ACCP and D is a ULT domain.*

Proof ($\Rightarrow$) Suppose D is a UFD. It is well known that D satisfies ACCP and since a UFD is a GCD domain, D is a ULT domain by Theorem 1 (4). ($\Leftarrow$) Since D satisfies ACCP, D is an RLT domain by Theorem 2. So D is an RLT domain and a ULT domain. By Theorem 1 (4), D is a GCD domain. But a GCD domain satisfying ACCP is a UFD.

We next give an example of an integral domain that is not an LT domain. We later use this example to give an example (Example 4) of an integrally closed atomic domain that is not an LT domain.

Example 1 (An Integral Domain That is Not an LT Domain) Let D be the integral domain $k[X, Y, Z][\{\frac{X}{Z^n}, \frac{Y}{Z^n}\}_{n \geq 1}]$ where k is a field and X, Y, Z are indeterminates over k. Then we cannot write $X/Y = a/b$ where $a, b \in D^*$ with $[a, b] = 1$. For suppose $X/Y = a/b$ for $a, b \in D^*$. We can write $a = f/Z^m$ and $b = g/Z^n$ where $f, g \in k[X, Y, Z]^*$ and $m, n \geq 0$. Then $XgZ^m = YfZ^n$. Then $X|f$ and $Y|g$ in $k[X, Y, Z]$ and hence $\frac{X}{Z^m}|a$ and $\frac{Y}{Z^n}|b$ in D. But $Z|\frac{X}{Z^m}$ and $Z|\frac{Y}{Z^n}$ in D, so $Z|a$ and $Z|b$ in D. Hence $[a, b] \neq 1$.

By Corollary 1, any integral domain satisfying ACCP that is not a UFD is an RLT domain that is not a ULT domain. We next examine a concrete example.

Example 2 (An RLT Domain That is Not a ULT Domain) Let $D = k[X^2, X^3]$ where k is a field and X is an indeterminate over k. Then D is Noetherian and hence satisfies ACCP and thus is an RLT domain. Here X^2 and X^3 are irreducible, with $X^2 \cdot X^2 \cdot X^2 = X^3 \cdot X^3$, so R is not a UFD and hence not a ULT domain. Indeed $X^4/X^3 = X^3/X^2$

where $[X^3, X^4] = [X^2, X^3] = 1$. Now $[X^4, X^5] = X^2$, but $[X^5, X^6]$ does not exist. In fact, both X^2 and X^3 are weak gcds for X^5 and X^6 since $\left[\frac{X^5}{X^2}, \frac{X^6}{X^2}\right] = [X^3, X^4] = 1$ and $\left[\frac{X^5}{X^3}, \frac{X^6}{X^3}\right] = [X^2, X^3] = 1$. Moreover, X^6/X^5 can be reduced to both X^4/X^3 and X^3/X^2, each of which is in lowest terms, but there does not exist a unit $u \in D$ with $X^3 = uX^4$ and $X^2 = uX^3$. Here X^4/X^3 can be put in lowest terms form X^3/X^2, but cannot be reduced to the lowest terms form X^3/X^2. Note that $[X^2, X^3] = 1$, but $(X^2, X^3)_v = (X^2, X^3) \neq D$ and $(X^2) \cap (X^3) = (X^5, X^6)$ is not principal. In fact, by Theorem 1 (5)(a) we cannot write $X^3/X^2 = f/g$ where $f, g \in D^*$ with $(f, g)_v = D$.

We have made a distinction between putting a fraction in lowest terms and reducing a fraction to lowest terms. We now give an example of a fraction that can be put in lowest terms but cannot be reduced to lowest terms.

Example 3 (A Fraction That Can Be Put in, But Not Reduced to, Lowest Terms) Let $D = k[X, Y, Z, T][\frac{X}{T}, \frac{Y}{T}, \{\frac{X}{Z^n}, \frac{Y}{Z^n}\}_{n \geq 1}]$ where k is a field and X, Y, Z, and T are indeterminates over k. Then T is a weak gcd for X and Y, so X/Y can be reduced to lowest terms $\left(\frac{X}{T}\right) / \left(\frac{Y}{T}\right)$. Now the set of divisors of $\frac{X}{Z}$ (resp., $\frac{Y}{Z}$) is $\left\{uZ^n, \frac{uX}{Z^{n+1}} \mid u \in k^*, n \geq 0\right\}$ (resp., $\left\{uZ^n, \frac{uY}{Z^{n+1}} \mid u \in k^*, n \geq 0\right\}$). So any common divisor of $\frac{X}{Z}$ and $\frac{Y}{Z}$ is of the form uZ^n where $n \geq 0$ and $u \in k^*$. It follows that $\frac{X}{Z}$ and $\frac{Y}{Z}$ do not have a weak gcd in D and hence $\left(\frac{X}{Z}\right) / \left(\frac{Y}{Z}\right)$ cannot be reduced to lowest terms in D. However, $\left(\frac{X}{Z}\right) / \left(\frac{Y}{Z}\right)$ can be put in lowest terms $\left(\frac{X}{T}\right) / \left(\frac{Y}{T}\right)$. It is interesting to note that in the localization $D[T^{-1}]$ of D, X/Y cannot be put in lowest terms.

Now LT domains and weak GCD domains were introduced in [3] in the context of atomic factorization. Let D be an integral domain. Then a nonzero nonunit element a of D is *irreducible*, or an *atom*, if $a = bc$ for $b, c \in D$ implies b or c is a unit. And D is *atomic* if every nonzero nonunit of D is a finite product of atoms. It is well known that if D satisfies ACCP, then D is atomic, but the converse is false. It is easily shown that if D satisfies ACCP, then so does $D[X]$. This raised the question of whether D atomic implies $D[X]$ is atomic which was answered in the negative in [6]. (It is easy to see that if $D[X]$ satisfies ACCP (resp., is atomic), then D satisfies ACCP (resp., is atomic)). Recall that an integral domain D is *strongly atomic* if for $a, b \in D^*$, there exist atoms $a_1, \ldots, a_s (s \geq 0)$ and $c, d \in D^*$ with $[c, d] = 1$ and $a = a_1 \cdots a_s c$ and $b = a_1 \cdots a_s d$. Note that D satisfies ACCP $\Rightarrow$ $D[X]$ is atomic $\Rightarrow$ D is strongly atomic. The following result links these various concepts.

Theorem 3 *For an integral domain D the following are equivalent.*

1. *D is an atomic RLT domain.*
2. *D is an atomic weak GCD domain.*
3. *D is strongly atomic.*
4. *Every linear polynomial in D[X] is a product of atoms.*

Proof (1)⇔(2) Theorem 1 (2). (2)⇒(3) Let $a, b \in D^*$. So D a weak GCD domain gives $a = a'c, b = b'c$ where $[a', b'] = 1$. Since D is atomic, either c is a unit or c is a product of atoms.

(3)⇒(4) Let $aX + b \in D[X]$ be a linear polynomial, so $a \in D^*$. Suppose $b \neq 0$. Then $a = a_1 \cdots a_s c$ and $b = a_1 \cdots a_s d$ where the a_i's are atoms ($s \geq 0$) and $[c, d] = 1$, $c, d \in D^*$. Then $aX + b = a_1 \cdots a_s(cX + d)$ is a product of atoms. So suppose $b = 0$. Then aX is a product of atoms if and only if a is, so it suffices to show that D is atomic, i.e., strongly atomic ⇒ atomic. Let $a \in D^*$ be a nonunit. Write $a = a_1 \cdots a_s c$ and $a^2 = a_1 \cdots a_s d$ where $[c, d] = 1$. Then $a_1^2 \cdots a_s^2 c^2 = a^2 = a_1 \cdots a_s d$. So canceling gives $d = a_1 \cdots a_s c^2$. Thus $c|d$ and hence c is a unit. So a is a product of atoms.

(4)⇒(2) For a nonunit $a \in D^*$, aX a product of atoms implies a is a product of atoms, so D is atomic. For $a, b \in D^*$, $aX + b$ is a product of atoms, so $aX + b = a_1 \cdots a_s(cX + d)$ where each a_i is an atom and $[c, d] = 1$. Put $e = a_1 \cdots a_s$. So $[\frac{a}{e}, \frac{b}{e}] = [c, d] = 1$ and hence e is a weak gcd for a and b. (We note that the equivalence of (2) and (3) is given in [3, Theorem 1.3]).

While an integral domain satisfying ACCP is an RLT domain, we next give an example of an atomic domain that is not even an LT domain.

Example 4 (An Integrally Closed Atomic Domain That is Not an LT Domain) Let D be the integral domain $k[X, Y, Z][\{\frac{X}{Z^n}, \frac{Y}{Z^n}\}_{n\geq 1}]$ where k is a field. From Example *1* we have $Z|a$ and $Z|b$ whenever $X/Y = a/b$ for $a, b \in D^*$. Let $A = \mathcal{A}^\infty(D)$ as in [6, Example 5.1]. There it is noted that A is integrally closed and atomic, in fact every reducible element of A is a product of two atoms. It is shown that X and Y do not have a weak gcd, so A is not a weak GCD domain, or equivalently, not an RLT domain. We prove the stronger result that if $X/Y = a/b$ where $a, b \in A$, then $Z|a$ and $Z|b$; so X/Y cannot be put in lowest terms in A. Thus A is not an LT domain. To prove this it suffices to prove the following. Let S be a subring of A containing D with the property that wherever $X/Y = a/b$ for $a, b \in S^*$, then $Z|a$ and $Z|b$. Then for $s \in S^*$ and indeterminate X_s, if $X/Y = a/b$ where $a, b \in S[X_s, s/X_s]^*$, then $Z|a$ and $Z|b$. With a change of notation, it suffices to prove the following. Let R be an integral domain and let $a, b \in R^*$. Suppose that $t \in R^*$ has the property that whenever $a/b = c/d$ for $c, d \in R^*$, then $t|c$ and $t|d$. Let $s \in R^*$ and X be an indeterminate over R. Suppose that $a/b = f/g$ for $f, g \in R[X, s/X]^*$. Then $t|f$ and $t|g$. Let $f = \frac{r_n' s^n}{X^n} + \cdots + \frac{r_1' s}{X} + r_0 + r_1 X + \cdots + r_m X^m$ and $g = \frac{t_n' s^n}{X^n} + \cdots + \frac{t_1' s}{X} + t_0 + t_1 X + \cdots + t_m X^m$. Then $a/b = f/g$ gives $ag = bf$. So equating coefficients gives $at_i' s^i = br_i' s^i$ and $at_i = br_i$. If $t_i' \neq 0$ (equivalently $r_i' \neq 0$), then $a/b = r_i'/t_i'$; so $t|r_i'$ and $t|t_i'$. If $t_i' = 0$ (equivalently $r_i' = 0$), then certainly $t|r_i'$ and $t|t_i'$. Likewise, $t|r_i$ and $t|t_i$. Hence $t|f$ and $t|g$.

We next consider the stability of the various types of "lowest terms" domains with respect to certain ring constructions. First, none of the "lowest term" domains except Bezout domains are preserved by homomorphic image. Indeed, for any set of indeterminates $\{X_a\}$, $\mathbb{Z}[\{X_a\}]$ is a UFD, but any integral domain is a homomorphic image of a suitable $\mathbb{Z}[\{X_a\}]$. Also, as a field satisfies all the lowest term properties, none of the various "lowest term" domains are preserved by subrings. Example 4 shows that none of the "lowest term" domains except Bezout domains are preserved by overrings. For $k[X, Y, Z]$ is a UFD while its overring $K[X, Y, Z][\{\frac{X}{Z^n}, \frac{Y}{Z^n}\}_{n\geq 1}]$ is not even an LT-domain. We next show that the LT and RLT properties are not preserved by polynomial extensions. In fact, we give an example of an atomic RLT domain A with $A[X]$ neither atomic nor even an LT domain.

Example 5 (An Atomic RLT Domain A So That A[X] is Neither Atomic Nor an LT Domain) Let D be the integral domain $k[X_1, X_2, X_3, Z][\{\frac{X_1}{Z^n}, \frac{X_2}{Z^n}, \frac{X_3}{Z^n}\}_{n\geq 1}]$ where X_1, X_2, X_3, and Z are indeterminates over the field k. Let $A = \mathcal{A}_{\omega,2}(D)$ as in [6, Example 5.2]. There it is shown that A is an atomic domain, in fact, every nonzero nonunit of A is either irreducible or a product of two irreducibles, and that A is a weak GCD (= RLT) domain. But it is also shown that the polynomial ring $A[X]$ is not atomic and is not a weak GCD domain. Indeed, $X_1X + X_2$, and X_3 do not have a weak GCD in $A[X]$ since X_1, X_2, and X_3 do not have an MCD in A (i.e., an element d with $[\frac{X_1}{d}, \frac{X_2}{d}, \frac{X_3}{d}] = 1$). Thus $A[X]$ is not an RLT domain. We prove the stronger result that if $(X_1X + X_2)/X_3 = a/b$ for $a, b \in A[X]$, then $Z|a$ and $Z|b$; so $(X_1X + X_2)/X_3$ cannot be put in lowest terms. Thus $A[X]$ is not an LT domain. To prove this it suffices to prove the following. Let S be a subring of A containing D with the property that whenever $(X_1X+X_2)/X_3 = a/b$ for $a, b \in S[X]^*$, then $Z|a$ and $Z|b$. Then for any ideal I of S and indeterminate Y over $S[X]$, if $(X_1X+X_2)/X_3 = a/b$ for $a, b \in S[Y, IY^{-1}][X]^*$, then $Z|a$ and $Z|b$. Since $S[Y, IY^{-1}][X] = S[X][Y, IS[X]Y^{-1}]$, it suffices to prove the following. Let R be an integral domain and let $a, b \in R^*$. Suppose that $t \in R^*$ has the property that whenever $a/b = c/d$ for $c, d \in R^*$, then $t|c$ and $t|d$. Let I be a nonzero ideal of R and X an indeterminate over R. Suppose that $a/b = f/g$ for $f, g \in R[X, IX^{-1}]^*$. Then $t|f$ and $t|g$. The proof of this follows mutatis mutandis from the proof given for $f, g \in R[X, s/X]^*$ in Example 4.

Suppose $a, b \in D^*$. We can consider $a, b \in D[X]^*$. As such it is possible to put a/b in lowest terms $f(X)/g(X)$ where $f(X), g(X)$ are positive degree polynomials of $D[X]$; see the paragraph after Theorem 4. However, if a/b is reduced to lowest terms $f(X)/g(X)$ in $D[X]$, then $f(X), g(X) \in D^*$. Thus, if $D[X]$ is an RLT domain so is D. Now $D[X]$ is an "absolute LT domain" if and only if $D[X]$ is Bezout, or equivalently, D is a field. And $D[X]$ is a "strong LT domain", equivalently a GCD domain, if and only if D is a GCD domain, equivalently, a "strong LT domain". We have been unable to determine if $D[X]$ an LT domain implies that D is an LT domain. However, if $D[X]$ is a ULT domain, so is D; in fact, D must be a GCD domain. This is our next theorem.

Theorem 4 *For an integral domain, the following are equivalent:* (1) *$D[X]$ is a ULT domain,* (2) *$D[X]$ is a GCD domain, and* (3) *D is a GCD domain.*

Proof It is well known that (2)⇔(3) and (2)⇒(1) by Theorem 1 (5)(b); so it suffices to show that (1)⇒(3). We first show that D is a ULT domain. Suppose $a, b \in D^*$, so $a/b = f(X)/g(X)$ where $f(X), g(X) \in D[X]^*$ with $[f(X), g(X)] = 1$. Suppose that $\deg f(X) > 0$. Then for each $r \in D, f(X)/g(X) = a/b = f(X - r)/g(X - r)$ and $[f(X-r), g(X-r)] = 1$. We may assume that D is infinite, so there exists an $r_0 \in D^*$ with $f(X)$ and $f(X - r_0)$ not associate. But then $f(X)/g(X)$ and $f(X - r_0)/g(X - r_0)$

are two lowest term representations for a/b, contradicting our assumption that $D[X]$ is a ULT domain. Hence $0 = \deg f(X) = \deg g(X)$. So D is an LT domain and hence a ULT domain. By Theorem 1 (5)(b), to show that D is a GCD domain it suffices to show that D is an RLT domain. Let $a, b \in D^*$; so $a/b = c/d$ where $c, d \in D^*$ with $[c, d] = 1$. Now $a/b = c/d = (c + aX)/(d + bX)$. Since $D[X]$ is a ULT domain, we must have $[c + aX,\ d + bX] \neq 1$. Let $f(X) \in D[X]^*$ be a nonunit divisor of $c + aX$ and $d + bX$. If $\deg f(X) = 0$, then $f(X)$ is a nonunit of D dividing both c and d, a contradiction. So $\deg f(X) = 1$ say $f(X) = \alpha - \beta X$. So $\alpha \mid c$ and $\alpha \mid d$; hence α must be a unit of D, so we can take $\alpha = 1$. Then $c + aX = c(1 + \beta X)$ and $d + bX = d(1 + \beta X)$. Thus $a = c\beta$ and $b = d\beta$. So $a/b = c/d = (\frac{a}{\beta})/(\frac{b}{\beta})$. Thus D is an RLT domain.

Now it is quite possible for $a, b \in D^*$, to have $a/b = f(X)/g(X)$ where $f(X), g(X) \in D[X]^*$ with $[f(X), g(x)] = 1$ and $\deg f(X) = \deg g(X) > 0$. Indeed, suppose that $a/b = c/d = e/f$ where $c, d, e, f \in D^*$ with $[c, d] = 1 = [e, f]$ and c and e are not associates. Then $a/b = (c + eX)/(d + fX)$ where $[c + eX,\ d + fX] = 1$. Suppose that we take $D = k[X^2, X^3]$ as in Example 2. Let T be an indeterminate over D. Then $X^3/X^2 = (X^3 + X^4T)/(X^2 + X^3T)$ where $[X^3 - X^4T,\ X^2 + X^3T] = 1$.

The following diagram shows the relationships among the various integral domains we have discussed. None of the implications can be reversed with the possible exceptions of RLT⇒LT and GCD⇒ULT.

	UFD			Bezout
⇙		⇘		⇙
ACCP			GCD	
⇓		⇙		⇘
strongly atomic	⇒ RLT			ULT
⇓		⇘		⇙
atomic			LT	

We end with the following two questions.

Question 1 Must an LT domain be an RLT domain?

Question 2 Must a ULT domain be a GCD domain?

References

1. Anderson, D.D.: GCD domains, Gauss' Lemma, and content of polynomials. In: Chapman, S. Glaz, S. (eds.) Non-Noetherian Commutative Ring Theory. Mathematics and Its Applications, vol. 520, pp. 1–31. Kluwer Academic Publishers, Dordrecht, Boston, London (2000)
2. Anderson, D.D., Quintero, R.: Some generalizations of GCD domains. In: Anderson, D.D. (ed.) Factorization in Integral Domains, pp. 189–195. Marcel Dekker, New York (1997)
3. Anderson, D.D., Anderson, D.F., Zafrullah, M.: Factorization in integral domains. J. Pure Appl. Algebra **69**, 1–19 (1990)

4. Gilmer, R.: Multiplicative Ideal Theory. In: Queen's Papers in Pure and Applied Mathematics, vol. 90. Queen's University, Kingston, ON (1992)
5. Mott, J.L., Zafrullah, M.: On Prüfer v-multiplication domains. Manuscripta Math. **35**, 1–26 (1981)
6. Roitman, M.: Polynomial extensions of atomic domains. J. Pure Appl. Algebra **87**, 187–199 (1993)

Unique Factorization in Torsion-Free Modules

Gerhard Angermüller

Abstract A generalization of unique factorization in integral domains to torsion-free modules ("factorial modules") has been proposed by A.-M. Nicholas in the 1960s and subsequently refined by D.L. Costa, C.-P. Lu and D.D. Anderson, S. Valdes-Leon. The aim of this note is to prove new results of this theory. In particular, it is shown that locally projective modules, flat Mittag-Leffler modules and torsion-free content modules are factorial modules. Moreover, factorially closed extensions of factorial domains are characterized with help of factorial modules.

Keywords Content module • Factorial domain • Factorial module • Inert extension • Locally projective module • Mittag-Leffler module

2000 Mathematics Subject Classification 13C13, 13F15, 13G05

1 Introduction

Unique factorization in integral domains plays a prominent role in algebra. It is explained in many books on basic algebra; moreover, the last decades have seen generalizations of this concept in several directions, see, e.g., [1] and the literature cited there. One of these directions is to introduce various types of factorizations of elements in domains (cf. also [2]); another one is the generalization to commutative rings with zero-divisors (cf. also [3, 4]). Further, in [17], a generalization to torsion-free modules over (factorial) domains has been proposed by A.-M. Nicholas and subsequently refined in [18–20] as well as by Costa [8], by Lu [16], and by Anderson and Valdes-Leon [4].

In this note it is proven that locally projective modules, flat Mittag-Leffler modules, and torsion-free content modules are factorial modules in the sense of Nicholas [17, 18]. Moreover, factorially closed extensions of factorial domains are characterized with the help of factorial modules.

G. Angermüller (✉)
e-mail: gerhard.angermueller@googlemail.com

M. Fontana et al. (eds.), *Rings, Polynomials, and Modules*,
https://doi.org/10.1007/978-3-319-65874-2_2

The content of this paper is organized as follows: In Sect. 2, preliminaries are proven to be used in the following sections; possibly, some of them are of independent interest, e.g. Proposition 1. The basic definitions and properties of factorable modules over commutative domains are contained in Sect. 3. The core of this paper is Sect. 4, where the theory is further developed in the special case of factorial base domains; moreover, in this section it is proven that locally projective modules, flat Mittag-Leffler modules or torsion-free content modules are factorial modules. In Sect. 5 ring extensions are considered which are factorable as modules; in particular, factorially closed extensions of factorial domains are characterized with factorial modules. The last section contains some hints to related literature.

Notation
The basics on (unique) factorization in domains can be found, e.g., in [13, 2.14 and 2.15]; basic concepts of Commutative Algebra used in this note are contained in [5], as well as our standard notation. For more advanced subjects we give detailed references to [6, 7, 11, 12, 14].

In the following sections R denotes a commutative domain with 1, $K = Q(R)$ the field of quotients of R and M a torsion-free R-module. M is identified with its image in $KM := K \otimes_R M$ under the map $1 \otimes id_M$; further, $\widehat{M} := \bigcap \{M_P | P \in Spec(R), ht(P) = 1\} \subseteq KM$. Moreover, $R^\times$ denotes the group of units of R.

2 Preliminaries

In this section we recall some definitions and prove some results to be used in the subsequent sections.

If R is a commutative ring and M an R-module, an element x of R is called a *zero-divisor on M*, if x annihilates some non-zero element of M; M is called *torsion-free*, if $0 \in R$ is the only zero-divisor on M. r, s is called a *(two-element) M-sequence*, if $r, s \in R$, r is not a zero-divisor on M and s is not a zero-divisor on M/rM. M is said to satisfy *accc*, if M satisfies the ascending chain condition on cyclic submodules (or equivalently: any non-empty family of cyclic submodules of M has a maximal element). R is said to satisfy *accp*, if R satisfies the ascending chain condition on principal ideals. A submodule N of M is called *torsion-closed in M* if M/N is torsion-free; N is called *pure in M*, if for all finite families $(x_i)_{i\in I}$, $(y_j)_{j\in J}$, $(r_{ij})_{i\in I, j\in J}$ of elements of N, M and R respectively such that for all $i \in I$, $x_i = \sum_{j\in J} r_{ij} y_j$, there is a family $(z_j)_{j\in J}$ of elements of N such that for all $i \in I$, $x_i = \sum_{j\in J} r_{ij} z_j$. Clearly, any pure submodule of M is torsion-closed in M. M^* denotes the R-module of R-linear forms on M. M is called *torsionless*, if for each $x \in M$ there is a $f \in M^*$ such that $f(x) \neq 0$; M is called *reflexive*, if the canonical homomorphism $m \mapsto (f(m))_{f\in M^*}$ is a bijection from M onto M^{**}. M is called *locally projective*, if for each surjective homomorphism $f : P \to Q$ of R-modules, each R-homomorphism $g : M \to Q$ and each finitely generated submodule N of M there is an R-homomorphism $h : M \to P$ such that $f \circ h(x) = g(x)$ for all $x \in N$; obviously, any projective module is locally projective. M is called a *content module*, if for every family $(I_\lambda)_{\lambda\in\Lambda}$ of ideals I_λ of R: $\bigcap_{\lambda\in\Lambda}(I_\lambda M) = (\bigcap_{\lambda\in\Lambda} I_\lambda)M$; if $x \in M$ and $(I_\lambda)_{\lambda\in\Lambda}$ is the family of all ideals I_λ of R such that $x \in I_\lambda M$, then $c(x) := \bigcap_{\lambda\in\Lambda} I_\lambda$ is called the *content of* x. M is called a *Mittag-Leffler-module*, if for every family $(Q_i)_{i\in I}$ of R-modules, the canonical map $M \otimes_R \prod_{i\in I} Q_i \to \prod_{i\in I} M \otimes_R Q_i$ is injective.

An element q of a commutative ring R is called an *atom*, if it is non-zero, not a unit and for all $r, s \in R$ such that $q = rs$, either r or s is a unit of R; q is called *prime*, if qR is a non-zero prime ideal of R. A domain R is called *atomic*, if any non-zero non-unit element of R can be expressed as a finite product of atoms. R is called a *factorial domain* (or a *UFD*), if any non-zero non-unit element of R can be expressed uniquely as a finite product of atoms up to units of R; as is well-known, e.g., by [13, Theorem 2.21], a domain R is factorial if and only if it satisfies accp (resp. is atomic) and every atom of R is prime. R is called a *GCD-domain*, if any two elements of R have a greatest common divisor. By [13, Theorem 2.22], a domain is factorial iff it is a GCD-domain satisfying accp. Concerning *Krull domains*, we refer to [7, Chapter VII].

The following technical argument is used in the proof of [4, Theorem 2.8].

Lemma 1 *Let $M \neq 0$ be such that for each $x \in M$, $x \neq 0$, there is a natural number $j(x)$ so that if $x = r_1 \cdots r_k y$, where r_i $(i = 1, \ldots, k)$ is a non-unit of R, and $y \in M$, then $k \leq j(x)$. Then R is atomic.*

Proof Choose $m \in M$, $m \neq 0$. Let r be a non-zero non-unit of R. If $r = r_1 \cdots r_k$ is a factorization of r into non-units r_i of R $(i = 1, \ldots, k)$, then $k \leq j(rm)$ by assumption on M; in particular, there are such factorizations of r with maximal k. By definition of an atom, any factorization $r = r_1 \cdots r_k$ of r into non-units r_i of R $(i = 1, \ldots, k)$ with maximal k is in fact a factorization into atoms. □

The following lemma is easily proved and probably known, but we could not find a reference.

Lemma 2 *Let $R \subseteq S$ be a ring extension.*

a) If S is a torsion-free R-module such that $S \cap Q(R) = R$, then $S^\times \cap R = R^\times$.
b) If S is a domain satisfying accp and $S^\times \cap R = R^\times$, then S satisfies accc as R-module.

Proof a) Let $r \in S^\times \cap R$, i.e. $rs = 1$ for some $s \in S$. Then $s = 1/r \in S \cap Q(R) = R$, whence $r \in R^\times$. b) Let $(Rx_i)_{i\in I}$ be any non-empty family of cyclic R-submodules of S. Then $(Sx_i)_{i\in I}$ is a non-empty family of principal ideals of S, whence has a maximal element Sx for some $x = x_j, j \in I$. Rx is a maximal element of $(Rx_i)_{i\in I}$: Let $i \in I$ and $Rx \subseteq Rx_i$. Then $x = rx_i$ for some $r \in R$; by choice of x, $Sx = Sx_i$. If $x = 0$, the assertion is obvious; otherwise, $rs = 1$ for some $s \in S$. Then $r \in S^{\times} \cap R = R^{\times}$ and in particular, $Rx = Rx_i$. □

The following assertion is more generally true for arbitrary (weakly) regular sequences, but for the sake of this note, the special case is sufficient.

Lemma 3 *If M is a flat M-module, then any two-element R-sequence is an M-sequence.*

Proof Let r, s be an R-sequence. As M is torsion-free, r is not a zero-divisor on M. Further, multiplication by s on R/rR is injective; as M is flat, multiplication by s on $M/rM \cong R/rR \otimes_R M$ is injective too. □

The following lemma is shown in the proof of [9, Proposition 1.5].

Lemma 4 *If R is a Krull domain, $M = \widehat{M}$ if and only if every two-element R-sequence is an M-sequence.*

Proof First, assume $M = \widehat{M}$. Let r, s be an R-sequence and $x, y \in M$ such that $rx = sy$. Then for any prime ideal P of R of height 1, $(r, s) \not\subseteq P$ (see, e.g., [14, Theorem 132]), i.e. $r \notin P$ or $s \notin P$. In the first case, $y \in M_P = rM_P$; in the second case, $y = r(x/s) \in rM_P$. Thus $y \in \bigcap\{rM_P | P \in Spec(R), ht(P) = 1\} = r\widehat{M} = rM$. Secondly, assume that every two-element R-sequence is an M-sequence and let $x \in \widehat{M}$. Then for every prime ideal P of R of height 1, $(M : x)_P = (M_P : x) = R_P$, whence $(M : x) \not\subseteq P$. Choose $r \in (M : x)$, $r \neq 0$, and denote by $P_1, \ldots, P_n$ the prime ideals of R of height 1 containing r. If $n = 0$, $r \in R^{\times}$ and thus $x \in M$; so, let us assume $n > 0$. R being a Krull domain, Rr is a decomposable ideal of R [11, Corollary 43.10], whence $\bigcup_{i=1,\ldots,n} P_i$ is the set Z of zero-divisors of R on R/Rr [5, Proposition 4.7]. By the above, $(M : x)$ is not contained in any P_i for $i = 1, \ldots, n$, and hence $(M : x) \not\subseteq Z$ by [5, Proposition 1.11 i)]. Thus we can choose $s \in (M : x) \setminus Z$. Then r, s is an R-sequence, whence an M-sequence by assumption. By choice of r, s: $rx, sx \in M$ and thus $r(sx) = s(rx)$ implies $rx \in rM$, i.e. $x \in M$. □

An extension $R \subseteq S$ of Krull domains is said to satisfy PDE, if for every prime ideal P of S of height 1, the prime ideal $P \cap R$ of R is zero or of height 1 (cf. [7, VII, §1.10]). The following lemma is proven in [9, Proposition 1.5].

Lemma 5 *If R and S are Krull domains, $R \subseteq S$ satisfies PDE if and only if every two-element R-sequence is an S-sequence.*

Proof Observe that in any Krull domain T, any principal ideal I, $0 \neq I \neq T$, has a primary decomposition, whose primary ideals belong to prime ideals of height 1 [11, Corollary 43.10 a)]; in particular, the set of zero-divisors on T/I is a finite union of prime ideals of T of height 1 [5, Proposition 4.7]. Assume first that $R \subseteq S$ satisfies PDE and let r, s be an R-sequence. Further, denote by Z the set of zero-divisors on S/rS; by the above, $Z = P_1 \cup \ldots \cup P_n$ for some height 1 prime ideals P_i of S. If s would be an element of Z, then $r, s \in P_i \cap R$ for some i by [5, Proposition 1.11 i)], whence $P_i \cap R$ would be of height > 1 [14, Theorem 132], contradicting PDE. Thus s is not a zero-divisor on S/rS, that is, r, s is an S-sequence. Assume now that any two-element R-sequence is an S-sequence and let P be a prime ideal of S of height 1; it has to be shown that the height of $P \cap R$ is 0 or 1 . Assuming the contrary, we can choose $r \in P \cap R$, $r \neq 0$ and, by the above, $s \in P \cap R$ such that s is not a zero-divisor on R/rR. Then r, s would be an R-sequence, whence an S-sequence contained in the height 1 prime ideal P, a contradiction [14, Theorem 132]. □

The conditions (3) and (4) of the following proposition are considered in [16, Theorem 2.1].

Proposition 1 *Let R be a factorial domain. The following conditions are equivalent:*

(1) $M = \widehat{M}$
(2) Every two-element R-sequence is an M-sequence
(3) For every prime element p of R, $rx \in pM$ for $r \in R, x \in M$ implies $r \in Rp$ or $x \in pM$.
(4) For all $r \in R$, $x \in M$, the submodule $rM \cap Rx$ is cyclic.

Proof (1)⇔(2) follows by Lemma 4, any factorial domain being a Krull domain. (2)⇒(3): Let p be a prime element of R, $r \in R$ and $x, y \in M$ such that $rx = py$. If p does not divide r, then p, r is an R-sequence, whence an M-sequence and thus $x \in pM$. (3)⇒(2): Let r, s be an R-sequence and $x, y \in M$ such that $rx = sy$. As R is factorial, r and s are relatively prime. We prove by induction on the number n of primes dividing r that $y \in rM$. If $n = 0$, r is a unit in R and the assertion is clear. Let now $n > 0$ and assume the assertion to be true for $n - 1$. Choose a prime element p of R dividing r, i.e. $r = pt$ for some $t \in R$. Then $sy = rx = ptx$, whence $y = pz$ for some $z \in M$ by assumption (3). Thus $psz = sy = ptx$, whence $sz = tx$ and thus $z = tw$ for some $w \in M$ by induction hypothesis. Putting all together, one obtains $y = pz = ptw = rw \in rM$. (3)⇒(4): Let $r \in R$, $x \in M$. We prove by induction on the number n of primes dividing r that $rM \cap Rx$ is cyclic. If $n = 0$, r is a unit in R and $rM \cap Rx = M \cap Rx = Rx$. Let now $n > 0$ and assume the assertion to be true for $n - 1$. Choose a prime element p of R dividing r, i.e. $r = ps$ for some $s \in R$. If $x = py$ for some $y \in M$, then $sM \cap Ry = Rz$ for some $z \in M$ by induction hypothesis, and thus $rM \cap Rx = psM \cap Rpy = p(sM \cap Ry) = Rpz$. If $x \notin pM$, we first observe that $pM \cap Rx = Rpx$: in fact, if $t \in R$, $w \in M$ are such that $pw = tx$, then (by assumption (3)) $t = up$ for some $u \in R$, whence $pw = tx = upx \in Rpx$. Moreover, by induction hypothesis, $sM \cap Rx = Rv$ for some $v \in M$. Then $rM \cap Rx = (rM \cap pM) \cap Rx = rM \cap (pM \cap Rx) = psM \cap Rpx = p(sM \cap Rx) = Rpv$. (4)⇒(3): Let p be a prime element of R, $r \in R$ and $x, y \in M$ such that $rx = py$. By assumption (4), $pM \cap Rx = Rz$ for some $z \in M$; in particular, $rx = py = sz$ for some $s \in R$ as well as $px = tz$ for some $t \in R$. Furthermore, $z = ux = pw$ for some $u \in R$, $w \in M$. If $x = 0$, the assertion is trivially true. If $x \neq 0$, then $px = tz = tux$, whence $p = tu$; as p is prime, either $t \in R^\times$ or $u \in R^\times$. If $t \in R^\times$, then $rx = sz = st^{-1}px$, whence $r = st^{-1}p \in Rp$; if $u \in R^\times$, then $x = u^{-1}z = u^{-1}pw \in pM$. □

3 Atomic and Factorable Modules

The following definitions are basic for this note.

Definition 1 Let $x \in M$.

An element $r \in R$ (resp. $m \in M$) is called *an R-divisor* of x (resp. *an M-divisor of x*), if $x = ry$ for some $y \in M$ (resp. $x = sm$ for some $s \in R$); r (resp. m) is called *a greatest R-divisor* (resp. *a smallest M-divisor*) of x, if any R-divisor of x divides r (resp. m is an M-divisor of any M-divisor of x).

x is called *irreducible* if any R-divisor of x is a unit of R.

x is called *primitive* if $x \neq 0$ and x is a smallest M-divisor of any non-zero element of Rx.

M is called *atomic*, if any non-zero element of M has an irreducible M-divisor.

M is called *factorable*, if any non-zero element of M has a smallest M-divisor.

M has the *finite divisor property* (or has *fdp*) if each non-zero element of M has, up to units, only a finite number of proper R-divisors.

A prime element p of R is called *prime for M* if $rx \in pM$ for $r \in R, x \in M$ implies $r \in Rp$ or $x \in pM$.

Remark 1

a) The irreducible elements of R—considered as an R-module—are the units of R. To avoid conflicts, we use the term "atom" for "irreducible elements of a domain" in the sense of [13, Section 2.14].
b) R—considered as an R-module—is atomic and factorable; the primitive elements are the units of R.
c) It is easily checked that the R-module $K = Q(R)$ has irreducible elements if and only if $K = R$; in particular, K is neither atomic nor factorable, if $K \neq R$.
d) If R is a factorial domain, then R has fdp when considered as an R-module; in this case, any prime element of R is prime for R.
e) If R is a field and x is any non-zero element of M, then x is a smallest M-divisor of x: in fact, if $x = ry$, where $r \in R$ and $y \in M$, then $r \neq 0$ and $y = r^{-1}x$. In particular, every vector space is factorable.
f) Let $R := k[X, Y]$ be a polynomial ring in two variables X, Y over a field k and $M := RX + RY$. The irreducible elements of M are the prime elements of the factorial domain R which are contained in M, thus showing, e.g., that the sets of irreducible elements of a module and that of a factorable module containing it can be disjoint (cf. a)). Moreover, M does not have primitive elements: otherwise there would exist an element x of M, $x \neq 0$, such that x is a smallest M-divisor of xX as well as of xY, i.e. a divisor of X and of Y, a contradiction to $M \neq R$. A similar argument shows that the element XY of M does not have a smallest divisor in M, thus proving that the submodule M of the factorable module R is not factorable. As M is a submodule of R, it has fdp too (cf. d); in Remark 3 there is an example of a module having fdp which is not a submodule of a factorable module.

More elementary facts related to Definition 1 are contained in the following lemma:

Lemma 6

a) *Greatest R-divisors (resp. smallest M-divisors) of any non-zero element of M are uniquely determined up to units of R.*
b) *Let $m, x \in M$, $r \in R$, $x \neq 0$ such that $x = rm$. Then r is a greatest R-divisor of x if and only if m is a smallest M-divisor of x.*
c) *A smallest M-divisor of any non-zero element of M is irreducible.*
d) *Every primitive element of M is irreducible.*
e) *If $m, n \in M$ are primitive and $r, s \in R$, $r, s \neq 0$ are such that $rm = sn$, then $m = un$ and $s = ur$ for some $u \in R^{\times}$.*
f) *For every $x \in M$ the following assertions are equivalent:*

   (1) *x is primitive*
   (2) *$x \neq 0$ and $Kx \cap M = Rx$*
   (3) *Rx is a maximal rank 1 submodule of M*
   (4) *$x \neq 0$ and Rx is torsion-closed in M*
   (5) *$x \neq 0$ and for every $y \in M$ either $Rx \cap Ry = 0$ or $Ry \subseteq Rx$*
   (6) *$x \neq 0$ and for every $r \in R$, $r \neq 0$, r is a greatest R-divisor of rx.*

Proof a): Let $x \in M$, $x \neq 0$. If $r, s \in R$ (resp. $m, n \in M$) are greatest R-divisors (resp. smallest M-divisors) of x, then r divides s (resp. m is an M-divisor of n) and vice versa. In any case, by the assumption on R and M, r and s (resp. m and n) differ by units of R. b) follows immediately from the definitions. c): Let $x \in M$, $x \neq 0$. Let $m \in M$ be a smallest M-divisor of x and let $r \in R$ and $n \in M$ be such that $m = rn$. By assumption on m, m is an M-divisor of n, whence r is a unit of R. d) follows from c). e) follows from the definition of primitive elements and a). f): Let $x \in M$. (1)⇒(2): If x is primitive and $(r/s)x = y$ for some $y \in M$ and some $r, s \in R$, $s \neq 0$, then $rx = sy$, whence $y = tx$ for some $t \in R$ and thus $r = st$ and $y = (r/s)x = tx \in Rx$. (2)⇒(3): Let N be a rank 1 submodule of M such that $Rx \subseteq N$. As $x \neq 0$ by assumption (2), one has $KN = Kx$ and thus $N \subseteq KN \cap M = Kx \cap M = Rx$. (3)⇒(4): Let $r, s \in R$, $r \neq 0$ and $y \in M$ be such that $ry = sx$. Then $y = 0 \in Rx$ or $Ky = Kx$ and thus $Rx \subseteq Ky \cap M$, whence $Ky \cap M = Rx$ by assumption (3); in particular, $y \in Rx$. (4)⇒(5): Let $y \in M$ and assume that there are $r, s \in R$ such that $rx = sy \neq 0$. By assumption (4), $y \in Rx$ and thus $Ry \subseteq Rx$. (5)⇒(1): Let $r, s \in R$, $r \neq 0$ and $y \in M$ be such that $rx = sy$. Then $Rx \cap Ry \neq 0$ and by assumption, $Ry \subseteq Rx$, whence x is an M-divisor of y. (1)⇔(6) follows from b). □

Lemma 7

a) *If M is factorable, then M is atomic.*
b) *If M satisfies accc, then any submodule of M is atomic.*
c) *If M is atomic (resp. factorable), then any torsion-closed submodule of M is atomic (resp. factorable); in particular, if M is atomic (resp. factorable), then any pure submodule of M is atomic (resp. factorable).*
d) *If $M = Rm$ is free of rank 1, then M is factorable; more precisely, m is a smallest M-divisor of any element of M and $R^{\times}m$ is the set of irreducible (resp. primitive) elements of M.*

Proof a) follows from Lemma 6 c). b): Assuming the contrary, a strictly ascending infinite chain of cyclic submodules of M can be easily constructed. c): Clearly, any irreducible (resp. smallest) M-divisor of any non-zero element of a torsion-closed submodule N of M is an N-divisor and is obviously irreducible in N too (resp. a smallest N-divisor). d) follows from the definitions (and Lemma 6 d)). □

The following proposition and its corollaries give some indication of the usefulness of factorable modules.

Proposition 2 *The following conditions are equivalent:*

(1) M is factorable
(2) Every non-zero element of M has a greatest R-divisor
(3) Every non-zero element $x \in M$ has a representation $x = ry$ with $r \in R$, y an irreducible element of M and this representation is unique up to a unit of R
(4) M is atomic and every irreducible element of M is primitive
(5) Every non-zero element of M has a primitive M-divisor
(6) Every non-zero element $x \in M$ has a representation $x = ry$ with r a greatest R-divisor of x and y a primitive M-divisor
(7) Every maximal rank 1 submodule of M is free.

Proof (1)⇔(2) follows from Lemma 6 b). (1)⇒(3): The first part of (3) follows from Lemma 6 c). Let $x \in M$, $x \neq 0$ be such that $x = ry = sz$ with $r, s \in R$ and irreducible elements y, z of M. Choose a smallest M-divisor w of x; then $y = ew$ and $z = fw$ with $e, f \in R^\times$. Further, $u := e^{-1}f \in R^\times$ and $r = us, z = uy$. (3)⇒(4): By the first part of (3), M is atomic. To prove the second part of (4), let $x \in M$ be irreducible and $r, s \in R, y \in M$ be such that $rx = sy$. Let $y = tz$ with $t \in R$ and $z \in M$ irreducible. By assumption, $r = ust, z = ux$ for some $u \in R^\times$, whence $y = tux$ showing x primitive. (4)⇒(5) is clear from the definition. (5)⇒(6): by assumption, x has a representation $x = ry$ with $r \in R$ and y a primitive M-divisor. By definition, y is a smallest M-divisor of x, whence the assertion follows by Lemma 6 b). (6)⇒(7): let N be a maximal rank 1 submodule of M. Choose $n \in N, n \neq 0$ and $r \in R, x \in M$ primitive such that $n = rx$. Maximality of N implies $N = KN \cap M$. In particular, $x = r^{-1}n \in N$, whence $Rx \subseteq N$ and thus $Rx = N$ by Lemma 6 f) (1)⇒(3). (7)⇒(1): let $x \in M, x \neq 0$. Then $Kx \cap M$ is a maximal rank 1 submodule of M; by assumption, $Kx \cap M = Ry$ for some $y \in Kx \cap M$. By Lemma 6 f) (3)⇒(1), y is primitive in M, whence a smallest divisor of x. □

Proposition 2 (7)⇒(1) yields another proof that any vector space is factorable (cf. Remark 1, e)). The next two corollaries shed some light upon modules of rank 1 and factorability:

Corollary 1 *If M has rank 1, then M is factorable if and only if M is free.*

Proof One direction by Proposition 2, (1)⇒(7), and the other by Lemma 7 d). □

Remark 2 An immediate consequence of Corollary 1 is that every non-principal ideal of any domain R is not factorable, although R is so (cf. Remark 1 b), f)).

Corollary 2 *If M is factorable, every non-zero cyclic submodule N of M is contained in a unique maximal rank 1 submodule of M, and this submodule is free.*

Proof Let $N = Rx$ for some $x \in M, x \neq 0$. By Proposition 2 (1)⇒(5), x has a primitive M-divisor y, i.e. $Rx \subseteq Ry$ and Ry is a maximal rank 1 submodule of M by Lemma 6 f) (1)⇒(3). If N is contained in a maximal rank 1 submodule N' of M, then $N' = Rz$ for some $z \in M$ by Proposition 2 (1)⇒(7) and z is primitive by Lemma 6 f) (3)⇒(1). As $x \in Ry \cap Rz$, one has $Ry = Rz$ by Lemma 6 f) (1)⇒(5). □

Corollary 3 *If M is factorable, then every R-sequence r, s is an M-sequence.*

Proof Let $x, y \in M$ be such that $rx = sy$. Then $x = r'u, y = s'v$ for some $r', s' \in R$ and some primitive elements u, v of M by Proposition 2 (1)⇒(6). This implies $rr'u = ss'v$, whence $rr't = ss'$ for some $t \in R^{\times}$ by Lemma 6 e). As r, s is an R-sequence, this yields $s' \in Rr$ and thus $y \in rM$. □

Corollary 4 *Let R be a Noetherian integrally closed domain and M a finitely generated factorable R-module. Then M is reflexive.*

Proof If M is factorable, then $M = \widehat{M}$ by Corollary 3 and Lemma 4. As M is finitely generated and R a Noetherian integrally closed domain, M is reflexive by [7, VII, §4.2, Theorem 2]. □

Corollary 5 *If M is factorable and $M \neq 0$, then:*

a) For all $r, s \in R$: $rM \subseteq sM$ if and only if s divides r.
b) For all $r \in R$: $rM = M$ if and only if $r \in R^{\times}$.
c) For every greatest divisor r of a non-zero element x of M and for every $s \in R$, sr is a greatest divisor of sx.

Proof a): Let $r, s \in R$ be such that $rM \subseteq sM$. By Proposition 2, (1)⇒(5), there is a primitive element x of M; in particular, $rx = sy$ for some $y \in M$. By Lemma 6 b), r is a greatest R-divisor of sy, whence s divides r. b) follows immediately from a). c): Let $y \in M$ be such that $x = ry$. By Lemma 6 b),c), y is irreducible, whence a primitive element of M by Proposition 2, (1)⇒(4). Thus y is a smallest divisor of $sx = sry$ and the assertion follows by Lemma 6 b). □

The next result shows that in general, factorable modules are far away from containing divisible modules. If R is not a field, it shows in particular that any torsion-free R-module containing $K = Q(R)$ as a submodule is not factorable.

Corollary 6 *If M is a factorable module containing an element $x \neq 0$ which has every non-zero element of R as R-divisor, then R is a field.*

Proof By Proposition 2 (1)⇒(2), x has a greatest R-divisor t. By assumption, t^2 divides t, whence $t \in R^{\times}$. As by assumption, any non-zero element of R divides t, $R \setminus 0 \subseteq R^{\times}$, i.e. R is a field. □

Factorability of modules have consequences for the base ring:

Corollary 7 *R^2 is factorable if and only if R is a GCD-domain. In particular, if R satisfies accp, R^2 is factorable if and only if R is factorial.*

Proof Let $r, s, t \in R$. It is easily seen that t is a greatest R-divisor of $(r, s) \in R^2$ if and only if t is a greatest common divisor of r and s. Thus the first assertion follows from Proposition 2 (1)⇔(2), the second from [13, Thm. 2.22]. □

Corollary 8 *If M is factorable and R satisfies accp, then M satisfies accc.*

Proof Let $(Rx_i)_{i\in I}$ be any non-empty family of cyclic submodules of M. By Proposition 2 (1)⇒(5), for each $i \in I$, $x_i = s_i y_i$ for some $s_i \in R$ and primitive $y_i \in M$. Then $(Rs_i)_{i\in I}$ is a non-empty family of principal ideals of R, whence has a maximal element Rs_j for some $j \in I$. Then $Rx_j = Rs_j y_j$ is a maximal element of $(Rx_i)_{i\in I}$: let $i \in I$ and $Rx_j \subseteq Rx_i$. Then $x_j = rx_i$ for some $r \in R$, whence $s_j y_j = rs_i y_i$. If $s_j = 0$, the assertion is clear; otherwise, $s_j = trs_i$, $y_i = ty_j$ for some $t \in R^\times$ by Lemma 6 e). By choice of j, $Rs_j = Rs_i$, whence $r \in R^\times$ and thus $Rx_j = Rx_i$. □

Corollary 9 *If M is factorable and R satisfies accp, then any submodule of M is atomic.*

Proof Immediate by Corollary 8 and Lemma 7 b). □

If M is factorable, then M is atomic (by Proposition 2) and every two-element R-sequence is an M-sequence (by Corollary 3). In case of GCD-domains there is a converse:

Corollary 10 *Let R be a GCD-domain. If M is atomic and every two-element R-sequence is an M-sequence, then M is factorable.*

Proof By Proposition 2 (4)⇒(1), it is sufficient to prove that every irreducible element of M is primitive. Let $x \in M$ be irreducible and $y \in M$, $r, s \in R \setminus 0$ be such that $rx = sy$. Let d be a greatest common divisor of r, s; then $r = r'd$, $s = s'd$ for some relatively prime elements $r', s' \in R$. Moreover, $r'x = s'y$ and s', r' is an R-sequence; by assumption, s', r' is an M-sequence, whence s' is an R-divisor of x. As x is irreducible, $s' \in R^\times$ and x is an M-divisor of y. □

4 Factorable Modules over Factorial Domains

The next definition reflects a straightforward approach to the generalization of UFDs to modules, cf. [17–20]. The subsequent propositions and its corollaries explain the role of factorial domains in this context.

Definition 2 M is called *factorial*, if $M \neq 0$ and every non-zero element x of M has a representation $x = r_1 \cdots r_n y$ with atoms r_i $(i = 1, \dots, n)$ of R, y an irreducible element of M and this representation is unique up to units of R; that is, if $x = s_1 \cdots s_m z$ is another representation with atoms s_i $(i = 1, \dots, m)$ of R and an irreducible element z of M, then $n = m$, there are units $u_i, u \in R^\times$ and a permutation σ of $\{1, \dots, n\}$ such that $r_i = u_i s_{\sigma(i)}$ $(i = 1, \dots, n)$ and $y = uz$.

Proposition 3 *The following conditions are equivalent:*

(1) M is factorial
(2) R is atomic and M is factorial
(3) R is a factorial domain, $M \neq 0$ and M is factorable

(4) *$M \neq 0$ and every non-zero element x of M has a representation $x = r_1 \cdots r_i y$ with atoms $r_1, \dots, r_i$ of R and an irreducible element y of M; moreover, every atom of R is a prime element of R and every irreducible element of M is primitive*
(5) *$M \neq 0$ and every non-zero element x of M has a representation $x = r_1 \cdots r_i y$ with atoms $r_1, \dots, r_i$ of R, y a primitive element of M and this representation is unique up to units of R.*

Proof (1)⇒(2): To prove that R is atomic, it suffices to show that M satisfies the assumptions of Lemma 1. Let $x \in M$, $x \neq 0$, and $x = r_1 \dots r_k y$ with non-units $r_1, \dots, r_k$ of R and $y \in M$. Further, by (1), write $x = s_1 \cdots s_m z$ with atoms s_i $(i = 1, \dots, m)$ of R and an irreducible element z of M; $j(x) := m$ is independent of the choice of atoms s_i $(i = 1, \dots, m)$ of R and irreducible element z of M. We prove now: $k \leq j(x)$. By (1), $r_k y = r_{k,1} \cdots r_{k,m_k} y_{k-1}$ with atoms $r_{k,1}, \dots, r_{k,m_k}$ of R and an irreducible element $y_{k-1}$of M. It follows $m_k \geq 1$; otherwise, $r_k y = y_{k-1}$ would be irreducible, i.e. r_k a unit of R, a contradiction. Continuing similarly with $r_{k-1} y_{k-1}$ etc., we obtain a representation of $x = r_1 \cdots r_k y = r_1 \cdots r_{k-1} r_{k,1} \cdots r_{k,m_k} y_{k-1} = r_{1,1} \cdots r_{1,m_1} \cdots r_{k-1,1} \cdots r_{k-1,m_{k-1}} r_{k,1} \cdots r_{k,m_k} y_1$ with atoms $r_{1,1}, \dots, r_{k,m_k}$ of R, an irreducible element y_1 of M and $m_j \geq 1$ for $j = 1, \dots, k$. Thus $k \leq \sum_{j=1}^{k} m_j = j(x)$ by assumption (1). (2)⇒(3): As $M \neq 0$, M contains an irreducible element m. R is a factorial domain: by assumption (2), any non-zero element r of R has a factorization into atoms and this factorization is unique up to units of R, because that is the case for the element rm of M. M is factorable by Proposition 2 (3)⇒(1). (3)⇒(4) follows from basic properties of factorial domains and Proposition 2 (1)⇒(3), (4). (4)⇒(5) is clear. (5)⇒(1) follows by Lemma 6 d). □

Proposition 4 *Let R be a factorial domain. The following conditions are equivalent:*

(1) *M is factorable*
(2) *M is a submodule of a factorable R-module and every two-element R-sequence is an M-sequence*
(3) *M has fdp and every two-element R-sequence is an M-sequence*
(4) *M satisfies accc and every two-element R-sequence is an M-sequence*
(5) *M is atomic and every two-element R-sequence is an M-sequence.*

Proof (1)⇒(2) follows by Corollary 3. (2)⇒(3) by Proposition 2 (1)⇒(2) and the fact that R is factorial. (3)⇒(4) by definition of accc. (4)⇒(5) by Lemma 7 b). (5)⇒(1) by Corollary 10. □

The condition "every two-element R-sequence is an M-sequence" has some remarkable equivalencies, cf. Proposition 1.

Corollary 11 *Let R be a factorial domain and M be a flat R-module. Then the following conditions are equivalent:*

(1) *M is factorable*
(2) *M is a submodule of a factorable R-module*

(3) M has fdp
(4) M satisfies accc
(5) M is atomic.

Proof If M is flat, every two-element R-sequence is an M-sequence by Lemma 3, whence Proposition 4 implies the assertion. □

The next two corollaries describe in some aspects submodules of factorable modules.

Corollary 12 *If R is a factorial domain and $(N_i)_{i\in I}$ a non-empty family of factorable submodules of M, then $N := \bigcap_{i\in I} N_i$ is factorable. In particular, any submodule of a factorable R-module is contained in a smallest factorable submodule.*

Proof By Proposition 4 (2)⇒(1), it is sufficient to show that any R-sequence r, s is an N-sequence. Let r, s be an R-sequence and $x, y \in N$ such that $rx = sy$. Then, by Proposition 4 (1)⇒(2), for each $i \in I$, $y = rz_i$ for some $z_i \in N_i$. M being torsion-free implies $z_i = z_j$ for each $j \in I$, i.e. r is an N-divisor of y. □

Corollary 13 *If R is a factorial domain, the following conditions are equivalent:*

(1) M is a submodule of a factorable R-module
(2) For each non-zero $x \in M$ there is a non-zero $r \in R$ such that for any $s \in R$ the R-divisors of sx divide sr
(3) For each prime ideal P of height 1 of R, M_P is a factorable R_P-module and each non-zero $x \in \widehat{M}$ is irreducible in all but a finite number of the M_P's
(4) $\widehat{M}$ is factorable
(5) $\widehat{M}$ has fdp
(6) $\widehat{M}$ satisfies accc
(7) $\widehat{M}$ is atomic.
Moreover, if any of these conditions hold, then $\widehat{M}$ is isomorphic to the smallest factorable submodule of any factorable R-module containing M.

Proof (1)⇒(2): Let N be a factorable R-module containing M and $x \in M$, $x \neq 0$. Then x has a greatest R-divisor r in N, and for every $s \in R$ the greatest R-divisor of sx in N is sr by Corollary 5 c), whence the assertion follows. (2)⇒(3): Let P be a prime ideal of R of height 1, i.e. $P = Rp$ for some prime element p of R. To show that M_P is a factorable R_P-module, let $x \in M$, $x \neq 0$ and $t \in R\backslash P$; moreover, choose $r \in R$, $r \neq 0$ fulfilling the condition in (2) for x. If $x/t = (u/t')(y/t'')$ for some $u \in R$, $y \in M$ and $t', t'' \in R\backslash P$, then u is an R-divisor of $xt't''$. Writing $u = vp^n$ for some $v \in R\backslash P$ and $n \geqq 0$, that implies that p^n divides $t't''r$, whence p^n divides r, thus proving that any non-zero element of M_P has a greatest R_P-divisor, and the first assertion follows by Proposition 2 (2)⇒(1). By the above, if x is not irreducible in M_P, then p divides r. In other words, any non-zero element of M is irreducible in all but a finite number of the M_P's, whence this is true for any non-zero element of $\widehat{M}$ too. (3)⇒(4): Let $x \in \widehat{M}$, $x \neq 0$ and $(p_i)_{i\in I}$ be a representative family of non-associate prime elements p_i of R and put $P_i := Rp_i$ for $i \in I$. Then for each $i \in I$, a greatest R_{P_i}-divisor of x in M_{P_i} is $p_i^{n_i}$ for some non-negative integer n_i; by assumption (3), $n_i = 0$ for all but finitely many $i \in I$. It is easily checked that $r := \prod_{i \in I} p_i^{n_i}$ is a greatest R-divisor of x in $\widehat{M}$, and the proof is finished by Proposition 2 (2)⇒(1). (4)⇒(1) is obvious. (4)⇔(5)⇔(6)⇔(7) follows by Propositions 4 and 1, observing that $(\widehat{M})_P = M_P$ for each prime ideal P of height 1 of R, i.e. $\widehat{\widehat{M}} = \widehat{M}$. To prove the last assertion, let N be a factorable R-module containing M and denote by Q the smallest factorable submodule of N containing M (see Corollary 12). Then for every prime ideal P of R of height 1, M_P can be identified with a submodule of Q_P, whence $M \subseteq \widehat{M} = \bigcap \{M_P | P \in Spec(R), ht(P) = 1\} \subseteq \bigcap \{Q_P | P \in Spec(R), ht(P) = 1\} = \widehat{Q} = Q$ by Propositions 4 and 1. By (1)⇒(4) above, $\widehat{M}$ is factorable, whence $\widehat{M} = Q$. □

Remark 3 Corollary 13 shows in particular that all submodules of factorial modules have fdp. The following example shows that not every module having fdp is a submodule of a factorial module (cf. [8, 2.3 Example]). Put $R := k[X, Y]$, a polynomial ring in two variables X, Y over a field k, and $M := R[X/Y] \subseteq K = Q(R)$. To prove that M is not a submodule of a factorable R-module, observe that for any $m > 0$, $1/Y^m = (X/Y)^m(1/X^m) \in \bigcap \{M_P | P \in Spec(R), ht(P) = 1\} = \widehat{M}$, whence the element $1/Y = Y^m(1/Y^{m+1})$ of $\widehat{M}$ does not have a greatest R-divisor; in particular, $\widehat{M}$ is not factorable and the assertion follows by Corollary 13. To prove that M has fdp, let $x \in M$, $x \neq 0$. Then $Y^m x \in R$ for some $m \geq 0$ and x has only finitely many R-divisors up to units if $Y^m x$ has so; in other words, we can assume $x \in R$. Let p be a prime element of R such that p is not associated to Y. If $m \geq 0$ and p^m is an R-divisor of x, then $x = p^m(a/Y^n)$ for some $n \geq 0$ and some $a \in R$, i.e. p^m divides $Y^n x$. By assumption on p, p^m divides x in R. Thus we have shown that the only prime powers of an R-divisor of x are the prime powers occurring in a factorization of x in prime elements of R and—possibly—prime powers of Y. To finish the proof, we have to show that not arbitrarily large powers of Y are R-divisors of x. Write $x = Y^n z$ with $n \geq 0$ and $z \in R$ such that Y does not divide z. Let $m > n$ and assume that Y^m is an R-divisor of x in M. Then $x/Y^m = z/Y^{m-n} = \sum_{i=0}^{j} r_i(X/Y)^i = \sum_{i=0}^{j} r_i X^i Y^{j-i}/Y^j$ for some $j > 0$ and some $r_i \in R$ for $i = 0, \ldots, j$ such that Y does not divide r_j. This implies $m - n = j$ and $z = \sum_{i=0}^{j} r_i X^i Y^{j-i} \in I^j$, where $I := RX + RY$, and thus $x = Y^n z \in I^n I^j = I^m$. Now, if arbitrarily large powers of Y would be R-divisors of x, that implies $x \in \bigcap_{m \geq 0} I^m$; by Krull's Intersection Theorem [5, Corollary 10.18] then $x = 0$, a contradiction.

Factorial modules behave well with respect to direct products or sums:

Proposition 5 *Let R be a factorial domain and $(M_i)_{i \in I}$ be any family of R-modules. Then the following assertions are equivalent:*

(1) M_i is factorable for each $i \in I$.
(2) The direct product $\prod_{i \in I} M_i$ is factorable.
(3) The direct sum $\bigoplus_{i \in I} M_i$ is factorable.

Proof The assertion is obvious, as far as the torsion-freeness of the modules in question is concerned. (1)⇒(2): Let $x = (x_i)_{i\in I}$ be an element of the direct product of the M_i. By Proposition 2 (1)⇒(2), for each $i \in I$ there is a greatest R-divisor r_i of x_i. Let d be a greatest common divisor of $\{r_i | i \in I\}$. Then d is a greatest R-divisor of x: obviously, d is an R-divisor of x and if $x = sy$ for some $s \in R$ and an element y of $\prod_{i\in I} M_i$, then s divides r_i for each $i \in I$, whence s divides d. (2)⇒(3): As the direct sum is torsion-closed in the direct product, the assertion follows by Lemma 7 c). (3)⇒(1): Lemma 7 c) applies again: each M_i is torsion-closed in $\bigoplus_{i\in I} M_i$. □

An immediate consequence is the following corollary:

Corollary 14 *If R is a factorial domain, for any set I, R^I and $R^{(I)}$ are factorable; moreover, any projective R-module is factorable.*
Even more is true:

Corollary 15 *If R is a factorial domain, any locally projective R-module is factorable.*

Proof As any locally projective R-module is isomorphic to a pure submodule of some R^I [12, Proposition 3.39], the assertion follows from Corollary 14 and Lemma 7 c). □

Proposition 6 *Let R be a factorial domain. The R-module $M^* = Hom_R(M, R)$ is factorable. More precisely, if $f \in M^*$ then:*

a) Any greatest common divisor of $f(M)$ is a greatest R-divisor of f.
b) f is irreducible in M^* if and only if the greatest common divisor of $f(M)$ is 1.

Proof Clearly, M^* is torsion-free. Let $f \in M^*$. To prove a), let r be a greatest common divisor of $f(M)$. Then for every $x \in M$ there is a unique $g(x) \in R$ such that $f(x) = rg(x)$; obviously, $g \in M^*$, whence r is an R-divisor of f. Moreover, if $f = sh$ for some $s \in R$ and $h \in M^*$, s divides r. b) follows from a), as well as the first assertion, taking Proposition 2 (2)⇒(1) into account. □

Corollary 16 *If R is a factorial domain, any reflexive R-module is factorable.*

Proof If M is reflexive, M is isomorphic to M^{**}, whence factorable by Proposition 6. □

Corollary 17 *Let R be a Noetherian factorial domain and M a finitely generated R-module. Then M is factorable if and only if it is reflexive.*

Proof If M is factorable, then M is reflexive by Corollary 4. The other direction is clear by Corollary 16. □

Proposition 7 *If R is a factorial domain, any torsion-free content R-module is factorable.*

Proof Let M be a torsion-free content module; denote for every $x \in M$ by $c(x)$ the content of x, i.e. $c(x)$ is the smallest ideal I of R such that $x \in IM$. We show that for each non-zero element x of M, a greatest common divisor r of $c(x)$ is a greatest

R-divisor of x: From $x \in c(x)M$, we conclude $x \in rM$; moreover, if $x = sy$ for some $s \in R$, $y \in M$, then $c(x) \subseteq sc(y)$ by definition of the content ideal, whence s divides r. The proof is finished by applying Proposition 2. □

Remark 4 If R is a factorial Hilbert domain, any content R-module is torsion-free [10, (2.17)]. Further, every locally projective module is torsion-free and a content module by [12, Proposition 3.38], whence Corollary 15 is a consequence of Proposition 7 too.

The next corollary is a generalization of Corollary 14:

Corollary 18 *If R is a factorial domain, any flat Mittag-Leffler R-module is factorable.*

Proof By Proposition 7, it is sufficient to show that any flat Mittag-Leffler-module is a content module. Let M be a flat Mittag-Leffler-module and $(I_\lambda)_{\lambda\in\Lambda}$ a family of ideals I_λ of R. As M is flat, the canonical exact sequence $\bigcap_{\lambda\in\Lambda} I_\lambda \to R \to \prod_{\lambda\in\Lambda} R/I_\lambda$ yields the exact sequence $\bigcap_{\lambda\in\Lambda} I_\lambda \otimes_R M \to M \to (\prod_{\lambda\in\Lambda} R/I_\lambda) \otimes_R M$; as M is a Mittag-Leffler-module, the canonical morphism $(\prod_{\lambda\in\Lambda} R/I_\lambda) \otimes_R M \to \prod_{\lambda\in\Lambda}((R/I_\lambda) \otimes_R M)$ is injective. Thus the kernels of the canonical maps $M \to (\prod_{\lambda\in\Lambda} R/I_\lambda) \otimes_R M$ and $M \to \prod_{\lambda\in\Lambda}((R/I_\lambda) \otimes_R M)$ coincide, whence $(\bigcap_{\lambda\in\Lambda} I_\lambda)M = \bigcap_{\lambda\in\Lambda}(I_\lambda M)$. □

5 Factorable Ring Extensions

In this section the results of the preceding sections are specialized to the case of an extension ring which is simultaneously a factorable module.

Definition 3 A *factorable extension* is any ring extension $R \subseteq S$ such that R is a domain and S is a torsion-free factorable R-module.

Proposition 8 *Let $R \subseteq S$ be an extension of domains. Consider the following conditions:*

(1) $R \subseteq S$ is a factorable extension
(2) $S \cap Q(R) = R$ and any two-element R-sequence is an S-sequence
(3) $S^\times \cap R = R^\times$ and any two-element R-sequence is an S-sequence.

Then:

a) (1) implies (2) and (2) implies (3).
b) If R is a GCD-domain and S satisfies accp, then (3) implies (1).

Proof a) (1)⇒(2): Let $x \in S \cap Q(R)$, i.e. $x = r/s$ for some $r, s \in R$, $s \neq 0$. Then $rS = sxS \subseteq sS$, whence s divides r by Corollary 5 a), i.e. $x \in R$. The second assertion follows by Corollary 3. (2)⇒(3) by Lemma 2 a). b) follows by Lemma 2 b), Lemma 7 b) and Corollary 10. □

The following corollaries are easily derived from the results of the subsequent sections.

Corollary 19 *Let $R \subseteq S$ be an extension of domains, R a GCD-domain, and S satisfying accp. If S is a flat R-module, the following conditions are equivalent:*

(1) $R \subseteq S$ is a factorable extension
(2) $S \cap Q(R) = R$
(3) $S^\times \cap R = R^\times$.

Proof As S is flat, any two-element R-sequence is an S-sequence by Lemma 3; thus Proposition 8 gives the result. □

Corollary 20 *Let $R \subseteq S$ be an extension of domains, R a GCD-domain, and S satisfying accp. If S is a faithfully flat R-module, then $R \subseteq S$ is a factorable extension.*

Proof By Corollary 19, only $S^\times \cap R = R^\times$ has to be shown. Let $r \in S^\times \cap R$, i.e. $rs = 1$ for some $s \in S$. Then $s \in R$ by the linear extension property [7, I, §3.7, Proposition 13]. □

Corollary 21 *Let $R \subseteq S$ be an extension of Krull domains such that R is factorial. If $R \subseteq S$ satisfies PDE, the following conditions are equivalent:*

(1) $R \subseteq S$ is a factorable extension
(2) $S \cap Q(R) = R$
(3) $S^\times \cap R = R^\times$.

Proof Immediate by Proposition 8 and Lemma 5. □

Corollary 22 *Let $R \subseteq S$ be an extension of Krull domains such that R is factorial. If S is integral over R, then $R \subseteq S$ is a factorable extension.*

Proof $R \subseteq S$ satisfies PDE by [7, VII, §1.10] and by Corollary 21 (2)⇒(1), the proof is finished. □

Recall that a subring R of a ring S is called *factorially closed in S*, if for all non-zero $x, y \in S$, the condition $xy \in R$ implies $x \in R$ and $y \in R$.

Lemma 8 *Let $R \subseteq S$ be an extension of domains such that S is factorial.*

a) The following conditions are equivalent:

(1) R is factorially closed in S
(2) R is factorial, $S^\times = R^\times$ and every prime element of R is a prime element of S.

b) If R is factorially closed in S, then $R \subseteq S$ is a factorable extension.

Proof a) (1)⇒(2): The first and the second assertion is clear. To prove the third one, observe that any prime element of R is by assumption (1) an atom in the factorial domain S, whence a prime element of S. (2)⇒(1): Let $x, y \in S$ be such that $xy \in R$. We prove by induction on the number n of the prime divisors of xy in R: $x \in R$ and $y \in R$. In case $n = 0$, the assertion follows from $S^\times = R^\times$. Assume now $n > 0$

and choose a prime element p of R dividing xy. By assumption (2), p divides x or y in S. Assume, e.g., $x = px'$ for some $x' \in S$. Then $xy = px'y$ and thus $x'y \in R$ by assumption on the factoriality of R. Applying the induction hypothesis yields $x' \in R$ and $y \in R$, whence $x = px' \in R$ too. b): By a), every prime element of R is a prime element of S, in particular, prime to the R-module S; by Proposition 1 (3)⇒(2), any two-element R-sequence is an S-sequence. Moreover, by a), R is factorial, $S^\times = R^\times$ and the assertion follows by [7, VII, §1.3, Theorem 2] and Proposition 8 b). □

The next proposition is an analogue to weak content algebras [22, Theorem 1.2]:

Proposition 9 *Let R be a factorial domain and $R \subseteq S$ a factorable extension. The following conditions are equivalent:*

(1) Every prime element of R is a prime element of S or is a unit of S
(2) The product of greatest R-divisors of any two elements x, y of S has the same prime divisors as any greatest R-divisor of xy
(3) The product of any two irreducible elements of the R-module S is irreducible.

Proof (1)⇒(2): Let $x, y \in S$ have greatest R-divisors r, s respectively; further, let t be a greatest R-divisor of xy. Clearly, rs is an R-divisor of xy, i.e. rs divides t; thus, it has only to be shown that any prime divisor p of t divides r or s, i.e. p divides x or y in S. But this is clear from assumption (1), either p being a prime element of S dividing xy in S or p being a unit in S. (2)⇒(3): Clearly, an element is irreducible if and only if it has 1 as greatest R-divisor. (3)⇒(1): Let p be a prime element of R, which is not a unit in S; further, let $x, y \in S$ be such that $xy \in pS$. Choose a greatest R-divisor r (resp. s) of x (resp. y). Then $x = ru$, $y = sv$ for some irreducible elements u, v of the R-module S by Proposition 2. By assumption (3), uv is irreducible, whence rs is a greatest R-divisor of xy by Proposition 2. Thus p divides rs, i.e. p divides r or s, whence $x \in pS$ or $y \in pS$. □

Corollary 23 *Let $R \subseteq S$ be an extension of domains such that S is factorial. The following conditions are equivalent:*

(1) R is factorially closed in S
(2) $S^\times = R^\times$ and S is a factorial R-module such that the product of any two irreducible elements of the R-module S is irreducible.

Proof Immediate by Proposition 3, Lemma 8, and Proposition 9. □

6 Notes

Historically, Nicolas in [17–20] first defined factorial modules and afterwards factorable ones. To clarify and to generalize ideas, we show in Sect. 3 that the assumption on the factoriality of the base domain is not always needed. Basically, the results proven in Sect. 3 seem to be known to Nicolas, Costa [8], and Lu [16] (at least for factorial domains). Their work was substantially complemented by Anderson and Valdes-Leon [4, Theorem 4.4], cf. Proposition 3. In Sect. 4, we follow Costas ideas and results [8] concerning *fdp*, and Lu [16] concerning prime elements *prime for a module*. Our applications of the theory developed so far to *content modules, flat Mittag-Leffler modules* or *locally projective modules* seem to be new. The basic results of Sect. 5 are due to Nicolas [19, 20]. Using that, Costa investigated symmetric algebras [8] and defined and analyzed Krull modules [9] (with J. L. Johnson). Other types of ring extensions, namely $M[X]$ or $M[[X]]$ are investigated by Lu [16]. For connections with faithful multiplication modules, cf. Kim, H. and Kim, M.O. [15]. The characterization of factorially closed extensions of factorial domains (Proposition 9, Corollary 23) seems to be new.

The topic of this note are torsion-free modules over domains; similarly to generalizations of *UFD*s, generalizations to arbitrary modules over arbitrary commutative rings have been considered by Anderson and Valdes-Leon [4, Theorem 2.8, Theorem 2.9]. This attempt has been continued by Nikseresht and Azizi [21].

Acknowledgements I thank the referee for detailed suggestions improving substantially the clearness of the paper.

References

1. Anderson, D.D.: Extensions of unique factorization: a survey. In: Dobbs, D.E., Fontana, M., Kabbaj, S.-E. (eds.) Advances in Commutative Ring Theory: Proceedings of the Third International Conference in Communication Ring Theory, Fez. Lecture Notes in Pure and Applied Mathematics, vol. 205, pp. 31–53. Marcel Dekker Inc., New York (1999)
2. Anderson, D.D., Frazier, A.M.: On a general theory of factorization in integral domains. Rocky Mt. J. Math. **61**, 663–705 (2011)
3. Anderson, D.D., Valdes-Leon, S.: Factorization in commutative rings with zero divisors. Rocky Mt. J. Math. **26**, 439–480 (1996)
4. Anderson, D.D., Valdes-Leon, S.: Factorization in commutative rings with zero divisors II. In: Anderson, D.D. (ed.) Factorization in Integral Domains. Lecture Notes in Pure and Applied Mathematics, vol. 189, pp. 197–219. Marcel Dekker Inc., New York (1997)
5. Atiyah, M.F., Macdonald, I.G.: Introduction to Commutative Algebra. Addison-Wesley Publishing Co., Boston (1969)
6. Bourbaki, N.: Algèbre I. Hermann, Paris (1970)
7. Bourbaki, N.: Commutative Algebra. Addison-Wesley Publishing Co., New York (1972)
8. Costa, D.L.: Unique factorization in modules and symmetric algebras Trans. Am. Math. Soc. **224**, 267–280 (1976)
9. Costa, D.L., Johnson, J.L.: Inert extensions of Krull domains. Proc. Am. Math. Soc. **59**, 189–194 (1976)
10. Eakin, P., Silver, J.: Rings which are almost polynomial rings. Trans. Am. Math. Soc. **174**, 425–449 (1972)
11. Gilmer, R.: Multiplicative Ideal Theory. Marcel Dekker Inc., New York (1972)
12. Göbel, R., Trlifaj, J.: Approximations and Endomorphism Algebras of Modules, 2nd Revised and Extended edn., vol. 1. De Gruyter, Berlin (2012)
13. Jacobson, N.: Basic Algebra I W.H. Freeman and Co., San Francisco (1974)
14. Kaplansky, I.: Commutative Rings. University of Chicago Press, Chicago (1974)
15. Kim, H., Kim, M.O.: Krull modules. Algebra Colloq. **20**, 463–474 (2013)
16. Lu, C.-P.: Factorial modules. Rocky Mt. J. Math. **224**, 125–139 (1976)

17. Nicolas, A.-M.: Modules factoriels, Sém. Dubreil-Pisot: Algèbre et Théorie des Nombres, 20e année (1966/1967) 10-01–10-12
18. Nicolas, A.-M.: Modules factoriels Bull. Sc. math. 2e série **95**, 33–52 (1971)
19. Nicolas, A.-M.: Extensions factorielles, Sém. Dubreil-Pisot: Algèbre et Théorie des Nombres, 27e année (1973/1974) 15-01–15-08
20. Nicolas, A.-M.: Extensions factorielles. Bull. Sc. math., 2e série **98**, 117–143 (1974)
21. Nikseresht, A., Azizi, A.: On factorization in modules. Commun. Algebra **39**, 292–311 (2011)
22. Rush, D.E.: Content algebras. Canad. Math. Bull. **21**, 329–334 (1978)

n-Absorbing Ideals of Commutative Rings and Recent Progress on Three Conjectures: A Survey

Ayman Badawi

Abstract Let R be a commutative ring with $1 \neq 0$. Recall that a proper ideal I of R is called a *2-absorbing ideal* of R if $a, b, c \in R$ and $abc \in I$, then $ab \in I$ or $ac \in I$ or $bc \in I$. A more general concept than 2-absorbing ideals is the concept of n-absorbing ideals. Let $n \geq 1$ be a positive integer. A proper ideal I of R is called an *n-absorbing ideal* of R if $a_1, a_2, \ldots, a_{n+1} \in R$ and $a_1 a_2 \cdots a_{n+1} \in I$, then there are n of the a_i's whose product is in I. The concept of n-absorbing ideals is a generalization of the concept of prime ideals (note that a prime ideal of R is a 1-absorbing ideal of R). In this survey article, we collect some old and recent results on n-absorbing ideals of commutative rings.

Keywords Prime • Primary • Weakly prime • Weakly primary • 2-Absorbing • n-Absorbing • Weakly 2-absorbing • Weakly n-absorbing • 2-Absorbing primary • Weakly 2-absorbing primary

Mathematics Subject Classification (2010) 13A15

Dedicated to Professors David F. Anderson and Luigi Salce on their retirement

1 Introduction

We assume throughout that all rings are commutative with $1 \neq 0$. Over the past several years, there has been considerable attention in the literature to n-absorbing ideals of commutative rings and their generalizations, for example see [1–62]. We recall from [6] that a proper ideal I of R is called a *2-absorbing ideal* of R if $a, b, c \in R$ and $abc \in I$, then $ab \in I$ or $ac \in I$ or $bc \in I$. A more general concept than 2-absorbing ideals is the concept of n-absorbing ideals. Let $n \geq 1$ be a positive integer. A proper ideal I of R is called an *n-absorbing ideal* of R as

A. Badawi (✉)
Department of Mathematics and Statistics, American University of Sharjah, P.O. Box 26666 Sharjah, United Arab Emirates
e-mail: abadawi@aus.edu

M. Fontana et al. (eds.), *Rings, Polynomials, and Modules*,
https://doi.org/10.1007/978-3-319-65874-2_3

in [3] if $a_1, a_2, \ldots, a_{n+1} \in R$ and $a_1a_2 \cdots a_{n+1} \in I$, then there are n of the a_i's whose product is in I. The concept of n-absorbing ideals is a generalization of the concept of prime ideals (note that a prime ideal of R is a 1-absorbing ideal of R).

Let R be a (commutative) ring. Then $\dim(R)$ denotes the Krull dimension of R, $Spec(R)$ denotes the set of prime ideals of R, $Max(R)$ denotes the set of maximal ideals of R, $T(R)$ denotes the total quotient ring of R, $qf(R)$ denotes the quotient field of R when R is an integral domain, and $Nil(R)$ denotes the ideal of nilpotent elements of R. If I is a proper ideal of R, then $Rad(I)$ and $Min_R(I)$ denote the radical ideal of I and the set of prime ideals of R minimal over I, respectively. We will often let 0 denote the zero ideal.

The purpose of this survey article is to collect some properties of n-absorbing ideals in commutative rings. In particular, we state some recent progresses on three outstanding conjectures (see Sect. 5). Our aim is to give the flavor of the subject, but not be exhaustive.

We recall some background material. A prime ideal P of a ring R is said to be a *divided prime ideal* if $P \subset xR$ for every $x \in R \setminus P$; thus, a divided prime ideal is comparable to every ideal of R. An integral domain R is said to be a *divided domain* if every prime ideal of R is a divided prime ideal.

An integral domain R is said to be a *valuation domain* if either $x|y$ or $y|x$ (in R) for all $0 \neq x, y \in R$ (a valuation domain is a divided domain). If I is a nonzero fractional ideal of a ring R, then $I^{-1} = \{ x \in T(R) \mid xI \subseteq R \}$. An integral domain R is called a *Dedekind* (resp., *Prüfer*) *domain* if $II^{-1} = R$ for every nonzero fractional ideal (resp., finitely generated fractional ideal) I of R. Moreover, an integral domain R is a Prüfer domain if and only if R_M is a valuation domain for every maximal ideal M of R.

Some of our examples use the $R(+)M$ construction. Let R be a ring and M an R-module. Then $R(+)M = R \times M$ is a ring with identity $(1, 0)$ under addition defined by $(r, m) + (s, n) = (r + s, m + n)$ and multiplication defined by $(r, m)(s, n) = (rs, rn + sm)$.

2 Basic Properties of n-Absorbing Ideals

Let I be a proper ideal of R. If I be an n-absorbing ideal of R for some positive integer n, then recall from [3] that $\omega_R(I) = \min\{ n \mid I \text{ is an } n\text{-absorbing ideal of } R \}$; otherwise, set $\omega_R(I) = \infty$. It is convenient to define $\omega_R(R) = 0$. We start by recalling some basic properties of n-absorbing ideals.

Theorem 2.1

1. *([6, Theorem 2.3]). Let I be a 2-absorbing ideal of a ring R. Then there are at most two prime ideals of R that are minimal over I (i.e., $|Min_R(I)| = 1$ or 2).*
2. *([6, Theorems 2.1 and 2.4]). Let I be a 2-absorbing ideal of a ring R. Then $Rad(I)$ is a 2-absorbing ideal of R and $(Rad(I))^2 \subseteq I$.*
3. *([3, Theorem 2.5]). Let I be an n-absorbing ideal of a ring R. Then there are at most n prime ideals of R minimal over I. Moreover, $|Min_R(I)| \leq \omega_R(I)$.*

4. *([3, Theorem 2.9]). Let $M_1, \ldots, M_n$ be maximal ideals of a ring R (not necessarily distinct). Then $I = M_1 \cdots M_n$ is an n-absorbing ideal of R. Moreover, $\omega_R(I) \leq n$.*
5. *([6, Theorem 2.4]). Let I be an 2-absorbing ideal of a ring R with exactly two minimal prime ideals P_1, P_2 over I. Then $P_1P_2 \subseteq I$.*
6. *([3, Theorem 2.14]). Let I be an n-absorbing ideal of a ring R such that I has exactly n minimal prime ideals, say $P_1, \ldots, P_n$. Then $P_1 \cdots P_n \subseteq I$. Moreover, $\omega_R(I) = n$.*
7. *([6, Theorem 2.5]). Let I be a 2-absorbing ideal of R such that $Rad(I) = P$ is a prime ideal of R and suppose that $I \neq P$. For each $x \in P \setminus I$ let $B_x = (I :_R x) = (I :_R x) = \{y \in R \mid yx \in I\}$. Then B_x is a prime ideal of R containing P. Furthermore, either $B_y \subseteq B_x$ or $B_x \subseteq B_y$ for every $x, y \in P \setminus I$.*
8. *([6, Theorem 2.6]). Let I be a 2-absorbing ideal of R such that $I \neq Rad(I) = P_1 \cap P_2$, where P_1 and P_2 are the only nonzero distinct prime ideals of R that are minimal over I. Then for each $x \in Rad(I) \setminus I$, $B_x = (I :_R x) = \{y \in R \mid xy \in I\}$ is a prime ideal of R containing P_1 and P_2. Furthermore, either $B_y \subseteq B_x$ or $B_x \subseteq B_y$ for every $x, y \in Rad(I) \setminus I$.*
9. *([3, Theorem 3.4]). Let I be an n-absorbing ideal of a ring R. Then $(I :_R x) = \{y \in R \mid yx \in I\}$ is an n-absorbing ideal of R containing I for all $x \in R \setminus I$. Moreover, $\omega_R(I_x) \leq \omega_R(I)$ for all $x \in R$.*
10. *([3, Theorem 3.5]). Let $n \geq 2$ and $I \subset Rad(I)$ be an n-absorbing ideal of a ring R. Suppose that $x \in Rad(I) \setminus I$, and let $m(\geq 2)$ be the least positive integer such that $x^m \in I$. Then $(I :_R x^{m-1}) = \{y \in R \mid yx^{m-1} \in I\}$ is an $(n-m+1)$-absorbing ideal of R containing I.*
11. *([3, Corollary 2.6]). Let $n \geq 2$ and $I \subset Rad(I)$ be an n-absorbing ideal of a ring R. Suppose that $x \in Rad(I) \setminus I$ and $x^n \in I$, but $x^{n-1} \notin I$. Then $(I :_R x^{n-1}) = \{y \in R \mid yx^{n-1} \in I\}$ is a prime ideal of R containing $Rad(I)$.*
12. *([3, Corollary 2.7]). Let $n \geq 2$ and I be an n-absorbing P-primary ideal of a ring R for some prime ideal P of R. If $x \in Rad(I) \setminus I$ and n is the least positive integer such that $x^n \in I$, then $(I :_R x^{n-1}) = \{y \in R \mid yx^{n-1} \in I\} = P$.*
13. *([3, Theorem 3.8]). Let $n \geq 2$ and $I \subset Rad(I)$ be an n-absorbing ideal of a ring R such that I has exactly n minimal prime ideals, say $P_1, \ldots, P_n$. Suppose that $x \in Rad(I) \setminus I$, and let $m(\geq 2)$ be the least positive integer such that $x^m \in I$. Then every product of $n - m + 1$ of the P_i's is contained in $(I :_R x^{m-1}) = \{y \in R \mid yx^{m-1} \in I\}$.*
14. *([3, Theorem 3.9]). Let I be a P-primary ideal of a ring R such that $P^n \subseteq I$ for some positive integer n (for example, if R is a Noetherian ring), and let $x \in P \setminus I$. If $x^m \notin I$ for some positive integer m, then $(I :_R x^m) = \{y \in R \mid yx^m \in I\}$ is an $(n - m)$-absorbing ideal of R.*

Assume that I is a proper ideal of R such that $I \neq Rad(I)$. The following two results give a characterization of 2-absorbing ideals in terms of $(I :_R x) = \{y \in R \mid yx \in I\}$, where $x \in Rad(I) \setminus I$.

Theorem 2.2 ([6, Theorem 2.8]) *Let I be an ideal of R such that $I \neq Rad(I)$ and $Rad(I)$ is a prime ideal of R. Then the following statements are equivalent:*

1. *I is a 2-absorbing ideal of R;*
2. $B_x = \{y \in R \mid yx \in I\}$ *is a prime ideal of R for each* $x \in Rad(I) \setminus I$.

Theorem 2.3 ([6, Theorem 2.9]) *Let I be an ideal of R such that* $I \neq Rad(I) = P_1 \cap P_2$, *where* P_1 *and* P_2 *are nonzero distinct prime ideals of R that are minimal over I. Then the following statement are equivalent:*

1. *I is a 2-absorbing ideal of R;*
2. $P_1P_2 \subseteq I$ *and* $B_x = \{y \in R \mid yx \in I\}$ *is a prime ideal of R for each* $x \in Rad(I) \setminus I$.
3. $B_x = \{y \in R \mid yx \in I\}$ *is a prime ideal of R for each* $x \in (P_1 \cup P_2) \setminus I$.

In view of Theorem 2.2, the following is an example of a prime ideal P of an integral domain R such that P^2 is not a 2-absorbing ideal of R.

Example 2.4 ([6, Example 3.9]) Let $R = \mathbb{Z} + 6X\mathbb{Z}[X]$ and $P = 6X\mathbb{Z}[X]$. Then P is a prime ideal of R. Since $6X^2 \in P \setminus P^2$ and $B_{6X^2} = \{y \in R \mid 6X^2y \in P^2\} = 6\mathbb{Z} + 6X\mathbb{Z}[X]$ is not a prime ideal of R, P^2 is not a 2-absorbing ideal of R by Theorem 2.2.
The following result characterizes all P-primary ideals that are 2-absorbing ideals.

Theorem 2.5 ([6, Theorem 3.1]) *Let I be a P-primary ideal of a ring R for some prime ideal P of R. Then I is a 2-absorbing ideal of R if and only if* $P^2 \subseteq I$. *In particular,* M^2 *is a 2-absorbing ideal of R for each maximal ideal M of R.*

The following is an example of a prime ideal P of an integral domain R such that P^2 is a 2-absorbing ideal of R, but P^2 is not a P-primary ideal of R.

Example 2.6 ([6, Example 3.11]) Let $R = \mathbb{Z} + 3x\mathbb{Z}[X]$ and let $P = 3X\mathbb{Z}[X]$ be a prime ideal of R. Since $3(3X^2) \in P^2$, we conclude that P^2 is not a P-primary ideal of R. It is easy to verify that if $d \in P \setminus P^2$, then either $B_d = \{y \in R \mid yd \in I\} = P$ or $B_d = 3\mathbb{Z} + 3X\mathbb{Z}[X]$ is a prime ideal of R. Hence P^2 is a 2-absorbing ideal by Theorem 2.2.

Let I be an ideal of R such that $Rad(I) = P$ is a nonzero divided prime ideal of R. The following result characterizes all such ideals that are 2-absorbing ideals.

Theorem 2.7 ([6, Theorem 3.6]) *Suppose that P is a nonzero divided prime ideal of R and I is an ideal of R such that* $Rad(I) = P$. *Then the following statements are equivalent:*

1. *I is a 2-absorbing ideal of R;*
2. *I is a P-primary ideal of R such that* $P^2 \subseteq I$.

Theorem 2.8 ([6, Theorem 3.7] and [3, Theorem 3.3]) *Let* $n \geq 1$ *be a positive integer. Suppose that Nil(R) and P are divided prime ideals of a ring R such that* $P \neq Nil(R)$. *Then* P^n *is a P-primary ideal of R, and thus* P^n *is an n-absorbing ideal of R with* $\omega_R(P^n) \leq n$. *Moreover,* $\omega_R(P^n) = n$ *if* $P^{n+1} \subset P^n$.

In view of Theorems 2.5, 2.7, and 2.8, for $n \geq 3$, we have the following two results.

Theorem 2.9 ([3, Theorem 3.1]) *Let P be a prime ideal of a ring R, and let I be a P-primary ideal of R such that $P^n \subseteq I$ for some positive integer n (for example, if R is a Noetherian ring). Then I is an n-absorbing ideal of R. Moreover, $\omega_R(I) \leq n$. In particular, if P^n is a P-primary ideal of R, then P^n is an n-absorbing ideal of R with $\omega_R(P^n) \leq n$, and $\omega_R(P^n) = n$ if $P^{n+1} \subset P^n$.*

Theorem 2.10 ([3, Theorem 3.2]) *Let P be a divided prime ideal of a ring R, and let I be an n-absorbing ideal of R with $Rad(I) = P$. Then I is a P-primary ideal of R.*

Mostafanasab and Darani in [50] proved the following result.

Theorem 2.11 ([50, Theorem 2.15]) *(n-absorbing avoidance theorem). Let $I_1, I_2, \ldots, I_m$ $(m \geq 2)$ be ideals of R such that I_i is an n_i-absorbing ideal of R for every $3 \leq i \leq m$. Suppose that $I_i \not\subseteq (I_j :_R x^{n_j - 1}) \subset R$ for every $x \in Rad(I_j) \setminus I_j$ with $i \neq j$. If I is an ideal of R such that $I \subseteq I_1 \cup I_2 \cup \cdots \cup I_m$, then $I \subseteq I_i$ for some $1 \leq i \leq m$.*

3 Extensions of n-Absorbing Ideals

The following results show the stability of n-absorbing ideals in various ring-theoretic constructions. These results generalize well-known results about prime ideals.

Theorem 3.1

1. *([3, Theorem 4.1]). Let I be an n-absorbing ideal of a ring R, and let S be a multiplicatively closed subset of R with $I \cap S = \emptyset$. Then I_S is an n-absorbing ideal of R_S. Moreover, $\omega_{R_S}(I_S) \leq \omega_R(I)$.*
2. *Let $f : R \longrightarrow T$ be a homomorphism of rings.*
   a. *([3, Theorem 4.1]). Let J be an n-absorbing ideal of T. Then $f^{-1}(J)$ is an n-absorbing ideal of R. Moreover, $\omega_R(f^{-1}(J)) \leq \omega_T(J)$.*
   b. *Let f be surjective and I be an n-absorbing ideal of R containing $ker(f)$. Then $f(I)$ is an n-absorbing ideal of T if and only if I is an n-absorbing ideal of R. Moreover, $\omega_T(f(I)) = \omega_R(I)$. In particular, this holds if f is an isomorphism.*

In the following result, we determine the n-absorbing ideals in the product of any two rings.

Theorem 3.2 ([3, Theorem 4.7]) *Let I_1 be an m-absorbing ideal of a ring R_1 and I_2 an n-absorbing ideal of a ring R_2. Then $I_1 \times I_2$ is an $(m+n)$-absorbing ideal of the ring $R_1 \times R_2$. Moreover, $\omega_{R_1 \times R_2}(I_1 \times I_2) = \omega_{R_1}(I_1) + \omega_{R_2}(I_2)$.*

Let R be a ring, M be an R-module, and $T = R(+)M$. If I is an n-absorbing ideal of R, then it is easy to show that $I(+)M$ is an n-absorbing ideal of T. In fact, $\omega_T(I(+)M) = \omega_R(I)$. We have the following result for the special case $T = R(+)R$, where R is an integral domain.

Theorem 3.3 ([3, Theorem 4.10]) *Let D be an integral domain, $R = D(+)D$, and I be an n-absorbing ideal of D that is not an $(n-1)$-absorbing ideal of D. Then $0(+)I$ is an $(n+1)$-absorbing ideal of R that is not an n-absorbing ideal of R; so $\omega_R(0(+)I) = \omega_D(I) + 1$. In particular, if P is a prime ideal of D, then $0(+)P$ is a 2-absorbing ideal of R.*

Let T be a ring extension of an integral domain D and P a prime ideal of D. Then $0(+)P$ need not be a 2-absorbing ideal of the ring $R = D(+)T$; so Theorem 3.3 does not extend to general R. We have the following example.

Example 3.4 ([3, Example 4.12]) Let $R = \mathbb{Z}(+)\mathbb{Q}$. Then $I = 0(+)2\mathbb{Z}$ is an ideal of R with $Rad(I) = 0(+)\mathbb{Q}$. Let $x = (0, \frac{1}{2}) \in Rad(I) \setminus I$. Then $B_x = (I :_R x) = (4\mathbb{Z})(+)\mathbb{Q}$ is not a prime ideal of R ($\omega_R(B_x) = 2$), and hence I is not a 2-absorbing ideal of R by Theorem 2.2. In fact, one can easily show that I is not an n-absorbing ideal of R for any positive integer n. For each positive integer n, let $x_i = (2, 0)$ for $1 \le i \le n$ and $x_{n+1} = (0, \frac{1}{2^{n-1}})$. Then $x_1 \cdots x_{n+1} = (0, 2) \in I$, but no proper subproduct of the x_i's is in I. Thus $\omega_R(I) = \infty$.

We next consider extensions of n-absorbing ideals of R in the polynomial ring $R[X]$ and the power series ring $R[[X]]$.

Theorem 3.5 *Let I be a proper ideal of a ring R. Then*

1. *([3, Theorem 4.13]). (I, X) is an n-absorbing ideal of $R[X]$ if and only if I is an n-absorbing ideal of R. Moreover, $\omega_{R[X]}((I, X)) = \omega_R(I)$.*
2. *([3, Theorem 4.15]). $I[X]$ is a 2-absorbing ideal of $R[X]$ if and only if I is a 2-absorbing ideal of R. (If $n \ge 3$ and I is an n-absorbing ideal of R, does it follow that $I[X]$ is an n-absorbing ideal of $R[X]$? (See Sect. 5.)*
3. *([44, Proposition 2.13]) $I[[X]]$ is a 2-absorbing ideal of $R[[X]]$ if and only if I is a 2-absorbing ideal of R (and therefore $I[X]$ is a 2-absorbing ideal of $R[X]$ if and only if I is a 2-absorbing ideal of R).*

Let K be a field. For rings of the form $D + XK[[X]]$, where D is a subring of K, we have the following result.

Theorem 3.6 ([3, Theorem 4.17]) *Let D be a subring of a field K and $R = D + XK[[X]]$.*

(a) If D is a field, then every proper ideal of R is an n-absorbing ideal of R for some positive integer n.
(b) If D is a proper subring of K with $qf(D) = K$, then the nonzero n-absorbing ideals of R have the form $I + XK[[X]]$, where I is an n-absorbing ideal of D, or $X^m K[[X]]$ for m an integer with $1 \le m \le n$. Moreover, $\omega_R(I + XK[[X]]) = \omega_D(I)$ and $\omega_R(X^m K[[X]]) = m$.

4 n-Absorbing Ideals in Specific Rings

If R is a Noetherian ring, then we have the following result.

Theorem 4.1 ([3, Theorem 5.3]) *Let R be a Noetherian ring. Then every proper ideal of R is an n-absorbing ideal of R for some positive integer n.*

A characterization of Dedekind domains in terms of 2-absorbing ideals is first given in [6, Theorem 3.15] and a similar characterization of Dedekind domains in terms of n-absorbing ideals ($n \geq 2$) is given in [3, Theorem 5.1].

Theorem 4.2 ([6, Theorem 3.15] and [3, Theorem 5.1]) *If R be a Noetherian integral domain. Then the following statements are equivalent.*

1. *R is a Dedekind domain.*
2. *If I is an n-absorbing ideal of R, then* $I = M_1 \cdots M_m$ *for maximal ideals* $M_1, \ldots, M_m$ *of R with* $1 \leq m \leq n$.

   Moreover, if $I = M_1 \cdots M_n$ *for maximal ideals* $M_1, \ldots, M_n$ *of a Dedekind domain R which is not a field, then* $\omega_R(I) = n$.

All 2-absorbing ideals of a valuation domain are determined in [6, Theorem 3.10]. If $n \geq 3$, then a similar result [3, Theorem 5.5] determines all n-absorbing ideals of a valuation domain.

Theorem 4.3 ([6, Theorem 3.10] and [3, Theorem 5.5]) *Let R be a valuation domain and n a positive integer. Then the following statements are equivalent for an ideal I of R.*

(1) *I is an n-absorbing ideal of R.*
(2) *I is a P-primary ideal of R for some prime ideal P of R and* $P^n \subseteq I$.
(3) $I = P^m$ *for some prime ideal* $P(= Rad(I))$ *of R and integer m with* $1 \leq m \leq n$.

Moreover, $\omega_R(P^n) = n$ *for P a nonidempotent prime ideal of R.*

Theorem 4.4 ([50, Proposition 2.10]) *Let V be a valuation domain with quotient field K, and let I be a proper ideal of V. Then I is an n-absorbing ideal of V if and only if whenever* $x_1x_2 \cdots x_{n+1} \in I$ *with* $x_1, x_2, \ldots, x_{n+1} \in K$, *then there are n of* $x_1, x_2, \ldots, x_{n+1}$ *whose product is in I.*

All 2-absorbing ideals of a Prüfer domain are determined in [6, Theorem 3.14].

Theorem 4.5 ([3, Theorem 3.14]) *Let R be a Prüfer domain and I be a nonzero ideal of R. Then the following statements are equivalent:*

1. *I is a 2-absorbing ideal of R;*
2. *I is a prime ideal of R or* $I = P^2$ *is a P-primary ideal of R or* $I = P_1 \cap P_2$, *where* P_1 *and* P_2 *are nonzero prime ideals of R.*

If n is a positive integer and R is a Prüfer domain, then we have the following result.

Theorem 4.6 ([6, Theorem 5.7]) *Let R be a Prüfer domain. Then an ideal I of R is an n-absorbing ideal of R for some positive integer n if and only if I is a product of prime ideals of R. Moreover, if* $P_1, \ldots, P_k$ *are incomparable prime ideals of R and* $n_1, \ldots, n_k$ *are positive integers with* $n_i = 1$ *if* P_i *is idempotent, then* $\omega_R(P_1^{n_1} \cdots P_k^{n_k}) = n_1 + \cdots + n_k$.

5 Strongly n-Absorbing Ideals and Recent Progresses on Three Conjectures

It is well known that a proper ideal I of a ring R is a prime ideal of R if and only if whenever $I_1 I_2 \subseteq I$ for ideals I_1, I_2 of R, then either $I_1 \subseteq I$ or $I_2 \subseteq I$. Let n be a positive integer. We recall from [3] that a proper ideal I of a ring R is a *strongly n-absorbing ideal* if whenever $I_1 \cdots I_{n+1} \subseteq I$ for ideals $I_1, \ldots, I_{n+1}$ of R, then the product of some n of the I_j's is in I. Thus a strongly 1-absorbing ideal is just a prime ideal, and the intersection of n prime ideals is a strongly n-absorbing ideal. It is clear that a strongly n-absorbing ideal of R is also an n-absorbing ideal of R, and in [6, Theorem 2.13], it was shown that these two concepts agree when $n = 2$.

Theorem 5.1 ([6, Theorem 2.13]) *Let I be a proper ideal of R. Then I is a 2-strongly absorbing ideal of R if and only if I is a 2-absorbing ideal of R.*

If R is a Prüfer domain and I is a proper ideal of R, it was shown in [3, Corollary 6.9] that I is an n-strongly absorbing ideal of R if and only if I is an n-absorbing ideal of R..

Theorem 5.2 ([3, Corollary 6.9]) *Let R be a Prüfer domain and n a positive integer. Then an ideal I of R is a strongly n-absorbing ideal of R if and only if I is an n-absorbing ideal of R.*

In view of Theorem 5.1, the following result is a generalization of Theorem 2.5 [6, Theorem 3.1].

Theorem 5.3 ([3, Theorem 6.6]) *Let I be a P-primary ideal of a ring R and n a positive integer. Then the following statements are equivalent.*

(1) I is an n-absorbing ideal of R and $P^n \subseteq I$.
(2) I is a strongly n-absorbing ideal of R.

In particular, if P^n is P-primary, then P^n is a strongly n-absorbing ideal of R.

For a Notherian ring R, we have the following result.

Theorem 5.4 ([3, Corollary 6.8]) *Let R be a Noetherian ring. Then every proper ideal of R is a strongly n-absorbing ideal of R for some positive integer n.*

Theorem 5.5 ([3, Corollary 6.7]) *Let $M_1, \ldots, M_n$ be maximal ideals of a ring R. Then $I = M_1 \cdots M_n$ is a strongly n-absorbing ideal of R.*

In view of Theorem 5.1, the following result is a generalization of [6, Theorem 2.4].

Theorem 5.6 ([3, Theorem 6.2]) *Let n be a positive integer and I a strongly n-absorbing ideal of a ring R such that I has exactly $m(\leq n)$ minimal prime ideals $P_1, \ldots, P_m$. Then $P_1^{n_1} \cdots P_m^{n_m} \subseteq I$ for positive integers $n_1, \ldots, n_m$ with $n = n_1 + \cdots + n_m$. In particular, if $Rad(I) = P$ is a prime ideal of R, then $P^n \subseteq I$.*

Theorem 5.7 ([50, Corollary 2.14]) *Let I_i be a strongly n_i-absorbing ideal of a ring R for every $1 \leq i \leq m$ $(m \geq 2)$. If I is an ideal of R such that $I \subseteq I_1 \cup I_2 \cdots \cup I_m$, then $I^{n_i} \subseteq I_i$ for some $1 \leq i \leq m$.*

Three outstanding conjectures on n-absorbing ideals are the following (see Anderson and Badawi [3] and also Cahen et al. [14, Problem 30]):

1. **Conjecture one**. If an ideal of R is n-absorbing, then it is strongly n-absorbing.
2. **Conjecture two**. If an ideal I of R is n-absorbing, then $I[X]$ is an n-absorbing ideal of $R[X]$.
3. **Conjecture three**. If an ideal I of R is n-absorbing, then $(Rad(I))^n \subseteq I$.

Laradji in [44] gave an affirmative answer for Conjecture three when $n = 3$. Note that an affirmative answer for Conjecture three was given in Theorem 2.1(2) when $n = 2$.

Theorem 5.8 ([44, Proposition 2.7]) *Let I be a 3-absorbing ideal of R. Then $(Rad(I))^3 \subseteq I$.*

Recently, Choi and Walker in [21, Theorem 1] gave an affirmative answer for Conjecture three for any positive integer n.

Theorem 5.9 ([21, Theorem 1]) *Let n be a positive integer and I be an n-absorbing ideal of R. Then $(Rad(I))^n \subseteq I$.*

It was shown [3, Theorem 6.1] that Conjecture one implies Conjecture three.

Theorem 5.10 ([3, Theorem 6.1]) *Let n be a positive integer and I be a strongly n-absorbing ideal of R. Then $(Rad(I))^n \subseteq I$.*

Laradji in [44] showed that Conjecture two implies Conjecture one.

Theorem 5.11 ([44, Proposition 2.9(i)]) *Let I be a proper ideal of R and n be a positive integer. If $I[X]$ is an n-absorbing ideal of $R[X]$, then I is a strongly n-absorbing ideal of R.*

Let $f(X) = a_m x^m + \cdots + a_0 \in R[X]$, for some positive integer m and for some $a_m, \ldots, a_0 \in R$. Then $c(f) = (a_m, \ldots, a_0)R$ is an ideal of R and it is called the *content* of $f(X)$. We recall that a ring R is called *Armandariz* if whenever $f(X)g(X) = 0 \in R[X]$ for some $f(X), g(X) \in R[X]$, then $c(f)c(g) = 0 \in R$.

Let I be a strongly n-absorbing ideal of R. The author in [44] showed that if R/I is Armandariz, then Conjecture one implies Conjecture two.

Theorem 5.12 ([44, Proposition 2.9(ii)]) *Let I be a strongly n-absorbing ideal of R for some positive integer n. If R/I is Armandariz, then $I[X]$ is an n-absorbing ideal of $R[X]$.*

Note that Theorem 5.2 gives an affirmative answer for Conjecture one when R is a Prüfer domain.

Let I be an n-absorbing ideal of R. Darani and Puczylowski in [27] proved that Conjecture one holds if the additive group of R/I is torsion-free.

Theorem 5.13 ([27, Theorem 4.2]) *Let I be a proper ideal of R and n be a positive integer. If I is an n-absorbing ideal of R such that the additive group of R/I is torsion-free, then I is a strongly n-absorbing ideal of R.*

Donadze in [31] proved that Conjecture one holds in the following case.

Theorem 5.14 ([31, Proposition 2.2]) *Let R be a ring and $n \geq 2$ be an integer. Suppose that R contains $n-1$ distinct invertible elements $u_1, \ldots, u_{n-1}$ such that $u_i - u_j$ is also invertible for all $i \neq j$, $1 \leq i, j \leq n-1$. Then every n-absorbing ideal of R is strongly n-absorbing.*

Laradji in [44] proved that Conjecture two holds in the following cases.

Theorem 5.15 ([44, Proposition 2.10]) *Let n be a positive integer, I be an n-absorbing ideal of R, and let $S = R/I$. Then $I[X]$ is an n-absorbing ideal of $R[X]$ in each of the following cases.*

1. *S is Armendariz and $|R/M| \geq n$ for each maximal ideal M of R containing I.*
2. *S is Armendariz and is $(n-1)!$-torsion free as an additive group.*
3. *S is torsion-free as an additive group.*

Donadze in [31] proved the following result.

Theorem 5.16 ([31, Corollary 2.10]) *If Conjecture two holds for $\mathbb{Z}[X_1, \ldots, X_m]$ for all $m \geq 1$, then Conjecture one holds for any commutative ring R.*

Recall that *R* is called *arithmetical* ring if the set of ideals of every localization of *R* by a prime ideal of *R* is totally ordered by inclusion.

Laradji in [44] proved that Conjecture two holds if *R* is arithmetical.

Theorem 5.17 ([44, Corollary 2.11]) *Let n be a positive integer and I be an n-absorbing ideal of an arithmetical ring R. Then $I[X]$ is an n-absorbing ideal of $R[X]$.*

In light of Theorems 5.17, 5.11, and 5.10, we conclude that all three Conjectures hold if *R* is arithmetical.

Theorem 5.18 *Let R be an arithmetical ring (for example, if R is a Prüfer domain). If I is an n-absorbing ideal of R for some positive integer n, then the following statements are true:*

1. *I is a strongly n-absorbing ideal of R;*
2. *$I[X]$ is an n-absorbing ideal of $R[X]$;*
3. *$(Rad(I))^n \subseteq I$.*

Laradji in [44] showed that when attempting to prove either Conjecture one, Conjecture two, or Conjecture three, it is enough to restrict our attention to the zero ideal of some total quotient rings.

Theorem 5.19 ([44, Proposition 2.15]) *Let I be a proper ideal of R and $T(R/I)$ be the total quotient ring of R/I. If Conjecture one, Conjecture two, or Conjecture three holds for the zero ideal of $T(R/I)$, then it holds for I.*

Let *I* be a proper ideal of *R*. Badawi and Anderson in [3] conjectured that $\omega_{R[X]}(I[X]) = \omega_R(I)$.

In view of Theorem 5.18, we have the following result.

Theorem 5.20 *Let R be an arithmatical ring (for example, if R is a Prüfer domain). Then $\omega_{R[X]}(I[X]) = \omega_R(I)$ for every proper ideal I of R.*

Nesehpour in [52, Corollary 10], independently, proved that $\omega_{R[X]}(I[X]) = \omega_R(I)$ for every proper ideal *I* of a Prüfer domain *R*.

6 n-Krull Dimension of Commutative Rings

From [46], we recall the following definitions.

Definition ([46])

1. Let R be a ring and n a positive integer. A chain of ideals: $I_0 \subset I_1 \cdots \subset I_m$, where $I_0, I_1, \ldots, I_m$ are distinct n-absorbing ideals of R, is called a chain of n-absorbing ideals of length m. The *n-Krull dimension* of R, denoted by $dim_n(R)$, is defined to be the supremum of the lengths of these chains. Thus $dim_1(R)$ is just the usual Krull dimension, dim(R), of R.
2. An n-absorbing ideal I of R is called a *minimal n-absorbing* ideal of the ideal J if $J \subseteq I$ and there is no n-absorbing ideal L such that $J \subseteq L \subset I$. An n-absorbing ideal I of R is called *a minimal n-absorbing ideal of R* if I is a minimal n-absorbing ideal of 0.
3. If I is an n-absorbing ideal of R, the n-height of I, denoted by $ht_n(I)$, is defined to be the supremum of lengths of chains $I_0 \subset I_1 \cdots \subset I_m$ of n-absorbing ideals of R for which $I_m = I$ if this supremum exists, and ∞ otherwise.
4. If I is a proper ideal of R (not necessarily an n-absorbing ideal) and n a positive integer, the n-height of I, denoted by $ht_n(I)$, is defined to be $min\{ht_n(J) \mid J$ is an n-absorbing ideal and $I \subseteq J\}$.

Remark 6.1 Although every prime ideal of R is an n-absorbing ideal for each $n \geq 1$, there exists a minimal prime ideal which is not a minimal n-absorbing ideal for each $n \geq 2$. For example, if $R = K[X]$ is the polynomial ring in one variable X over a field K, the minimal prime ideal $P = RX$ of (0) is not a minimal 2-absorbing ideal of 0, since by [3, Lemma 2.8], RX^2 is a 2-absorbing ideal of R.

Let $l(R)$ denote the length of a composition series for a ring R which is of finite length. We recall the following results.

Theorem 6.2 *Let R be a ring. Then*

1. *([46, Theorem 2.1]). For each positive integer n, there is an n-absorbing ideal of R which is minimal among all n-absorbing ideals of R.*
2. *([46, Theorem 2.1]. If I a proper ideal of R, then for each positive integer n, there is an n-absorbing ideal of R which is minimal among all n-absorbing ideals of R containing I.*
3. *([46, Theorem 2.7]). Let n be a positive integer. If $dim_n(R)$ is finite, then $dim_n(R) = sup\{ht_n(M) \mid M$ is a maximal ideal of $R\}$.*
4. *([46, Theorem 2.8]). If R is an Artinian ring, then $dim_n(R)$ is finite for each positive integer n.*
5. *([46, Theorem 2.9]). If (R, M) is a quasilocal Noetherian domain with maximal ideal M such that $dim_1(R) = 1$, then $dim_2(R)$ is finite.*
6. *([46, Theorem 2.12]). If (R, M) is a quasilocal Artinian ring and n is the smallest positive integer such that $M^n = 0$, then $dim_k R = l(R) - 1$ for each $k \geq n$.*
7. *([46, Theorem 2.13]). If R is an Artinian ring with k maximal ideals, then there exists a positive integer n such that $dim_n(R) = l(R) - k$.*

It was shown [46, Theorem 2.10] that if Conjecture three holds (see Sect. 5), then Theorem 6.2(5) can be extended to any positive integer n. Hence in view of Theorem 5.9, we have the following result.

Theorem 6.3 ([21, Theorem 1] and [46, Theorem 2.10]) *If (R, M) is a quasilocal Noetherian domain with maximal ideal M such that $dim_1(R) = 1$, then $dim_n(R)$ is finite for every positive integer n.*

In light of Theorem 4.2, the following result provides a characterization of Dedekind domains in terms of n-Krull dimension.

Theorem 6.4 ([46, Theorem 2.13]) *Let R be a Noetherian integral domain which is not a field. Then the following statements are equivalent.*

1. *R is a Dedekind domain.*
2. *$dim_n(R) = n$ for every positive integer n.*
3. *$dim_2(R) = 2$.*

Theorem 6.5 ([46, Theorem 2.21]) *Let (R, M) be a discrete valuation ring and I an ideal of R. Then*

1. *I is an n-absorbing ideal for some positive integer n and $\omega_R(I) = l_R(R/I)$.*
2. *For every positive integer n, $dim_n(R) = l_R(R/M^n) = n$.*

7 (m, n)-Closed Ideals and Quasi-n-Absorbing Ideals

We start by recalling some definitions.

Definition Let I be a proper ideal I of R. Then

1. ([2]). I is called a *semi-n-absorbing* ideal of R if $x^{n+1} \in I$ for $x \in R$ implies $x^n \in I$. More generally, for positive integers m and n, I is said to be an *(m, n)-closed* ideal of R if $x^m \in I$ for $x \in R$ implies $x^n \in I$ (observe that I is a semi-n-absorbing ideal of R if and only if I is a $(n + 1, n)$-closed ideal of R).
2. ([2]). For positive integers m and n, I is said to be an *strongly (m, n)-closed* ideal of R if $J^m \subseteq I$ for some ideal J of R implies $J^n \subseteq I$.
3. ([50]). I is called a *quasi-n-absorbing* ideal if whenever $a^n b \in I$ for some $a, b \in R$, then $a^n \in I$ or $a^{(n-1)} b \in I$.
4. [50]. I is called a *strongly quasi-n-absorbing* ideal if whenever $I_1^n I_2 \subseteq I$ for some ideals I_1, I_2 of R, then $I_1^n \subseteq I$ or $I_1^{(n-1)} I_2 \subseteq I$.

Remark 7.1 Note that Mostafanasab and Darani in [50] called a proper ideal I of R to be a *semi-(m, n)-absorbing* ideal if I is an (m, n)-closed ideal.

The following examples show that for every integer $n \geq 2$, there is a semi-n-absorbing ideal (i.e., $(n + 1, n)$-closed ideal) that is neither a radical ideal nor an n-absorbing ideal, and that there is an ideal that is not a semi-n-absorbing ideal (i.e., $(n + 1, n)$-closed ideal) for any positive integer n.

Example 7.2 ([2, Example 2.2])

1. Let $R = \mathbb{Z}$, $n \geq 2$ an integer, and $I = 2 \cdot 3^n\mathbb{Z}$. Then I is a semi-n-absorbing ideal (i.e., $(n+1, n)$-closed ideal) of R. In fact, I is a semi-m-absorbing ideal for every integer $m \geq n$. However, $(2 \cdot 3^{n-1})^2 \in I$ and $2 \cdot 3^{n-1} \notin I$; so I is not a radical ideal of R. Moreover, $2 \cdot 3^n \in I$, $3^n \notin I$, and $2 \cdot 3^{n-1} \notin I$; so I is not an n-absorbing ideal of R (but, I is an $(n+1)$-absorbing ideal of R). Note that for $n = 1$, $I = 6\mathbb{Z}$ is a semi-1-absorbing ideal (i.e., radical ideal) of R, but not a 1-absorbing ideal (i.e., prime ideal) of R.
2. Let $R = \mathbb{Q}[\{X_n\}_{n\in\mathbb{N}}]$ and $I = (\{X_n^n\}_{n\in\mathbb{N}})$. Then $X_{n+1}^{n+1} \in I$ and $X_{n+1}^n \notin I$ for every positive integer n; so I is not a semi-n-absorbing ideal (i.e., $(n+1, n)$-closed ideal) for any positive integer n. Thus I is (m, n)-closed if and only if $1 \leq m \leq n$.
3. Let R be a commutative Noetherian ring. Then every proper ideal of R is an n-absorbing ideal of R, and hence a semi-n-absorbing ideal of R, for some positive integer n (Theorem 4.1). Thus, for every proper ideal I of R, there is a positive integer n such that I is (m, n)-closed for every positive integer m. Note that the ring in (2) is not Noetherian.
4. Clearly, an n-absorbing ideal of R is also an $(n+1)$-absorbing ideal of R. However, this need not be true for semi-n-absorbing ideals. For example, it is easily seen that $I = 16\mathbb{Z}$ is a semi-2-absorbing ideal (i.e., $(3, 2)$-closed ideal) of $\mathbb{Z}$, but not a semi-3-absorbing ideal (i.e., $(4, 3)$-closed ideal) of $\mathbb{Z}$.
5. Let R be a valuation domain. Then it is known that a radical ideal of R is also a prime ideal of R, i.e., a semi-1-absorbing ideal of R is a 1-absorbing ideal of R. However, a semi-n-absorbing ideal of R need not be an n-absorbing ideal of R for $n \geq 2$. For example, let $R = \mathbb{Z}_{(2)}$ and $I = 16\mathbb{Z}_{(2)}$. Then R is a DVR, and it is easily verified that I is a semi-2-absorbing ideal (i.e., $(3, 2)$-closed ideal) of R, but not a 2-absorbing ideal of R.

It was conjectured (see Conjecture one in Sect. 5) that a proper ideal I of R is an n-absorbing ideal of R if and only if I is a strongly n-absorbing ideal of R. However, an (m, n)-closed ideal of R need not be a strongly (m, n)-closed ideal of R; we have the following example.

Example 7.3 ([2, Example 2.5]) Let $R = \mathbb{Z}[X, Y]$, $I = (X^2, 2XY, Y^2)$, and $J = \sqrt{I} = (X, Y)$. Suppose that $a^m \in I$ for $a \in R$ and m a positive integer. Then $a \in \sqrt{I}$, and thus $a = bX + cY$ for some $b, c \in R$. Hence $a^2 = (bX + cY)^2 = b^2X^2 + 2bcXY + c^2Y^2 \in I$, and thus I is an $(m, 2)$-closed ideal of R for every positive integer m. It is easily checked that $J^m \subseteq I$ for every integer $m \geq 3$. However, $J^2 \not\subseteq I$ since $XY \notin I$; so I is not a strongly $(m, 2)$-closed ideal of R for any integer $m \geq 3$.
In view of Example 7.3, we have the following result.

Theorem 7.4 ([2, Theorem 2.6]) *Let R be a commutative ring, m a positive integer, I an $(m, 2)$-closed ideal of R, and J an ideal of R.*

1. *If $J^m \subseteq I$, then $2J^2 \subseteq I$.*
2. *Suppose that $2 \in U(R)$. Then I is a strongly $(m, 2)$-closed ideal of R.*

In view of Theorem 7.4(2), we have the following result.

Theorem 7.5 ([50, Corollary 4.11]) *Let R be a ring and n be a positive integer such that $n!$ is a unit in R. Then every semi-n-absorbing ideal of R is strongly semi-n-absorbing.*

We have the following result.

Theorem 7.6 ([50, Proposition 4.6]) *Let I be an ideal of a ring R and n be a positive integer. If for every ideal J of R, we have $J^{n+1} \subseteq I \subseteq J$ implies $J^n \subseteq I$, then I is a strongly semi-n-absorbing ideal of R.*

The following result is a characterization of zero-dimensional rings in terms of (m, n)-closed ideals.

Theorem 7.7 ([2, Theorem 2.15]) *Let R be a commutative ring and n a positive integer. Then the following statements are equivalent.*

1. *Every proper ideal of R is (m, n)-closed for every positive integer m.*
2. *There is an integer $m > n$ such that every proper ideal of R is (m, n)-closed.*
3. *For every proper ideal I of R, there is an integer $m_I > n$ such that I is (m_I, n)-closed.*
4. *Every proper ideal of R is a semi-n-absorbing ideal (i.e., $(n+1, n)$-closed ideal) of R.*
5. *$dim(R) = 0$ and $w^n = 0$ for every $w \in nil(R)$.*

Let R be an integral domain and m, k be fixed positive integers. The next result determines the smallest positive integer n such that $I = p^k R$ is (m, n)-closed. As usual, $\lfloor x \rfloor$ is the greatest integer, or floor function.

Theorem 7.8 ([2, Theorem 3.10]) *Let R be an integral domain and $I = p^k R$, where p is a prime element of R and k is a positive integer. Let m be a positive integer and n be the smallest positive integer such that I is (m, n)-closed.*

1. *If $m \geq k$, then $n = k$.*
2. *Let $m < k$ and write $k = ma + r$, where a is a positive integer and $0 \leq r < m$.*
   a. *If $r = 0$, then $n = m$.*
   b. *If $r \neq 0$ and $a \geq m$, then $n = m$.*
   c. *If $r \neq 0$, $a < m$, and $(a+1)|k$, then $n = k/(a+1)$.*
   d. *If $r \neq 0$, $a < m$, and $(a+1) \nmid k$, then $n = \lfloor k/(a+1) \rfloor + 1$.*

Let R be an integral domain and n, k be fixed positive integers. The next result determines the largest positive integer m such that $I = p^k R$ is (m, n)-closed.

Theorem 7.9 ([2, Theorem 3.11]) *Let R be an integral domain, n a positive integer, and $I = p^k R$, where p is a prime element of R and k is a positive integer.*

1. *If $n \geq k$, then I is (m, n)-closed for every positive integer m.*
2. *Let $n < k$ and write $k = na + r$, where a is a positive integer and $0 \leq r < n$. Let m be the largest positive integer such that I is (m, n)-closed.*
   a. *If $a > n$, then $m = n$.*
   b. *If $a = n$ and $r = 0$, then $m = n + 1$.*

c. *If* $a = n$ *and* $r \neq 0$*, then* $m = n$.
d. *If* $a < n$*,* $r = 0$*, and* $(a-1)|k$*, then* $m = k/(a-1) - 1$.
e. *If* $a < n$*,* $r = 0$*, and* $(a-1) \nmid k$*, then* $m = \lfloor k/(a-1) \rfloor$.
f. *If* $a < n$*,* $r \neq 0$*, and* $a|k$*, then* $m = k/a - 1$.
g. *If* $a < n$*,* $r \neq 0$*, and* $a \nmid k$*, then* $m = \lfloor k/a \rfloor$.

In view of Theorems 7.8 and 7.9, let I be a proper ideal of a commutative ring R and m and n positive integers. Anderson and Badawi in [2] defined $f_I(m) = \min\{n \mid I \text{ is } (m,n) - closed\} \in \{1, \ldots, m\}$ and $g_I(n) = sup\{m \mid I \text{ is } (m,n) - closed\} \in \{n, n+1, \ldots\} \cup \{\infty\}$. We have the following example.

Example 7.10 Let R be an integral domain and $I = p^{30}R$ for p a prime element of R. By Theorem 7.8, one may easily calculate that $f_I(m) = m$ for $1 \leq m \leq 6$, $f_I(7) = 6, f_I(8) = f_I(9) = 8, f_I(m) = 10$ for $10 \leq m \leq 14, f_I(m) = 15$ for $15 \leq m \leq 29$, and $f_I(m) = 30$ for $m \geq 30$. Using Theorem 7.9, one may easily calculate that $g_I(n) = n$ for $1 \leq n \leq 5, g_I(6) = g_I(7) = 7, g_I(8) = g_I(9) = 9, g_I(n) = 14$ for $10 \leq n \leq 14, g_I(n) = 29$ for $15 \leq n \leq 29$, and $g_I(n) = \infty$ for $n \geq 30$.

If R is a Prüfer domain, we have the following result.

Theorem 7.11 ([50, Corollary 3.26]) *Let R be a Prüfer domain, n be a positive integer, and I be an ideal of R.*

1. *If I is a strongly quasi-n-absorbing (resp. strongly semi-n-absorbing) ideal of R, then I[X] is a quasi-n-absorbing (resp. semi-n-absorbing) ideal of R[X].*
2. *If I[X] is a quasi-n-absorbing (resp. semi-n-absorbing) ideal of R[X], then I is a quasi-n-absorbing (resp. semi-n-absorbing) ideal of R.*

The following result determines the quasi-n-absorbing ideals in the product of any two rings.

Theorem 7.12 ([50, Proposition 4.20]) *Let* $n \geq 2$ *be an integer,* R_1, R_2 *be rings,* $R = R_1 \times R_2$*, and L be a quasi-n-absorbing ideal of R. Then either* $L = I_1 \times R_2$*, where* I_1 *is a quasi-n-absorbing ideal of* R_1 *or* $L = R_1 \times I_2$*, where* I_2 *is a quasi-n-absorbing ideal of* R_2 *or* $L = I_1 \times I_2$*, where* I_1 *is a semi-*$(n-1)$*-absorbing ideal of* R_1 *and* I_2 *is a semi-*$(n-1)$*-absorbing ideal of* R_2.

8 2-Absorbing Primary Ideals of Commutative Rings

We recall the following definition from [9] which is a generalization of primary ideal. A proper ideal I of R is said to be a 2-*absorbing primary ideal* of R if whenever $a, b, c \in R$ with $abc \in I$, then $ab \in I$ or $ac \in Rad(I)$ or $bc \in Rad(I)$.

In the following result, we collect some basic properties of 2-absorbing primary ideals of commutative rings.

Theorem 8.1

1. *([9, Theorem 2.2]). If I is a 2-absorbing primary ideal of R, then $Rad(I)$ is a 2-absorbing ideal of R.*
2. *([9, Theorem 2.3]). Suppose that I is a 2-absorbing primary ideal of R. Then one of the following statements must hold.*

   a. *$Rad(I) = P$ is a prime ideal,*
   b. *$Rad(I) = P_1 \cap P_2$, where P_1 and P_2 are the only distinct prime ideals of R that are minimal over I.*

3. *([9, Corollary 2.5]). Let R be a commutative ring with $1 \neq 0$, and let P_1, P_2 be prime ideals of R. If P_1^n is a P_1-primary ideal of R for some positive integer $n \geq 1$ and P_2^m is a P_2-primary ideal of R for some positive integer $m \geq 1$, then $P_1^n P_2^m$ and $P_1^n \cap P_2^m$ are 2-absorbing primary ideals of R. In particular, P_1P_2 is a 2-absorbing primary ideal of R.*
4. *([9, Theorem 2.8]). Let I be an ideal of R. If $Rad(I)$ is a prime ideal of R, then I is a 2-absorbing primary ideal of R. In particular, if P is a prime ideal of R, then P^n is a 2-absorbing primary ideal of R for every positive integer $n \geq 1$.*
5. *([9, Theorem 2.10]). Let R be a commutative divided ring with $1 \neq 0$ (for example, if R is a valuation domain). Then every proper ideal of R is a 2-absorbing primary ideal of R.*
6. *([9, Theorem 2.20]). Let $f : R \to R'$ be a homomorphism of commutative rings. Then the following statements hold.*

   a. *If I' is a 2-absorbing primary ideal of R', then $f^{-1}(I')$ is a 2-absorbing primary ideal of R.*
   b. *If f is an epimorphism and I is a 2-absorbing primary ideal of R containing $Ker(f)$, then $f(I)$ is a 2-absorbing primary ideal of R'.*

7. *([9, Theorem 2.22]). Let R be a commutative ring with $1 \neq 0$, S be a multiplicatively closed subset of R, and I be a proper ideal of R. Then the following statements hold.*

   a. *If I is a 2-absorbing primary ideal of R such that $I \cap S = \varnothing$, then $S^{-1}I$ is a 2-absorbing primary ideal of $S^{-1}R$.*
   b. *If $S^{-1}I$ is a 2-absorbing primary ideal of $S^{-1}R$ and $S \cap Z_I(R) = \varnothing$, then I is a 2-absorbing primary ideal of R.*

The following result is a characterization of Dedekind domains in terms of 2-absorbing primary ideals.

Theorem 8.2 ([9, Theorem 2.11]) *Let R be a Noetherian integral domain with $1 \neq 0$ that is not a field. Then the following statements are equivalent.*

1. *R is a Dedekind domain.*
2. *A nonzero proper ideal I of R is a 2-absorbing primary ideal of R if and only if either $I = M^n$ for some maximal ideal M of R and some positive integer $n \geq 1$ or $I = M_1^n M_2^m$ for some maximal ideals M_1, M_2 of R and some positive integers $n, m \geq 1$.*

3. *If I is a nonzero proper 2-absorbing primary ideal of R, then either $I = M^n$ for some maximal ideal M of R and some positive integer $n \geq 1$ or $I = M_1^n M_2^m$ for some maximal ideals M_1, M_2 of R and some positive integers $n, m \geq 1$.*
4. *A nonzero proper ideal I of R is a 2-absorbing primary ideal of R if and only if either $I = P^n$ for some prime ideal P of R and some positive integer $n \geq 1$ or $I = P_1^n P_2^m$ for some prime ideals P_1, P_2 of R and some positive integers $n, m \geq 1$.*
5. *If I is a nonzero proper 2-absorbing primary ideal of R, then either $I = P^n$ for some prime ideal P of R and some positive integer $n \geq 1$ or $I = P_1^n P_2^m$ for some prime ideals P_1, P_2 of R and some positive integers $n, m \geq 1$.*

The following result determines the 2-absorbing primary ideals in the product of any finite number of rings.

Theorem 8.3 ([9, Theorem 2.24]) *Let $R = R_1 \times R_2 \times \cdots \times R_n$, where $2 \leq n < \infty$, and $R_1, R_2, \ldots, R_n$ are commutative rings with $1 \neq 0$. Let J be a proper ideal of R. Then the following statements are equivalent.*

1. *J is a 2-absorbing primary ideal of R.*
2. *Either $J = \times_{t=1}^n I_t$ such that for some $k \in \{1, 2, \ldots, n\}$, I_k is a 2-absorbing primary ideal of R_k, and $I_t = R_t$ for every $t \in \{1, 2, \ldots, n\} \setminus \{k\}$ or $J = \times_{t=1}^n I_t$ such that for some $k, m \in \{1, 2, \ldots, n\}$, I_k is a primary ideal of R_k, I_m is a primary ideal of R_m, and $I_t = R_t$ for every $t \in \{1, 2, \ldots, n\} \setminus \{k, m\}$.*

A proper ideal I of R is said to be a *strongly 2-absorbing primary ideal* of R if whenever I_1, I_2, I_3 are ideals of R with $I_1 I_2 I_3 \subseteq I$, then $I_1 I_2 \subseteq I$ or $I_1 I_3 \subseteq I$ or $I_2 I_3 \subseteq I$. We have the following result.

Theorem 8.4 ([9, Theorem 2.19]) *Let I be a proper ideal of R Then I is a 2-absorbing primary ideal of R if and only if I is a strongly 2-absorbing primary ideal of R.*

Remark 8.5 Many topics related to the concept of n-absorbing ideals have been left untouched; the interested reader may consult the many articles mentioned in the references and MathSciNet. In the following, we will outline some of the related topics.

1. For topics on 2-absorbing preradicals, see [24–26]
2. For topics related to 2-absorbing commutative semigroups, see [27].
3. For topics related to (weakly) n-absorbing ideals of commutative rings, see [4, 5, 7–9, 11, 12], [15–17], [20, 30], and [36–38].
4. For topics related to n-absorbing ideals in semirings, see [18, 22, 32, 42, 43, 57, 58], and [61].
5. For topics related to (weakly) n-absorbing submodules, see [19], [25, 28, 29], [32–35], [47, 48, 51], [53, 55, 59], and [62].

Acknowledgements I would like to thank the referee for the great effort in proofreading the manuscript.

References

1. Akca, F.K., Ulucak, G.: Uniformly primary elements of multiplicative lattices. Int. Math. Forum **10**, 157–163 (2015)
2. Anderson, D.F., Badawi, A.: On (m, n)-closed ideals of commutative rings. J. Algebra Appl. **16**, 21pp. (2017)
3. Anderson, D.F., Badawi, A.: On n-absorbing ideals of commutative rings. Commun. Algebra **39**, 1646–1672 (2011)
4. Anderson, D.D., Bataineh, M.: Generalizations of prime ideals. Commun. Algebra **36**, 686–696 (2008)
5. Anderson, D.D., Smith, E.: Weakly prime ideals. Houston J. Math. **29**, 831–840(2003)
6. Badawi, A.: On 2-absorbing ideals of commutative rings. Bull. Aust. Math. Soc. **75**, 417–429 (2007)
7. Badawi, A.: On weakly semiprime ideals of commutative rings. In: Beitrge zur Algebra und Geometri/Contributions to Algebra and Geometry, vol. 57, pp. 589–597. Springer, Berlin (2016)
8. Badawi, A., Yousef Darani, A.: On weakly 2-absorbing ideals of commutative rings. Houston J. Math. **39**, 441–452 (2013)
9. Badawi, A., Tekir, U., Yeikin, E.: On 2-absorbing primary ideals of commutative rings. Bull. Korean Math. Soc. **51**, 1163–1173 (2014)
10. Badawi, A., Tekir, U., Yeikin, E.: On weakly 2-absorbing primary ideals of commutative rings. J. Korean Math. Soc. **52**, 97–111 (2015)
11. Badawi, A., Tekir, U., Yeikin, E.: Generalizations of 2-absorbing primary ideals of commutative rings. Turk. J. Math. **40**, 703–717 (2016)
12. Ballal, S., Kharat, V.: On generalization of prime, weakly prime and almost prime elements in multiplicative lattices. Int. J. Algebra **8**, 439–449 (2014)
13. Becker, A.E.: Results on n-absorbing ideals of commutative rings. M.S. thesis, University of Wisconsin-Milwaukee, Milwaukee (2015)
14. Cahen, P.J., Fontana, M., Frisch, S., Glaz, S.: Open problems in commutative ring theory. In: Commutative Algebra, pp. 353–375. Springer, New York (2014)
15. Callialp, F., Yetkin, E., Tekir, U.: On 2-absorbing primary and weakly 2-absorbing elements in multiplicative lattices. Ital. J. Pure Appl. Math. **34**, 263–276 (2015)
16. Celikel, E.Y., Ugurlu, E.A., Ulucak, G.: On ϕ-2-absorbing elements in multiplicative lattices. Palest. J. Math. **5**, 127–135 (2016)
17. Celikel, E.Y., Ugurlu, E.A., Ulucak, G.: On ϕ-2-absorbing primary elements in multiplicative lattices. Palest. J. Math. **5**, 136–146 (2016)
18. Chaudhari, J.N.: 2-absorbing ideals in semirings. Int. J. Algebra **6**, 265–270 (2012)
19. Chaudhari, J.N., Bendale, H.P.: Weakly prime ternary subsemimodules of ternary semimodules. J. Algebra Relat. Top. **2**, 63–72 (2014)
20. Chinwarakorn, S., Pianskool, S.: On almost generalized 2-absorbing and weakly almost generalized 2-absorbing structures. ScienceAsia **41**, 64–72 (2015)
21. Choi, H.S., Walker, A.: The radical of an n-absorbing ideal. arXiv:1610.10077 [math.AC] (2016), to appear in J. Comm. Algebra
22. Darani, A.Y.: On 2-absorbing and weakly 2-absorbing ideals of commutative semirings. Kyungpook Math. J. **52**, 91–97 (2012)
23. Darani, A.Y., Ghasemi, G.: On L-fuzzy 2-absorbing ideals. Ital. J. Pure Appl. Math. **36**, 147–154 (2016)
24. Darani, A.Y., Mostafanasab, H.: On 2-absorbing preradicals. J. Algebra Appl. **14**, 22pp. (2015)
25. Darani, A.Y., Mostafanasab, H.: Co-2-absorbing preradicals and submodules. J. Algebra Appl. **14**, 23pp. (2015)
26. Darani, A.Y., Mostafanasab, H.: Generalizations of semicoprime preradicals. Algebra and Discrete Math. **21**, 214–238 (2016)

27. Darani, A.Y., Puczyłowski, E.R.: On 2-absorbing commutative semigroups and their applications to rings. Semigroup Forum **86**, 83–91 (2013)
28. Darani, A.Y., Soheilnia, F.: 2-absorbing and weakly 2-absorbing submodules. Thai J. Math. **9**, 577–584 (2011)
29. Darani, A.Y., Soheilnia, F.: On n-absorbing submodules. Math. Commun. **17**, 547–557 (2012)
30. Davvaz, B., Fahid, B.: Rough fuzzy n-absorbing (n-absorbing primary) ideal in commutative rings. Ital. J. Pure Appl. Math. **36**, 473–482 (2016)
31. Donadze, G.: The Anderson-Badawi conjecture for commutative algebras over infinite fields. Indian J. Pure Appl. Math. **47**, 691–696 (2016)
32. Dubey, M.K., Aggarwal, P.: On 2-absorbing primary submodules of modules over commutative ring with unity. Asian-Eur. J. Math. **8**, 12pp. (2015)
33. Dubey, M.K., Aggarwal, P.: On n-absorbing submodules of modules over commutative rings. In: Beitrge zur Algebra und Geometrie/Contributions to Algebra and Geometry, vol. 57, pp. 679–690. Springer, Berlin (2016)
34. Dubey, M.K., Sarohe, P.: On 2-absorbing semimodules. Quasigroups Relat. Sys. **21**, 175–184 (2013)
35. Dubey, M.K., Sarohe, P.: On $(n1, n) - \phi$-Prime ideals in semirings. Mat. Vesn. **67**, 222–232 (2015)
36. Ebrahimpour, M.: On generalization of almost prime and weakly prime ideals. Bull. Iranian Math. Soc. **40**, 531–540 (2014)
37. Ebrahimpour, M.: On generalizations of prime ideals (II). Commun. Algebra **42**, 3861–3875 (2014)
38. Ebrahimpour, M., Nekooei, R.: On generalization of prime ideals. Commun. Algebra **40**, 1268–1279 (2012)
39. Farzalipour, F., Ghiasvand, P.: On graded semiprime and graded weakly semiprime ideals. Int. Electron. J. Algebra **13**, 15–32 (2013)
40. Hamoda, M., Ashour, A.E.: On graded n-absorbing submodules. Le Matematiche **70**, 243–254 (2015)
41. Jayaram, C., Tekir, U., Yetkin, E.: 2-absorbing and weakly 2-absorbing elements in multiplicative lattices. Commun. Algebra **42**, 2338–2353 (2014)
42. Kumar, P., Dubey, M.K., Sarohe, P.: Some results on 2-absorbing ideals in commutative semirings. J. Math. Appl. **38**, 5–13 (2015)
43. Kumar, P., Dubey, M.K., Sarohe, P.: On 2-absorbing primary ideals in commutative semirings. Eur. J. Pure Appl. Math. **9**, 186–195 (2016)
44. Laradji, A.: On n-absorbing rings and ideals. Colloq. Math. **147**, 265–273 (2017)
45. Manjarekar, C.S., Bingi, A.V.: Absorbing elements in lattice modules. Int. Electron. J. Algebra **19**, 58–76 (2016)
46. Moghimi, H.F., Naghani, S.R.: On n-absorbing ideals and the n-Krull dimension of a commutative ring. J. Korean Math. Soc. **53**, 1225–1236 (2016)
47. Moradi, S., Azizi, A.: 2-absorbing and n-weakly prime submodules. Miskolc Math. Notes **13**, 75–86 (2012)
48. Moradi, S., Azizi, A.: Weakly 2-absorbing submodules of modules. Turk. J. Math. **40**, 350–364 (2016)
49. Moradi, R., Ebrahimpour, M.: On ϕ-2-absorbing primary submodules. Acta Math. Vietnam [First Online] (2016). doi:10.1007/s40306-016-0181–0
50. Mostafanasab, H., Darani, A.Y.: On n-absorbing ideals and two generalizations of semiprime ideals. Thai J. Math. **15**, 772–794 (2017)
51. Mostafanasab, H., Yetkin, E., Tekir, U., Darani, A.Y.: On 2-absorbing primary submodules of modules over commutative rings. An. St. Univ. Ovidius Constanta **24**, 335–351 (2016)
52. Nasehpour, P.: On the Anderson-Badawi $\omega_{R[X]}(I[X]) = \omega_R(I)$ conjecture. Archivum Mathematicum (BRNO) **52**, 71–78(2016)
53. Payrovi, S., Babaei, S.: On 2-absorbing submodules over commutative rings. Lobachevskii J. Math. **36**, 58–64 (2015)

54. Payrovi, S., Babaei, S.: On the 2-absorbing ideals and zero divisor graph of equivalence classes of zero divisors. J. Hyperstructures **3**, 1–9 (2014)
55. Payrovi, S., Babaei, S.: On 2-absorbing submodules. Algebra Colloq. **19**, 913–920 (2012)
56. Payrovi, S., Babaei, S.: On the 2-absorbing ideals. Int. Math. Forum **7**, 265–271 (2012)
57. Petchkaew, P., Wasanawichit, A., Pianskool, S.: Generalizations of n-absorbing ideals of commutative semirings. Thai J. Math. **14**, 477–489 (2016)
58. Soheilnia, F.: On 2-absorbing and weakly 2-absorbing primary ideals of a commutative semiring. Kyungpook Math. J. **56**, 107–120 (2016)
59. Tekir, U., Mostafanasab, H., Oral, K.H.: Classical 2-absorbing submodules of modules over commutative rings. Eur. J. Pure Appl. Math. **8**, 417–430 (2015)
60. Tekir, U., Ko, S., Oral, K.H., Shum, K.P.: On 2-absorbing quasi-primary ideals in commutative rings. Commun. Math. Stat. **4**, 55–62 (2016)
61. Tessema, Y.Y., Belayneh, B.B., Venkateswarlu, K.: Certain generalized prime ideals in boolean like semirings. Int. J. Algebra **8**, 663–669 (2014)
62. Yasein, M., Abu-Dawwas, R.: On almost 2-absorbing submodules. Ital. J. Pure Appl. Math. **36**, 923–928 (2016)

Embedding Dimension and Codimension of Tensor Products of Algebras over a Field

S. Bouchiba and S. Kabbaj

To David Dobbs on the occasion of his 70th birthday

Abstract Let k be a field. This paper investigates the embedding dimension and codimension of Noetherian local rings arising as localizations of tensor products of k-algebras. We use results and techniques from prime spectra and dimension theory to establish an analogue of the "special chain theorem" for the embedding dimension of tensor products, with effective consequence on the transfer or defect of regularity as exhibited by the (embedding) codimension given by $\mathrm{codim}(R) := \mathrm{embdim}(R) - \dim(R)$.

Keywords Tensor product of k-algebras • Regular ring • Embedding dimension • Krull dimension • Embedding codimension • Separable extension

Mathematics Subject Classification 13H05, 13F20, 13B30, 13E05, 13D05, 14M05, 16E65

1 Introduction

Throughout, all rings are commutative with identity elements, ring homomorphisms are unital, and k stands for a field. The embedding dimension of a Noetherian local ring $(R, \mathfrak{m})$, denoted by $\mathrm{embdim}(R)$, is the least number of generators of

S. Bouchiba
Department of Mathematics, University of Meknes, Meknes 50000, Morocco
e-mail: bouchibasamir@gmail.com

S. Kabbaj (✉)
Department of Mathematics and Statistics, King Fahd University of Petroleum and Minerals, Dhahran 31261, Saudi Arabia
e-mail: kabbaj@kfupm.edu.sa

M. Fontana et al. (eds.), *Rings, Polynomials, and Modules*,
https://doi.org/10.1007/978-3-319-65874-2_4

$\mathfrak{m}$ or, equivalently, the dimension of $\mathfrak{m}/\mathfrak{m}^2$ as an $R/\mathfrak{m}$-vector space. The ring R is regular if its Krull dimension and embedding dimensions coincide. The (embedding) codimension of R measures the defect of regularity of R and is given by the formula $\text{codim}(R) := \text{embdim}(R) - \dim(R)$. The concept of regularity was initially introduced by Krull and became prominent when Zariski showed that a local regular ring corresponds to a smooth point on an algebraic variety. Later, Serre proved that a ring is regular if and only if it has finite global dimension. This allowed to see that regularity is stable under localization and then the definition got globalized as follows: a Noetherian ring is regular if its localizations with respect to all prime ideals are regular. The ring R is a complete intersection if its $\mathfrak{m}$-completion is the quotient ring of a local regular ring modulo an ideal generated by a regular sequence; R is Gorenstein if its injective dimension is finite; and R is Cohen–Macaulay if the grade and height of $\mathfrak{m}$ coincide. All these algebro-geometric notions are globalized by carrying over to localizations.

These concepts transfer to tensor products of algebras over a field under suitable assumptions. It has been proved that a Noetherian tensor product of algebras (over a field) inherits the notions of (locally) complete intersection ring, Gorenstein ring, and Cohen–Macaulay ring [3, 19, 31, 34]. In particular, a Noetherian tensor product of any two extension fields is a complete intersection ring. As to regularity and unlike the above notions, a Noetherian tensor product of two extension fields of k is not regular in general. In 1965, Grothendieck proved a positive result in case one of the two extension fields is a finitely generated separable extension [18]. Recently, we have investigated the possible transfer of regularity to tensor products of algebras over a field k. If A and B are two k-algebras such that A is geometrically regular; i.e., $A \otimes_k F$ is regular for every finite extension F of k (e.g., A is a separable extension field over k), we proved that $A \otimes_k B$ is regular if and only if B is regular and $A \otimes_k B$ is Noetherian [4, Lemma 2.1]. As a consequence, we established necessary and sufficient conditions for a Noetherian tensor product of two extension fields of k to inherit regularity under (pure in)separability conditions [4, Theorem 2.4]. Also, Majadas' relatively recent paper tackled questions of regularity and complete intersection of tensor products of commutative algebras via the homology theory of André and Quillen [24]. Finally, it is worthwhile recalling that tensor products of rings subject to the above concepts were recently used to broaden or delimit the context of validity of some homological conjectures; see, for instance, [20, 22]. Suitable background on regular, complete intersection, Gorenstein, and Cohen–Macaulay rings is [14, 18, 23, 25]. For a geometric treatment of these properties, we refer the reader to the excellent book of Eisenbud [15].

Throughout, given a ring R, I an ideal of R and p a prime ideal of R, when no confusion is likely, we will denote by I_p the ideal IR_p of the local ring R_p and by $\kappa_R(p)$ the residue field of R_p. One of the cornerstones of dimension theory of polynomial rings in several variables is *the special chain theorem*, which essentially asserts that the height of any prime ideal of the polynomial ring can always be realized via a special chain of prime ideals passing by the extension of its contraction over the basic ring; namely, if R is a Noetherian ring and P is a prime ideal of $R[X_1, \ldots, X_n]$ with $p := P \cap R$, then

$$\dim(R[X_1,\dots,X_n]_P) = \dim(R_p) + \dim\left(\kappa_R(p)[X_1,\dots,X_n]_{\frac{P_p}{pR_p[X_1,\dots,X_n]}}\right)$$

An analogue of this result for Noetherian tensor products, established in [3], states that, for any prime ideal P of $A \otimes_k B$ with $p := P \cap A$ and $q := P \cap B$, we have

$$\dim(A \otimes_k B)_P = \dim(A_p) \; + \; \dim\left(\left(\kappa_A(p) \otimes_k B\right)_{\frac{P_p}{pA_p \otimes_k B}}\right)$$

which also comes in the following extended form

$$\dim(A \otimes_k B)_P = \dim(A_p) + \dim(B_q) + \dim\left(\left(\kappa_A(p) \otimes_k \kappa_B(q)\right)_{\frac{P(A_p \otimes_k B_q)}{pA_p \otimes_k B_q + A_p \otimes_k qB_q}}\right).$$

This paper investigates the embedding dimension of Noetherian local rings arising as localizations of tensor products of k-algebras. We use results and techniques from prime spectra and dimension theory to establish satisfactory analogues of the "special chain theorem" for the embedding dimension in various contexts of tensor products, with effective consequences on the transfer or defect of regularity as exhibited by the (embedding) codimension. The paper traverses four sections along with an introduction.

In Sect. 2, we introduce and study a new invariant which allows to correlate the embedding dimension of a Noetherian local ring B with the fiber ring $B/\mathfrak{m}B$ of a local homomorphism $f : A \longrightarrow B$ of Noetherian local rings. This enables us to provide an analogue of the special chain theorem for the embedding dimension as well as to generalize the known result that "*if f is flat and A and $B/\mathfrak{m}B$ are regular rings, then B is regular*."

Section 3 is devoted to the special case of polynomial rings which will be used in the investigation of tensor products. The main result (Theorem 3.1) states that, for a Noetherian ring R and $X_1,\dots,X_n$ indeterminates over R, for any prime ideal P of $R[X_1,\dots,X_n]$ with $p := P \cap R$, we have:

$$\begin{aligned}\operatorname{embdim}(R[X_1,\dots,X_n]_P) &= \operatorname{embdim}(R_p) + \operatorname{ht}\left(\frac{P}{p[X_1,\dots,X_n]}\right)\\ &= \operatorname{embdim}(R_p) + \operatorname{embdim}\left(\kappa_R(p)[X_1,\dots,X_n]_{\frac{P_p}{pR_p[X_1,\dots,X_n]}}\right)\end{aligned}$$

Then, Corollary 3.2 asserts that

$$\operatorname{codim}(R[X_1,\dots,X_n]_P) = \operatorname{codim}(R_p)$$

and recovers a well-known result on the transfer of regularity to polynomial rings; i.e., *$R[X_1,\dots,X_n]$ is regular if and only if so is R* (this result was initially proved via Serre's result on finite global dimension and Hilbert Theorem on syzygies). Then Corollary 3.3 characterizes regularity in general settings of localizations of polynomial rings and, in the particular cases of Nagata rings and Serre conjecture rings, it states that $R(X_1, \dots, X_n)$ *is regular if and only if* $R\langle X_1, \dots, X_n\rangle$ *is regular if and only if* R *is regular*.

Let A and B be two k-algebras such that $A \otimes_k B$ is Noetherian and let P be a prime ideal of $A \otimes_k B$ with $p := P \cap A$ and $q := P \cap B$. Due to known behavior of tensor products of k-algebras subject to regularity (cf. [4, 18, 19, 31, 34]), Sect. 4 investigates the case when A (or B) is a separable (not necessarily algebraic) extension field of k. The main result (Theorem 4.2) asserts that, if K is a separable extension field of k, then

$$\text{embdim}(K \otimes_k A)_P = \text{embdim}(A_p) + \text{embdim}\left(\left(K \otimes_k \kappa_A(p)\right)_{\frac{P_p}{K \otimes_k pA_p}}\right).$$

In particular, if K is separable algebraic over k, then

$$\text{embdim}(K \otimes_k A)_P = \text{embdim}(A_p).$$

Then, Corollary 4.5 asserts that

$$\text{codim}(K \otimes_k A)_P = \text{codim}(A_p)$$

and hence $K \otimes_k A$ is regular if and only if so is A. This recovers Grothendieck's result on the transfer of regularity to tensor products issued from finite extension fields [18, Lemma 6.7.4.1].

Section 5 examines the more general case of tensor products of k-algebras with separable residue fields. The main theorem (Theorem 5.1) states that if $\kappa_B(q)$ is a separable extension field of k, then

$$\begin{aligned}\text{embdim}(A \otimes_k B)_P &= \text{embdim}(A_p) + \text{embdim}(B_q)\\ &\quad + \text{embdim}\left(\left(\kappa_A(p) \otimes_k \kappa_B(q)\right)_{\frac{P(A_p \otimes_k B_q)}{pA_p \otimes_k B_q + A_p \otimes_k qB_q}}\right)\end{aligned}$$

Then, Corollary 5.2 contends that

$$\text{codim}(A \otimes_k B)_P = \text{codim}(A_p) + \text{codim}(B_q)$$

recovering known results on the transfer of regularity to tensor products over perfect fields [31, Theorem 6(c)] and, more generally, to tensor products issued from residually separable extension fields [4, Theorem 2.11].

The four aforementioned main results are connected as follows:

$$\begin{array}{ccccc} & & \text{Proposition 4.1} & & \\ & & & \searrow & \\ \text{Theorem 3.1} & & \Downarrow & & \text{Theorem 5.1} \\ & \searrow & & \nearrow & \\ & & \text{Theorem 4.2} & & \end{array}$$

Of relevance to this study is Bouchiba, Conde-Lago, and Majadas' recent preprint [8] where the authors prove some of our results via the homology theory of André and Quillen. In the current paper, we offer direct and self-contained proofs using techniques and basic results from commutative ring theory. Early and recent developments on prime spectra and dimension theory are to be found in [3, 5–7, 28–30, 32, 33] for the special case of tensor products of k-algebras, and in [1, 11, 17, 22, 23, 25, 26] for the general case. Any unreferenced material is standard, as in [23, 25].

2 Embedding Dimension of Noetherian Local Rings

In this section, we discuss the relationship between the embedding dimensions of Noetherian local rings connected by a local ring homomorphism. To this purpose, we introduce a new invariant μ which allows to relate the embedding dimension of a local ring to that of its fiber ring.

Throughout, let $(A, \mathfrak{m}, K)$ and $(B, \mathfrak{n}, L)$ be local Noetherian rings, $f : A \longrightarrow B$ a local homomorphism (i.e., $\mathfrak{m}B := f(\mathfrak{m})B \subseteq \mathfrak{n}$), and I a proper ideal of A. Let

$$\mu_A(I) := \dim_K\left(\frac{I + \mathfrak{m}^2}{\mathfrak{m}^2}\right).$$

Note that $\mu_A(I)$ equals the maximal number of elements of I which are part of a minimal basis of $\mathfrak{m}$; so that $0 \leq \mu_A(I) \leq \text{embdim}(A)$ and $\mu_A(\mathfrak{m}) = \text{embdim}(A)$. Next, let $\mu^f_B(I)$ denote the maximal number of elements of $IB := f(I)B$ which are part of a minimal basis of $\mathfrak{n}$; that is,

$$\mu^f_B(I) := \mu_B(IB) = \dim_L\left(\frac{IB + \mathfrak{n}^2}{\mathfrak{n}^2}\right).$$

It is easily seen that if $x_1, \ldots, x_r$ are elements of $\mathfrak{m}$ such that $f(x_1), \ldots, f(x_r)$ are part of a minimal basis of $\mathfrak{n}$, then $x_1, \ldots, x_r$ are part of a minimal basis of $\mathfrak{m}$ as well. That is, $0 \leq \mu^f_B(I) \leq \mu_A(I)$. Moreover, if J is a proper ideal of B and $\pi : B \twoheadrightarrow \dfrac{B}{J}$ is the canonical surjection, then the natural linear map of L-vector spaces $\dfrac{IB + \mathfrak{n}^2}{\mathfrak{n}^2} \twoheadrightarrow \dfrac{IB + \mathfrak{n}^2 + J}{\mathfrak{n}^2 + J}$ yields $\mu^{\pi\circ f}_{B/J}(I) \leq \mu^f_B(I)$.

Proposition 2.1 *Under the above notation, we have:*

$$\mathrm{embdim}(B) = \mu_B^f(I) + \mathrm{embdim}\left(\frac{B}{IB}\right).$$

In particular,

$$\mathrm{embdim}(A) = \mu_A(I) + \mathrm{embdim}\left(\frac{A}{I}\right).$$

Proof The first statement follows easily from the following exact sequence of L-vector spaces

$$0 \longrightarrow \frac{IB + n^2}{n^2} \longrightarrow \frac{n}{n^2} \longrightarrow \frac{n}{IB + n^2} = \frac{n/IB}{(n/IB)^2} \longrightarrow 0.$$

The second statement holds since $\mu_A(I) = \mu_A^{\mathrm{id}_A}(I)$. □

Recall that, under the above notation, the following inequality always holds: $\dim(B) \leq \dim(A) + \dim\left(\frac{B}{\mathfrak{m}B}\right)$. The first corollary provides an analogue for the embedding dimension.

Corollary 2.2 *Under the above notation, we have:*

$$\mathrm{embdim}(B) \leq \mathrm{embdim}(A) - \mathrm{embdim}(A/I) + \mathrm{embdim}\left(\frac{B}{IB}\right).$$

In particular,

$$\mathrm{embdim}(B) \leq \mathrm{embdim}(A) + \mathrm{embdim}\left(\frac{B}{\mathfrak{m}B}\right).$$

It is well known that if f is flat and both A and $\frac{B}{\mathfrak{m}B}$ are regular, then B is regular. The second corollary generalizes this result to homomorphisms subject to going-down. Recall that a ring homomorphism $h : R \longrightarrow S$ satisfies going-down (henceforth abbreviated GD) if for any pair $p \subseteq q$ in $\mathrm{Spec}(R)$ such that there exists $Q \in \mathrm{Spec}(S)$ lying over q, then there exists $P \in \mathrm{Spec}(S)$ lying over p with $P \subseteq Q$. Any flat ring homomorphism satisfies GD.

Corollary 2.3 *Under the above notation, assume that f satisfies GD. Then:*

(a) $\mathrm{codim}(B) = \left(\mu_B^f(\mathfrak{m}) - \dim(A)\right) + \mathrm{codim}\left(\frac{B}{\mathfrak{m}B}\right)$.

(b) $\mathrm{codim}(B) + \left(\mathrm{embdim}(A) - \mu_B^f(\mathfrak{m})\right) = \mathrm{codim}(A) + \mathrm{codim}\left(\frac{B}{\mathfrak{m}B}\right)$.

(c) B is regular and $\mu_B^f(\mathfrak{m}) = \mathrm{embdim}(A) \Longleftrightarrow A$ and $\frac{B}{\mathfrak{m}B}$ are regular.

Proof The proof is straightforward via a combination of Proposition 2.1 and [25, Theorem 15.1]. □

Corollary 2.4 *Under the above notation, assume that f satisfies GD. Then:*

(a) $\operatorname{codim}(B) \leq \operatorname{codim}(A) + \operatorname{codim}\left(\frac{B}{\mathfrak{m}B}\right)$.

(b) If $\frac{B}{\mathfrak{m}B}$ *is regular, then* $\operatorname{codim}(B) \leq \operatorname{codim}(A)$.

Proof The proof is direct via a combination of Corollary 2.2 and the known fact that $\dim(B) = \dim(A) + \dim\left(\frac{B}{\mathfrak{m}B}\right)$. □

3 Embedding Dimension and Codimension of Polynomial Rings

This section is devoted to the special case of polynomial rings which will be used, later, for the investigation of tensor products. The main result of this section (Theorem 3.1) settles a formula for the embedding dimension for the localizations of polynomial rings over Noetherian rings. It recovers (via Corollary 3.2) a well-known result on the transfer of regularity to polynomial rings; that is, $R[X_1, \ldots, X_n]$ is regular if and only if so is R. Moreover, Theorem 3.1 leads to investigate the regularity of two famous localizations of polynomial rings in several variables; namely, the Nagata ring $R(X_1, X_2, \ldots, X_n)$ and Serre conjecture ring $R\langle X_1, X_2, \ldots, X_n\rangle$. We show that the regularity of these two constructions is entirely characterized by the regularity of R (Corollary 3.3).

Recall that one of the cornerstones of dimension theory of polynomial rings in several variables is *the special chain theorem*, which essentially asserts that the height of any prime ideal P of $R[X_1, \ldots, X_n]$ can always be realized via a special chain of prime ideals passing by the extension $(P \cap R)[X_1, \ldots, X_n]$. This result was first proved by Jaffard in [22] and, later, Brewer, Heinzer, Montgomery, and Rutter reformulated it in the following simple way [12, Theorem 1]: Let P be a prime ideal of $R[X_1, \ldots, X_n]$ with $p := P \cap R$. Then $\operatorname{ht}(P) = \operatorname{ht}(p[X_1, \ldots, X_n]) + \operatorname{ht}\left(\frac{P}{p[X_1, \ldots, X_n]}\right)$. In a Noetherian setting, this formula becomes:

$$\begin{aligned} \dim(R[X_1, \ldots, X_n]_P) &= \dim(R_p) + \operatorname{ht}\left(\frac{P}{p[X_1, \ldots, X_n]}\right) \\ &= \dim(R_p) + \dim\left(\kappa_R(p)[X_1, \ldots, X_n]_{\frac{P_p}{pR_p[X_1, \ldots, X_n]}}\right) \end{aligned} \tag{1}$$

where the second equality holds on account of the basic fact $\frac{P}{p[X_1, \ldots, X_n]} \cap \frac{R}{p} = 0$. The main result of this section (Theorem 3.1) features a "special chain theorem" for the embedding dimension with effective consequence on the codimension.

Theorem 3.1 *Let R be a Noetherian ring and $X_1, \ldots, X_n$ be indeterminates over R. Let P be a prime ideal of $R[X_1, \ldots, X_n]$ with $p := P \cap R$. Then:*

$$\begin{aligned}\mathrm{embdim}(R[X_1,\ldots,X_n]_P) &= \mathrm{embdim}(R_p) + \mathrm{ht}\left(\frac{P}{p[X_1,\ldots,X_n]}\right)\\ &= \mathrm{embdim}(R_p) + \mathrm{embdim}\left(\kappa_R(p)[X_1,\ldots,X_n]_{\frac{P_p}{pR_p[X_1,\ldots,X_n]}}\right)\end{aligned}$$

Proof We use induction on n. Assume $n = 1$ and let P be a prime ideal of $R[X]$ with $p := P \cap R$ and $r := \mathrm{embdim}(R_p)$. Then $p_p = (a_1,\ldots,a_r)R_p$ for some $a_1,\ldots,a_r \in p$. We envisage two cases; namely, either P is an extension of p or an upper to p. For both cases, we will use induction on r.

Case 1 P is an extension of p (i.e., $P = pR[X]$). We prove that $\mathrm{embdim}(R[X]_P) = r$. Indeed, we have $P_P = pR_p[X]_{pR_p[X]} = (a_1,\ldots,a_r)R_p[X]_{pR_p[X]} = (a_1,\ldots,a_r)R[X]_P$. So, obviously, if $p_p = (0)$, then $P_P = 0$. Next, we may assume $r \geq 1$. One can easily check that the canonical ring homomorphism $\varphi : R_p \longrightarrow R[X]_P$ is injective with $\varphi(p_p) \subseteq P_P$. This forces $\mathrm{embdim}(R[X]_P) \geq 1$. Hence, there exists $j \in \{1,\ldots,n\}$, say $j = 1$, such that $a := a_1 \in p$ with $\frac{a}{1} \in P_P \setminus P_P^2$ and, a fortiori, $\frac{a}{1} \in p_p \setminus p_p^2$. By [23, Theorem 159], we get

$$\begin{cases}\mathrm{embdim}(R[X]_P) = 1 + \mathrm{embdim}\left(\frac{R}{(a)}[X]_{\frac{P}{aR[X]}}\right)\\ \mathrm{embdim}(R_p) = 1 + \mathrm{embdim}\left(\left(\frac{R}{(a)}\right)_{\frac{p}{(a)}}\right)\end{cases} \tag{2}$$

Therefore $\mathrm{embdim}\left(\left(\frac{R}{(a)}\right)_{\frac{p}{(a)}}\right) = r - 1$ and then, by induction on r, we obtain

$$\mathrm{embdim}\left(\frac{R}{(a)}[X]_{\frac{P}{aR[X]}}\right) = \mathrm{embdim}\left(\left(\frac{R}{(a)}\right)_{\frac{p}{(a)}}\right). \tag{3}$$

A combination of (2) and (3) leads to $\mathrm{embdim}(R[X]_P) = r$, as desired.

Case 2 P is an upper to p (i.e., $P \neq pR[X]$). We prove that $\mathrm{embdim}(R[X]_P) = r + 1$. Note that $PR_p[X]$ is also an upper to p_p and then there exists a (monic) polynomial $f \in R[X]$ such that $\overline{\frac{f}{1}}$ is irreducible in $\kappa_R(p)[X]$ and $PR_p[X] = pR_p[X] + fR_p[X]$. Notice that $pR[X] + fR[X] \subseteq P$ and we have

$$\begin{aligned}P_P &= PR_p[X]_{PR_p[X]} = (pR_p[X] + fR_p[X])_{PR_p[X]}\\ &= (p[X] + fR[X])R_p[X]_{PR_p[X]} = (p[X] + fR[X])_P\\ &= p[X]_P + fR[X]_P = (a_1,\ldots,a_r,f)R[X]_P.\end{aligned}$$

Assume $r = 0$. Then P is an upper to zero with $P_P = fR[X]_P$. So that $\mathrm{embdim}(R[X]_P) \leq 1$. Further, by the principal ideal theorem [23, Theorem 152], we have

$$\mathrm{embdim}(R[X]_P) \geq \dim(R[X]_P) = \mathrm{ht}(P) = 1.$$

It follows that $\text{embdim}(R[X]_P) = 1$, as desired.

Next, assume $r \geq 1$. We claim that $pR[X]_P \nsubseteq P_P^2$. Deny and suppose that $pR[X]_P \subseteq P_P^2$. This assumption combined with the fact $P_P = p[X]_P + fR[X]_P$ yields $\frac{P_P}{P_P^2} = \overline{f}R[X]_P$ as $R[X]_P$-modules and hence $P_P = fR[X]_P$ by [23, Theorem 158]. Next, let $a \in p$. Then, as $\frac{a}{1} \in P_P = fR[X]_P$, there exist $g \in R[X]$ and $s, t \in R[X] \setminus P$ such that $t(sa - fg) = 0$. So that $tfg \in p[X]$, whence $tg \in p[X] \subset P$ as $f \notin p[X]$. It follows that $tsa = tfg \in P^2$ and thus $\frac{a}{1} \in P_P^2 = f^2R[X]_P$. We iterate the same process to get $\frac{a}{1} \in P_P^n = f^nR[X]_P$ for each integer $n \geq 1$. Since $R[X]_P$ is a Noetherian local ring, $\bigcap P_P^n = (0)$ and thus $\frac{a}{1} = 0$ in $R[X]_P$. By the canonical injective homomorphism $R_p \hookrightarrow R[X]_P$, $\frac{a}{1} = 0$ in R_p. Thus $p_p = (0)$, the desired contradiction.

Consequently, $pR[X]_P = (a_1, \ldots, a_r)R[X]_P \nsubseteq P_P^2$. So, there exists $j \in \{1, \ldots, n\}$, say $j = 1$, such that $a := a_1 \in P_P \setminus P_P^2$ and, a fortiori, $a \in p_p \setminus p_p^2$. Similar arguments as in Case 1 lead to the same two formulas displayed in (2). Therefore $\text{embdim}\left(\left(\frac{R}{(a)}\right)_{\frac{p}{(a)}}\right) = r - 1$ and then, by induction on r, we obtain

$$\text{embdim}\left(\frac{R}{(a)}[X]_{\frac{P}{aR[X]}}\right) = 1 + \text{embdim}\left(\left(\frac{R}{(a)}\right)_{\frac{p}{(a)}}\right). \tag{4}$$

A combination of (2) and (4) leads to $\text{embdim}(R[X]_P) = r + 1$, as desired.

Now, assume that $n \geq 2$ and set $R[k] := R[X_1, \ldots, X_k]$ and $p[k] = p[X_1, \ldots, X_k]$ for $k := 1, \ldots, n$. Let $P' := P \cap R[n-1]$. We prove that $\text{embdim}(R[n]_P) = r + \text{ht}\left(\frac{P}{p[n]}\right)$. Indeed, by virtue of the case $n = 1$, we have

$$\text{embdim}(R[n]_P) = \text{embdim}(R[n-1]_{P'}) + \text{ht}\left(\frac{P}{P'[X_n]}\right). \tag{5}$$

Moreover, by induction hypothesis, we get

$$\text{embdim}(R[n-1]_{P'}) = r + \text{ht}\left(\frac{P'}{p[n-1]}\right). \tag{6}$$

Note that the prime ideals $\frac{P'[X_n]}{p[n]}$ and $\frac{P}{p[n]}$ both survive in $\kappa_R(p)[n]$, respectively. Hence, as $\kappa_R(p)[n]$ is catenarian and $(R/p)[n-1]$ is Noetherian, we obtain

$$\text{ht}\left(\frac{P}{p[n]}\right) = \text{ht}\left(\frac{P'[X_n]}{p[n]}\right) + \text{ht}\left(\frac{P}{P'[X_n]}\right) = \text{ht}\left(\frac{P'}{p[n-1]}\right) + \text{ht}\left(\frac{P}{P'[X_n]}\right). \tag{7}$$

Further, the fact that $\kappa_R(p)[X_1, \ldots, X_n]$ is regular yield

$$\mathrm{ht}\left(\frac{P}{p[X_1,\ldots,X_n]}\right) = \mathrm{embdim}\left(\kappa_R(p)[X_1,\ldots,X_n]_{\frac{P_p}{pR_p[X_1,\ldots,X_n]}}\right). \tag{8}$$

So (5)–(8) lead to the conclusion, completing the proof of the theorem. □

As a first application of Theorem 3.1, we get the next corollary on the (embedding) codimension. In particular, it recovers a well-known result on the transfer of regularity to polynomial rings (initially proved via Serre's result on finite global dimension and Hilbert Theorem on syzygies [27, Theorem 8.37]. See also [23, Theorem 171]).

Corollary 3.2 *Let R be a Noetherian ring and $X_1,\ldots,X_n$ be indeterminates over R. Let P be a prime ideal of $R[X_1,\ldots,X_n]$ with $p := P\cap R$. Then:*

$$\mathrm{codim}(R[X_1,\ldots,X_n]_P) = \mathrm{codim}(R_p).$$

In particular, $R[X_1,\ldots,X_n]$ is regular if and only if R is regular.

Theorem 3.1 allows us to characterize the regularity for two famous localizations of polynomial rings; namely, Nagata rings and Serre conjecture rings. Let R be a ring and $X, X_1,\ldots,X_n$ indeterminates over R. Recall that $R(X_1,\ldots,X_n) = S^{-1}R[X_1,\ldots,X_n]$ is the Nagata ring, where S is the multiplicative set of $R[X_1,\ldots,X_n]$ consisting of the polynomials whose coefficients generate R. Let $R\langle X\rangle := U^{-1}R[X]$, where U is the multiplicative set of monic polynomials in $R[X]$, and $R\langle X_1,\cdots,X_n\rangle := R\langle X_1,\ldots,X_{n-1}\rangle\langle X_n\rangle$. Then $R\langle X_1,\ldots,X_n\rangle$ is called the Serre conjecture ring and is a localization of $R[X_1,\ldots,X_n]$.

Corollary 3.3 *Let R be a Noetherian ring and $X_1,\ldots,X_n$ indeterminates over R. Let S be a multiplicative subset of $R[X_1,\ldots,X_n]$. Then:*

(a) $S^{-1}R[X_1,\ldots,X_n]$ is regular if and only if R_p is regular for each prime ideal p of R such that $p[X_1,\ldots,X_n]\cap S = \emptyset$.
(b) In particular, $R(X_1,\ldots,X_n)$ is regular if and only if $R\langle X_1,\ldots,X_n\rangle$ is regular if and only if $R[X_1,\ldots,X_n]$ is regular if and only if R is regular.

Proof (a) Let $Q = S^{-1}P$ be a prime ideal of $S^{-1}R[X_1,\ldots,X_n]$, where P is the inverse image of Q by the canonical homomorphism $R[X_1,\ldots,X_n]\to S^{-1}R[X_1,\ldots,X_n]$ and let $p := P\cap R$. Notice that $S^{-1}R[X_1,\ldots,X_n]_Q \cong R[X_1,\ldots,X_n]_P$ and $\dfrac{Q}{S^{-1}p[X_1,\ldots,X_n]} \cong \overline{S}^{-1}\dfrac{P}{p[X_1,\ldots,X_n]}$ where $\overline{S}$ denotes the image of S via the natural homomorphism $R[X_1,\ldots,X_n]\to \frac{R}{p}[X_1,\ldots,X_n]$. Therefore, by (1), we obtain

$$\dim(S^{-1}R[X_1,\ldots,X_n]_Q) = \dim(R[X_1,\ldots,X_n]_P) = \dim(R_p) + \mathrm{ht}\left(\frac{Q}{S^{-1}p[X_1,\ldots,X_n]}\right) \tag{9}$$

and, by Theorem 3.1, we have

$$\begin{aligned}\operatorname{embdim}(S^{-1}R[X_1,\dots,X_n]_Q) &= \operatorname{embdim}(R[X_1,\dots,X_n]_P)\\ &= \operatorname{embdim}(R_p) + \operatorname{ht}\left(\frac{Q}{S^{-1}p[X_1,\dots,X_n]}\right).\end{aligned} \tag{10}$$

Now, observe that the set $\{Q\cap R \mid Q \text{ is a prime ideal of } S^{-1}R[X_1,\dots,X_n]\}$ is equal to the set $\{p \mid p \text{ is a prime ideal of } R \text{ such that } p[X_1,\dots,X_n] \cap S = \emptyset\}$. Therefore, (9) and (10) lead to the conclusion.

(b) Combine (a) with the fact that the extension of any prime ideal of R to $R[X_1,\dots,X_n]$ does not meet the multiplicative sets related to the rings $R(X_1,\dots,X_n)$ and $R\langle X_1,\dots,X_n\rangle$. □

4 Embedding Dimension and Codimension of Tensor Products Issued from Separable Extension Fields

This section establishes an analogue of the "special chain theorem" for the embedding dimension of Noetherian tensor products issued from separable extension fields, with effective consequences on the transfer or defect of regularity. Namely, due to known behavior of a tensor product $A \otimes_k B$ of two k-algebras subject to regularity (cf. [4, 18, 19, 25, 31, 34]), we will investigate the case where A or B is a separable (not necessarily algebraic) extension field of k.

Throughout, let A and B be two k-algebras such that $A \otimes_k B$ is Noetherian and let P be a prime ideal of $A \otimes_k B$ with $p := P \cap A$ and $q := P \cap B$. Recall that A and B are Noetherian too; and the converse is not true, in general, even if $A = B$ is an extension field of k (cf. [16, Corollary 3.6] or [32, Theorem 11]). We assume familiarity with the natural isomorphisms for tensor products and their localizations as in [9, 10, 27]. In particular, we identify A and B with their respective images in $A \otimes_k B$ and we have $\dfrac{A\otimes_k B}{p\otimes_k B + A\otimes_k q} \cong \dfrac{A}{p} \otimes_k \dfrac{B}{q}$ and $A_p \otimes_k B_q \cong S^{-1}(A \otimes_k B)$ where $S := \{s\otimes t \mid s \in A\setminus p, t \in B \setminus q\}$. Throughout this and next sections, we adopt the following simplified notation for the invariant μ:

$$\mu_P(pA_p) := \mu^i_{(A\otimes_k B)_P}(pA_p) \ \text{ and } \ \mu_P(qA_q) := \mu^j_{(A\otimes_k B)_P}(qB_q)$$

where $i : A_p \longrightarrow (A \otimes_k B)_P$ and $j : B_q \longrightarrow (A \otimes_k B)_P$ are the canonical (local flat) ring homomorphisms.

Recall that $A \otimes_k B$ is Cohen–Macaulay (resp., Gorenstein, locally complete intersection) if and only if so are A and the fiber rings $\kappa_A(p) \otimes_k B$ (for each prime ideal p of A) [3, 31]. Also if A and the fiber rings $\kappa_A(p) \otimes_k B$ are regular then so is $A \otimes_k B$ [25, Theorem 23.7(ii)]. However, the converse does not hold in general; precisely, if $A \otimes_k B$ is regular then so is A [25, Theorem 23.7(i)] but the fiber rings are not necessarily regular (see [4, Example 2.12(iii)]).

From [3, Proposition 2.3] and its proof, recall an analogue of the special chain theorem (recorded in (1)) for the tensor products which correlates the dimension of $(A \otimes_k B)_P$ to the dimension of its fiber rings; namely,

$$\begin{aligned}\dim(A \otimes_k B)_P &= \dim(A_p) + \operatorname{ht}\left(\frac{P}{p \otimes_k B}\right)\\ &= \dim(A_p) + \dim\left(\left(\kappa_A(p) \otimes_k B\right)_{\frac{P_p}{pA_p \otimes_k B}}\right)\end{aligned} \tag{11}$$

Our first result reformulates Proposition 2.1 and thus gives an analogue of the special chain theorem for the embedding dimension in the context of tensor products of algebras over a field.

Proposition 4.1 *Let A and B be two k-algebras such that $A \otimes_k B$ is Noetherian and let P be a prime ideal of $A \otimes_k B$ with $p := P \cap A$ and $q := P \cap B$. Then:*

(a) $\operatorname{embdim}(A \otimes_k B)_P = \mu_P(pA_p) + \operatorname{embdim}\left(\left(\kappa_A(p) \otimes_k B\right)_{\frac{P_p}{pA_p \otimes_k B}}\right)$.

(b) $\operatorname{codim}(A \otimes_k B)_P + \left(\operatorname{embdim}(A_p) - \mu_P(pA_p)\right) =$
$$\operatorname{codim}(A_p) + \operatorname{codim}\left(\left(\kappa_A(p) \otimes_k B\right)_{\frac{P_p}{pA_p \otimes_k B}}\right).$$

(c) *$(A \otimes_k B)_P$ is regular and $\mu_P(pA_p) = \operatorname{embdim}(A_p)$ if and only if both A_p and $\left(\kappa_A(p) \otimes_k B\right)_{\frac{P_p}{pA_p \otimes_k B}}$ are regular.*

Recall that an extended form of the special chain theorem [3] states that

$$\dim(A \otimes_k B)_P = \dim(A_p) + \dim(B_q) + \dim\left(\left(\kappa_A(p) \otimes_k \kappa_B(q)\right)_{\frac{P(A_p \otimes_k B_q)}{pA_p \otimes_k B_q + A_p \otimes_k qB_q}}\right).$$

In this vein, notice that, via Proposition 4.1(a), we always have the following inequalities:

$$\begin{aligned}\operatorname{embdim}(A \otimes_k B)_P &\leq \operatorname{embdim}(A_p) + \operatorname{embdim}\left(\left(\kappa_A(p) \otimes_k B\right)_{\frac{P_p}{pA_p \otimes_k B}}\right)\\ &\leq \operatorname{embdim}(A_p) + \operatorname{embdim}(B_q)\\ &\quad + \operatorname{embdim}\left(\left(\kappa_A(p) \otimes_k \kappa_B(q)\right)_{\frac{P(A_p \otimes_k B_q)}{pA_p \otimes_k B_q + A_p \otimes_k qB_q}}\right).\end{aligned}$$

Let us state the main theorem of this section.

Theorem 4.2 *Let K be a separable extension field of k and A a k-algebra such that $K \otimes_k A$ is Noetherian. Let P be a prime ideal of $K \otimes_k A$ with $p := P \cap A$. Then:*

$$\begin{aligned}\text{embdim}(K \otimes_k A)_P &= \text{embdim}(A_p) + \text{ht}\left(\frac{P}{K \otimes_k p}\right)\\ &= \text{embdim}(A_p) + \text{embdim}\left(\left(K \otimes_k \kappa_A(p)\right)_{\frac{P_p}{K \otimes_k pA_p}}\right)\end{aligned}$$

If, in addition, K is algebraic over k, then $\text{embdim}(K \otimes_k A)_P = \text{embdim}(A_p)$.

The proof of this theorem requires the following two preparatory lemmas; the first of which determines a formula for the embedding dimension of the tensor product of two k-algebras A and B localized at a special prime ideal P with no restrictive conditions on A or B.

Lemma 4.3 *Let A and B be two k-algebras such that $A \otimes_k B$ is Noetherian and let P be a prime ideal of $A \otimes_k B$ with $p := P \cap A$ and $q := P \cap B$. Assume that $P_P = (p \otimes_k B + A \otimes_k q)_P$. Then:*

(a) $\mu_P(pA_p) = \text{embdim}(A_p)$ *and* $\mu_P(qB_q) = \text{embdim}(B_q)$.
(b) $\text{embdim}(A \otimes_k B)_P = \text{embdim}(A_p) + \text{embdim}(B_q)$.

Proof We proceed through two steps.

Step 1. Assume that $K := B$ is an extension field of k. Then $q = (0)$ and $P_P = p_p(A_p \otimes_k K)_{P_p}$. Let $n := \text{embdim}(A_p)$ and let $a_1, \ldots, a_n$ be elements of p such that $p_p = \left(\frac{a_1}{1}, \ldots, \frac{a_n}{1}\right) A_p$. Our argument uses induction on n. If $n = 0$, then A_p is a field and $p_p = (0)$; hence $P_P = (0)$, whence $\text{embdim}(A \otimes_k K)_P = 0$, as desired. Next, suppose $n \geq 1$. We have $P_P = \left(\frac{a_1}{1}, \ldots, \frac{a_n}{1}\right)(A \otimes_k K)_P$. If $\text{embdim}(A \otimes_k K)_P = 0$, $(A \otimes_k K)_P$ is regular and so is A_p by [25, Theorem 23.7(i)]. Hence, $n = \dim(A_p) = 0$ by (11). Absurd. So, necessarily, $\text{embdim}(A \otimes_k K)_P \geq 1$. Without loss of generality, we may assume that $\frac{a_1}{1} \in P_P \setminus P_P^2$. Note that we already have $\frac{a_1}{1} \in p_p \setminus p_p^2$. Now, $\frac{P}{(a_1) \otimes_k K}$ is a prime ideal of $\frac{A}{(a_1)} \otimes_k K$ with $\frac{P}{(a_1) \otimes_k K} \cap \frac{A}{(a_1)} = \frac{p}{(a_1)}$. By [23, Theorem 159], we obtain $\text{embdim}\left(\left(\frac{A}{(a_1)}\right)_{\frac{p}{(a_1)}}\right) = n-1$.
By induction, we get

$$\text{embdim}\left(\left(\frac{A}{(a_1)} \otimes_k K\right)_{\frac{P}{(a_1)\otimes_k K}}\right) = \text{embdim}\left(\left(\frac{A}{(a_1)}\right)_{\frac{p}{(a_1)}}\right).$$

We conclude, via [23, Theorem 159], to get

$$\text{embdim}(A \otimes_k K)_P = 1 + \text{embdim}\left(\left(\frac{A}{(a_1)} \otimes_k K\right)_{\frac{P}{(a_1)\otimes_k K}}\right) = n.$$

Moreover, observe that $\left(\kappa_A(p)\otimes_k K\right)_{\frac{P_p}{pp\otimes_k K}}$ is a field as $P_P=(p\otimes_k K)_P$. By Proposition 4.1, we have

$$\mu_P(pA_p)=\operatorname{embdim}(A\otimes_k K)_P=\operatorname{embdim}(A_p). \tag{12}$$

Step 2. Assume that B is an arbitrary k-algebra. Since $P_P=(p\otimes_k B+A\otimes_k q)_P$, then $P(A_p\otimes_k B_q)=pA_p\otimes_k B_q+A_p\otimes_k qB_q$, hence $\left(\kappa_A(p)\otimes_k \kappa_B(q)\right)_{\frac{P(A_p\otimes_k B_q)}{pA_p\otimes_k B_q+A_p\otimes_k qB_q}}$ is an extension field of k. So, apply Proposition 4.1 twice to get

$$\operatorname{embdim}(A\otimes_k B)_P=\mu_P(qB_q)+\mu_{\frac{P_q}{A\otimes_k qB_q}}(pA_p). \tag{13}$$

Further, notice that

$$\begin{aligned}\left(\frac{P_q}{A\otimes_k qB_q}\right)_{\frac{P_q}{A\otimes_k qB_q}} &=\frac{(P_q)_{P_q}}{(A\otimes_k qB_q)_{P_q}}=\frac{P_P}{(A\otimes_k q)_P}=\frac{(p\otimes_k B+A\otimes_k q)_P}{(A\otimes_k q)_P}\\ &=\left(\frac{p\otimes_k B+A\otimes_k q}{A\otimes_k q}\right)_{\frac{P}{A\otimes_k q}}\\ &\cong\left(p\otimes_k\frac{B}{q}\right)_{\frac{P}{A\otimes_k q}}=\left(p\otimes_k\kappa_B(q)\right)_{\frac{P_q}{A\otimes_k qB_q}}.\end{aligned}$$

Therefore, by (12), we get

$$\mu_{\frac{P_q}{A\otimes_k qB_q}}(pA_p)=\operatorname{embdim}\left(\left(A\otimes_k\kappa_B(q)\right)_{\frac{P_q}{A\otimes_k qB_q}}\right)=\operatorname{embdim}(A_p).$$

Similar arguments yield

$$\mu_{\frac{P_p}{pA_p\otimes_k B}}(qB_q)=\operatorname{embdim}\left(\left(\kappa_A(p)\otimes_k B\right)_{\frac{P_p}{pA_p\otimes_k B}}\right)=\operatorname{embdim}(B_q)$$

and, by the facts $0\leq\mu_P(pA_p)\leq\operatorname{embdim}(A_p)$ and $\mu_{\frac{P_p}{pA_p\otimes_k B}}(qB_q)\leq\mu_P(qB_q)$, we obtain

$$\mu_P(pA_p)=\operatorname{embdim}(A_p)\ \text{ and }\ \mu_P(qB_q)=\operatorname{embdim}(B_q)$$

completing the proof of the lemma via (13). □

The second lemma will allow us to reduce our investigation to tensor products issued from finite extension fields.

Lemma 4.4 *Let K be an extension field of k and A a k-algebra such that $K\otimes_k A$ is Noetherian. Let P be a prime ideal of $K\otimes_k A$. Then, there exists a finite extension field E of k contained in K such that*

$$\mathrm{embdim}(K \otimes_k A)_P = \mathrm{embdim}(F \otimes_k A)_Q$$

for each intermediate field F between E and K and $Q := P \cap (F \otimes_k A)$.

Proof Let $z_1, \ldots, z_t \in K \otimes_k A$ such that $P = (z_1, \ldots, z_t) K \otimes_k A$; and for each $i = 1, \ldots, t$, let $z_i := \sum_{j=1}^{n_i} \alpha_{ij} \otimes_k a_j$ with $\alpha_{ij} \in K$ and $a_j \in A$. Let $E := k\left(\{\alpha_{ij} \mid i = 1, \ldots, t\,;\, j = 1, \ldots, n_i\}\right)$ and $Q := P \cap (E \otimes_k A)$. Clearly, $z_1, \ldots, z_t \in Q$ and hence $P = Q(K \otimes_k A) = K \otimes_E Q$. Apply Lemma 4.3 to $K \otimes_k A \cong K \otimes_E (E \otimes_k A)$ to obtain $\mathrm{embdim}(K \otimes_k A)_P = \mathrm{embdim}(E \otimes_k A)_Q$. Now, let F be an intermediate field between E and K and $Q' := P \cap (F \otimes_k A)$. Then

$$P = Q'(K \otimes_k A) = K \otimes_E Q' \tag{14}$$

since $Q' \cap (E \otimes_k A) = Q$. As above, Lemma 4.3 leads to the conclusion. □

Next, we give the proof of the main theorem.

Proof of Theorem 4.2 We proceed through three steps.

Step 1. Assume that K is an algebraic separable extension field of k. We claim that

$$P_P = (K \otimes_k p)_P. \tag{15}$$

Indeed, set $S_p := \frac{A}{p} \setminus \{\bar{0}\}$. The basic fact $\frac{P}{K \otimes_k p} \cap \frac{A}{p} = (\bar{0})$ yields

$$\frac{(K \otimes_k A)_P}{(K \otimes_k p)_P} \cong \left(K \otimes_k \frac{A}{p}\right)_{\frac{P}{K \otimes_k p}} = \left(K \otimes_k \kappa_A(p)\right)_{S_p^{-1}\left(\frac{P}{K \otimes_k p}\right)}$$

where $K \otimes_k \kappa_A(p)$ is a zero-dimensional ring [30, Theorem 3.1], reduced [35, Chap. III, §15, Theorem 39], and hence von Neumann regular [23, Ex. 22, p. 64]. It follows that $\left(K \otimes_k \kappa_A(p)\right)_{S_p^{-1}\left(\frac{P}{K \otimes_k p}\right)}$ is a field. Consequently, $(K \otimes_k p)_P = P_P$, the unique maximal ideal of $(K \otimes_k A)_P$, proving our claim. By (15) and Lemma 4.3, we get $\mathrm{embdim}(K \otimes_k A)_P = \mathrm{embdim}(A_p)$.

Step 2. Assume that K is a finitely generated separable extension field of k. Let $T = \{x_1, \ldots, x_t\}$ be a finite separating transcendence base of K over k; that is, K is algebraic separable over $k(T) := k(x_1, \ldots, x_t)$. Let $S := k[T] \setminus \{0\}$ and notice that

$$K \otimes_k A \cong K \otimes_{k(T)} (k(T) \otimes_k A) \cong K \otimes_{k(T)} S^{-1}A[T].$$

Let $P \cap S^{-1}A[T] = S^{-1}P'$ for some prime ideal P' of $A[T]$. Note that $P' \cap A = p$. Then, we have

$$
\begin{aligned}
\operatorname{embdim}(K \otimes_k A)_P &= \operatorname{embdim}\left(K \otimes_{k(T)} S^{-1}A[T]\right)_P \\
&= \operatorname{embdim}\left(S^{-1}A[T]_{S^{-1}P'}\right), \text{ by Step 1} \\
&= \operatorname{embdim}\left(A[T]_{P'}\right) \\
&= \operatorname{embdim}(A_p) + \operatorname{ht}\left(\frac{P'}{p[T]}\right), \text{ by Theorem 3.1.}
\end{aligned}
$$

Moreover, note that

$$
\begin{aligned}
K \otimes_k \frac{A}{p} &\cong K \otimes_{k(T)} \left(k(T) \otimes_k \frac{A}{p}\right) \\
&\cong K \otimes_{k(T)} \frac{S^{-1}A[T]}{S^{-1}p[T]}
\end{aligned}
$$

and

$$
\frac{P}{K \otimes_k p} \cap \frac{S^{-1}A[T]}{S^{-1}p[T]} = \frac{S^{-1}P'}{S^{-1}p[T]}
$$

as $K \otimes_k p \cong K \otimes_{k(T)} S^{-1}p[T]$ so that $(K \otimes_k p) \cap S^{-1}A[T] = S^{-1}p[T]$. Therefore the integral extension $\frac{S^{-1}A[T]}{S^{-1}p[T]} \hookrightarrow K \otimes_k \frac{A}{p}$ is flat and hence satisfies the Going-down property; that is, $\operatorname{ht}\left(\frac{P'}{p[T]}\right) = \operatorname{ht}\left(\frac{S^{-1}P'}{S^{-1}p[T]}\right) = \operatorname{ht}\left(\frac{P}{K \otimes_k p}\right)$. It follows that $\operatorname{embdim}(K \otimes_k A)_P = \operatorname{embdim}(A_p) + \operatorname{ht}\left(\frac{P}{K \otimes_k p}\right)$.

Step 3. Assume that K is an arbitrary separable extension field of k. Then, by Lemma 4.4, there exists a finite extension field E of k contained in K such that

$$
\operatorname{embdim}(K \otimes_k A)_P = \operatorname{embdim}(E \otimes_k A)_Q
$$

where $Q := P \cap (E \otimes_k A)$. Let Ω denote the set of all intermediate fields between E and K. For each $F \in \Omega$, note that $P = Q'(K \otimes_k A)$, where $Q' := P \cap (F \otimes_k A)$, as seen in (14); and by Lemma 4.4 and Step 2, we obtain

$$
\operatorname{embdim}(K \otimes_k A)_P = \operatorname{embdim}(F \otimes_k A)_{Q'} = \operatorname{embdim}(A_p) + \operatorname{ht}\left(\frac{Q'}{F \otimes_k p}\right). \tag{16}
$$

Further, as the ring extension $F \otimes_k \frac{A}{p} \hookrightarrow K \otimes_k \frac{A}{p}$ satisfies the Going-down property, we get

$$
\operatorname{ht}\left(\frac{Q'}{F \otimes_k p}\right) \leq \operatorname{ht}\left(\frac{P}{K \otimes_k p}\right). \tag{17}
$$

Next let $K \otimes_k p \subseteq P_0 \subsetneq P_1 \subsetneq \ldots \subsetneq P_n = P$ be a chain of distinct prime ideals of $K \otimes_k A$ such that $n := \operatorname{ht}\left(\frac{P}{K \otimes_k p}\right)$. Let $t_i \in P_i \setminus P_{i-1}$ for each $i = 1, \ldots, n$. One readily checks that there exists a finite extension field G of k contained in

K such that, for each $i = 1, \ldots, n$, $t_i \in G \otimes_k A$ and thus $t_i \in Q'_i \setminus Q'_{i-1}$, where $Q'_i := P_i \cap (G \otimes_k A)$. Let $H := k(E, G) \in \Omega$ and $Q_i := P_i \cap (H \otimes_k A)$ for each $i = 1, \ldots, n$. Then $t_i \in Q_i \setminus Q_{i-1}$ for each $i = 1, \ldots, n$. Therefore, we get the following chain of distinct prime ideals in $H \otimes_k A$

$$H \otimes_k p \subseteq Q_0 \subsetneq Q_1 \subsetneq \ldots \subsetneq Q_n = Q' := P \cap (H \otimes_k A).$$

It follows that $\mathrm{ht}\left(\frac{Q'}{H \otimes_k p}\right) \geq n$ and then (17) yields $\mathrm{ht}\left(\frac{Q'}{F \otimes_k p}\right) = \mathrm{ht}\left(\frac{P}{K \otimes_k p}\right)$. Further, $K \otimes_k \kappa_A(p)$ is regular since K is separable over K [18, Lemma 6.7.4.1]. Consequently, by (16), we get

$$\begin{aligned}\mathrm{embdim}(K \otimes_k A)_P &= \mathrm{embdim}(A_p) + \mathrm{ht}\left(\frac{P}{K \otimes_k p}\right) \\ &= \mathrm{embdim}(A_p) + \mathrm{embdim}\left(\left(K \otimes_k \kappa_A(p)\right)_{\frac{P_p}{K \otimes_k p A_p}}\right)\end{aligned}$$

completing the proof of the theorem.

As a direct application of Theorem 4.2, we obtain the next corollary on the (embedding) codimension which extends Grothendieck's result on the transfer of regularity to tensor products issued from finite extension fields [18, Lemma 6.7.4.1]. See also [4].

Corollary 4.5 *Let K be a separable extension field of k and A a k-algebra such that $K \otimes_k A$ is Noetherian. Let P be a prime ideal of $K \otimes_k A$ with $p := P \cap A$. Then:*

$$\mathrm{codim}(K \otimes_k A)_P = \mathrm{codim}(A_p).$$

In particular, $K \otimes_k A$ is regular if and only if A is regular.

Proof Combine Theorem 4.2 and (11). □

5 Embedding Dimension and Codimension of Tensor Products of Algebras with Separable Residue Fields

This section deals with a more general setting (than in Sect. 4); namely, we compute the embedding dimension of localizations of the tensor product of two k-algebras A and B at prime ideals P such that the residue field $\kappa_B(P \cap B)$ is a separable extension of k. The main result establishes an analogue for an extended form of the "special chain theorem" for the Krull dimension which asserts that

$$\begin{aligned}\dim(A \otimes_k B)_P &= \dim(A_p) + \dim(B_q) + \mathrm{ht}\left(\frac{P}{p \otimes_k B + A \otimes_k q}\right)\\ &= \dim(A_p) + \dim(B_q) + \dim\left(\left(\kappa_A(p) \otimes_k \kappa_B(q)\right)_{\frac{P(A_p \otimes_k B_q)}{pA_p \otimes_k B_q + A_p \otimes_k qB_q}}\right).\end{aligned} \tag{18}$$

As an application, we formulate the (embedding) codimension of these constructions with direct consequence on the transfer or defect of regularity.

Here is the main result of this section.

Theorem 5.1 *Let A and B be two k-algebras such that $A \otimes_k B$ is Noetherian and let P be a prime ideal of $A \otimes_k B$ with $p := P \cap A$ and $q := P \cap B$. Assume $\kappa_B(q)$ is separable over k. Then:*

$$\begin{aligned}\mathrm{embdim}(A \otimes_k B)_P &= \mathrm{embdim}(A_p) + \mathrm{embdim}(B_q) + \mathrm{ht}\left(\frac{P}{p \otimes_k B + A \otimes_k q}\right)\\ &= \mathrm{embdim}(A_p) + \mathrm{embdim}(B_q)\\ &\quad + \mathrm{embdim}\left(\left(\kappa_A(p) \otimes_k \kappa_B(q)\right)_{\frac{P(A_p \otimes_k B_q)}{pA_p \otimes_k B_q + A_p \otimes_k qB_q}}\right)\end{aligned}$$

Proof Notice first that, as $\kappa_B(q)$ is separable over k, $\kappa_A(p) \otimes_k \kappa_B(q)$ is a regular ring and hence

$$\begin{aligned}\mathrm{embdim}\left(\left(\kappa_A(p) \otimes_k \kappa_B(q)\right)_{\frac{P(A_p \otimes_k B_q)}{pA_p \otimes_k B_q + A_p \otimes_k qB_q}}\right) &= \mathrm{ht}\left(\frac{P(A_p \otimes_k B_q)}{pA_p \otimes_k B_q + A_p \otimes_k qB_q}\right)\\ &= \mathrm{ht}\left(\frac{P}{p \otimes_k B + A \otimes_k q}\right).\end{aligned}$$

So, we only need to prove the first equality in the theorem and, without loss of generality, we may assume that $(A, \mathfrak{n})$ and $(B, \mathfrak{m})$ are local k-algebras such that $A \otimes_k B$ is Noetherian, $\frac{B}{\mathfrak{m}}$ is a separable extension field of k, and P is a prime ideal of $A \otimes_k B$ with $P \cap A = \mathfrak{n}$ and $P \cap B = \mathfrak{m}$. Similar arguments used in the proof of Lemma 4.4 show that there exists a finite extension field K of k contained in $\frac{B}{\mathfrak{m}}$ such that

$$\frac{P}{A \otimes_k \mathfrak{m}} = Q\left(A \otimes_k \frac{B}{\mathfrak{m}}\right) \cong Q \otimes_K \frac{B}{\mathfrak{m}}$$

where $Q := \frac{P}{A \otimes_k \mathfrak{m}} \cap (A \otimes_k K)$. Since $\frac{B}{\mathfrak{m}}$ is separable over k and K is a finitely generated intermediate field, then K is separably generated over k (cf. [21, Chap. VI, Theorem 2.10 & Definition 2.11]). Let t denote the transcendence degree of K over k and let $c_1, \ldots, c_t \in B$ such that $\{\overline{c_1}, \ldots, \overline{c_t}\}$ is a separating transcendence base of K over k; i.e., K is separable algebraic over $\Omega := k(\overline{c_1}, \ldots, \overline{c_t})$. Also $c_1, \ldots, c_t$ are algebraically independent over k with

$$\mathfrak{m} \cap k[c_1, \ldots, c_t] = (0). \tag{19}$$

So one may view $A \otimes_k k[c_1, \dots, c_t] \cong A[c_1, \dots, c_t]$ as a polynomial ring in t indeterminates over A. Set $S := k[c_1, \dots, c_t] \setminus \{0\}$; $k(t) := k(c_1, \dots, c_t)$; $A[t] := A[c_1, \dots, c_t]$. Then, we have

$$P \cap S = \mathfrak{m} \cap S = \emptyset \text{ and } A \otimes_k S^{-1}B \cong S^{-1}A[t] \otimes_{k(t)} S^{-1}B. \tag{20}$$

Next, let $T := \frac{P}{A\otimes_k \mathfrak{m}} \cap (A \otimes_k \Omega) = Q \cap (A \otimes_k \Omega)$ and consider the following canonical isomorphisms of k-algebras $\theta_1 : A \otimes_k \dfrac{S^{-1}B}{S^{-1}\mathfrak{m}} \longrightarrow \big(A \otimes_k k(t)\big) \otimes_{k(t)} \dfrac{S^{-1}B}{S^{-1}\mathfrak{m}}$ and $\theta_2 : A \otimes_k \dfrac{B}{\mathfrak{m}} \longrightarrow (A \otimes_k \Omega) \otimes_\Omega \dfrac{B}{\mathfrak{m}}$. As $A \otimes_k K \cong (A \otimes_k \Omega) \otimes_\Omega K$, by (15) we obtain $Q_Q = (T \otimes_\Omega K)_Q = T(A \otimes_k K)_Q$ and hence

$$\begin{aligned}
\left(\frac{P}{A \otimes_k \mathfrak{m}}\right)_{\frac{P}{A\otimes_k \mathfrak{m}}} &= Q\left(A \otimes_k \frac{B}{\mathfrak{m}}\right)_{\frac{P}{A\otimes_k \mathfrak{m}}} = Q_Q\left(A \otimes_k \frac{B}{\mathfrak{m}}\right)_{\left(\frac{P}{A\otimes_k \mathfrak{m}}\right)_Q} \\
&= T(A \otimes_k K)_Q\left(A \otimes_k \frac{B}{\mathfrak{m}}\right)_{\left(\frac{P}{A\otimes_k \mathfrak{m}}\right)_Q} \\
&= T(A \otimes_k K)\left(A \otimes_k \frac{B}{\mathfrak{m}}\right)_{\frac{P}{A\otimes_k \mathfrak{m}}} = T\left(A \otimes_k \frac{B}{\mathfrak{m}}\right)_{\frac{P}{A\otimes_k \mathfrak{m}}} \\
&= \left(\theta_2^{-1}\left(\theta_2\left(T\left(A \otimes_k \frac{B}{\mathfrak{m}}\right)\right)\right)\right)_{\frac{P}{A\otimes_k \mathfrak{m}}} = \left(\theta_2^{-1}\left(T \otimes_\Omega \frac{B}{\mathfrak{m}}\right)\right)_{\frac{P}{A\otimes_k \mathfrak{m}}}.
\end{aligned} \tag{21}$$

Moreover, on account of (19) and by considering the natural surjective homomorphism of k-algebras $k[c_1, \dots, c_t] \xrightarrow{\varphi} k[\overline{c_1}, \dots, \overline{c_t}]$ defined by $\varphi(c_i) = \overline{c_i}$ for each $i = 1, \dots, t$, we get $k[c_1, \dots, c_t] \overset{\varphi}{\cong} k[\overline{c_1}, \dots, \overline{c_t}]$ inducing the following natural isomorphism of extension fields $\phi := S^{-1}\varphi : k(t) \longrightarrow k(\overline{c_1}, \dots, \overline{c_t}) = \Omega$. Then, ϕ induces a structure of $k(t)$-algebras on Ω and thus on $\dfrac{B}{\mathfrak{m}}$. We adopt a second structure of $k(t)$-algebras on $\dfrac{B}{\mathfrak{m}}$, inherited from the canonical injection $k(t) \overset{i}{\hookrightarrow} S^{-1}B$. Indeed, consider the following k-algebra homomorphisms $k(t) \xrightarrow{\bar{i}} \frac{S^{-1}B}{S^{-1}\mathfrak{m}} \xrightarrow{\gamma} \frac{B}{\mathfrak{m}}$ defined by $\bar{i}(\alpha) = \overline{\alpha}$ for each $\alpha \in k(t)$, and where γ is the isomorphism of k-algebras defined by $\gamma\left(\frac{\overline{b}}{s}\right) = \frac{\overline{b}}{\overline{s}}$ for each $b \in B$ and each $s \in S$. It is easy to see that these two structures of $k(t)$-algebras coincide on $\dfrac{B}{\mathfrak{m}}$. This is due to the commutativity of the following diagram of homomorphisms of k-algebras

$$\begin{array}{ccc}
k(t) & \xrightarrow{\bar{i}} & \dfrac{S^{-1}B}{S^{-1}\mathfrak{m}} \\
 & \phi \searrow & \downarrow \gamma \\
 & & \dfrac{B}{\mathfrak{m}}
\end{array}$$

since, for each $\alpha := \frac{f}{s} \in k(t)$ with $f \in k[c_1, \dots, c_t]$ and $s \in S$, we have

$$(\gamma \circ \bar{i})(\alpha) = \gamma(\overline{\alpha}) = \frac{\bar{f}}{\bar{s}} = \frac{\varphi(f)}{\varphi(s)} = \phi(\alpha).$$

Now, consider the following isomorphism of k-algebras

$$\psi := \theta_2 \circ (1_A \otimes_k \gamma) \circ \theta_1^{-1} : \big(A \otimes_k k(t)\big) \otimes_{k(t)} \frac{S^{-1}B}{S^{-1}\mathfrak{m}} \longrightarrow (A \otimes_k \Omega) \otimes_\Omega \frac{B}{\mathfrak{m}}$$

where, for each $a \in A$, $\alpha \in k(t)$, $b \in B$, and $s \in S$, we have

$$\begin{aligned}\psi\left((a \otimes_k \alpha) \otimes_{k(t)} \overline{\tfrac{b}{s}}\right) &= \theta_2\left((1_A \otimes_k \gamma)\left(a \otimes_k \overline{\alpha}\overline{\tfrac{b}{s}}\right)\right) = \theta_2\left(a \otimes_k \gamma\left(\overline{\alpha}\overline{\tfrac{b}{s}}\right)\right) \\ &= \theta_2\left(a \otimes_k \left((\gamma \circ \bar{i})(\alpha)\gamma\left(\overline{\tfrac{b}{s}}\right)\right)\right) = \theta_2\left(a \otimes_k \left(\phi(\alpha)\gamma\left(\overline{\tfrac{b}{s}}\right)\right)\right) \\ &= (a \otimes_k \bar{1}) \otimes_\Omega \phi(\alpha)\gamma\left(\overline{\tfrac{b}{s}}\right) = (a \otimes_k \phi(\alpha)) \otimes_\Omega \gamma\left(\overline{\tfrac{b}{s}}\right) \\ &= (1_A \otimes_k \phi)(a \otimes_k \alpha) \otimes_\Omega \gamma\left(\overline{\tfrac{b}{s}}\right).\end{aligned}$$

Next, let $\delta : A \otimes_k S^{-1}B \longrightarrow S^{-1}A[t] \otimes_{k(t)} S^{-1}B$ denote the canonical isomorphism of k-algebras mentioned in (20) and let $S^{-1}H := S^{-1}P \cap S^{-1}A[t]$ where H is a prime ideal of $A[t]$ with $H \cap S = \emptyset$. Therefore

$$\begin{aligned}\psi\left(S^{-1}H \otimes_{k(t)} \frac{S^{-1}B}{S^{-1}\mathfrak{m}}\right) &= (1_A \otimes_k \phi)(S^{-1}H) \otimes_\Omega \gamma\left(\frac{S^{-1}B}{S^{-1}\mathfrak{m}}\right) \\ &= (1_A \otimes_k \phi)(S^{-1}H) \otimes_\Omega \frac{B}{\mathfrak{m}}.\end{aligned} \tag{22}$$

Claim $\delta(S^{-1}P)_{\delta(S^{-1}P)} = \left(S^{-1}H \otimes_{k(t)} S^{-1}B + S^{-1}A[t] \otimes_{k(t)} S^{-1}\mathfrak{m}\right)_{\delta(S^{-1}P)}$.

Indeed, consider the following commutative diagram (as $\phi = \gamma \circ \bar{i}$)

$$\begin{array}{ccc} S^{-1}(A \otimes_k B) = A \otimes_k S^{-1}B & \overset{1_A \otimes_k (\gamma\circ\pi)}{\longrightarrow} & A \otimes_k \frac{B}{\mathfrak{m}} \\ 1_A \otimes_k i \uparrow & & \uparrow \\ A \otimes_k k(t) & \overset{1_A \otimes_k \phi}{\longrightarrow} & A \otimes_k \Omega \end{array}$$

where $\pi : S^{-1}B \longrightarrow \frac{S^{-1}B}{S^{-1}\mathfrak{m}}$ denotes the canonical surjection (with $\pi \circ i = \bar{i}$) and the vertical maps are the canonical injections. Also, it is worth noting that $1_A \otimes_k \phi$ is an isomorphism of k-algebras. Hence

$$
\begin{aligned}
T &= \frac{P}{A \otimes_k \mathfrak{m}} \cap (A \otimes_k \Omega) \\
&= (1_A \otimes_k \phi)\left(\left(\left(1_A \otimes_k (\gamma \circ \pi)\right)^{-1}\left(\frac{P}{A \otimes_k \mathfrak{m}}\right)\right) \cap \left(A \otimes_k k(t)\right)\right) \\
&= (1_A \otimes_k \phi)\left(S^{-1}P \cap \left(A \otimes_k k(t)\right)\right) = (1_A \otimes_k \phi)(S^{-1}P \cap S^{-1}A[t]) \\
&= (1_A \otimes_k \phi)(S^{-1}H).
\end{aligned}
\tag{23}
$$

It follows, via (21), (23), and (22), that

$$
\begin{aligned}
\left(\frac{P}{A \otimes_k \mathfrak{m}}\right)_{\frac{P}{A\otimes_k \mathfrak{m}}} &= \theta_2^{-1}\left(T \otimes_\Omega \frac{B}{\mathfrak{m}}\right)_{\frac{P}{A\otimes_k \mathfrak{m}}} \\
&= \theta_2^{-1}\left((1_A \otimes_k \phi)(S^{-1}H) \otimes_\Omega \frac{B}{\mathfrak{m}}\right)_{\frac{P}{A\otimes_k \mathfrak{m}}} \\
&= \theta_2^{-1}\left(\psi\left(S^{-1}H \otimes_{k(t)} \frac{S^{-1}B}{S^{-1}\mathfrak{m}}\right)\right)_{\frac{P}{A\otimes_k \mathfrak{m}}} \\
&= (1_A \otimes_k \gamma)\left(\theta_1^{-1}\left(S^{-1}H \otimes_{k(t)} \frac{S^{-1}B}{S^{-1}\mathfrak{m}}\right)\right)_{\frac{P}{A\otimes_k \mathfrak{m}}}.
\end{aligned}
$$

Further, notice that $\frac{P}{A \otimes_k \mathfrak{m}} = (1_A \otimes_k \gamma)\left(\frac{S^{-1}P}{A \otimes_k S^{-1}\mathfrak{m}}\right)$. Then the isomorphism $1_A \otimes_k \gamma$ yields the canonical isomorphism of local k-algebras

$$
(1_A \otimes_k \gamma)_P : \left(A \otimes_k \frac{S^{-1}B}{S^{-1}\mathfrak{m}}\right)_{\frac{S^{-1}P}{A\otimes_k S^{-1}\mathfrak{m}}} \longrightarrow \left(A \otimes_k \frac{B}{\mathfrak{m}}\right)_{\frac{P}{A\otimes_k \mathfrak{m}}} \quad \text{with}
$$

$$
\begin{aligned}
(1_A \otimes_k \gamma)_P &\left(\left(\frac{S^{-1}P}{A \otimes_k S^{-1}\mathfrak{m}}\right)_{\frac{S^{-1}P}{A\otimes_k S^{-1}\mathfrak{m}}}\right) = \left(\frac{P}{A \otimes_k \mathfrak{m}}\right)_{\frac{P}{A\otimes_k \mathfrak{m}}} \\
&= (1_A \otimes_k \gamma)_P\left(\theta_1^{-1}\left(S^{-1}H \otimes_{k(t)} \frac{S^{-1}B}{S^{-1}\mathfrak{m}}\right)_{\frac{S^{-1}P}{A\otimes_k S^{-1}\mathfrak{m}}}\right).
\end{aligned}
$$

Therefore

$$
\theta_1^{-1}\left(S^{-1}H \otimes_{k(t)} \frac{S^{-1}B}{S^{-1}\mathfrak{m}}\right)_{\frac{S^{-1}P}{A\otimes_k S^{-1}\mathfrak{m}}} = \left(\frac{S^{-1}P}{A \otimes_k S^{-1}\mathfrak{m}}\right)_{\frac{S^{-1}P}{A\otimes_k S^{-1}\mathfrak{m}}}. \tag{24}
$$

Moreover, consider the following commutative diagram

$$\begin{array}{ccc} A \otimes_k S^{-1}B & \xrightarrow{\delta} & S^{-1}A[t] \otimes_{k(t)} S^{-1}B \\ \pi_1 \downarrow & & \downarrow \pi_2 \\ A \otimes_k \dfrac{S^{-1}B}{S^{-1}\mathfrak{m}} & \xrightarrow{\theta_1} & S^{-1}A[t] \otimes_{k(t)} \dfrac{S^{-1}B}{S^{-1}\mathfrak{m}} \end{array}$$

where $\pi_1 = 1_A \otimes_k \pi$ and $\pi_2 = 1_{S^{-1}A[t]} \otimes_k \pi$ are the canonical surjective homomorphisms of k-algebras. Hence

$$\begin{aligned} \pi_1^{-1}\left(\theta_1^{-1}\left(S^{-1}H \otimes_{k(t)} \frac{S^{-1}B}{S^{-1}\mathfrak{m}}\right)\right) &= (\theta_1 \circ \pi_1)^{-1}\left(S^{-1}H \otimes_{k(t)} \frac{S^{-1}B}{S^{-1}\mathfrak{m}}\right) \\ &= (\pi_2 \circ \delta)^{-1}\left(S^{-1}H \otimes_{k(t)} \frac{S^{-1}B}{S^{-1}\mathfrak{m}}\right) \\ &= \delta^{-1}\left(\pi_2^{-1}\left(S^{-1}H \otimes_{k(t)} \frac{S^{-1}B}{S^{-1}\mathfrak{m}}\right)\right) \\ &= \delta^{-1}\left(S^{-1}H \otimes_{k(t)} S^{-1}B + S^{-1}A[t] \otimes_{k(t)} S^{-1}\mathfrak{m}\right) \end{aligned}$$

so that

$$\begin{aligned} \theta_1^{-1}\left(S^{-1}H \otimes_{k(t)} \frac{S^{-1}B}{S^{-1}\mathfrak{m}}\right) &= \pi_1\left(\delta^{-1}\left(S^{-1}H \otimes_{k(t)} S^{-1}B + S^{-1}A[t] \otimes_{k(t)} S^{-1}\mathfrak{m}\right)\right) \\ &= \frac{\delta^{-1}\left(S^{-1}H \otimes_{k(t)} S^{-1}B + S^{-1}A[t] \otimes_{k(t)} S^{-1}\mathfrak{m}\right)}{A \otimes_k S^{-1}\mathfrak{m}}. \end{aligned}$$

It follows, via (24), that

$$\begin{aligned} \frac{S^{-1}P_{S^{-1}P}}{(A \otimes_k S^{-1}\mathfrak{m})_{S^{-1}P}} &= \left(\frac{S^{-1}P}{A \otimes_k S^{-1}\mathfrak{m}}\right)_{\frac{S^{-1}P}{A\otimes_k S^{-1}\mathfrak{m}}} = \theta_1^{-1}\left(S^{-1}H \otimes_{k(t)} \frac{S^{-1}B}{S^{-1}\mathfrak{m}}\right)_{\frac{S^{-1}P}{A\otimes_k S^{-1}\mathfrak{m}}} \\ &= \left(\frac{\delta^{-1}\left(S^{-1}H \otimes_{k(t)} S^{-1}B + S^{-1}A[t] \otimes_{k(t)} S^{-1}\mathfrak{m}\right)}{A \otimes_k S^{-1}\mathfrak{m}}\right)_{\frac{S^{-1}P}{A\otimes_k S^{-1}\mathfrak{m}}} \\ &= \frac{\delta^{-1}\left(S^{-1}H \otimes_{k(t)} S^{-1}B + S^{-1}A[t] \otimes_{k(t)} S^{-1}\mathfrak{m}\right)_{S^{-1}P}}{(A \otimes_k S^{-1}\mathfrak{m})_{S^{-1}P}} \end{aligned}$$

and thus $S^{-1}P_{S^{-1}P} = \delta^{-1}\left(S^{-1}H \otimes_{k(t)} S^{-1}B + S^{-1}A[t] \otimes_{k(t)} S^{-1}\mathfrak{m}\right)_{S^{-1}P}$. Also, note that the isomorphism of k-algebras δ induces the isomorphism of local k-algebras $\delta_P : (A \otimes_k S^{-1}B)_{S^{-1}P} \longrightarrow (S^{-1}A[t] \otimes_{k(t)} S^{-1}B)_{\delta(S^{-1}P)}$. Hence

$$\begin{aligned} \delta_P^{-1}\left(\delta(S^{-1}P)_{\delta(S^{-1}P)}\right) &= S^{-1}P_{S^{-1}P} \\ &= \delta_P^{-1}\left(\left(S^{-1}H \otimes_{k(t)} S^{-1}B + S^{-1}A[t] \otimes_{k(t)} S^{-1}\mathfrak{m}\right)_{\delta(S^{-1}P)}\right) \end{aligned}$$

so that $\delta(S^{-1}P)_{\delta(S^{-1}P)} = \left(S^{-1}H \otimes_{k(t)} S^{-1}B + S^{-1}A[t] \otimes_{k(t)} S^{-1}\mathfrak{m}\right)_{\delta(S^{-1}P)}$ proving the claim.

It follows, by Lemma 4.3 applied to $S^{-1}A[t] \otimes_{k(t)} S^{-1}B$, that

$$\mu_{\delta(S^{-1}P)}(S^{-1}\mathfrak{m}S^{-1}B_{S^{-1}\mathfrak{m}}) = \text{embdim}(S^{-1}B_{S^{-1}\mathfrak{m}}) = \text{embdim}(B)$$

so that, by Proposition 4.1, we have

$$\begin{aligned}
\text{embdim}(A \otimes_k B)_P &= \text{embdim}\left((A \otimes_k S^{-1}B)_{S^{-1}P}\right) \\
&= \text{embdim}\left((S^{-1}A[t] \otimes_{k(t)} S^{-1}B)_{\delta(S^{-1}P)}\right) \\
&= \mu_{\delta(S^{-1}P)}(S^{-1}\mathfrak{m}S^{-1}B_{S^{-1}\mathfrak{m}}) \\
&\quad + \text{embdim}\left(\left(S^{-1}A[t] \otimes_{k(t)} \frac{S^{-1}B}{S^{-1}\mathfrak{m}}\right)_{\frac{\delta(S^{-1}P)}{S^{-1}A[t]\otimes_{k(t)}S^{-1}\mathfrak{m}}}\right) \\
&= \text{embdim}(B) + \text{embdim}\left(\left(S^{-1}A[t] \otimes_{k(t)} \frac{S^{-1}B}{S^{-1}\mathfrak{m}}\right)_{\frac{\delta(S^{-1}P)}{S^{-1}A[t]\otimes_{k(t)}S^{-1}\mathfrak{m}}}\right) \\
&= \text{embdim}(B) + \text{embdim}\left(\left(A \otimes_k \frac{S^{-1}B}{S^{-1}\mathfrak{m}}\right)_{\frac{S^{-1}P}{A\otimes_k S^{-1}\mathfrak{m}}}\right).
\end{aligned}$$

Finally, as $\frac{S^{-1}B}{S^{-1}\mathfrak{m}} \cong \frac{B}{\mathfrak{m}}$ is a separable extension field of k, we get, by Theorem 4.2, that

$$\begin{aligned}
\text{embdim}(A \otimes_k B)_P &= \text{embdim}(A) + \text{embdim}(B) + \text{ht}\left(\frac{S^{-1}P/(A \otimes_k S^{-1}\mathfrak{m})}{\mathfrak{n} \otimes_k (S^{-1}B/S^{-1}\mathfrak{m})}\right) \\
&= \text{embdim}(A) + \text{embdim}(B) + \text{ht}\left(\frac{S^{-1}P}{\mathfrak{n} \otimes_k S^{-1}B + A \otimes_k S^{-1}\mathfrak{m}}\right) \\
&= \text{embdim}(A) + \text{embdim}(B) + \text{ht}\left(\frac{P}{\mathfrak{n} \otimes_k B + A \otimes_k \mathfrak{m}}\right)
\end{aligned}$$

completing the proof of the theorem. □

As a direct application of Theorem 5.1, we obtain the next corollary on the (embedding) codimension which recovers known results on the transfer of regularity to tensor products over perfect fields [31, Theorem 6(c)] and, more generally, to tensor products issued from residually separable extension fields [4, Theorem 2.11]. Recall that a k-algebra R is said to be residually separable, if $\kappa_R(p)$ is separable over k for each prime ideal p of R.

Corollary 5.2 *Let A and B be two k-algebras such that $A \otimes_k B$ is Noetherian and let P be a prime ideal of $A \otimes_k B$ with $p := P \cap A$ and $q := P \cap B$. Assume $\kappa_B(q)$ is separable over k. Then:*

$$\text{codim}(A \otimes_k B)_P = \text{codim}(A_p) + \text{codim}(B_q).$$

Proof Combine Theorem 5.1 and (18). □

Note that if k is perfect, then every k-algebra is residually separable. Now, if k is an arbitrary field, one can easily provide original examples of residually separable k-algebras through localizations of polynomial rings or pullbacks [2, 13]. For instance, let X be an indeterminate over k and $K \subseteq L$ two separable extensions of k. Then, the one-dimensional local k-algebras $R := K + XL[X]_{(X)} \subseteq S := L[X]_{(X)}$ are residually separable since the extensions $k \subseteq \kappa_R\left(XL[X]_{(X)}\right) = K \subseteq \kappa_S\left(XL[X]_{(X)}\right) = L \subset \kappa_R(0) = \kappa_S(0) = L(X)$ are separable over k by Mac Lane's Criterion and transitivity of separability. Also, similar arguments show that the two-dimensional local k-algebra $R' := R + YL(X)[Y]_{(Y)}$ is residually separable, where Y is an indeterminate over k. Therefore, one may reiterate the same process to build residually separable k-algebras of arbitrary Krull dimension.

Corollary 5.3 *Let A be a finitely generated k-algebra and B a residually separable k-algebra. Let P be a prime ideal of $A \otimes_k B$ with $p := P \cap A$ and $q := P \cap B$. Then:*

$$\operatorname{codim}(A \otimes_k B)_P = \operatorname{codim}(A_p) + \operatorname{codim}(B_q).$$

In particular, $A \otimes_k B$ is regular if and only if so are A and B.

Corollary 5.4 *Let k be an algebraically closed field, A a finitely generated k-algebra, p a maximal ideal of A, and B an arbitrary k-algebra. Let P be a prime ideal of $A \otimes_k B$ such that $P \cap A = p$ and set $q := P \cap B$. Then:*

$$\operatorname{codim}(A \otimes_k B)_P = \operatorname{codim}(A_p) + \operatorname{codim}(B_q).$$

Acknowledgements This work was funded by King Fahd University of Petroleum & Minerals under DSR Research Grant # RG1212.

References

1. Anderson, D.F., Bouvier, A., Dobbs, D.E., Fontana, M., Kabbaj, S.: On Jaffard domains. Expo. Math. **6**(2), 145–175 (1988)
2. Bastida, E., Gilmer, R.: Overrings and divisorial ideals of rings of the form $D + M$. Mich. Math. J. **20**, 79–95 (1973)
3. Bouchiba, S., Kabbaj, S.: Tensor products of Cohen-Macaulay rings. Solution to a problem of Grothendieck. J. Algebra **252**, 65–73 (2002)
4. Bouchiba, S., Kabbaj, S.: Regularity of tensor products of k-algebras. Math. Scand. **115**(1), 5–19 (2014)
5. Bouchiba, S., Girolami, F., Kabbaj, S.: The dimension of tensor products of AF-rings. Lecture Notes in Pure Applied Mathematics, vol. 185, pp. 141–154. Dekker, New York (1997)
6. Bouchiba, S., Girolami, F., Kabbaj, S.: The dimension of tensor products of k-algebras arising from pullbacks. J. Pure Appl. Algebra **137**, 125–138 (1999)
7. Bouchiba, S., Dobbs, D.E., Kabbaj, S.: On the prime ideal structure of tensor products of algebras. J. Pure Appl. Algebra **176**, 89–112 (2002)
8. Bouchiba, S., Conde-Lago, J., Majadas, J.: Cohen-Macaulay, Gorenstein, complete intersection and regular defect for the tensor product of algebras. J. Pure Appl. Algebra (to appear)

9. Bourbaki, N.: Algèbre, Chapitres 4–7, Masson, Paris (1981)
10. Bourbaki, N.: Algèbre Commutative, Chapitres 8–9, Masson (1981)
11. Bouvier, A., Dobbs, D.E., Fontana, M.: Universally catenarian integral domains. Adv. Math. **72**(2), 211–238 (1988)
12. Brewer, J., Montgomery, P., Rutter, E., Heinzer, W.: Krull dimension of polynomial rings. In: Conference on Commutative Algebra. Lecture Notes in Mathematics, vol. 311, pp. 26–45. Springer, Berlin (1973)
13. Brewer, J.W., Rutter, E.A.: $D + M$ constructions with general overrings. Mich. Math. J. **23**, 33–42 (1976)
14. Bruns, W., Herzog, J.: Cohen-Macaulay Rings. Cambridge University Press, Cambridge (1993)
15. Eisenbud, D.: Commutative Algebra with a View Toward Algebraic Geometry. Graduate Texts in Mathematics, vol. 150. Springer, New York (1995)
16. Ferrand, D.: Monomorphismes et morphismes absolument plats. Bull. Soc. Math. France **100**, 97–128 (1972)
17. Fontana, M., Kabbaj, S.: Essential domains and two conjectures in dimension theory. Proc. Am. Math. Soc. **132**(9), 2529–2535 (2004)
18. Grothendieck, A.: Eléments de géométrie algébrique. Institut des Hautes Etudes Sci. Publ. Math. No. 24, Bures-sur-yvette (1965)
19. Haghighi, H., Tousi, M., Yassemi, S.: Tensor product of algebras over a field. In: Commutative Algebra. Noetherian and non-Noetherian Perspectives, pp. 181–202. Springer, New York (2011)
20. Huneke, C., Jorgensen, D.A.: Symmetry in the vanishing of Ext over Gorenstein rings. Math. Scand. **93**(2), 161–184 (2003)
21. Hungerford, T.: Algebra. Springer, New York (1974)
22. Jaffard, P.: Théorie de la dimension dans les anneaux de polynômes. Mém. Sc. Math. vol. 146. Gauthier-Villars, Paris (1960)
23. Kaplansky, I.: Commutative Rings. University of Chicago Press, Chicago (1974)
24. Majadas, J.: On tensor products of complete intersections. Bull. Lond. Math. Soc. **45**(6), 1281–1284 (2013)
25. Matsumura, H.: Commutative Ring Theory. Cambridge University Press, Cambridge (1986)
26. Nagata, M.: Local Rings. Robert E. Krieger Publishing Co., Huntington, NY (1975)
27. Rotman, J.J.: An Introduction to Homological Algebra, 2nd edn. Universitext. Springer, New York (2009)
28. Sharp, R.Y.: The effect on associated prime ideals produced by an extension of the base field. Math. Scand. **38**, 43–52 (1976)
29. Sharp, R.Y.: Simplifications in the theory of tensor products of field extensions. J. Lond. Math. Soc. **15** 48–50 (1977)
30. Sharp, R.Y.: The dimension of the tensor product of two field extensions. Bull. Lond. Math. Soc. **9**, 42–48 (1977)
31. Tousi, M., Yassemi, S.: Tensor products of some special rings. J. Algebra **268**, 672–676 (2003)
32. Vamos, P.: On the minimal prime ideals of a tensor product of two fields. Math. Proc. Camb. Philos. Soc. **84**, 25–35 (1978)
33. Wadsworth, A.R.: The Krull dimension of tensor products of commutative algebras over a field. J. Lond. Math. Soc. **19**, 391–401 (1979)
34. Watanabe, K., Ishikawa, T., Tachibana, S., Otsuka, K.: On tensor products of Gorenstein rings. J. Math. Kyoto Univ. **9**, 413–423 (1969)
35. Zariski, O., Samuel, P.: Commutative Algebra, vol. I. Van Nostrand, Princeton (1960)

Minimal Generating Sets for the D-Algebra $\mathrm{Int}(S, D)$

Jacques Boulanger and Jean-Luc Chabert

Abstract We are looking for minimal generating sets for the D-algebra $\mathrm{Int}(S, D)$ of integer-valued polynomials on any infinite subset S of a Dedekind domain D. For instance, the binomial polynomials $\binom{X}{p^r}$, where p is a prime number and r is any nonnegative integer, form a minimal generating set for the classical $\mathbb{Z}$-algebra $\mathrm{Int}(\mathbb{Z}) = \{f \in \mathbb{Q}[X] \mid f(\mathbb{Z}) \subseteq \mathbb{Z}\}$. In the local case, when D is a valuation domain and S is a regular subset of D, we are able to construct minimal generating sets, but we are not always able to extract from a generating set a minimal one. In particular, we prove that, in local fields, the generating set of integer-valued polynomials obtained by de Shalit and Iceland by means of Lubin-Tate formal group laws is minimal. In our proofs we make an extensive use of Bhargava's notion of p-ordering.

Keywords Integer-valued polynomials • Bhargava's factorials • Minimal generating sets • Lubin-Tate formal group laws • Dirichlet series

Subject Classifications Codes: Primary: 13F20, Secondary: 11S31, 11B65, 11R42

1 Introduction

When studying the ring $\mathrm{Int}(D)$ of integer-valued polynomials on a domain D, one of the first things we are looking for is the existence of bases of $\mathrm{Int}(D)$ as a D-module. Here, we consider $\mathrm{Int}(D)$ where D is a Dedekind domain, or more generally $\mathrm{Int}(S, D)$ where S is an infinite subset of D, as a D-algebra.

The origin of our study comes from a statement of de Shalit and Iceland [11] that we recall now. Let K be a local field and V be the corresponding valuation domain. De Shalit and Iceland obtained by a very interesting and surprising way,

J. Boulanger • J.-L. Chabert (✉)
LAMFA/CNRS-UMR 7352, Université de Picardie, 80039 Amiens, France
e-mail: jaboulanger@wanadoo.fr; jean-luc.chabert@u-picardie.fr

M. Fontana et al. (eds.), *Rings, Polynomials, and Modules*,
https://doi.org/10.1007/978-3-319-65874-2_5

namely by means of Lubin-Tate formal groups, a generating set of the V-algebra of integer-valued polynomials on V, that is,

$$\mathrm{Int}(V) = \{f \in K[X] \mid f(V) \subseteq V\}.$$

More precisely, if $F(t_1, t_2)$ denotes a Lubin-Tate formal group law on V, one knows that, for every $x \in V$, there is a unique power series

$$[x](t) = \sum_{n=1}^{\infty} c_n(x)t^n$$

such that $c_1(x) = x$ and $F([x](t_1), [x](t_2)) = [x](F(t_1, t_2))$. It turns out that the $c_n(x)$'s are polynomials of $\mathrm{Int}(V)$ of degree $\leq n$. Denoting by q the cardinality of the residue field of V, the authors proved [11, Thm 3.1] that the set $\{c_{q^m}(x) \mid m \geq 0\}$ is a generating set for the V-algebra $\mathrm{Int}(V)$. They also state that this set is a minimal generating set, but they did not give really a proof. Thus, the aim of this paper is to give a proof of this statement (Theorem 3 below). By the way, we study the question of the existence of minimal generating sets for the D-algebra of integer-valued polynomials $\mathrm{Int}(S, D)$ in a more general framework, namely, when D is a Dedekind domain and S is an infinite subset of D.

For instance, in the particular case where $K = \mathbb{Q}_p$ and $V = \mathbb{Z}_p$, following [8, §5.1], we know a minimal generating set of the $\mathbb{Z}_p$-algebra $\mathrm{Int}(\mathbb{Z}_p)$ obtained by a more direct way, namely the set formed by the polynomials $F_{p^m}(X)$ $(m \geq 0)$ defined inductively by $F_1(X) = X$, $F_p(X) = \frac{X^p - X}{p}$ and, for $m \geq 1$, $F_{p^m}(X) = F_p(F_{p^{m-1}}(X))$. In fact, the author gives no proof of minimality (which is true however, see Proposition 3 below). Let us consider another example, the classical ring of integer-valued polynomials:

$$\mathrm{Int}(\mathbb{Z}) = \{f(X) \in \mathbb{Q}[X] \mid f(\mathbb{Z}) \subseteq \mathbb{Z}\}.$$

It is well known that the binomial polynomials

$$\binom{X}{n} = \frac{X(X-1)\dots(X-n+1)}{n!} \quad (n \geq 0)$$

form a basis of the $\mathbb{Z}$-module $\mathrm{Int}(\mathbb{Z})$. The set $\{\binom{X}{n}\}_{n \geq 0}$ is then a generating set, and in fact a minimal generating set, for the $\mathbb{Z}$-module $\mathrm{Int}(\mathbb{Z})$. If we try to extract from this set a minimal generating set for the $\mathbb{Z}$-algebra $\mathrm{Int}(\mathbb{Z})$, we see that there exists one and only one such subset, namely (see Proposition 12): $\left\{\binom{X}{p^k} \;\middle|\; p \in \mathbb{P}, k \in \mathbb{N}\right\}$.

We will prove below analogous results when replacing $\mathbb{Z}$ by an integral domain.

Notation Let D denote an integral domain with quotient field K, and consider the D-algebra of *integer-valued polynomials on D*, that is,

$$\mathrm{Int}(D) = \{f(X) \in K[X] \mid f(D) \subseteq D\}.$$

More generally, for every infinite subset S of D, consider the D-algebra of *integer-valued polynomials on S with respect to D*, that is,

$$\mathrm{Int}(S, D) = \{f(X) \in K[X] \mid f(S) \subseteq D\}.$$

Recall that, for every $n \in \mathbb{N}$, the leading coefficients of the polynomials of $\mathrm{Int}(S, D)$ with degree $\leq n$ form a nonzero fractional ideal of D denoted by $\mathfrak{I}_n(S, D)$ and called the n-th *characteristic ideal* of S (cf. [5, §II.1]). We also know:

Proposition 1 ([5, II.1.5]) *Let $\mathcal{G} \subseteq \mathrm{Int}(S, D)$. If, for every $n \geq 0$, the leading coefficients of the polynomials of $\mathcal{G}$ with degree n generate the fractional ideal $\mathfrak{I}_n(S, D)$, then $\mathcal{G}$ is a generating set for the D-module* $\mathrm{Int}(S, D)$.

In the sequel, we will look for generating sets of the D-algebra $\mathrm{Int}(S, D)$ which are extracted from sets $\mathcal{G}$ obtained by this way. In particular, when the characteristic ideals are principal, there exists bases of $\mathrm{Int}(S, D)$ having one and only one polynomial of each degree. Following Pólya [16], such a basis is called a *regular basis*. Proposition 1 shows that there exists regular bases for the D-module $\mathrm{Int}(S, D)$ if and only if all the characteristic ideals $\mathfrak{I}_n(S, D)$ are principal.

We begin our study with the local case: in the next section, we consider first the easiest case, namely, the $\mathbb{Z}_p$-algebra $\mathrm{Int}(\mathbb{Z}_p)$. Then, in Sect. 3, we consider generating sets for the V-algebra extracted from regular bases of the V-module $\mathrm{Int}(S, V)$ where V is any rank-one valuation domain and S any infinite subset of V. To obtain minimal generating sets with Theorem 2 we need to assume that the subset S is regular (we recall there the notion of regular subset). Finally, in Sect. 5, we globalize the previous results to Dedekind domains.

2 Minimal Generating Sets for the $\mathbb{Z}$-Algebra $\mathrm{Int}(\mathbb{Z}_p)$

Let p be a fixed prime number. We denote by $\mathbb{Q}_p$ the field of p-adic numbers, by $\mathbb{Z}_p$ the ring of p-adic integers, and by v_p the p-adic valuation on $\mathbb{Q}_p$.

2.1 Extracted from the Regular Basis Formed by the $\binom{X}{n}$

Lemma 1 *The $\mathbb{Z}_p$-algebra* $\mathrm{Int}(\mathbb{Z}_p)$ *admits the generating set:* $\left\{\binom{X}{p^r} \,\middle|\, r \geq 0\right\}$.

Proof Assume that $n \geq 2$ is not of the form p^r with $r \geq 0$. Then, $n = mp^r$ with $m \geq 2$, and $p \nmid m$. By Legendre formula [15]: $v_p(n!) = \sum_{k\geq 1} \left[\frac{n}{p^k}\right]$, we have

$$v_p(n!) = \sum_{k=1}^{r} m\left[\frac{p^r}{p^k}\right] + \sum_{k\geq 1}\left[\frac{m}{p^k}\right]$$

$$= \sum_{k=1}^{r}\left[\frac{p^r}{p^k}\right] + \sum_{k=1}^{r}(m-1)\left[\frac{p^r}{p^k}\right] + \sum_{k\geq 1}\left[\frac{m-1}{p^k}\right] = v_p(p^r!) + v_p((m-1)p^r!)\,.$$

Consequently, $n! = u \times p^r! \times ((m-1)p^r)!$ where u is invertible in $\mathbb{Z}_p$. Thus, the degree of the polynomial $\binom{X}{n} - \frac{1}{u}\binom{X}{p^r}\binom{X}{(m-1)p^r}$ is strictly less than n. Since the polynomial $\binom{X}{n}$ is generated by the binomials $\binom{X}{m}$ where $m < n$, it may be deleted from the generating set.

Remark 1 If n is of the form p^r for some $r \geq 1$, the previous reasoning does not hold since

$$\forall j \in \{1, \ldots, p^r - 1\} \quad v_p(j!) + v_p((p^r - j)!) < v_p(p^r!)\,.$$

This is a consequence of Legendre formula written in the following way:

$$v_p(n!) = \frac{n - \sigma_p(n)}{p-1} \quad [15]$$

where $\sigma_p(n)$ denotes the sum of the digits of n in base p. Indeed,

$$v_p\left(\binom{p^r}{j}\right) = v_p(p^r!) - v_p(j!) - v_p((p^r - j)!) = \frac{\sigma_p(j) + \sigma_p(p^r - j) - \sigma_p(p^r)}{p-1}\,.$$

Since each carry in the addition of j and $p^r - j$ decreases the sum of the digits by $p-1$, p divides $\binom{p^r}{j}$.

We show now that the generating set given in Lemma 1 is minimal.

Proposition 2 *The set* $\left\{\binom{X}{p^r} \,\middle|\, r \geq 0\right\}$ *is a minimal generating set for the* $\mathbb{Z}_p$*-algebra* $\mathrm{Int}(\mathbb{Z}_p)$*. Moreover, it is the only minimal generating set that one may extract from* $\left\{\binom{X}{n} \,\middle|\, n \in \mathbb{N}\right\}$.

Proof First, the polynomial $X = \binom{X}{p^0}$ cannot be deleted since if we have a relation of the form

$$X = c_0 + \sum_{\underline{\alpha}\neq \underline{0}} c_{\underline{\alpha}} \binom{X}{k_1}^{\alpha_1} \cdots \binom{X}{k_s}^{\alpha_s} \quad \text{where } c_{\underline{\alpha}} \in \mathbb{Z}_p \text{ and } k_i \neq 0, 1\,, \tag{1}$$

the substitution of 0 and 1 for X would lead to a contradiction.

Now assume that there exists some $r \geq 1$ such that $\binom{X}{p^r}$ could be deleted, that is, that there exists a relation of the form

$$\binom{X}{p^r} = c_0 + \sum_{\underline{\alpha} \neq \underline{0}} c_{\underline{\alpha}} \binom{X}{k_1}^{\alpha_1} \cdots \binom{X}{k_s}^{\alpha_s} \quad \text{where } c_{\underline{\alpha}} \in \mathbb{Z}_p \text{ and } k_i \neq 0, p^r . \tag{2}$$

The classical formula (see, for instance, [10, T. 1, Ch. 1, Ex. 23]):

$$\binom{X}{m}\binom{X}{n} = \sum_{l=\max(m,n)}^{m+n} \frac{l!}{(l-m)!(l-n)!(m+n-l)!} \binom{X}{l} \tag{3}$$

shows that the coefficient of $\binom{X}{p^r}$ in the development of the product $\binom{X}{m}\binom{X}{n}$ where $m, n \neq p^r$ is always divisible by p since if $m > p^r$, or $n > p^r$, or $m + n < p^r$, $\binom{X}{p^r}$ does not appear, and if $0 < m, n < p^r$ and $m + n \geq p^r$,

$$\frac{p^r!}{(p^r - m)!(p^r - n)!(m + n - p^r)!} = \binom{p^r}{m}\binom{m}{p^r - n},$$

we may conclude with Remark 1.

Thus, if in the right side of Eq. (2), we replace successively all the products $\binom{X}{m}\binom{X}{n}$ by means of Eq. (3) until we obtain a sum which is linear with respect to the $\binom{X}{l}$'s then, if $\binom{X}{p^r}$ appears, its coefficient is divisible by p, and this property remains until the end of the process. So that, we will obtain an equality of the form:

$$\binom{X}{p^r} = \sum_l b_l \binom{X}{l} \quad \text{with } p | b_{p^r} .$$

Since, the $\binom{X}{l}$'s form a basis of the $\mathbb{Z}_p$-module Int$(\mathbb{Z}_p)$, we have a contradiction. Moreover, this is the only minimal generating subset that can be extracted from $\{\binom{X}{n} \mid n \in \mathbb{N}\}$ since $\binom{X}{p^r}$ cannot be obtained from all of the others.

2.2 *Extracted from the Basis Formed by the Fermat Polynomials*

As previously said in the introduction, we know another natural basis of the $\mathbb{Z}_p$-module Int$(\mathbb{Z}_p)$ constructed from the Fermat binomial

$$F_p(X) = \frac{X^p - X}{p} . \tag{4}$$

Consider the sequence formed by the iterates of F_p:

$$F_{p^0}(X) = F_1(X) = X \text{ and, for } k \geq 2, F_{p^k}(X) = F_p(F_{p^{k-1}}(X)) . \quad (5)$$

Now, if n may be written $n = n_k n_{k-1} \cdots n_1 n_0$ in base p, one let

$$F_n(X) = \prod_{j=0}^{k} (F_{p^j})^{n_j} . \quad (6)$$

One knows [5, §II.2] that the set $\{F_n(X)\}_{n\geq 0}$ is a regular basis of the $\mathbb{Z}_p$-module $\text{Int}(\mathbb{Z}_p)$. It follows from Eq.(6) that the polynomials F_{p^k} $(k \in \mathbb{N})$ form a generating set of the $\mathbb{Z}_p$-algebra $\text{Int}(\mathbb{Z}_p)$. In fact, this generating set is minimal:

Proposition 3 *[8, §5.1] The polynomials F_{p^k} $(k \in \mathbb{N})$ defined by Formulas (4) and (5) form a minimal generating set for the $\mathbb{Z}_p$-algebra* $\text{Int}(\mathbb{Z}_p)$.

Proof In order to prove that we cannot delete any F_{p^k} from this generating set, we consider their images f_{p^k} in $\text{Int}(\mathbb{Z}_p)/p\text{Int}(\mathbb{Z}_p)$. Clearly, the f_{p^k}'s generate the $\mathbb{F}_p$-algebra $\text{Int}(\mathbb{Z}_p)/p\text{Int}(\mathbb{Z}_p)$ and satisfy the relations $f_{p^k}^p - f_{p^k} = 0$ since $(F_{p^k})^p - F_{p^k} = p\, F_{p^{k+1}}$. We prove now that all the relations between the f_{p^k}'s are generated by the previous ones. Let

$$\psi : \mathbb{F}_p[Y_0, Y_1, \ldots, Y_k, \ldots] \to \text{Int}(S, \mathbb{Z}_p)/p\text{Int}(S, \mathbb{Z}_p)$$

be the homomorphism of $\mathbb{F}_p$-algebra defined by $\psi(Y_k) = f_{p^k}$. Clearly, ψ is onto and $\ker(\psi)$ contains the ideal $\mathfrak{I}$ generated by the elements $Y_k^p - Y_k$ $(k \geq 0)$. Then, every $q \in \mathbb{F}_p[Y_0, \ldots, Y_k, \ldots]$ is congruent modulo $\mathfrak{I}$ to a polynomial:

$$q_0 = \sum_{\underline{\alpha}} d_{\underline{\alpha}} Y_0^{\alpha_0} Y_1^{\alpha_1} \cdots Y_r^{\alpha_r} \text{where } d_{\underline{\alpha}} \in \mathbb{F}_p \text{ and } 0 \leq \alpha_k < p .$$

Let

$$Q_0 = \sum_{\underline{\alpha}} c_{\underline{\alpha}} Y_0^{\alpha_0} Y_1^{\alpha_1} \cdots Y_r^{\alpha_r} \text{ where } c_{\underline{\alpha}} \in \mathbb{Z} \text{ is a representative of } d_{\underline{\alpha}} .$$

Then,

$$Q_0(F_1, F_p, \ldots, F_{p^k}, \ldots) = \sum_{\underline{\alpha}} c_{\underline{\alpha}} F_1^{\alpha_0} F_p^{\alpha_1} \cdots F_{p^r}^{\alpha_r} = \sum_{\underline{\alpha}} c_{\underline{\alpha}} F_{|\underline{\alpha}|}$$

where $|\underline{\alpha}| = \alpha_0 + \alpha_1 p + \cdots + \alpha_r p^r$.

If $\psi(q) = 0$, then $\psi(q_0) = 0$ and $Q_0(F_1, F_p, \ldots, F_{p^k}, \ldots) \in p\,\text{Int}(\mathbb{Z}_p)$. Consequently, p divides all the $c_{\underline{\alpha}}$ since the F_n's form a basis of the $\mathbb{Z}_p$-module $\text{Int}(\mathbb{Z}_p)$. Thus, $q_0 = 0$ and $q \in \mathfrak{I}$. We then may conclude that the f_{p^k}'s form an irredundant set of generators of $\text{Int}(\mathbb{Z}_p)/p\text{Int}(\mathbb{Z}_p)$, and hence, the same assertion holds for the polynomials F_{p^k}'s in $\text{Int}(\mathbb{Z}_p)$.

3 Generating Sets in the Local Case

Hypotheses and Notation for the Section

Let K be a valued field. We denote by v the valuation of K and by V the corresponding valuation domain. By definition, v is a rank-one valuation, that is, $v : K^* \to \mathbb{R}$.

Let S be an infinite subset of V which is assumed to be precompact, that is, such that its completion with respect to the topology defined by v is compact.

3.1 Gaps and Generating Sets

Let w_S denote the 'characteristic function' of S which is associated to the sequence of characteristic ideals of S :

$$w_S : n \in \mathbb{N} \mapsto -v(\mathfrak{I}_n(S, V)) \in \mathbb{R}\,.$$

The function w_S is super-additive, that is,

$$w_S(i + j) \geq w_S(i) + w_S(j) \text{ for all } i, j \geq 0$$

since clearly $\mathfrak{I}_i(S, V) \times \mathfrak{I}_j(S, V) \subseteq \mathfrak{I}_{i+j}(S, V)$. We are led to consider the indices for which the function is strictly super-additive.

Definition 1 The set of *indices of gaps* of w_S is defined by

$$\mathfrak{g}_V(S) = \{n > 0 \mid \forall i \in \{1, \ldots, n-1\}\ \ w_S(i) + w_S(n-i) < w_S(n)\,\}\,.$$

By definition, we always have $1 \in \mathfrak{g}_V(S)$. It follows from the previous section (see, for example, Remark 1) that, for every prime number p

$$\mathfrak{g}(\mathbb{Z}_p) = \mathfrak{g}_{\mathbb{Z}_p}(\mathbb{Z}_p) = \{p^r \mid r \geq 0\}\,. \tag{7}$$

Recall that the inverse of the characteristic ideals $\mathfrak{I}_n(S, V)$ is the n-th factorial ideal $(n!)_S^V$ of S as defined by Bhargava [2]. Consequently, $w_S(n) = v((n!)_S^V)$.

The following proposition is a slight generalization of [14, Prop. 4.3].

Proposition 4 *If* $\{g_n\}_{n\geq 0}$ *is a regular basis of* $\mathrm{Int}(S, V)$, *then the following set is a generating set for the* V*-algebra* $\mathrm{Int}(S, V)$:

$$\{\, g_n \mid n \in \mathfrak{g}_V(S)\,\}\,.$$

Proof It follows from Proposition 1 that the set $\{\, g_n \mid n \geq 0\}$ is a generating set of the module. We may forget $g_0 = 1$ for the generating set of the algebra. Let

$n > 0$ and assume that $n \notin \mathfrak{g}_V(S)$: there exist $i > 0$ and $j > 0$ such that $i + j = n$ and $w_S(i) + w_S(j) = w_S(n)$. Denoting by $lc(g)$ the leading coefficient of every polynomial g, we have $v(lc(g_k)) = w_S(k)$ for all k. Thus, in the valuation domain V we have the relation:

$$lc(g_n) = u \times lc(g_i) \times lc(g_j) \quad \text{where } v(u) = 0\,.$$

Consequently,

$$\deg\big(g_n(X) - ug_i(X)g_j(X)\big) < n,$$

and hence, $g_n(X) - ug_i(X)g_j(X)$ is a linear combination of the g_m's where $m < n$. Thus, we may delete g_n from the generating set.

In other words, every polynomial $f \in \mathrm{Int}(S, V)$ of degree n may be written:

$$f(X) = P(g_{i_1}(X), \ldots, g_{i_k}(X)) \text{ where } i_1, \ldots, i_k \in \mathfrak{g}_V(S) \cap \{1, \ldots, n\}$$

and $P(T_1, \ldots, T_k) \in V[T_1, \ldots, T_k]$. Moreover, the previous proof shows that we may add: the total degree of P is ≤ 2.

We will prove next that some of these generating sets are minimal generating sets for the V-algebra $\mathrm{Int}(S, V)$ if the subset S itself is sufficiently regular.

Remark 2 Take care that there are generating sets extracted from regular bases which do not necessarily contain all the g_n's where $n \in \mathfrak{g}_V(S)$. Indeed, consider the $\mathbb{Z}_3$-algebra $\mathrm{Int}(\mathbb{Z}_3)$ and the regular basis $\{g_n\}_{n\geq 0}$ where $g_n(X) = \binom{X}{n}$ for $n \neq 5$ and $g_5(X) = \binom{X}{5} + \binom{X}{3}$. The $\mathbb{Z}_3$-algebra $\mathrm{Int}(\mathbb{Z}_3)$ is generated by the g_n's where $n \in \{1, 4, 5\} \cup \{3^r \mid r \geq 2\}$ since $\binom{X}{3} = g_5(X) - \frac{1}{5}(g_1(X) - 4) \times g_4(X)$, although $3 \in \mathfrak{g}(\mathbb{Z}_3)$.

3.2 Structural Constants

As said in the introduction, the V-module $\mathrm{Int}(S, V)$ is free. Thus, if $\{f_n\}_{n\geq 0}$ denotes a basis, we may consider the corresponding structural constants $c_k(n, m)$ of the V-algebra $\mathrm{Int}(S, V)$ defined by the relations:

$$f_n(X)f_m(X) = \sum_{k\geq 0} c_k(n, m)f_k(X) \quad (m, n, k \in \mathbb{N}). \tag{8}$$

The $c_k(n, m)$ are unique and belong to V since the f_k's form a basis of the V-module $\mathrm{Int}(S, V)$. Recall the following result due to Elliott [12, Prop. 2.2] in the case where the basis is a regular basis associated to a v-ordering (see Sect. 3.3):

Proposition 5 *Let $\{f_n\}_{n\geq 0}$ be a basis and consider the $c_k(n, m)$ defined by (8). Then, all the relations between the generators f_n are generated by relations (8).*

Proof Let $\varphi : V[T_1, \ldots, T_n, \ldots] \to \text{Int}(S, V)$ be the homomorphism of V-algebra defined by $\varphi(T_n) = f_n$. Clearly, φ is onto and $\ker(\varphi)$ contains the ideal $\mathfrak{I}$ generated by the elements $T_nT_m - \sum_{k\geq 0} c_k(n, m)T_k$. It is easy to see that every $P \in V[T_1, \ldots, T_n, \ldots]$ is congruent modulo $\mathfrak{I}$ to a linear form $\sum_{0\leq k\leq n} \lambda_k T_k$, and then, $\varphi(P) = \sum_{k=0}^{n} \lambda_k f_k$. Moreover, this linear form is unique because of the fact that the f_k's form a basis. Consequently, the morphism $\overline{\varphi} : V[\ldots, T_n, \ldots]/\mathfrak{I} \to \text{Int}(S, V)$ deduced from φ is a bijection.

In order to generalize the previous results obtained for $\text{Int}(\mathbb{Z}_p)$, we need a formula analogous to Eq. (3) which allowed us to say that the coefficient of $\binom{X}{p^r}$ is divisible by p. This is the hypothesis of the next lemma.

Lemma 2 *Fix some $l \in \mathbb{N}^*$. If for all $n, m \neq l$, either $c_l(n, m) = 0$ or $v(c_l(n, m)) > 0$, then f_l does not belong to the V-algebra generated by the f_n's where $n \in \mathbb{N} \setminus \{l\}$.*

Proof Assume that there exists a relation of the form:

$$f_l(X) = \sum_{\underline{\alpha}} d_{\underline{\alpha}} f_{k_1}^{\alpha_1}(X) \ldots f_{k_s}^{\alpha_s}(X) \text{ where } d_{\underline{\alpha}} \in V \text{ and } k_i \neq l\,.$$

Replacing successively every product of two polynomials f_{k_i} and f_{k_j}, by means of relations (8), until we obtain a linear combination of the $f_k(X)$'s, we see that, if $f_l(X)$ appears at some step, then this is always with a coefficient whose valuation is > 0, and this property is preserved at the following steps. Thus, we obtain an equality of the form:

$$f_l(X) = \sum_k b_k f_k(X) \text{ with } v(b_l) > 0\,.$$

We have a contradiction since the $f_k(X)$'s form a basis of the V-module $\text{Int}(S, V)$.

The following proposition is an obvious consequence of Proposition 4 and Lemma 2.

Proposition 6 *Let $\{f_n\}_{n\geq 0}$ be a regular basis of* $\text{Int}(S, V)$. *If for every $l \in \mathcal{G}_V(S)$ and for every $n, m \neq l$, either $c_l(n, m) = 0$ or $v(c_l(n, m)) > 0$, then the set $\{f_l \mid l \in \mathcal{G}_V(S)\}$ is a minimal generating set for the V-algebra* $\text{Int}(S, V)$. *Moreover, it is the only minimal generating set that one may extract from $\{f_n \mid n \geq 0\}$.*

3.3 Regular Bases Associated to v-Orderings

Let us recall the notion of v-ordering introduced by Manjul Bhargava [2]. A v-*ordering* of the subset S is a sequence $\{a_n\}_{n\geq 0}$ of elements of S such that, for every $n \geq 1$, a_n satisfies

$$v\left(\prod_{k=0}^{n-1}(a_n-a_k)\right)=\inf_{x\in S} v\left(\prod_{k=0}^{n-1}(x-a_k)\right).$$

Such sequences exist thanks to the precompactness of S [6].

Clearly, if $\{a_n\}_{n\geq 0}$ is a v-ordering of S, then the following sequence of polynomials is a regular basis of $\mathrm{Int}(S,V)$ (see [4]):

$$f_0(X)=1 \text{ and, for } n\geq 1,\ f_n(X)=\prod_{k=0}^{n-1}\frac{X-a_k}{a_n-a_k}. \tag{9}$$

Lemma 3 *Let $\mathcal{B}=\{f_n \mid n\geq 0\}$ be a regular basis of* $\mathrm{Int}(S,V)$ *constructed by means of a v-ordering $\{a_n\}_{n\geq 0}$ of S. Any subset $\mathcal{G}$ of $\mathcal{B}$ which is a generating set for the V-algebra* $\mathrm{Int}(S,V)$ *contains* $f_1(X)=\frac{X-a_0}{a_1-a_0}$.

Proof Otherwise, analogously to the beginning of the proof of Proposition 2, we would obtain a contradiction by substituting a_0 and a_1 to X.

Lemma 4 *[12, Prop. 2.2] Let $\{f_n\}_{n\geq 0}$ be a regular basis associated to a v-ordering $\{a_n\}_{n\geq 0}$ of S and consider the structural constants $c_k(n,m)$ associated to this basis. Then, for $k<\max\{n,m\}$ or $k>n+m$, $c_k(n,m)=0$. Moreover, $c_m(0,m)=1$ and $c_m(m,m)=1$.*

Proof By considering the degree, we see that $c_k(n,m)=0$ for $k>n+m$. We may prove by induction on k that $c_k(n,m)=0$ for $k<\max(n,m)$ since $f_k(a_h)=0$ for $0\leq h\leq k-1$ and $f_k(a_k)=1$. In particular,

$$f_m(X)f_m(X)=c_m(m,m)f_m(X)+\ldots+c_k(m,m)f_k(X)+\ldots+c_{2m}(m,m)f_{2m}(X).$$

By substituting a_m to X, we obtain $c_m(m,m)=1$. The equality $c_m(0,m)=1$ is obvious.

Despite Elliott's work [12] on the structural constants of the V-algebra $\mathrm{Int}(S,V)$ with respect to regular bases associated to v-orderings, we need to add a hypothesis on the subset S : we assume that S is regular in a sense which generalizes the regular compact subsets of local fields considered by Amice [1].

4 Minimal Generating Sets in the case of a Regular Subset

Hypotheses and Notation for the Section

Let K be a valued field, let V be the corresponding valuation domain, and let S be an infinite precompact subset of V.

For every $\gamma\in\mathbb{R}$ and every $x\in S$, consider the class in S of x modulo γ, that is, the S-ball

$$S(x, \gamma) = \{y \in S \mid v(x - y) \geq \gamma\}.$$

Denote by q_γ the number of classes of S modulo γ, that is, the number of distinct nonempty S-balls $S(x, \gamma)$.

The fact that S is precompact is equivalent to the fact that all the q_γ's are finite (see, for instance, [6, Lemma 3.1]). Moreover, there is a strictly increasing sequence of non-negative numbers $\{\gamma_k\}_{k\geq 0}$, the *critical valuations* of S, which tends to $+\infty$ such that

$$\gamma_0 = \min_{x,y\in S\,,\; x\neq y} v(x - y)$$

and

$$q_{\gamma_k} < q_\gamma \leq q_{\gamma_{k+1}} \Leftrightarrow \gamma_k < \gamma \leq \gamma_{k+1} \qquad \text{[7, Prop. 5.1]}. \tag{10}$$

Note that $q_{\gamma_0} = 1$.

4.1 Regular Subsets, Gaps, and Strong v-Orderings

Definition 2 The precompact subset S of the valued field K is said to be *regular* if, whatever $\gamma < \delta$, all nonempty S-balls $S(x, \gamma)$ contain the same number of S-ball $S(y, \delta)$.

Consequently, with notation (10), if S is regular there exist a sequence of positive integers $\{\alpha_k\}_{k\geq 0}$ such that each nonempty S-ball $S(x, \gamma_k)$ contains α_k non-empty distinct S-balls $S(y, \gamma_{k+1})$. In particular, for every $k \geq 0$, we have:

$$q_{\gamma_{k+1}} = \alpha_k q_{\gamma_k}.$$

When S is regular, the characteristic function w_S satisfies the following generalization of Legendre formula:

$$w_S(n) = n\gamma_0 + \sum_{k\geq 1} \left[\frac{n}{q_{\gamma_k}}\right] (\gamma_k - \gamma_{k-1}) \qquad \text{[9, Thm 1.5]}. \tag{11}$$

Lemma 5 *If S is regular, then the indices of gaps of S are the cardinalities q_γ :*

$$\mathfrak{g}_V(S) = \{q_{\gamma_k} \mid k \geq 0\}.$$

Proof Fix some $n \neq 0, 1$ which is not of the form q_{γ_k} and let r be the largest integer such that q_{γ_r} divides n. Then, $n = mq_{\gamma_r}$ where $m \geq 2$ and $\alpha_r \nmid m$. It follows from (11) that

$$w_S(n) = mq_{\gamma_r}\gamma_0 + \sum_{k=1}^{r} m\left[\frac{q_{\gamma_r}}{q_{\gamma_k}}\right](\gamma_k - \gamma_{k-1}) + \sum_{k\geq r+1}\left[\frac{mq_{\gamma_r}}{q_{\gamma_k}}\right](\gamma_k - \gamma_{k-1})$$

$$= (m-1)q_{\gamma_r}\gamma_0 + \sum_{k=1}^{r}(m-1)\left[\frac{q_{\gamma_r}}{q_{\gamma_k}}\right](\gamma_k - \gamma_{k-1}) + \sum_{k\geq r+1}\left[\frac{(m-1)q_{\gamma_r}}{q_{\gamma_k}}\right](\gamma_k - \gamma_{k-1})$$

$$+q_{\gamma_r}\gamma_0 + \sum_{k=1}^{r}\left[\frac{q_{\gamma_r}}{q_{\gamma_k}}\right](\gamma_k - \gamma_{k-1}) = w_S(n - q_{\gamma_r}) + w_S(q_{\gamma_r})\,.$$

On the other hand, consider some n of the form q_{γ_r} with $r \geq 1$. Then, for every $j \in \{1, \ldots, n-1\}$, we have:

$$w_S(j) + w_S(n-j) = j\gamma_0 + \sum_{k=1}^{r-1}\left[\frac{j}{q_{\gamma_k}}\right](\gamma_k - \gamma_{k-1}) + (n-j)\gamma_0 + \sum_{k=1}^{r-1}\left[\frac{n-j}{q_{\gamma_k}}\right](\gamma_k - \gamma_{k-1})$$

$$\leq n\gamma_0 + \sum_{k=1}^{r-1}\left[\frac{n}{q_{\gamma_k}}\right](\gamma_k - \gamma_{k-1}) < w_S(n)\,.$$

The strict inequality follows from the fact that $\left[\frac{n}{q_{\gamma_r}}\right](\gamma_k - \gamma_{k-1}) = \gamma_k - \gamma_{k-1}$ is missing.

We also know that:

Proposition 7 *[9, Theorem 1.5] Any regular subset S admits* strong v-orderings, *that is, sequences $\{a_n\}_{n\geq 0}$ of elements of S such that, for every $k \geq 0$, $\{a_n\}_{n\geq k}$ is a v-ordering of S.*

For instance, the sequence $\{0, 1, 2, \ldots\}$ is a strong p-ordering of $\mathbb{Z}_p$.

4.2 Minimal Generating Sets Associated to Strong v-Orderings

Both following lemmas show clearly why regular subsets allow a generalization of what happens for $\mathbb{Z}_p$.

Lemma 6 *Assume that $\{a_n\}_{n\geq 0}$ is a strong v-ordering of S and let $\{f_n\}_{n\geq 0}$ be the regular basis of* Int(S, V) *associated to this strong v-ordering. Then, for every $l \in \mathfrak{g}_V(S)$ and every $k \in \{1, \ldots, l-1\}$, one has $v(f_k(a_l)) > 0$.*

Proof

$$f_k(a_l) = \prod_{j=0}^{k-1}\frac{a_l - a_j}{a_k - a_j} = \frac{\prod_{j=0}^{l-1}(a_l - a_j)}{\prod_{j=0}^{k-1}(a_k - a_j)\prod_{j=k}^{l-1}(a_l - a_j)}\,.$$

Since, the sequence $\{a_n\}$ is a strong v-ordering, we have

$$v\left(\prod_{j=k}^{l-1}(a_l - a_j)\right) = v\left(\prod_{j=0}^{l-k-1}(a_{l-k} - a_j)\right) = w_S(l-k).$$

Finally, $l \in \mathfrak{g}_V(S)$ implies:

$$v(f_k(a_l)) = w_S(l) - w_S(k) - w_S(l-k) > 0.$$

Lemma 4 may be completed by the following:

Lemma 7 *Let $\{f_n\}_{n\geq 0}$ be a regular basis associated to a strong v-ordering $\{a_n\}_{n\geq 0}$ of the regular subset S and let $c_k(n,m)$ be the corresponding structural constants. Then, for every $l \in \mathfrak{g}_V(S)$:*

$$[n \cdot m \neq 0 \text{ and } (n,m) \neq (l,l)] \Rightarrow [c_l(n,m) = 0 \text{ or } v(c_l(n,m)) > 0].$$

Proof By Lemma 4, we may assume that $1 \leq n \leq m \leq l \leq n+m$. Thus we have:

$$f_n(X)f_m(X) = c_m f_m(X) + \ldots + c_l f_l(X) + \ldots + c_{n+m} f_{n+m}(X).$$

Consequently,

$$f_n(a_l)f_m(a_l) = c_m f_m(a_l) + \ldots + c_{l-1} f_{l-1}(a_l) + c_l.$$

If $n < l$, it follows from Lemma 6 that $v(c_l) > 0$.

Lemma 7 allows us to apply Proposition 6 and we obtain:

Theorem 1 *Let K be a valued field, V be its valuation domain, and S be a precompact and regular subset of V. Let $\mathfrak{g}_V(S)$ denote the set of indices of gaps, that is, the set formed by the cardinalities q_γ of the subsets S mod γ. Let $\{a_n\}_{n\geq 0}$ be a strong v-ordering of S and let $\mathcal{B} = \{f_n\}_{n\geq 0}$ be the regular basis of the V-module* Int(S, V) *associated to this strong v-ordering, that is, defined by $f_n(X) = \prod_{k=0}^{n-1} \frac{X-a_k}{a_n-a_k}$. Then, there is one and only one subset $\mathcal{G}$ of $\mathcal{B}$ which is a minimal generating set of* Int(S, V) *as a V-algebra, namely*

$$\mathcal{G} = \{f_n \mid n \in \mathfrak{g}_V(S)\}.$$

4.3 *Other Minimal Generating Sets when S is a Regular Subset*

To extend Theorem 1 to other regular bases, we are looking for conditions which will allow us to use Lemma 7:

Proposition 8 *Let $\{a_n\}_{n\geq 0}$ be a strong v-ordering of the regular subset S, let $\{g_n\}_{n\geq 0}$ be a regular basis of* $\mathrm{Int}(S,V)$*, and let $l \in \mathfrak{g}_V(S)$. If $v(g_n(a_l)-g_n(a_0))>0$ for all $n>l$ then, $g_l(X)$ does not belong to the V-algebra generated by the g_n's where $n\in\mathbb{N}\setminus\{l\}$.*

Proof If $\{f_n\}_{n\geq 0}$ denotes the regular basis associated to the strong v-ordering $\{a_n\}_{n\geq 0}$, for every $g(X)=\sum_{k=0}^{n} d_k f_k(X)\in \mathrm{Int}(S,V)$, we have

$$g(a_l)=g(a_0)+\sum_{k=1}^{l-1} d_k f_k(a_l)+d_l,$$

and hence, by Lemma 6, $v(d_l-(g(a_l)-g(a_0)))>0$.

Consequently, by hypothesis, for every $n>l$, if $g_n(X)=\sum_{k=0}^{n} d_{n,k}f_k(X)$, then $v(d_{n,l})>0$. Assume now that we have

$$g_l(X)=\sum_{\underline{\alpha}} c_{\underline{\alpha}}\, g_{k_1}^{\alpha_1}(X)\ldots g_{k_s}^{\alpha_s}(X) \text{ where } c_{\underline{\alpha}}\in V \text{ and } k_i\neq l\,.$$

Replacing the g_{k_i}'s by the f_k's in the right-hand side by means of the equality $g_{k_i}=\sum_k d_{k_i,k}f_k$, we note that if f_l appears in a monomial the valuation of the corresponding coefficient is >0. Then, we compute all the products of the f_k's by means of Lemma 7 and we obtain in the right-hand side a sum of the form $\sum_k b_k f_k$ where $v(b_l)>0$, that is, we obtain an equality of the form:

$$g_l(X)=\sum_k b_k f_k(X) \text{ with } v(b_l)>0\,.$$

We have a contradiction since the f_k's and the g_h's form regular bases of the V-module $\mathrm{Int}(S,V)$.

Now, we are able to state and prove our main theorem in the local case.

Theorem 2 *Let K be a valued field, V be its valuation domain, and S be a precompact and regular subset of V. Let $\mathfrak{g}_V(S)$ denote the set of indices of gaps and let $\{a_n\}_{n\geq 0}$ denote a strong v-ordering of S. Let $\mathscr{B}=\{g_n\mid n\geq 0\}$ be a regular basis of the V-module* $\mathrm{Int}(S,V)$ *such that*

$$\forall l\in \mathfrak{g}_V(S)\ \forall n>l\ \ v(g_n(a_l)-g_n(a_0))>0\,. \tag{12}$$

Then, there is one and only one subset $\mathscr{G}$ of $\mathscr{B}$ which is a minimal generating set of $\mathrm{Int}(S,V)$ *as a V-algebra, namely $\mathscr{G}=\{g_n\mid n\in\mathfrak{g}_V(S)\}$.*

Proof By Proposition 4, $\mathscr{G}$ is a generating set. The fact that this is a minimal generating set and that this is the only minimal generating set that can be extracted from $\mathscr{B}$ is an obvious consequence of Proposition 8.

Remark 3 Condition (12) for all $n > l$ is not necessary to be able to extract a minimal generating set, as shown by the following corollary, but it is useful for us to be sure that this is the unique extracted minimal set as shown by the example of Remark 2.

Theorem 2 may be formulated in a slightly different way when starting with a generating set which is not a regular basis.

Corollary 1 *Let K be a valued field, V be its valuation domain, and S be a precompact and regular subset of V. Let $\{a_n\}_{n\geq 0}$ denote a strong v-ordering of S. For each $k \in \mathfrak{g}_V(S)$, let g_k be a polynomial of* Int(S, V) *of degree k such that the valuation of its leading coefficient is equal to $-w_S(k)$. If*

$$\forall\, k, l \in \mathfrak{g}_V(S)\ \ [\, l < k\ \Rightarrow\ v(g_k(a_l) - g_k(a_0)) > 0\,]\,, \tag{13}$$

then $\{\, g_n \mid n \in \mathfrak{g}_V(S)\,\}$ is a minimal generating set of Int(S, V) *as a V-algebra.*

Proof Let $\{f_n\}_{n\geq 0}$ be the regular basis associated to the strong v-ordering $\{a_n\}_{n\geq 0}$. For every $n \geq 0$, let $h_n = g_n$ if $n \in \mathfrak{g}_V(S)$ and $h_n = f_n$ if $n \notin \mathfrak{g}_V(S)$. Then, $\{h_n\}_{n\geq 0}$ is a regular basis which satisfies Eq. (12) of Theorem 2.

The following lemma is a complement to Corollary 1 that will be useful to obtain minimality in Sect. 5 concerning globalization (cf. Proposition 16).

Lemma 8 *With the hypotheses and notation of Corollary 1, let $\{h_j\}_{j\in J}$ be a set of polynomials of* Int(S, V) *such that, for every $j \in J$ and every $l \in \mathfrak{g}_V(S)$:*

$$\text{either } \deg(h_j) < l \text{ or } v(h_j(a_l) - h_j(a_0)) > 0\,. \tag{14}$$

Then, there is one and only one subset of $\{g_n \mid n \in \mathfrak{g}_V(S)\} \cup \{h_j \mid j \in J\}$ which is a minimal generating set of Int(S, V) *as a V-algebra, namely $\{g_n \mid n \in \mathfrak{g}_V(S)\}$.*

Proof The first argument of the proof of Proposition 8 shows that:

$$\forall l \in \mathfrak{g}_V(S)\ \ \forall h \in \mathrm{Int}(S, V)\ [\deg(h) < l \Rightarrow v(h(a_l) - h(a_0)) > 0].$$

Fix some $l \in \mathfrak{g}_V(S)$. Then, for every $j \in J$, one has $v(h_j(a_l) - h_j(a_0)) > 0$.

Let us prove now that $v(g_l(a_l) - g_l(a_0)) = 0$. Let $\{f_n\}$ be the regular basis associated to the sequence $\{a_n\}_{n\geq 0}$. Then, $g_l(X) = \sum_{m=0}^{l} c_m f_m(X)$ where $c_m \in V$ and $v(c_l) = 0$. In particular, $g_l(a_l) - g_l(a_0) = c_l + \sum_{m=1}^{l-1} c_m f_m(a_l)$. By Lemma 6, $v(g_l(a_l) - g_l(a_0)) = v(c_l) = 0$.

Lemma 8 is then a straightforward consequence of the following obvious lemma:

Lemma 9 *Without any hypotheses on S, let $\{g\} \cup \{h_j \mid j \in J\} \subset$* Int$(S, V)$. *If there exist a and b in S such that $v(g(a)-g(b)) = 0$ and, for every $j \in J$, $v(h_j(a)-h_j(b)) > 0$, then g does not belong to the V-algebra generated by the h_j's.*

Remark 4 Condition (13) is not necessary to have a minimal generating set: for instance, by Proposition 3, $\{F_{p^k} \mid k \in \mathbb{N}\}$ is a minimal generating set of $\mathrm{Int}(\mathbb{Z}_p)$ while $\mathfrak{g}(\mathbb{Z}_3) = \{3^k \mid k \geq 0\}$, $\{n\}_{n\geq 0}$ is a strong p-ordering of $\mathbb{Z}_p$, and $3 \not| F_{3^3}(3)$.

This remark leads us to suppose that the following conjecture could be true.

Conjecture For every regular subset S of V and every regular basis $\{f_n\}_{n\geq 0}$ of $\mathrm{Int}(S, V)$, the generating set $\{f_n \mid n \in \mathfrak{g}(S)\}$ is minimal.

4.4 The Generating Set Associated to a Lubin-Tate Formal Group Law

In the particular case where $S = V$ is a discrete valuation domain, Corollary 1 becomes:

Proposition 9 *Let V be the ring of a discrete valuation v with uniformizer π and finite residue field of cardinality q. Let $\{f_{q^m}(X) \mid m \geq 0\}$ be a set of polynomials of* $\mathrm{Int}(V)$ *such that* $\deg(f_{q^m}) = q^m$ *and the valuation of the leading coefficient of f_{q^m} is equal to $-\frac{q^m-1}{q-1}$. If for every $m > l \geq 1$ $v(f_{q^m}(\pi^l) - f_{q^m}(0)) > 0$, then the f_{q^m}'s form a minimal generating set of the V-algebra* $\mathrm{Int}(V)$.

Proof Recall first how we can construct a strong v-ordering $\{a_n\}_{n\geq 0}$ of V (cf. [5, §II.2]): let $\{a_0 = 0, a_1, \ldots, a_{q-1}\}$ be a system of representatives of the residue field of V then, for $n = n_r q^r + \ldots + n_1 q + n_0$ where $0 \leq n_j < q$, let $a_n = a_{n_r}\pi^r + \ldots a_{n_1}\pi + a_{n_0}$. We then have $w_V(n) = \sum_{k>0} \left[\frac{n}{q^k}\right]$, and hence, $\mathfrak{g}(V) = \{q^m \mid m \geq 0\}$.

Assume now that the discrete valuation domain V is the ring of integers of a local field K, that is, K is complete with respect to the topology defined by the valuation v and the residue field of V is finite with cardinality q. Recall that a commutative formal group law over V is a formal power series $F(X, Y) \in V[[X, Y]]$ with the following properties:

$$F(X, Y) \equiv X + Y \pmod{\deg 2},$$

$$F(X, F(Y, Z)) = F(F(X, Y), Z),$$

$$F(X, Y) = F(Y, X).$$

The formal group law F is said to be a Lubin-Tate formal group law if it admits an endomorphism f, that is a power series $f(T) \in V[[T]]$ such that

$$f(F(X, Y)) = F(f(X), f(Y)),$$

which satisfies

$$f(T) \equiv \pi T \pmod{\deg 2},$$

$$f(T) \equiv T^q \pmod{\pi},$$

Theorem 3 *Let V be the valuation domain of a local field K, whose residue field has cardinality q, and let F be a Lubin-Tate formal group law on V associated to a power series f. For each $x \in V$, let $[x](T) = \sum_{n\geq 1} c_n(x)T^n$ be the unique power series such that $c_1(x) = x$ and $f \circ [x] = [x] \circ f$. Then, for each $n \geq 1$, $c_n(x)$ is an integer-valued polynomial of degree $\leq n$. Moreover, for $m \geq 0$, the $c_{q^m}(x)$'s are integer-valued polynomials on V with degree exactly q^m which form a minimal generating set of* Int(V).

Proof Recall that $\log_F(T)$ is the unique isomorphism between $F(X, Y)$ and the formal group law $\mathbb{G}_a = X + Y$ (see, for example, [13, I.5.4]). Consequently, for every $x \in V$, the multiplication by x with respect to F, that is, $[x]_F(T)$ clearly satisfies the following formula

$$[x](T) = \exp_F(x \log_F(T)) . \tag{15}$$

Since $\log_F(T)$ and $\exp_F(T)$ are power series belonging to $TV[[T]]$, Formula (15) shows that the $c_n(x)$'s are polynomials of degree $\leq n$. That the $c_n(x)$'s are integer-valued follows clearly from the well-known fact that $[x](T) \in V[[T]]$. That the c_{q^m}'s are of degree q^m and form a generating set for the V-algebra Int(V) is already proved by de Shalit and Iceland [11, Theorem 3.1]. That this is a minimal generating set follows easily from Proposition 9: we just have to verify that, denoting by π a uniformizer such that $f(T) \equiv \pi T \pmod{\deg 2}$, we have:

$$\forall\ m > l \geq 1 \quad c_{q^m}(\pi^l) \equiv c_{q^m}(0) \pmod{\pi}.$$

First, it follows from (15) and $\exp_F(T) \equiv T \pmod{\deg 2}$ that $[0](T) = 0$ and $c_n(0) = 0$ for every n. Moreover, it is known that $[x] \circ [y] = [xy]$ for all $x, y \in V$, and that $[\pi] = f$. Finally, since by definition $f \equiv T^q \pmod{\pi}$, we have $[\pi^l](T) = f(T) \circ \cdots \circ f(T) \equiv T^{q^l} \pmod{\pi}$, and hence, $c_{q^m}(\pi^l) \equiv 0 \pmod{\pi}$.

5 Globalization when D is a Dedekind Domain

In this section, we assume that D is a Dedekind domain and S is an infinite subset of D. Recall that, in a Dedekind domain, every ideal $\mathfrak{I}$ is generated by two elements, moreover the first generator may be any nonzero element of $\mathfrak{I}$. Noticing that, for every n, $X^n \in$ Int(S, D), we then have:

Proposition 10 *If, for every $n \geq 0$, the polynomial $g_n \in$ Int(S, D) of degree n is chosen in such a way that its leading coefficient generates with 1 the fractional ideal $\mathfrak{I}_n(S, D)$, then the set $\{X^n \mid n \in \mathbb{N}\} \cup \{g_n \mid n \in \mathbb{N}^*\}$ is a generating set for the D-module* Int(S, D), *and the set $\{X\} \cup \{g_n \mid n \in \mathbb{N}^*\}$ is a generating set for the D-algebra* Int(S, D).

To obtain minimal generating subsets, we begin by deleting some generators that are useless. To do this we have to consider the 'gaps' of the factorial ideals of S.

5.1 The Factorial Ideals and Their Gaps

Bhargava [2–4] associate to the subset S of the Dedekind domain D a sequence of ideals called *factorial ideals of* S, denoted by $n!_S^D$ (or $n!_S$) which could be defined by:

$$n!_S^D = \mathfrak{I}_n(S,D)^{-1} \quad (n \geq 0).$$

The ideals $n!_S^D$ are entire ideals (for instance, $n!_{\mathbb{Z}} = n!\,\mathbb{Z}$). They form a decreasing sequence for the inclusion and satisfy:

$$0!_S = D \quad \text{and} \quad n!_S \times m!_S \text{ divides } (n+m)!_S\,.$$

Remark 5 One could think that, analogously to Definition 1, the gaps of the factorial ideals of S with respect to D should correspond to integers n such that, for every $j \in \{1,\dots,n-1\}$, $j!_S \times (n-j)!_S \neq n!_S$. With such a definition, the gaps of the factorials of $\mathbb{Z}$ would correspond to all nonnegative integers since $j(n-j) \neq 0$ implies $\frac{n!}{j!(n-j)!} = \binom{n}{j} \neq 1$. The following equality shows that this definition would be too extensive in view of a global statement of Theorem 2:

$$\binom{X}{6} = X\binom{X}{5} + \binom{X}{2}\binom{X}{4} - \binom{X}{3}^2 + \binom{X}{3} - 6\binom{X}{4} + 5\binom{X}{5}.$$

Thus, the set $\left\{\binom{X}{n}\right\}_{n\geq 0}$ is not a minimal generating set for the $\mathbb{Z}$-algebra $\mathrm{Int}(\mathbb{Z})$. We are then led to consider the following definition which in fact extends also Definition 1.

Definition 3 The set of *indices of gaps* of the factorial ideals $n!_S^D$ is

$$\mathfrak{g}_D(S) = \{n \geq 1 \mid n!_S^D \neq \cap_{1\leq j\leq n-1} \left(j!_S^D \times (n-j)!_S^D\right)\}\,.$$

In other words, the index n corresponds to a gap if the ideal $n!_S^D$ is not the least common multiple of the ideals $j!_S^D(n-j)!_S^D$ for $1 \leq j \leq n-1$. Equivalently,

$$n \in \mathfrak{g}_D(S) \;\Leftrightarrow\; \exists \mathfrak{m} \in \mathrm{Max}(D)\ \forall j \in \{1,\dots,n-1\}\ [\,w_{\mathfrak{m}}(n) > w_{\mathfrak{m}}(j) + w_{\mathfrak{m}}(n-j)\,]\,,$$

where $w_{\mathfrak{m}}(n) = w_{\mathfrak{m},S}(n) = -v_{\mathfrak{m}}(\mathfrak{I}_n(S,D))$ (denoting by $v_{\mathfrak{m}}$ the $\mathfrak{m}$-adic valuation associated to the maximal ideal $\mathfrak{m}$ of D). Consequently,

$$\mathfrak{g}_D(S) = \cup_{\mathfrak{m}\in\mathrm{Max}(D)}\, \mathfrak{g}_{D_{\mathfrak{m}}}(S)\,.$$

5.2 Localization and Globalization

Recall that, since D is Noetherian, for each maximal ideal $\mathfrak{m}$ of D [5, §I.2] :

$$\text{Int}(S, D)_{\mathfrak{m}} = \text{Int}(S, D_{\mathfrak{m}}) \,. \tag{16}$$

These equalities allow us to globalize the previous results obtained in the local case. In view of this globalization the following proposition will be useful.

Proposition 11 *Let* $\{h_j\}_{j\in J}$ *be a set of elements of* $\text{Int}(S, D)$*. The set* $\{h_j \mid j \in J\}$ *is a generating set for the D-algebra* $\text{Int}(S, D)$ *if and only if, for every* $\mathfrak{m} \in \text{Max}(D)$*, this set is a generating set for the* $D_{\mathfrak{m}}$*-algebra* $\text{Int}(S, D_{\mathfrak{m}})$.

Proof Clearly, it follows from Formula (16) that the condition is necessary. Conversely, assume that $\{h_j\}_{j\in J}$ is a generating set for all the localizations and consider some $g(X) \in \text{Int}(S, D)$. By hypothesis, for every maximal ideal $\mathfrak{m}$ of D, there exists a polynomial $P_{\mathfrak{m}} \in D_{\mathfrak{m}}[T_j \mid j \in J]$ such that $g = P_{\mathfrak{m}}((h_j)_{j\in J})$. Let $s_{\mathfrak{m}} \in D \setminus \mathfrak{m}$ be such that $s_{\mathfrak{m}} P_{\mathfrak{m}} \in D[T_j \mid j \in J]$. As the $s_{\mathfrak{m}}$'s generate the ideal D, there exist finitely many maximal ideals $\mathfrak{m}_1, \ldots, \mathfrak{m}_l$ and elements $t_1, \ldots, t_l \in D$ such that $t_1 s_{\mathfrak{m}_1} + \ldots + t_l s_{\mathfrak{m}_l} = 1$. Consequently,

$$g(X) = \sum_{i=1}^{l} t_i\, s_{\mathfrak{m}_i}\, g(X) = \sum_{i=1}^{l} t_i\, \left(s_{\mathfrak{m}_i} P_{\mathfrak{m}_i}((h_j)_{j\in J})\right) .$$

Let

$$Q(T_{j_1}, \ldots, T_{j_r}) = \sum_{i=1}^{l} t_i\, \left(s_{\mathfrak{m}_i} P_{\mathfrak{m}_i}((T_j)_{j\in J})\right) .$$

Then,

$$g(X) = Q(h_{j_1}(X), \ldots, h_{j_r}(X)) \quad \text{where} \quad Q \in D[T_{j_1}, \ldots, T_{j_r}] \,.$$

Corollary 2 *If the factorial ideals* $n!_S$ *are principal, that is, if* $\text{Int}(S, D)$ *admits a regular basis* $\{g_n\}_{n\geq 0}$*, then* $\{\, g_n \mid n \in \mathfrak{g}_D(S) \,\}$ *is a generating set for the D-algebra* $\text{Int}(S, D)$.

This is an obvious consequence of Propositions 4 and 11.

Remark 6 If $\{g_j \mid j \in J\}$ is a generating set for the D-algebra $\text{Int}(S, D)$, then the characteristic ideal $\mathfrak{I}_1(S, D)$ is generated as a fractional ideal by the leading coefficients of the polynomials g_j of degree 1. Thus, if $\mathfrak{I}_1(S, D)$ is not principal, a generating set of $\text{Int}(S, D)$ necessarily contains two polynomials of degree 1 (as in the following example).

Example 1 Let $K = \mathbb{Q}(\sqrt{-5})$, $D = \mathcal{O}_K = \mathbb{Z}[\sqrt{-5}]$. Then, $3\mathcal{O}_K = \mathfrak{p}\mathfrak{q}$ where $\mathfrak{p} = (3, 1 + \sqrt{-5})$ and $\mathfrak{q} = (3, 1 - \sqrt{-5})$. We consider $\mathrm{Int}(\mathfrak{p}, \mathbb{Z}[\sqrt{-5}])$. The characteristic ideal $\mathfrak{I}_1(\mathfrak{p}, \mathbb{Z}[\sqrt{-5}])$ is equal to $\mathfrak{p}^{-1} = (1, \frac{1-\sqrt{-5}}{3})$ and is not principal. Consequently, any generating set of $\mathrm{Int}(\mathfrak{p}, \mathbb{Z}[\sqrt{-5})$ contains two polynomials of degree 1, for instance X and $\frac{1-\sqrt{-5}}{3}X$.

Remark 7 Why do we assume that S is infinite? Because, if S is finite, there does no exist in general any minimal generating set for the D-algebra $\mathrm{Int}(S, D)$.

Let us look, for instance, to the $\mathbb{Z}$-algebra $\mathrm{Int}(\{0\}, \mathbb{Z}) = \mathbb{Z} + X\mathbb{Q}[X]$. Let $\{h_j\}_{j\in J}$ be the elements of degree one of a generating set, we may assume that $h_j(0) = 0$ for each j. By Remark 6, there exist $j_1, \ldots, j_r \in J$ such that $X \in \sum_{i=1}^r \mathbb{Z}h_{j_i}$. Noticing that, if $h = \frac{a}{b}X$ with $(a, b) = 1$, then there exist $u, v \in \mathbb{Z}$ such that $\frac{1}{b}X = uh + vX$, it is then easy to see that every h_{j_0} where $j_0 \in J \setminus \{j_1, \ldots, j_r\}$ belongs to the $\mathbb{Z}$-module generated by the h_j's where $j \in J \setminus \{j_0\}$, and hence, that no generating set can be minimal.

On the other hand, if S is empty, there may exist minimal generating sets for the D-algebra $\mathrm{Int}(\emptyset, D) = K[X]$ as shown by the set $\{\frac{1}{p} \mid p \in \mathbb{P}\} \cup \{X\}$ for the $\mathbb{Z}$-algebra $\mathbb{Q}[X]$.

5.3 Classical Examples

Proposition 12 *The set* $\{\binom{X}{p^r} \mid p \in \mathbb{P}, r \geq 0\}$ *is a generating set for the* $\mathbb{Z}$-*algebra* $\mathrm{Int}(\mathbb{Z})$. *Moreover, it is the only subset of* $\left\{\binom{X}{n} \mid n \in N\right\}$ *which is a minimal generating set.*

This is a consequence of Propositions 2 and 11. We may also note that the $\binom{X}{n}$'s are constructed by means of the sequence $\{n\}_{n\geq 0}$ which is a strong p-ordering for every p. With respect to Fermat polynomials, the next theorem follows from Propositions 3 and 11, and also from the fact that, if $p \neq p'$, the polynomials $F_{p'^k}$ belong to $\mathbb{Z}_{(p')}[X]$ and then are useless to generate $\mathrm{Int}(\mathbb{Z}_{(p)})$.

Proposition 13 *The set* $\{F_{p^k} \mid p \in \mathbb{P}, k \in \mathbb{N}\}$ *where* $F_1 = X$, $F_p = \frac{X^p - X}{p}$ *and, for* $k \geq 2$, $F_{p^k} = F_p(F_{p^{k-1}})$ *is a minimal generating set for the* $\mathbb{Z}$-*algebra* $\mathrm{Int}(\mathbb{Z})$.

With respect to subsets, the following result shows that a generating set for the $\mathbb{Z}$-module $\mathrm{Int}(S, \mathbb{Z})$ can be, just by deleting 1, a minimal generating set for the Z-algebra $\mathrm{Int}(S, \mathbb{Z})$.

Proposition 14 *Let q be an integer ≥ 2 and let $S = \{q^n \mid n \geq 0\}$. The set*

$$\left\{ \begin{bmatrix} X \\ n \end{bmatrix}_q = \prod_{k=0}^{n-1} \frac{X - q^k}{q^n - q^k} \mid n \geq 1 \right\}$$

is a minimal generating set of the $\mathbb{Z}$-*algebra* $\mathrm{Int}(S, \mathbb{Z})$.

Proof It is well known that the polynomials $\begin{bmatrix} X \\ n \end{bmatrix}_q$ $(n \geq 1)$ form with 1 a regular basis of the $\mathbb{Z}$-module Int$(S, \mathbb{Z})$ (cf. [5, Exercise II.15]) since the sequence $\{q^n\}_{n\geq 0}$ is a simultaneous ordering of S. In particular, $\mathfrak{I}_n(S, \mathbb{Z})^{-1} = \prod_{k=0}^{n-1}(q^n - q^k)\mathbb{Z} = q^{\frac{n(n-1)}{2}} \prod_{h=1}^{n}(q^h - 1)\mathbb{Z}$, and hence, $\mathcal{G}(S) = \mathbb{N}^*$ since, for every prime p dividing q, $w_{p,S}(n) = -v_p(\mathfrak{I}_n(S, \mathbb{Z})) = \frac{n(n-1)}{2} v_p(q)$. The proposition is then a consequence of Proposition 6 thanks to the following result due to Elliott, Adams, DeMoss, Freaney and Mostowa:

Proposition 15 *[12, Theorem 1.5] For all $m, n \in \mathbb{N}$*

$$\begin{bmatrix} X \\ m \end{bmatrix}_q \begin{bmatrix} X \\ n \end{bmatrix}_q = \sum_{l=\max(m,n)}^{m+n} q^{(l-m)(l-n)} \begin{pmatrix} l \\ l-m, l-n, m+n-l \end{pmatrix} \begin{bmatrix} X \\ l \end{bmatrix}_q$$

5.4 *Globalization of the Sets Given by Lubin-Tate Formal Group Laws*

At the end of their paper [11, § 4.2], de Shalit and Iceland described a globalization to number fields of their results in local fields. This globalization works in particular for number fields of class number one. We recall here this globalization with some slight changes so that it works in a more general framework, namely for Pólya fields.

Recall that a number field K is called a *Pólya field* [17] if Int$(\mathcal{O}_K)$ admits a regular basis. It is known [5, II.3.9] that this is equivalent to the fact that the Pólya group of $\mathcal{O}_K$ is trivial where the *Pólya group* $\mathcal{P}o(D)$ of a Dedekind domain D is the subgroup of the class group generated by the classes of the ideals $\Pi_q(D)$ where $\Pi_q(D)$ denotes the product of all the prime ideals $\mathfrak{p}$ of D with the same norm q :

$$\Pi_q(D) = \prod_{\mathfrak{p}\in \mathrm{Max}(D),\, |D/\mathfrak{p}|=q} \mathfrak{p}\,.$$

Now, let K be a number field with ring of integers $\mathcal{O}_K$. Let T be a finite (possibly empty) set of primes such that the Pólya group of $R = \mathcal{O}_{K,T} = \cap_{\mathfrak{p}\notin T}\mathcal{O}_{K,\mathfrak{p}}$ is trivial. Denote by $\mathcal{Q}$ the set of integers $\{q \mid q = N_{K/\mathbb{Q}}(\mathfrak{p}), \mathfrak{p} \notin T\}$ and, for each $q \in \mathcal{Q}$, let π_q be a generator of the principal ideal $\Pi_q(R)$. Now, consider the formal Dirichlet series

$$L(s) = \prod_{q\in\mathcal{Q}} \frac{1}{1 - \frac{1}{\pi_q q^s}} = \sum_{n=1}^{\infty} \frac{a_n}{n^s}\,, \tag{17}$$

Clearly, $a_1 = 1$, $a_n \in K$, and, for every $\mathfrak{p} \notin T$ with norm q, the Dirichlet series $(1-\pi_q^{-1} q^{-s}) L(s)$ has $\mathfrak{p}$-integral coefficients. [This is exactly de Shalit and Iceland's proof where the set $\{\mathfrak{p} \mid \mathfrak{p} \notin T\}$ is replaced by the set $\mathcal{Q}$.]

Consider the formal power series

$$f(X) = \sum_{n=1}^{\infty} a_n X^n \tag{18}$$

and the group law

$$F(X,Y) = f^{-1}(f(X) + f(Y)) \tag{19}$$

for which f is a logarithm. A priori F is defined over K, let us show that it is defined over R. Let $\mathfrak{p} \notin T$ with norm q. One has:

$$(1 - \pi_q^{-1} q^{-s}) L(s) = \sum_{n \geq 1} \left(a_n - \frac{1}{\pi_q} a_{\frac{n}{q}} \right) \frac{1}{n^s} \tag{20}$$

where $a_{\frac{n}{q}} = 0$ when $q \nmid n$. From the fact that the Dirichlet series (20) has $\mathfrak{p}$-integral coefficients, it follows that the corresponding power series has $\mathfrak{p}$-integral coefficients:

$$g(X) = f(X) - \frac{1}{\pi_q} f(X^q) \in \mathcal{O}_{K,\mathfrak{p}}[[X]] \,. \tag{21}$$

Now Hazewinkel's functional equation lemma [13, I.2.2 (i)] implies that the coefficients of the formal group law F defined by (19) are also in $\mathcal{O}_{K,\mathfrak{p}}$, and hence, in $R = \cap_{\mathfrak{p} \notin T} \mathcal{O}_{K,\mathfrak{p}}$.

Furthermore, by [13, I.8.3.6], F is a Lubin-Tate formal group law associated with the prime π_q and the corresponding power series $[\pi]_F = f^{-1}(\pi f(X))$. Then, for every $x \in \mathcal{O}_{K,\mathfrak{p}}$, $[x]_F(t) = f^{-1}(xf(t)) \in \mathcal{O}_{K,\mathfrak{p}}[[t]]$. Finally,

$$\forall x \in R \quad [x]_F(t) = \sum_{n=1}^{\infty} c_n(x) t^n \in R[[t]] \,, \tag{22}$$

and the $c_n(x)$'s belongs to $\mathrm{Int}(R)$. In the particular case where K is a Pólya field, we may state the following:

Proposition 16 *Let K be a Pólya field. Consider the Dirichlet series (17) where $\mathcal{Q}$ denotes the set formed by the norms of all the primes of K and the formal group law on $\mathcal{O}_K$ defined by means of equations (18) and (19). Then the functions $c_n(x)$ defined by (22) belong to* $\mathrm{Int}(\mathcal{O}_K)$ *and the subset $\{c_n(x) \mid n \in \mathfrak{g}(R)\}$ where $\mathfrak{g}(R) = \{q^m \mid q \in \mathcal{Q}, m \in \mathbb{N}\}$ is a minimal generating set for the $\mathcal{O}_K$-algebra* $\mathrm{Int}(\mathcal{O}_K)$.

Proof That $\{c_n(x) \mid n \in \mathfrak{g}(R)\}$ is a generating set follows from Theorem 3 in the local case. That it is minimal follows from Lemma 8: we just have to verify that, for every q^m and every $n > q^m$, π_q divides $c_n(\pi_q^m)$. In fact, $c_n(\pi_q^m)$ is the coefficient of T^n in $[\pi_q^m](T) \equiv T^{q^m} \pmod{\pi_q}$.

References

1. Amice, Y.: Interpolation p-adique. Bull. Soc. Math. Fr. **92**, 117–180 (1964)
2. Bhargava, M.: P-orderings and polynomial functions on arbitrary subsets of Dedekind rings. J. Reine Angew. Math. **490**, 101–127 (1997)
3. Bhargava, M.: Generalized factorials and fixed divisors over subsets of a Dedekind domain. J. Number Theory **72**, 67–75 (1998)
4. Bhargava, M.: The factorial function and generalizations. Am. Math. Mon. **107**, 783–799 (2000)
5. Cahen, P.-J., Chabert, J.-L.: In: Integer-Valued Polynomials. Mathematical Surveys and Monographs, vol. 48. American Mathematical Society, Providence (1997)
6. Cahen, P.-J., Chabert, J.-L.: On the ultrametric Stone-Weierstrass theorem and Mahler's expansion. J. Théor. Nombres de Bordeaux **14**, 43–57 (2002)
7. Chabert, J.-L.: Integer-valued polynomials in valued fields with an application to discrete dynamical systems. In: Commutative Algebra and Applications, pp. 103–134. de Gruyter, Berlin (2009)
8. Chabert, J.-L.: Integer-valued polynomials: Looking for regular bases (a survey). In: Commutative Algebra: Recent Advances in Commutative Rings, Integer-Valued Polynomials, and Polynomial Functions, pp. 83–111. Springer, New York (2014)
9. Chabert, J.-L., Evrard, S., Fares, Y.: Regular subsets of valued fields and Bhargava's v-orderings. Math. Z. **274**, 263–290 (2013)
10. Comtet, L.: Analyse Combinatoire. Presses Universitaires de France, Paris (1970)
11. de Shalit, E., Iceland, E.: Integer valued polynomials and Lubin-Tate formal groups. J. Number Theory **129**, 632–639 (2009)
12. Elliott, J.: Newton basis relations and applications to integer-valued polynomials and q-binomial coefficients. Integers **14**, A38 (2014)
13. Hazewinkel, M.: Formal Groups and Applications. Academic Press, New York (1978)
14. Johnson, K.: Super-additive sequences and algebras of polynomials. Proc. Am. Math. Soc. **139**, 3431–3443 (2011)
15. Legendre, A.-M.: Essai sur la Théorie des Nombres, 2nd ed. Courcier, Paris (1808)
16. Pólya, G., Über ganzwertige Polynome in algebraischen Zahlkörpern. J. Reine Angew. Math. **149**,97–116 (1919)
17. Zantema, H.: Integer valued polynomials over a number field. Manuscr. Math. **40**, 155–203 (1982)

Algebraic Entropy in Locally Linearly Compact Vector Spaces

Ilaria Castellano and Anna Giordano Bruno

Dedicated to the 70th birthday of Luigi Salce

Abstract We introduce algebraic entropy for continuous endomorphisms of locally linearly compact vector spaces over a discrete field, as a natural extension of the algebraic entropy for endomorphisms of discrete vector spaces studied in Giordano Bruno and Salce (Arab J Math 1:69–87, 2012). We show that the main properties continue to hold in the general context of locally linearly compact vector spaces, in particular we extend the Addition Theorem.

Keywords Linearly compact vector space • Locally linearly compact vector space • Algebraic entropy • Continuous linear transformation • Continuous endomorphism • Algebraic dynamical system

2010 Mathematics Subject Classification: 15A03; 15A04; 22B05; 20K30; 37A35

1 Introduction

In [1] Adler, Konheim, and McAndrew introduced the notion of topological entropy h_{top} for continuous self-maps of compact spaces, and they concluded the paper by sketching a definition of the algebraic entropy h_{alg} for endomorphisms of abelian groups. This notion of algebraic entropy, which is appropriate for torsion abelian groups and vanishes on torsion-free abelian groups, was later reconsidered by Weiss

I. Castellano
University of Southampton, Building 54, Salisbury Road, SO17 1BJ Southampton, UK
e-mail: ilaria.castellano88@gmail.com

A. Giordano Bruno (✉)
Università degli Studi di Udine, Via delle Scienze 2016, 33100 Udine, Italy
e-mail: anna.giordanobruno@uniud.it

M. Fontana et al. (eds.), *Rings, Polynomials, and Modules*,
https://doi.org/10.1007/978-3-319-65874-2_6

in [26], who proved all the basic properties of h_{alg}. Recently, h_{alg} was deeply investigated by Dikranjan, Goldsmith, Salce and Zanardo for torsion abelian groups in [10], where they proved in particular the Addition Theorem and the Uniqueness Theorem.

Later on, Peters suggested another definition of algebraic entropy for automorphisms of abelian groups in [19]; here, we denote Peters' entropy still by h_{alg}, since it coincides with Weiss' notion on torsion abelian groups; on the other hand, Peters' entropy is not vanishing on torsion-free abelian groups. In [9] h_{alg} was extended to all endomorphisms and deeply investigated, in particular the Addition Theorem and the Uniqueness Theorem were proved in full generality. In [20] Peters gave a further generalization of his notion of entropy for continuous automorphisms of locally compact abelian groups, which was recently extended by Virili in [25] to continuous endomorphisms.

Weiss in [26] connected the algebraic entropy h_{alg} for endomorphisms of torsion abelian groups with the topological entropy h_{top} for continuous endomorphisms of totally disconnected compact abelian groups by means of Pontryagin duality. Moreover, the same connection was shown by Peters in [19] between h_{alg} for topological automorphisms of countable abelian groups and h_{top} for topological automorphisms of metrizable compact abelian groups. These results, known as Bridge Theorems, were recently extended to endomorphisms of abelian groups in [6], to continuous endomorphisms of locally compact abelian groups with totally disconnected Pontryagin dual in [8], and to topological automorphisms of locally compact abelian groups in [24] (in the latter two cases on the Potryagin dual one considers an extension of h_{top} to locally compact groups based on a notion of entropy introduced by Hood in [15] as a generalization of Bowen's entropy from [3]—see also [14]).

A generalization of Weiss' entropy in another direction was given in [22], where Salce and Zanardo introduced the i-entropy ent_i for endomorphisms of modules over a ring R and an invariant i of $\mathrm{Mod}(R)$. For abelian groups (i.e., $\mathbb{Z}$-modules) and $i = \log|-|$, ent_i coincides with Weiss' entropy. Moreover, the theory of the entropies ent_L where L is a length function was pushed further in [21, 23].

In [12] the easiest case of ent_i was studied, namely, the case of vector spaces with the dimension as invariant, as an introduction to the algebraic entropy in the most convenient and familiar setting. The *dimension entropy* $\mathrm{ent}_{\dim}$ is defined for an endomorphism $\phi : V \to V$ of a vector space V as

$$\mathrm{ent}_{\dim}(\phi) = \sup\{H_{\dim}(\phi, F) : F \leq V,\ \dim F < \infty\},$$

where

$$H_{\dim}(\phi, F) = \lim_{n\to\infty} \frac{1}{n} \dim(F + \phi F + \ldots + \phi^{n-1} F).$$

All the basic properties of $\mathrm{ent}_{\dim}$ were proved in [12], namely, Invariance under conjugation, Monotonicity for linear subspaces and quotient vector spaces, Logarithmic Law, Continuity on direct limits, weak Addition Theorem (see Sect. 4 for the precise meaning of these properties).

Moreover, compared to the Addition Theorem for h_{alg} and other entropies, a simpler proof was given in [12, Theorem 5.1] of the Addition Theorem for $\mathrm{ent}_{\dim}$, which states that if V is a vector space, $\phi : V \to V$ an endomorphism and W a ϕ-invariant (i.e., $\phi W \leq W$) linear subspace of V, then

$$\mathrm{ent}_{\dim}(\phi) = \mathrm{ent}_{\dim}(\phi \restriction_W) + \mathrm{ent}_{\dim}(\overline{\phi}),$$

where $\phi \restriction_W$ is the restriction of ϕ to W and $\overline{\phi} : V/W \to V/W$ is the endomorphism induced by ϕ.

Also the Uniqueness Theorem was proved for the dimension entropy (see [12, Theorem 5.3]). Namely, $\mathrm{ent}_{\dim}$ is the unique collection of functions

$$\mathrm{ent}^V_{\dim} : \mathrm{End}(V) \to \mathbb{N} \cup \{\infty\}, \quad \phi \mapsto \mathrm{ent}_{\dim}(\phi),$$

satisfying for every vector space V: Invariance under conjugation, Continuity on direct limits, Addition Theorem and $\mathrm{ent}_{\dim}(\beta_F) = \dim F$ for any finite-dimensional vector space F, where $\beta_F : \bigoplus_{\mathbb{N}} F \to \bigoplus_{\mathbb{N}} F$, $(x_0, x_1, x_2, \ldots) \mapsto (0, x_0, x_1, \ldots)$ is the right Bernoulli shift.

Inspired by the extension of h_{alg} from the discrete case to the locally compact one, and by the approach used in [11] to define the intrinsic algebraic entropy (based on the concept of inert subgroup with respect to an endomorphism—see [2, 5]), we extend the dimension entropy to continuous endomorphisms of locally linearly compact vector spaces. Recall that a linearly topologized vector space V over a discrete field $\mathbb{K}$ is *locally linearly compact* (briefly, l.l.c.) if it admits a local basis at 0 consisting of linearly compact open linear subspaces; we denote by $\mathcal{B}(V)$ the family of all linearly compact open linear subspaces of V (see [16, 17]). Clearly, linearly compact and discrete vector spaces are l.l.c.. (See Sect. 2 for some background on linearly compact and locally linearly compact vector spaces.)

Let V be an l.l.c. vector space and $\phi: V \to V$ a continuous endomorphism. The *algebraic entropy of ϕ with respect to $U \in \mathcal{B}(V)$* is

$$H(\phi, U) = \lim_{n\to\infty} \frac{1}{n} \dim \frac{U + \phi U + \ldots + \phi^{n-1} U}{U}, \tag{1}$$

and the *algebraic entropy* of ϕ is

$$\mathrm{ent}(\phi) = \sup\{H(\phi, U) \mid U \in \mathcal{B}(V)\}.$$

In Sect. 3 we show that the limit in (1) exists. Moreover, we see that ent is always zero on linearly compact vector spaces (see Corollary 2). On the other hand, if V is a discrete vector space, then $\mathrm{ent}(\phi)$ turns out to coincide with $\mathrm{ent}_{\dim}(\phi)$ (see Lemma 1). Moreover, if V is an l.l.c. vector space over a finite field $\mathbb{F}$, then V is a totally disconnected locally compact abelian group and $h_{alg}(\phi) = \mathrm{ent}(\phi) \cdot \log|\mathbb{F}|$ (see Lemma 5).

In Sect. 4 we prove all of the general properties that the algebraic entropy is expected to satisfy, namely, Invariance under conjugation, Monotonicity for linear subspaces and quotient vector spaces, Logarithmic Law, Continuity on direct limits, weak Addition Theorem. As a consequence of the computation of the algebraic entropy for the Bernoulli shifts (see Example 2), we find in particular that the algebraic entropy for continuous endomorphisms of l.l.c. vector spaces takes all values in $\mathbb{N} \cup \{\infty\}$.

In Sect. 5 we prove the so-called Limit-free Formula for the computation of the algebraic entropy, that permits to avoid the limit in the definition in (1) (see Proposition 12). Indeed, taken V an l.l.c. vector space and $\phi : V \to V$ a continuous endomorphism, for every $U \in \mathcal{B}(V)$ we construct an open linear subspace U^- of V (see Definition 1) such that $\phi^{-1}U^-$ is an open linear subspace of U^- of finite codimension and

$$H(\phi, U) = \dim \frac{U^-}{\phi^{-1}U^-}.$$

A first Limit-free Formula for h_{alg} in the case of injective endomorphisms of torsion abelian groups was sketched by Yuzvinski in [28] and was later proved in a slightly more general setting in [7]; this result was extended in [13, Lemma 5.4] to a Limit-free Formula for the intrinsic algebraic entropy of automorphisms of abelian groups. In [7] one can find also a Limit-free Formula for the topological entropy of surjective continuous endomorphisms of totally disconnected compact groups, which was extended to continuous endomorphisms of totally disconnected locally compact groups in [14, Proposition 3.9], using ideas by Willis in [27]. Our Limit-free Formula is inspired by all these results, mainly by ideas from the latter one.

The Limit-free Formula is one of the main tools that we use in Sect. 6 to extend the Addition Theorem from the discrete case (i.e., the Addition Theorem for $\mathrm{ent}_{\mathrm{dim}}$ [12, Theorem 5.1]) to the general case of l.l.c vector spaces (see Theorem 2). If V is an l.l.c. vector space, $\phi : V \to V$ a continuous endomorphism and W a closed ϕ-invariant linear subspace of V, consider the following commutative diagram

$$\begin{array}{ccccccccc}
0 & \longrightarrow & W & \longrightarrow & V & \longrightarrow & V/W & \longrightarrow & 0 \\
 & & \big\downarrow{\scriptstyle \phi\restriction_W} & & \big\downarrow{\scriptstyle \phi} & & \big\downarrow{\scriptstyle \overline{\phi}} & & \\
0 & \longrightarrow & W & \longrightarrow & V & \longrightarrow & V/W & \longrightarrow & 0
\end{array}$$

of continuous endomorphisms of l.l.c. vector spaces, where $\phi \restriction_W$ is the restriction of ϕ to W and $\overline{\phi}$ is induced by ϕ; we say that the Addition Theorem holds if

$$\mathrm{ent}(\phi) = \mathrm{ent}(\phi \restriction_W) + \mathrm{ent}(\overline{\phi}).$$

While it is known that h_{alg} satisfies the Addition Theorem for endomorphisms of discrete abelian groups (see [9]), it is still an open problem to establish whether h_{alg} satisfies the Addition Theorem in the general case of continuous endomorphisms of locally compact abelian groups; from the Addition Theorem for the topological entropy in [14] and the Bridge Theorem in [8] one can only deduce that the Addition Theorem holds for h_{alg} in the case of topological automorphisms of locally compact abelian groups which are compactly covered (i.e., they have totally disconnected Pontryagin dual). Here, Theorem 2 shows in particular that the Addition Theorem holds for h_{alg} on the small subclass of compactly covered locally compact abelian groups consisting of all locally linearly compact spaces over finite fields.

With respect to the Uniqueness Theorem for $\mathrm{ent}_{\dim}$ mentioned above, we leave open the following question.

Question 1 Does a Uniqueness Theorem hold also for the algebraic entropy ent on locally linearly compact vector spaces?

In other words, we ask whether ent turns out to be the unique collection of functions $\mathrm{ent}^V : \mathrm{End}(V) \to \mathbb{N} \cup \{\infty\}$, $\phi \mapsto \mathrm{ent}(\phi)$, satisfying for every l.l.c. vector space V: Invariance under conjugation, Continuity on direct limits, Addition Theorem and $\mathrm{ent}(\beta_F) = \dim F$ for any finite-dimensional vector space F, where $V = \bigoplus_{n=-\infty}^{0} F \oplus \prod_{n=1}^{\infty} F$ is endowed with the topology inherited from the product topology of $\prod_{n\in\mathbb{Z}} F$, and $\beta_F : V \to V$, $(x_n)_{n\in\mathbb{Z}} \mapsto (x_{n-1})_{n\in\mathbb{Z}}$ is the right Bernoulli shift (see Example 2).

We end by remarking that in [4] we introduce a topological entropy for l.l.c. vector spaces and connect it to the algebraic entropy studied in this paper by means of Lefschetz Duality, by proving a Bridge Theorem in analogy to the ones recalled above for h_{alg} and h_{top} in the case of locally compact abelian groups and their continuous endomorphisms.

2 Background on Locally Linearly Compact Vector Spaces

Fix an arbitrary field $\mathbb{K}$ endowed always with the discrete topology. A topological vector space V over $\mathbb{K}$ is said to be *linearly topologized* if it is Hausdorff and it admits a neighborhood basis at 0 consisting of linear subspaces of V. Clearly, a discrete vector space V is linearly topologized, and if V has finite dimension then the vice-versa holds as well (see [17, p.76, (25.6)]).

If W is a linear subspace of a linearly topologized vector space V, then W with the induced topology is a linearly topologized vector space; if W is also closed in V, then V/W with the quotient topology is a linearly topologized vector space as well.

Given a linearly topologized vector space V, a *linear variety* M of V is a coset $v + W$, where $v \in V$ and W is a linear subspace of V. A linear variety $M = v + W$ is said to be *open* (respectively, *closed*) in V if W is open (respectively, closed) in V.

A linearly topologized vector space V is *linearly compact* if any collection of closed linear varieties of V with the finite intersection property has non-empty intersection (equivalently, any collection of open linear varieties of V with the finite intersection property has non-empty intersection) (see [17]).

For reader's convenience, we collect in the following proposition all those properties concerning linearly compact vector spaces that we use further on.

Proposition 1 *Let V be a linearly topologized vector space.*

(a) *If W is a linearly compact subspace of V, then W is closed.*
(b) *If V is linearly compact and W is a closed linear subspace of V, then W is linearly compact.*
(c) *If V is linearly compact, W a linearly topologized vector space, and $\phi : V \to W$ is a surjective continuous homomorphism, then W is linearly compact.*
(d) *If V is discrete, then V is linearly compact if and only if it has finite dimension (hence, if V has finite dimension then V is linearly compact).*
(e) *If W is a closed linear subspace of V, then V is linearly compact if and only if W and V/W are linearly compact.*
(f) *The direct product of linearly compact vector spaces is linearly compact.*
(g) *An inverse limit of linearly compact vector spaces is linearly compact.*
(h) *A linearly compact vector space is complete.*

Proof A proof for (a), (b), (c), and (d) can be found in [17, page 78]. Properties (e) and (f) are proved in [18, Propositions 2 and 9]. Finally, (g) follows from (b) and (f). Let $\iota: V \to \tilde{V}$ be the topological dense embedding of V into its completion $\tilde{V}$, thus (a) implies (h). □

A linearly topologized vector space V is *locally linearly compact* (briefly, l.l.c.) if there exists an open linear subspace of V that is linearly compact (see [17]). Thus V is l.l.c. if and only if it admits a neighborhood basis at 0 consisting of linearly compact linear subspaces of V. Clearly, linearly compact and discrete vector spaces are l.l.c.. The structure of an l.l.c. vector space can be characterized as follows.

Theorem 1 ([17, (27.10), page 79]) *If V is an l.l.c. vector space, then V is topologically isomorphic to $V_c \oplus V_d$, where V_c is a linearly compact open linear subspace of V and V_d is a discrete linear subspace of V.*

By Proposition 1 and Theorem 1, one may prove that an l.l.c. vector space verifies the following properties.

Proposition 2 *Let V be a linearly topologized vector space.*

(a) *If V is l.l.c., then V is complete.*
(b) *If W is an l.l.c. linear subspace of V, then W is closed.*
(c) *If W is a closed linear subspace of V, then V is l.l.c. if and only if W and V/W are l.l.c.*

Given an l.l.c. vector space V, for the computation of the algebraic entropy we are interested in the neighborhood basis $\mathcal{B}(V)$ at 0 of V consisting of all linearly compact open linear subspaces of V. We see now how the local bases $\mathcal{B}(W)$ and $\mathcal{B}(V/W)$ of a closed linear subspace W of V and the quotient V/W depend on $\mathcal{B}(V)$.

Proposition 3 *Let V be an l.l.c. vector space and W a closed linear subspace of V. Then:*

(a) $\mathcal{B}(W) = \{U \cap W \mid U \in \mathcal{B}(V)\}$;
(b) $\mathcal{B}(V/W) = \{(U + W)/W \mid U \in \mathcal{B}(V)\}$.

Proof (a) Clearly, $\{U \cap W \mid U \in \mathcal{B}(V)\} \subseteq \mathcal{B}(W)$. Conversely, let $U_W \in \mathcal{B}(W)$. Since U_W is open in W, there exists an open subset $A \subseteq V$ such that $U_W = A \cap W$. As A is a neighborhood of 0, there exists $U' \in \mathcal{B}(V)$ such that $U' \subseteq A$. In particular, $U' \cap W \subseteq U_W$ is an open linear subspace of the linearly compact space U_W, and so $U_W/(U' \cap W)$ has finite dimension by Proposition 1(d,e). Therefore, there exists a finite-dimensional subspace $F \leq U_W$ such that $U_W = F + (U' \cap W)$. Finally, let $U := F + U' \in \mathcal{B}(V)$. Hence, for $F \leq W$ we have $U_W = F + (U' \cap W) = (F + U') \cap W = U \cap W$.

(b) Since the canonical projection $\pi : V \to V/W$ is continuous and open, the set $\{\pi(U) \mid U \in \mathcal{B}(V)\}$ is contained in $\mathcal{B}(V/W)$.

To prove that $\mathcal{B}(V/W) \subseteq \{(U + W)/W \mid U \in \mathcal{B}(V)\}$, let $\overline{U} \in \mathcal{B}(V/W)$. Then $\pi^{-1}\overline{U}$ is an open linear subspace of V, hence it contains some $U \in \mathcal{B}(V)$. Then $\pi U \leq \overline{U}$ and πU has finite codimension in $\overline{U}$ by Proposition 1(d,e). Therefore, there exists a finite-dimensional linear subspace $\overline{F}$ of V/W such that $\overline{F} \leq \overline{U}$ and $\overline{U} = \pi U + \overline{F}$. Let F be a finite-dimensional linear subspace of V such that $F \leq \pi^{-1}\overline{U}$ and $\pi F = \overline{F}$. Now $\pi(U + F) = \overline{U}$ and $U + F \in \mathcal{B}(V)$ by Proposition 1(c). □

As consequence of Lefschetz Duality Theorem, every linearly compact vector space is topologically isomorphic to a direct product of one-dimensional vector spaces (see [17, Theorem 32.1]). From this result, we derive the known properties that if a linearly topologized vector space V over a finite discrete field is linearly compact then it is compact, and if V is l.l.c. then it is locally compact.

Proposition 4 *Let V be a linearly compact vector space over a discrete field $\mathbb{K}$. Then V is compact if and only if $\mathbb{K}$ is finite.*

Proof Write $V = \prod_{i \in I} \mathbb{K}_i$ with $\mathbb{K}_i = \mathbb{K}$ for all $i \in I$. If $\mathbb{K}$ is finite, then $\mathbb{K}_i$ is compact for all $i \in I$, and so V is compact. Conversely, if V is compact, then each $\mathbb{K}_i$ is compact as well, hence $\mathbb{K}$ is a compact discrete field, so $\mathbb{K}$ is finite. □

Corollary 1 *An l.l.c. vector space V over a finite discrete field $\mathbb{F}$ is a totally disconnected locally compact abelian group.*

Proof By Proposition 4, $\mathcal{B}(V)$ is a local basis at 0 of V consisting of compact open subgroups of V, so V is a totally disconnected locally compact abelian group. □

3 Existence of the Limit and Basic Properties

Let V be an l.l.c. vector space, $\phi : V \to V$ a continuous endomorphism and $U \in \mathcal{B}(V)$. For $n \in \mathbb{N}_+$ and a linear subspace F of V, the n-th *ϕ-trajectory* of F is

$$T_n(\phi, F) = F + \phi F + \phi^2 F + \ldots + \phi^{n-1} F.$$

If $U \in \mathcal{B}(V)$, notice that for every $n \in \mathbb{N}_+$, $T_n(\phi, U) \in \mathcal{B}(V)$ as well, since it is open being the union of cosets of U, and linearly compact by Proposition 1(c,f).

Moreover, $T_n(\phi, U) \leq T_{n+1}(\phi, U)$ for all $n \in \mathbb{N}_+$, thus we obtain an increasing chain of linearly compact open linear subspaces of V, namely

$$U = T_1(\phi, U) \leq T_2(\phi, U) \leq \ldots \leq T_n(\phi, U) \leq T_{n+1}(\phi, U) \leq \ldots .$$

The *ϕ-trajectory* of U is $T(\phi, U) = \bigcup_{n\in\mathbb{N}_+} T_n(\phi, U)$, which is the smallest ϕ-invariant open linear subspace of V containing U.

Hence, the algebraic entropy of ϕ with respect to U introduced in (1) can be written as

$$H(\phi, U) = \lim_{n\to\infty} \frac{1}{n} \dim \frac{T_n(\phi, U)}{U}. \tag{2}$$

Notice that since $T_n(\phi, U)$ is linearly compact and U is open, U has finite codimension in $T_n(\phi, U)$, that is, $\frac{T_n(\phi,U)}{U}$ has finite dimension by Proposition 1(d,e). Moreover, the following result shows that the limit in (2) exists.

Proposition 5 *Let V be an l.l.c. vector space and $\phi : V \to V$ a continuous endomorphism. For every $n \in \mathbb{N}_+$ let*

$$\alpha_n = \dim \frac{T_{n+1}(\phi, U)}{T_n(\phi, U)}.$$

Then the sequence of non-negative integers $\{\alpha_n\}_n$ is stationary and $H(\phi, U) = \alpha$ where α is the value of the stationary sequence $\{\alpha_n\}_n$ for n large enough.

Proof For every $n > 1$, $T_{n+1}(\phi, U) = T_n(\phi, U) + \phi^n U$ and $\phi T_{n-1}(\phi, U) \leq T_n(\phi, U)$. Thus,

$$\frac{T_{n+1}(\phi, U)}{T_n(\phi, U)} \cong \frac{\phi^n U}{T_n(\phi, U) \cap \phi^n U}$$

is a quotient of

$$B_n = \frac{\phi^n U}{\phi T_{n-1}(\phi, U) \cap \phi^n U}.$$

Therefore, $\alpha_n \leq \dim B_n$. Moreover, since $\phi T_n(\phi, U) = \phi T_{n-1}(\phi, U) + \phi^n U$,

$$B_n \cong \frac{\phi T_{n-1}(\phi, U) + \phi^n U}{\phi T_{n-1}(\phi, U)} = \frac{\phi T_n(\phi, U)}{\phi T_{n-1}(\phi, U)} \cong \frac{T_n(\phi, U)}{T_{n-1}(\phi, U) + (T_n(\phi, U) \cap \ker \phi)};$$

the latter vector space is a quotient of $T_n(\phi, U)/T_{n-1}(\phi, U)$, so $\dim B_n \leq \alpha_{n-1}$. Hence $\alpha_n \leq \alpha_{n-1}$. Thus $\{\alpha_n\}_n$ is a decreasing sequence of non-negative integers, therefore stationary. Since $U \leq T_n(\phi, U) \leq T_{n+1}(\phi, U)$,

$$\alpha_n = \dim \frac{T_{n+1}(\phi, U)}{U} - \dim \frac{T_n(\phi, U)}{U}. \tag{3}$$

As $\{\alpha_n\}_n$ is stationary, there exist $n_0 > 0$ and $\alpha \geq 0$ such that $\alpha_n = \alpha$ for every $n \geq n_0$. If $\alpha = 0$, equivalently $\dim \frac{T_{n+1}(\phi,U)}{U} = \dim \frac{T_n(\phi,U)}{U}$ for every $n \geq n_0$, and hence $H(\phi, U) = 0$. If $\alpha > 0$, by (3) we have that for every $n \in \mathbb{N}$

$$\dim \frac{T_{n_0+n}(\phi, U)}{U} = \dim \frac{T_{n_0}(\phi, U)}{U} + n\alpha.$$

Thus,

$$H(\phi, U) = \lim_{n\to\infty} \frac{1}{n+n_0} \dim \frac{T_{n_0+n}(\phi, U)}{U} = \lim_{n\to\infty} \frac{\dim \frac{T_{n_0}(\phi,U)}{U} + n\alpha}{n+n_0} = \alpha.$$

This concludes the proof. □

Proposition 5 yields that the value of $\mathrm{ent}(\phi)$ is either a non-negative integer or ∞. Moreover, Example 2 below witnesses that ent takes all values in $\mathbb{N} \cup \{\infty\}$.

We see now that the algebraic entropy ent coincides with $\mathrm{ent}_{\dim}$ on discrete vector spaces.

Lemma 1 *Let V be a discrete vector space and $\phi\colon V \to V$ an endomorphism. Then*

$$\mathrm{ent}(\phi) = \mathrm{ent}_{\dim}(\phi).$$

Proof Note that $\mathcal{B}(V) = \{F \leq V : \dim F < \infty\}$. Let now $F \in \mathcal{B}(V)$. Then

$$H(\phi, F) = \lim_{n\to\infty} \frac{1}{n} \dim \frac{T_n(\phi, F)}{F} = \lim_{n\to\infty} \frac{1}{n} (\dim T_n(\phi, F) - \dim F) =$$

$$= \lim_{n\to\infty} \frac{1}{n} \dim T_n(\phi, F) = H_{\dim}(\phi, F).$$

It follows from the definitions that $\mathrm{ent}(\phi) = \mathrm{ent}_{\dim}(\phi)$. □

We compute now the algebraic entropy in the easiest case of the identity automorphism.

Example 1

(a) Let $\phi\colon V \to V$ be a continuous endomorphism of an l.l.c. vector space V. Then $H(\phi, U) = 0$ for every $U \in \mathcal{B}(V)$ which is ϕ-invariant.
(b) Let $\phi = \mathrm{id}_V$. Since every element of $\mathcal{B}(V)$ is ϕ-invariant, (a) easily implies $\mathrm{ent}(\mathrm{id}_V) = 0$.

Inspired by the above example we provide now the general case of when the algebraic entropy is zero.

Proposition 6 *Let V be an l.l.c. vector space, $\phi : V \to V$ a continuous endomorphism and $U \in \mathcal{B}(V)$. Then the following conditions are equivalent:*

(a) $H(\phi, U) = 0$;
(b) there exists $n \in \mathbb{N}_+$ such that $T(\phi, U) = T_n(\phi, U)$;
(c) $T(\phi, U)$ is linearly compact.

In particular, $\mathrm{ent}(\phi) = 0$ *if and only if $T(\phi, U)$ is linearly compact for all $U \in \mathcal{B}(V)$.*

Proof (a)⇒(b) If $H(\phi, U) = 0$, then $\dim \frac{T_{n+1}(\phi,U)}{T_n(\phi,U)} = 0$ eventually by Proposition 5. Therefore, the chain of linearly compact open linear subspaces $\{T_n(\phi, U)\}_{n\in\mathbb{N}}$ is stationary.

(b)⇒(c) is clear from the definition.

(c)⇒(a) If $T(\phi, U)$ is linearly compact, by Proposition 1(d,e) we have that $\frac{T(\phi,U)}{U}$ is finite-dimensional. Since $T(\phi, U) = \bigcup_{n\in\mathbb{N}_+} T_n(\phi, U)$, it follows that

$$\frac{T(\phi, U)}{U} = \bigcup_{n\in\mathbb{N}_+} \frac{T_n(\phi, U)}{U}$$

and so the chain $\left\{\frac{T_n(\phi,U)}{U}\right\}_{n\in\mathbb{N}}$ is stationary. Therefore, $H(\phi, U) = 0$. □

As a consequence we see that ent always vanishes on linearly compact vector spaces.

Corollary 2 *If V is a linearly compact vector space and $\phi: V \to V$ a continuous endomorphism, then* $\mathrm{ent}(\phi) = 0$. *In particular, if V is a finite dimensional vector space, then* $\mathrm{ent}(\phi) = 0$.

The next result shows that when $\mathrm{ent}(\phi)$ is finite, this value is realized on some $U \in \mathcal{B}(V)$.

Lemma 2 *Let V be an l.l.c. vector space and $\phi : V \to V$ a continuous endomorphism. If* $\mathrm{ent}(\phi)$ *is finite, then there exists $U \in \mathcal{B}(V)$ such that* $\mathrm{ent}(\phi) = H(\phi, U)$.

Proof Since $\mathrm{ent}(\phi)$ is finite and $H(\phi, U) \in \mathbb{N}$ for every $U \in \mathcal{B}(V)$ by Proposition 5, the subset $\{H(\phi, U) : U \in \mathcal{B}(V)\}$ of $\mathbb{N}$ is bounded, hence finite. Therefore,

$$\mathrm{ent}(\phi) = \sup\{H(\phi, U) \mid U \in \mathcal{B}(V)\} = \max\{H(\phi, U) \mid U \in \mathcal{B}(V)\};$$

in other words, $\mathrm{ent}(\phi) = H(\phi, U)$ for some $U \in \mathcal{B}(V)$ as required. □

We prove now the monotonicity of $H(\phi, -)$ on the family $\mathcal{B}(V)$ ordered by inclusion.

Lemma 3 *Let V be an l.l.c. vector space and $\phi : V \to V$ a continuous endomorphism. If $U, U' \in \mathcal{B}(V)$ are such that $U' \leq U$, then $H(\phi, U') \leq H(\phi, U)$.*

Proof For $n \in \mathbb{N}_+$, since $T_n(\phi, U') + U$ is a linear subspace of $T_n(\phi, U)$, we have

$$\frac{T_n(\phi, U')/U'}{(T_n(\phi, U') \cap U)/U'} \cong \frac{T_n(\phi, U')}{T_n(\phi, U') \cap U} \cong \frac{T_n(\phi, U') + U}{U} \leq \frac{T_n(\phi, U)}{U}.$$

Thus,

$$\dim \frac{T_n(\phi, U')}{U'} \leq \dim \frac{T_n(\phi, U)}{U} + \dim \frac{T_n(\phi, U') \cap U}{U'}.$$

Finally, since $\dim \frac{T_n(\phi,U')\cap U}{U'} \leq \dim \frac{U}{U'}$, which is constant, for $n \to \infty$ we obtain the thesis. □

Let $(I, \leq)$ be a poset. A subset $J \subseteq I$ is said to be *cofinal* in I if for every $i \in I$ there exists $j \in J$ such that $i \leq j$. The following consequence of Lemma 3 permits to compute the algebraic entropy on a cofinal subset of $\mathscr{B}(V)$ ordered by inclusion.

Corollary 3 *Let V be an l.l.c. vector space and $\phi : V \to V$ a continuous endomorphism.*

(a) If $\mathscr{B}$ is a cofinal subset of $\mathscr{B}(V)$, then $\mathrm{ent}(\phi) = \sup\{H(\phi, U) \mid U \in \mathscr{B}\}$.
(b) If $U_0 \in \mathscr{B}(V)$ and $\mathscr{B} = \{U \in \mathscr{B}(V) : U_0 \leq U\}$, then $\mathrm{ent}(\phi) = \sup\{H(\phi, U) \mid U \in \mathscr{B}\}$.

Proof

(a) follows immediately from Lemma 3 and the definition.
(b) Since $U_0 + U \in \mathscr{B}$ for every $U \in \mathscr{B}(V)$, it follows that $\mathscr{B}$ is cofinal in $\mathscr{B}(V)$, so item (a) gives the thesis.

□

The following result simplifies the computation of the algebraic entropy in several cases.

Lemma 4 *Let V be an l.l.c. vector space, $\phi\colon V \to V$ a continuous endomorphism and $U \in \mathscr{B}(V)$. Then there exists a finite-dimensional linear subspace F of U such that, for every $n \in \mathbb{N}_+$,*

$$T_n(\phi, U) = U + T_n(\phi, F).$$

Proof We proceed by induction on $n \in \mathbb{N}_+$. For $n = 1$ it is obvious. Since U has finite codimension in $T_2(\phi, U) = U + \phi U$, there exists a finite-dimensional linear subspace F of V contained in U and such that $T_2(\phi, U) = U + \phi F = U + T_2(\phi, F)$. Assume now that $T_n(\phi, U) = U + T_n(\phi, F)$ for some $n \in \mathbb{N}_+$, $n \geq 2$. Then

$$\begin{aligned} T_{n+1}(\phi, U) = U + \phi T_n(\phi, U) = U + \phi(U) + \phi T_n(\phi, F) = \\ = U + \phi F + \phi T_n(\phi, F) = U + T_{n+1}(\phi, F). \end{aligned}$$

This concludes the proof. □

We end this section by discussing the relation of ent with h_{alg}. Recall that a topological abelian group G is *compactly covered* if each element of G is contained in some compact subgroup of G (equivalently, the Pontryagin dual of G is totally disconnected). If G is a compactly covered locally compact abelian group, $\phi\colon G \to G$ a continuous endomorphism and $U \in \mathscr{B}_{gr}(V) = \{U \leq G \mid$ compact open subgroup$\}$, then (see [8, Theorem 2.3])

$$h_{alg}(\phi) = \sup\{H_{alg}(\phi, U) \mid U \in \mathscr{B}_{gr}(V)\}$$

where

$$H_{alg}(\phi, U) = \lim_{n\to\infty} \frac{1}{n} \log \left| \frac{T_n(\phi, U)}{U} \right|.$$

If V is an l.l.c. vector space over a finite field $\mathbb{F}$, by Corollary 1 it is a totally disconnected locally compact abelian group. In particular V is compactly covered, since V is a torsion abelian group for $\mathbb{F}$ is finite.

Lemma 5 *Let V be an l.l.c. vector space over a finite field $\mathbb{F}$ and let $\phi : V \to V$ be a continuous endomorphism. Then*

$$h_{alg}(\phi) = \mathrm{ent}(\phi) \cdot \log |\mathbb{F}|.$$

Proof Let $\mathbb{F} = \{f_1, \ldots, f_{|\mathbb{F}|}\}$. Since every $U \in \mathscr{B}(V)$ is compact by Proposition 4, we have that $U \in \mathscr{B}_{gr}(V)$; hence, $\mathscr{B}(V) \subseteq \mathscr{B}_{gr}(V)$.

We show that $\mathscr{B}(V)$ is cofinal in $B_{gr}(V)$. Let $U \in \mathscr{B}_{gr}(V)$ and $U' = \sum_{i=1}^{|\mathbb{F}|} f_i U$. Since V is a topological vector space, $f_i U$ is compact for all $i = 1, \ldots, |\mathbb{F}|$, so U' is compact as well. Clearly, U' is contained in the linear subspace $\langle U \rangle$ of V generated by U. We see that actually $U' = \langle U \rangle$. Indeed, let

$$x = f_{i_1} u_1 + \ldots + f_{i_k} u_k, \quad u_1, \ldots, u_k \in U, \quad f_{i_1}, \ldots, f_{i_k} \in \mathbb{F},$$

be an arbitrary element in $\langle U \rangle$. Rearranging the summands, that is, letting for every $j \in \{1, \ldots, |\mathbb{F}|\}$, $u^j_{l_1\ldots l_j} = u_{l_1} + \ldots + u_{l_j} \in U$ for $l_1, \ldots, l_j \in \{1, \ldots, k\}$ such that $f_{i_{l_1}} = \ldots = f_{i_{l_j}} = f_j$, we obtain that

$$x = \sum_{j=1}^{|\mathbb{F}|} f_j u^j_{l_1\ldots l_j} \in U'.$$

Hence $U' = \langle U \rangle$. Therefore, $U' \in \mathscr{B}(V)$ and U' contains U. This proves that $\mathscr{B}(V)$ is cofinal in $\mathscr{B}_{gr}(V)$ as claimed.

Thus, by [25, Corollary 2.3], $h_{alg}(\phi) = \sup\{H_{alg}(\phi, U) \mid U \in \mathcal{B}(V)\}$. Since for every $U \in \mathcal{B}(V)$,

$$\left|\frac{T_n(\phi, U)}{U}\right| = |\mathbb{F}|^{\dim \frac{T_n(\phi,U)}{U}}$$

for all $n \in \mathbb{N}_+$, we obtain

$$H_{alg}(\phi, U) = \lim_{n\to\infty} \frac{1}{n} \log \left|\frac{T_n(\phi, U)}{U}\right| = \lim_{n\to\infty} \frac{1}{n} \dim \frac{T_n(\phi, U)}{U} \log |\mathbb{F}| = H(\phi, U) \log |\mathbb{F}|,$$

and so the thesis follows. □

4 General Properties and Examples

In this section we prove the general basic properties of the algebraic entropy. These properties extend their counterparts for discrete vector spaces proved for $\mathrm{ent}_{\dim}$ in [12]. Moreover, our proofs follow those of the same properties for the intrinsic algebraic entropy given in [11].

We start by proving the invariance of ent under conjugation by a topological isomorphism.

Proposition 7 (Invariance Under Conjugation) *Let V be an l.l.c. vector space and $\phi : V \to V$ a continuous endomorphism. If $\alpha: V \to W$ is a topological isomorphism of l.l.c. vector spaces, then* $\mathrm{ent}(\phi) = \mathrm{ent}(\alpha\phi\alpha^{-1})$.

Proof Let $U \in \mathcal{B}(W)$; then $\alpha^{-1}U \in \mathcal{B}(V)$. For $n \in \mathbb{N}_+$ we have $\alpha T_n(\phi, \alpha^{-1}U) = T_n(\alpha\phi\alpha^{-1}, U)$. As α induces an isomorphism $\frac{V}{\alpha^{-1}U} \to \frac{W}{U}$, and furthermore through this isomorphism $\frac{T_n(\phi,\alpha^{-1}U)}{\alpha^{-1}U}$ is isomorphic to $\frac{T_n(\alpha\phi\alpha^{-1},U)}{U}$, by applying the definition we have $H(\phi, \alpha^{-1}U) = H(\alpha\phi\alpha^{-1}, U)$. Now the thesis follows, since α induces a bijection between $\mathcal{B}(V)$ and $\mathcal{B}(W)$. □

The next lemma is useful to prove the monotonicity of the algebraic entropy in Proposition 8.

Lemma 6 *Let V be an l.l.c. vector space, $\phi : V \to V$ a continuous endomorphism and W a closed ϕ-invariant linear subspace of V. Then:*

$$\mathrm{ent}(\phi\restriction_W) = \sup\{H(\phi\restriction_W, U \cap W) \mid U \in \mathcal{B}(V)\} \textit{ and}$$
$$\mathrm{ent}(\overline{\phi}) = \sup\{H(\overline{\phi}, (U + W)/W) \mid U \in \mathcal{B}(V)\},$$

where $\overline{\phi} : V/W \to V/W$ is the continuous endomorphism induced by ϕ.

Proof Apply Proposition 3. □

Next we see that the algebraic entropy is monotone under taking invariant linear subspaces and quotient vector spaces.

Proposition 8 (Monotonicity) *Let V be an l.l.c. vector space, $\phi : V \to V$ a continuous endomorphism, W a ϕ-invariant closed linear subspace of V, and $\overline{\phi}$ is the continuous endomorphism of V/W induced by ϕ. Then:*

(a) $\mathrm{ent}(\phi) \geq \mathrm{ent}(\phi \restriction_W)$;
(b) $\mathrm{ent}(\phi) \geq \mathrm{ent}(\overline{\phi})$.

Proof (a) Let $U \in \mathcal{B}(V)$ and $n \in \mathbb{N}_+$. Since

$$\frac{T_n(\phi, U)}{U} \geq \frac{U + T_n(\phi \restriction_W, U \cap W)}{U} \cong \frac{T_n(\phi \restriction_W, U \cap W)}{T_n(\phi \restriction_W, U \cap W) \cap U},$$

and $T_n(\phi \restriction_W, U \cap W) \cap U = U \cap W$, it follows that

$$\dim \frac{T_n(\phi \restriction_W, U \cap W)}{U \cap W} \leq \dim \frac{T_n(\phi, U)}{U}.$$

Hence, $H(\phi \restriction_W, U \cap W) \leq H(\phi, U) \leq \mathrm{ent}(\phi)$. Finally, Lemma 6 yields the thesis.

(b) For $U \in \mathcal{B}(V)$ and $n \in \mathbb{N}_+$, we have that

$$\frac{T_n(\overline{\phi}, \frac{U+W}{W})}{\frac{U+W}{W}} \cong \frac{T_n(\phi, U + W)}{U + W} = \frac{T_n(\phi, U) + W}{U + W} \cong \frac{T_n(\phi, U)}{T_n(\phi, U) \cap (U + W)}, \tag{4}$$

where the latter vector space is clearly a quotient of $\frac{T_n(\phi,U)}{U}$. Therefore,

$$H\left(\overline{\phi}, \frac{U + W}{W}\right) \leq H(\phi, U) \leq \mathrm{ent}(\phi).$$

Now Lemma 6 concludes the proof. □

Note that equality holds in item (b) of the above proposition if W is also linearly compact. In fact, in this case for every $U \in \mathcal{B}(V)$ we have $U + W \in \mathcal{B}(V)$ by Proposition 1(c), and hence Lemma 3 and the first isomorphism in (4) yield $H(\phi, U) \leq H(\phi, U + W) = H\left(\overline{\phi}, \frac{U+W}{W}\right)$; therefore, $\mathrm{ent}(\phi) \leq \mathrm{ent}(\overline{\phi})$ and so $\mathrm{ent}(\phi) = \mathrm{ent}(\overline{\phi})$ by Lemma 8(b).

Proposition 9 (Logarithmic Law) *Let V be an l.l.c. vector space and $\phi : V \to V$ a continuous endomorphism. Then* $\mathrm{ent}(\phi^k) = k \cdot \mathrm{ent}(\phi)$ *for every* $k \in \mathbb{N}$.

Proof For $k = 0$, it is enough to note that $\mathrm{ent}(\mathrm{id}_V) = 0$ by Example 1. So let $k \in \mathbb{N}_+$ and $U \in \mathcal{B}(V)$. For every $n \in \mathbb{N}_+$,

$$T_{nk}(\phi, U) = T_n(\phi^k, T_k(\phi, U)) \quad \text{and} \quad T_n(\phi, T_k(\phi, U)) = T_{n+k-1}(\phi, U).$$

Let $E = T_k(\phi, U) \in \mathscr{B}(V)$. By Lemma 3,

$$k \cdot H(\phi, U) \leq k \cdot H(\phi, E) = k \cdot \lim_{n\to\infty} \frac{1}{nk} \dim \frac{T_{nk}(\phi, E)}{E} = \lim_{n\to\infty} \frac{1}{n} \dim \frac{T_{(n+1)k-1}(\phi, U)}{E}$$
$$\leq \lim_{n\to\infty} \frac{1}{n} \dim \frac{T_{(n+1)k}(\phi, U)}{E} = \lim_{n\to\infty} \frac{1}{n} \dim \frac{T_{n+1}(\phi^k, E)}{E} = H(\phi^k, E);$$

consequently, $k \cdot \mathrm{ent}(\phi) \leq \mathrm{ent}(\phi^k)$.

Conversely, as $U \leq E \leq T_{nk}(\phi, U)$,

$$\mathrm{ent}(\phi) \geq H(\phi, U) = \lim_{n\to\infty} \frac{1}{nk} \dim \frac{T_{nk}(\phi, U)}{U} = \lim_{n\to\infty} \frac{1}{nk} \dim \frac{T_n(\phi^k, E)}{U}$$
$$\geq \lim_{n\to\infty} \frac{1}{nk} \dim \frac{T_n(\phi^k, E))}{E} = \frac{1}{k} \cdot H(\phi^k, E).$$

By Lemma 3, it follows that $H(\phi^k, E) \geq H(\phi^k, U)$, and so $k \cdot \mathrm{ent}(\phi) \geq \mathrm{ent}(\phi^k)$. □

The next property shows that the algebraic entropy behaves well with respect to direct limits.

Proposition 10 (Continuity on Direct Limits) *Let V be an l.l.c. vector space and $\phi\colon V \to V$ a continuous endomorphism. Assume that V is the direct limit of a family $\{V_i \mid i \in I\}$ of closed ϕ-invariant linear subspaces of V, and let $\phi_i = \phi \restriction_{V_i}$ for all $i \in I$. Then* $\mathrm{ent}(\phi) = \sup_{i\in I} \mathrm{ent}(\phi_i)$.

Proof By Proposition 8(a), $\mathrm{ent}(\phi) \geq \mathrm{ent}(\phi_i)$ for every $i \in I$. So $\mathrm{ent}(\phi) \geq \sup_{i\in I} \mathrm{ent}(\phi_i)$.

Conversely, let $U \in \mathscr{B}(V)$. By Lemma 4, there exists a finite dimensional subspace F of U such that for all $n \in \mathbb{N}_+$

$$T_n(\phi, U) = U + T_n(\phi, F). \tag{5}$$

As F is finite dimensional, $F \leq V_i$ for some $i \in I$. In particular,

$$T_n(\phi_i, U \cap V_i) = (U \cap V_i) + T_n(\phi, F). \tag{6}$$

Indeed, since $F \leq U \cap V_i$, the inclusion $(U \cap V_i) + T_n(\phi, F) \leq T_n(\phi_i, U \cap V_i)$ follows easily. On the other hand, since $T_n(\phi, F) \leq V_i$,

$$T_n(\phi_i, U \cap V_i) \leq T_n(\phi, U) \cap V_i = (U + T_n(\phi, F)) \cap V_i = (U \cap V_i) + T_n(\phi, F).$$

Therefore, (6) yields

$$\frac{T_n(\phi_i, U \cap V_i)}{U \cap V_i} \cong \frac{(U \cap V_i) + T_n(\phi, F)}{U \cap V_i} \cong \frac{T_n(\phi, F)}{U \cap T_n(\phi, F)}.$$

At the same time, (5) implies

$$\frac{T_n(\phi, U)}{U} \cong \frac{U + T_n(\phi, F)}{U} \cong \frac{T_n(\phi, F)}{U \cap T_n(\phi, F)}.$$

Hence, $H(\phi, U) = H(\phi_i, U \cap V_i) \leq \sup_{i \in I} \mathrm{ent}(\phi_i)$, and so $\mathrm{ent}(\phi) \leq \sup_{i \in I} \mathrm{ent}(\phi_i)$. □

We end this list of properties of the algebraic entropy with the following simple case of the Addition Theorem.

Proposition 11 (Weak Addition Theorem) *For $i = 1, 2$, let V_i be an l.l.c. vector space and $\phi_i : V_1 \to V_1$ a continuous endomorphism. Let $\phi = \phi_1 \times \phi_2 : V \to V$, where $V = V_1 \times V_2$. Then* $\mathrm{ent}(\phi) = \mathrm{ent}(\phi_1) + \mathrm{ent}(\phi_2)$.

Proof Notice that $\mathscr{B} = \{U_1 \times U_2 \mid U_i \in \mathscr{B}(V_i), i = 1, 2\}$ is cofinal in $\mathscr{B}(V)$. Indeed, let $U \in \mathscr{B}(V)$; for $i = 1, 2$, since the canonical projection $\pi_i \colon V \to V_i$ is an open continuous map, $U_i = \pi_i U \in \mathscr{B}(V_i)$, and $U \leq U_1 \times U_2$.

Now, for $U_1 \times U_2 \in \mathscr{B}$ and for every $n \in \mathbb{N}_+$,

$$\frac{T_n(\phi, U_1 \times U_2)}{U_1 \times U_2} \cong \frac{T_n(\phi_1, U_1)}{U_1} \times \frac{T_n(\phi_2, U_2)}{U_2};$$

hence,

$$H(\phi, U_1 \times U_2) = H(\phi_1, U_1) + H(\phi_2, U_2). \tag{7}$$

By Corollary 3(a) we conclude that $\mathrm{ent}(\phi) \leq \mathrm{ent}(\phi_1) + \mathrm{ent}(\phi_2)$.

If $\mathrm{ent}(\phi) = \infty$, the thesis holds true. So assume that $\mathrm{ent}(\phi)$ is finite; then $\mathrm{ent}(\phi_1)$ and $\mathrm{ent}(\phi_2)$ are finite as well by Proposition 8(a). Hence, for $i = 1, 2$ by Lemma 2 there exists $U_i \in \mathscr{B}(V_i)$ such that $\mathrm{ent}(\phi_i) = H(\phi_i, U_i)$. By (7) we obtain

$$\mathrm{ent}(\phi_1) + \mathrm{ent}(\phi_2) = H(\phi_1, U_1) + H(\phi_2, U_2) = H(\phi, U_1 \times U_2) \leq \mathrm{ent}(\phi),$$

where the latter inequality holds because $U_1 \times U_2 \in \mathscr{B}(V)$. Therefore, $\mathrm{ent}(\phi_1) + \mathrm{ent}(\phi_2) \leq \mathrm{ent}(\phi)$ and this concludes the proof. □

In the case of a discrete vector space V and an automorphism $\phi : V \to V$, we have that $\mathrm{ent}_{\dim}(\phi^{-1}) = \mathrm{ent}_{\dim}(\phi)$ (see [12]). This property does not extend to the general case of an l.l.c. vector space V and a topological automorphism $\phi : V \to V$; in fact, the next example shows that $\mathrm{ent}(\phi)$ could not coincide with $\mathrm{ent}(\phi^{-1})$.

Let F be a finite dimensional vector space and let $V = V_c \oplus V_d$, with

$$V_c = \prod_{n=-\infty}^{0} F \quad \text{and} \quad V_d = \bigoplus_{n=1}^{\infty} F,$$

be endowed with the topology inherited from the product topology of $\prod_{n\in\mathbb{Z}} F$, so V_c is linearly compact and V_d is discrete.

The *left (two-sided) Bernoulli shift* is

$$_F\beta\colon V \to V, \quad (x_n)_{n\in\mathbb{Z}} \mapsto (x_{n+1})_{n\in\mathbb{Z}},$$

while the *right (two-sided) Bernoulli shift* is

$$\beta_F\colon V \to V, \quad (x_n)_{n\in\mathbb{Z}} \mapsto (x_{n-1})_{n\in\mathbb{Z}}.$$

Clearly, β_F and $_F\beta$ are topological automorphisms such that $_F\beta^{-1} = \beta_F$.

Let us compute their algebraic entropies.

Example 2

(a) Consider the case $F = \mathbb{K}$, i.e., $V_c = \prod_{n=-\infty}^{0} \mathbb{K}$ and $V_d = \bigoplus_{n=1}^{\infty} \mathbb{K}$, and let $\phi \in \{_\mathbb{K}\beta, \beta_\mathbb{K}\}$. By Corollary 3(b),

$$\mathrm{ent}(\phi) = \sup\{H(\phi, U) \mid U \in \mathscr{B}(V),\ V_c \le U\}.$$

Let $U \in \mathscr{B}(V)$ such that $V_c \le U$. Since V_c has finite codimension in U by Proposition 1(d,e), there exists $k \in \mathbb{N}_+$ such that

$$U \le U' := \prod_{n=-\infty}^{0} \mathbb{K} \times \bigoplus_{n=1}^{k} \mathbb{K} \in \mathscr{B}(V),$$

hence $H(\phi, U) \le H(\phi, U')$ by Lemma 3. Clearly,

$$\ldots \le {_\mathbb{K}\beta^n}(U') \le \ldots \le {_\mathbb{K}\beta}(U') \le U' \le \beta_\mathbb{K}(U') \le \ldots \le \beta_\mathbb{K}^n(U') \le \ldots.$$

So, for all $n \in \mathbb{N}_+$, $T_n({_\mathbb{K}\beta}, U') = U'$, while

$$\dim \frac{T_{n+1}(\beta_\mathbb{K}, U')}{T_n(\beta_\mathbb{K}, U')} = \dim \frac{\beta_\mathbb{K}^{n+1}(U')}{\beta_\mathbb{K}^n U'} = \dim \frac{\beta_\mathbb{K}(U')}{U'} = 1.$$

By Corollary 3(a), we can conclude that

$$\mathrm{ent}({_\mathbb{K}\beta}) = 0 \quad \text{and} \quad \mathrm{ent}(\beta_\mathbb{K}) = 1.$$

In particular, $\mathrm{ent}(\phi) \ne \mathrm{ent}(\phi^{-1})$ for $\phi \in \{_\mathbb{K}\beta, \beta_\mathbb{K}\}$.

(b) It is possible, slightly modifying the computations in item (a), to find that, for F a finite dimensional vector space,

$$\mathrm{ent}({_F\beta}) = 0 \quad \text{and} \quad \mathrm{ent}(\beta_F) = \dim F.$$

5 Limit-Free Formula

The aim of this subsection is to prove Proposition 12 that provides a formula for the computation of the algebraic entropy avoiding the limit in the definition. This formula is a fundamental ingredient in the proof of the Addition Theorem presented in the last section.

Definition 1 Let V be an l.l.c. vector space, $\phi : V \to V$ a continuous endomorphism and $U \in \mathcal{B}(V)$. Let:

- $U^{(0)} = U$;
- $U^{(n+1)} = U + \phi^{-1}U^{(n)}$ for every $n \in \mathbb{N}$;
- $U^{-} = \bigcup_{n\in\mathbb{N}} U^{(n)}$.

It can be proved by induction that $U^{(n)} \leq U^{(n+1)}$ for every $n \in \mathbb{N}$. Since U is open, clearly every $U^{(n)}$ is open as well, so also U^{-} and $\phi^{-1}U^{-}$ are open linear subspaces of V.

We see now that U^{-} is the smallest linear subspace of V containing U and inversely ϕ-invariant (i.e., $\phi^{-1}U^{-} \leq U^{-}$). Note that U^{-} coincides with $T(\phi^{-1}, U)$ when ϕ is an automorphism, otherwise it could be strictly smaller.

Lemma 7 *Let V be an l.l.c. vector space, $\phi : V \to V$ a continuous endomorphism and $U \in \mathcal{B}(V)$. Then:*

(a) $\phi^{-1}U^{-} \leq U^{-}$;
(b) if W is a linear subspace of V such that $U \leq W$ and $\phi^{-1}W \leq W$, then $U^{-} \leq W$.

Proof (a) follows from the fact that $\phi^{-1}U^{(n)} \leq U^{(n+1)}$ for every $n \in \mathbb{N}$.

(b) By the hypothesis, one can prove by induction that $U^{(n)} \leq W$ for every $n \in \mathbb{N}$; hence, $U^{-} \leq W$. □

In the next lemma we collect some other properties that we use in the sequel.

Lemma 8 *Let V be an l.l.c. vector space, $\phi : V \to V$ a continuous endomorphism and $U \in \mathcal{B}(V)$. Then:*

(a) $U^{-} = U + \phi^{-1}U^{-}$;
(b) $\frac{U^{-}}{\phi^{-1}U^{-}}$ *has finite dimension.*

Proof (a) follows from the equalities

$$U+\phi^{-1}U^{-} = U+\phi^{-1}\bigcup_{n\in\mathbb{N}} U^{(n)} = U+\bigcup_{n\in\mathbb{N}}\phi^{-1}U^{(n)} = \bigcup_{n\in\mathbb{N}}(U+\phi^{-1}U^{(n)}) = U^{-}.$$

(b) By Proposition 1(d,e), the quotient $\frac{U}{U\cap\phi^{-1}U^{-}}$ has finite dimension, since the linear subspace $U\cap\phi^{-1}U^{-}$ is open in the linearly compact space U. In view of item (a) we have the isomorphism

$$\frac{U^{-}}{\phi^{-1}U^{-}} = \frac{U+\phi^{-1}U^{-}}{\phi^{-1}U^{-}} \cong \frac{U}{U\cap\phi^{-1}U^{-}},$$

so we conclude that also $\frac{U^-}{\phi^{-1}U^-}$ has finite dimension. □

The next lemma is used in the proof of Proposition 12.

Lemma 9 *Let V be an l.l.c. vector space, $\phi : V \to V$ a continuous endomorphism and $U \in \mathcal{B}(V)$. Then, for every $n \in \mathbb{N}_+$,*

$$\phi^{-n}T_n(\phi, U) = \phi^{-1}U^{(n-1)}.$$

Proof We proceed by induction on $n \in \mathbb{N}_+$. We write simply $T_n = T_n(\phi, U)$.

If $n = 1$ we have $\phi^{-1}T_1 = \phi^{-1}U = \phi^{-1}U^{(0)}$. Assume now that the property holds for $n \in \mathbb{N}_+$, we prove it for $n + 1$, that is, we verify that

$$\phi^{-(n+1)}T_{n+1} = \phi^{-1}U^{(n)}. \tag{8}$$

Let $x \in \phi^{-1}U^{(n)}$. Then, by inductive hypothesis,

$$\phi(x) \in U^{(n)} = U + \phi^{-1}U^{(n-1)} = U + \phi^{-n}T_n.$$

Consequently,

$$\phi^{n+1}(x) = \phi^n(\phi(x)) \in \phi^n U + T_n = T_{n+1};$$

this shows that $x \in \phi^{-(n+1)}T_{n+1}$. Therefore, $\phi^{-1}U^{(n)} \leq \phi^{-(n+1)}T_{n+1}$.

Conversely, let $x \in \phi^{-(n+1)}T_{n+1}$. Then

$$\phi^{n+1}(x) \in T_{n+1} = T_n + \phi^n U,$$

and so $\phi^{n+1}(x) = y + \phi^n(u)$, for some $y \in T_n$ and $u \in U$. Therefore, $\phi^n(\phi(x) - u) = y \in T_n$, that is,

$$\phi(x) - u \in \phi^{-n}T_n = \phi^{-1}U^{(n-1)}$$

by inductive hypothesis. Hence, $\phi(x) \in U + \phi^{-1}U^{(n-1)} = U^{(n)}$, and we can conclude that $x \in \phi^{-1}U^{(n)}$. Thus, (8) is verified. So, the induction principle gives the thesis. □

We are now in position to prove the Limit-free Formula, where clearly we use that dim $\frac{U^-}{\phi^{-1}U^-}$ has finite dimension by Lemma 8(b).

Proposition 12 (Limit-Free Formula) *Let V be an l.l.c. vector space, $\phi : V \to V$ a continuous endomorphism and $U \in \mathcal{B}(V)$. Then*

$$H(\phi, U) = \dim \frac{U^-}{\phi^{-1}U^-}.$$

Proof We write simply $T_n = T_n(\phi, U)$ for every $n \in \mathbb{N}_+$. By Proposition 5, there exist $n_0 \in \mathbb{N}_+$ and $\alpha \in \mathbb{N}$, such that for every $n \geq n_0$, $H(\phi, U) = \alpha$, where $\alpha = \dim \frac{T_{n+1}}{T_n}$. So, our aim is to prove that $\alpha = \dim \frac{U^-}{\phi^{-1}U^-}$.

For every $n \in \mathbb{N}$, the quotient $U \cap \phi^{-1}U^{(n)} \leq U$ is open in the linearly compact space U, thus $\frac{U}{U \cap \phi^{-1}U^{(n)}}$ has finite dimension, by Proposition 1(d,e); moreover, since $U \cap \phi^{-1}U^{(n)} \leq U \cap \phi^{-1}U^{(n+1)}$, the space $\frac{U}{U\cap\phi^{-1}U^{(n+1)}}$ is a quotient of $\frac{U}{U\cap\phi^{-1}U^n}$. The decreasing sequence of finite-dimensional vector spaces $\left\{\frac{U}{U\cap\phi^{-1}U^n} \mid n \in \mathbb{N}\right\}$ must stabilize; this means that there exists $n_1 \in \mathbb{N}$ such that $U \cap \phi^{-1}U^{(n)} = U \cap \phi^{-1}U^{(n_1)}$ for every $n \geq n_1$. Hence, for every $m \geq n_1$,

$$U \cap \phi^{-1}U^{(m)} = \bigcup_{n\in\mathbb{N}} (U \cap \phi^{-1}U^{(n)}) = U \cap \bigcup_{n\in\mathbb{N}} \phi^{-1}U^{(n)} =$$
$$= U \cap \phi^{-1} \bigcup_{n\in\mathbb{N}} U^{(n)} = U \cap \phi^{-1}U^-.$$

Fix now $m \geq \max\{n_0, n_1\}$; since $\frac{U^-}{\phi^{-1}U^-} = \frac{U+\phi^{-1}U}{\phi^{-1}U} \cong \frac{U}{U\cap\phi^{-1}U^-}$ by Lemma 8(a), we have

$$\dim \frac{U^-}{\phi^{-1}U^-} = \dim \frac{U}{U \cap \phi^{-1}U^-} = \dim \frac{U}{U \cap \phi^{-1}U^{(m)}}$$
$$= \dim \frac{U + \phi^{-1}U^{(m)}}{\phi^{-1}U^{(m)}} = \dim \frac{U^{(m+1)}}{\phi^{-1}U^{(m)}}.$$

We see now that

$$\dim \frac{U^{(m)}}{\phi^{-1}U^{(m-1)}} = \dim \frac{T_{m+1}}{T_m} = \alpha$$

and this concludes the proof. To this end, noting that

$$\frac{U^{(m)}}{\phi^{-1}U^{(m-1)}} = \frac{U + \phi^{-1}U^{(m-1)}}{\phi^{-1}U^{(m-1)}} \quad \text{and} \quad \frac{T_{m+1}}{T_m} = \frac{\phi^{m+1}U + T_m}{T_m},$$

define

$$\Phi : \frac{U + \phi^{-1}U^{(m-1)}}{\phi^{-1}U^{(m-1)}} \longrightarrow \frac{\phi^{m+1}U + T_m}{T_m}$$
$$x + \phi^{-1}U^{(m-1)} \mapsto \phi^m(x) + T_m.$$

Then Φ is a surjective homomorphism by construction and it is well defined and injective since $\phi^{-m}T_m = \phi^{-1}U^{(m-1)}$ by Lemma 9. □

6 Addition Theorem

This section is devoted to the proof of the Addition Theorem for the algebraic entropy ent for l.l.c. vector spaces (see Theorem 2).

Let V be an l.l.c. vector space and $\phi : V \to V$ a continuous endomorphism. Theorem 1 allows us to decompose V into the direct sum of a linearly compact open linear subspace V_c and a discrete linear subspace V_d of V, namely $V \cong V_c \oplus V_d$ topologically. So, assume that $V = V_c \oplus V_d$ and let

$$\iota_*: V_* \to V, \quad p_*: V \to V_*, \quad * \in \{c, d\}, \tag{9}$$

be respectively the canonical embeddings and projections. Accordingly, we may associate to ϕ the following decomposition

$$\phi = \begin{pmatrix} \phi_{cc} & \phi_{dc} \\ \phi_{cd} & \phi_{dd} \end{pmatrix}, \tag{10}$$

where $\phi_{\bullet *} : V_\bullet \to V_*$ is the composition $\phi_{\bullet *} = p_* \circ \phi \circ \iota_\bullet$ for $\bullet, * \in \{c, d\}$. Therefore, each $\phi_{\bullet *}$ is continuous as it is composition of continuous homomorphisms.

Lemma 10 *In the above notations, consider $\phi_{cd}: V_c \to V_d$. Then:*

(a) $\mathrm{Im}(\phi_{cd}) \in \mathscr{B}(V_d)$;
(b) $\ker(\phi_{cd}) \in \mathscr{B}(V_c) \subseteq \mathscr{B}(V)$.

Proof (a) Since V_d is discrete, by Proposition 1(c,d) we have that $\mathrm{Im}(\phi_{cd}) \leq V_d$ has finite dimension, hence $\mathrm{Im}(\phi_{cd}) \in \mathscr{B}(V_d) = \{F \leq V_d \mid \dim F < \infty\}$.

(b) As $\ker(\phi_{cd})$ is a closed linear subspace of V_c, which is linearly compact, then $\ker(\phi_{cd})$ is linearly compact as well by Proposition 1(b). Since $V_c/\ker(\phi_{cd}) \cong \mathrm{Im}(\phi_{cd})$ is finite dimensional by item (a), $V_c/\ker(\phi_{cd})$ is discrete and so $\ker(\phi_{cd})$ is open in V_c; therefore, $\ker(\phi_{cd}) \in \mathscr{B}(V_c)$. □

We show now that the only positive contribution to the algebraic entropy of ϕ comes from its "discrete component" ϕ_{dd}.

Proposition 13 *In the above notations,* $\mathrm{ent}(\phi) = \mathrm{ent}(\phi_{dd})$.

Proof By Lemma 10(a), $\mathrm{Im}(\phi_{cd}) \in \mathscr{B}(V_d)$; hence, letting

$$\mathscr{B}_d = \{F \leq V_d \mid \mathrm{Im}(\phi_{cd}) \leq F, \dim F < \infty\} \subseteq \mathscr{B}(V_d),$$

Corollary 3(b) implies

$$\mathrm{ent}(\phi_{dd}) = \sup\{H(\phi_{dd}, F) \mid F \in \mathscr{B}_d\}. \tag{11}$$

Let $\mathscr{B} = \{U \in \mathscr{B}(V) \mid V_c \leq U\}$, which is cofinal in $\mathscr{B}(V)$. For $U \in \mathscr{E}$, since V_c has finite codimension in U by Proposition 1(d,e), there exists a finite dimensional linear subspace $F \leq V_d$ such that $U = V_c \oplus F$. Conversely, $V_c \oplus F \in \mathscr{B}$ for every finite dimensional linear subspace $F \leq V_d$. Hence, $\mathcal{B} = \{V_c \oplus F \mid F \in \mathcal{B}(V_d)\}$. Moreover, $\mathcal{B}' = \{V_c \oplus F \mid F \in \mathcal{B}_d\}$ is cofinal in $\mathcal{B}$ and so in $\mathcal{B}(V)$. Thus, Corollary 3(b) yields

$$\mathrm{ent}(\phi) = \sup\{H(\phi, U) \mid U \in \mathcal{B}'\}. \tag{12}$$

For $U = V_c \oplus F \in \mathcal{B}'$, as in Definition 1 let, for every $n \in \mathbb{N}$,

$$\begin{aligned} U^{(0)} &= U \quad \text{and} \quad F^{(0)} = F, \\ U^{(n)} &= U + \phi^{-1}U^{(n-1)} \quad \text{and} \quad F^{(n)} = F + \phi_{dd}^{-1}F^{(n-1)}, \\ U^{-} &= \bigcup_{n\in\mathbb{N}} U^{(n)} \quad \text{and} \quad F^{-} = \bigcup_{n\in\mathbb{N}} F^{(n)}. \end{aligned}$$

Proposition 12, together with (12) and (11) respectively, implies that

$$\mathrm{ent}(\phi) = \sup\left\{\dim \frac{U^-}{\phi^{-1}U^-} \mid U \in \mathcal{B}'\right\}, \tag{13}$$

$$\mathrm{ent}(\phi_{dd}) = \sup\left\{\dim \frac{F^-}{\phi_{dd}^{-1}F^-} \mid F \in \mathcal{B}_d\right\}. \tag{14}$$

Let $U = V_c \oplus F \in \mathcal{B}'$. We show by induction on $n \in \mathbb{N}$ that

$$U^{(n)} = V_c \oplus F^{(n)} \quad \text{for every } n \in \mathbb{N}. \tag{15}$$

For $n = 0$, we have $U^{(0)} = U = V_c \oplus F = V_c \oplus F^{(0)}$. Assume now that $n \in \mathbb{N}$ and that $U^{(n)} = V_c \oplus F^{(n)}$. First note that $U^{(n+1)} = U + \phi^{-1}U^{(n)} = U + \phi^{-1}(V_c \oplus F^{(n)})$. Moreover, since $\mathrm{Im}(\phi_{cd}) \leq F \leq F^{(n)}$,

$$\begin{aligned} \phi^{-1}(V_c \oplus F^{(n)}) &= \{(x, y) \in V_c \oplus V_d \mid \phi_{cd}(x) + \phi_{dd}(y) \in F^{(n)}\} \\ &= \{(x, y) \in V_c \oplus V_d \mid \phi_{dd}(y) \in F^{(n)}\} \\ &= V_c \oplus \phi_{dd}^{-1}F^{(n)}. \end{aligned}$$

Thus, $U^{(n+1)} = V_c \oplus F^{(n+1)}$ as required in (15).

Now (15) implies that $U^- = V_c \oplus F^-$; moreover, since $\mathrm{Im}(\phi_{cd}) \leq F \leq F^-$,

$$\phi^{-1}U^- = \{(x, y) \in V_c \oplus V_d \mid \phi_{dd}(y) \in F^-\} = V_c \oplus \phi_{dd}^{-1}F^-.$$

Therefore, $\frac{U^-}{\phi^{-1}U^-} = \frac{V_c \oplus F^-}{V_c \oplus \phi_{dd}^{-1}F^-} = \frac{F^-}{\phi_{dd}^{-1}F^-}$, so the thesis follows from (13) and (14).

□

We are now in position to prove the Addition Theorem.

Theorem 2 (Addition Theorem) *Let V be an l.l.c. vector space, $\phi\colon V \to V$ a continuous endomorphism and W a closed ϕ-invariant linear subspace of V. Then*

$$\mathrm{ent}(\phi) = \mathrm{ent}(\phi \restriction_W) + \mathrm{ent}(\overline{\phi}),$$

where $\overline{\phi} : V/W \to V/W$ is the continuous endomorphism induced by ϕ.

Proof Let $V_c \in \mathcal{B}(V)$ and $W_c = W \cap V_c \in \mathcal{B}(W)$. By Theorem 1, there exists a discrete linear subspace W_d of W such that $W = W_c \oplus W_d$. Let V_d be a linear subspace of V such that $V = V_c \oplus V_d$ and $W_d \leq V_d$. Clearly, V_d is discrete, since V_c is open and $V_c \cap V_d = 0$. By construction, the diagram

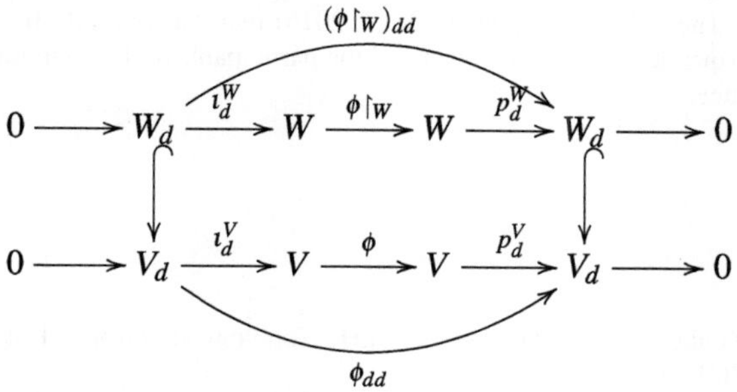

commutes, where $\iota_d^W, \iota_d^V, p_d^W, p_d^V$ are the canonical embeddings and projections of W and V, respectively. This yields that W_d is a ϕ_{dd}-invariant linear subspace of V_d and that

$$(\phi \restriction_W)_{dd} = \phi_{dd} \restriction_{W_d} .$$

Now, let $\pi\colon V \to V/W$ be the canonical projection and let $\overline{V} = V/W$. Let $\overline{V}_c = \pi(V_c)$ and $\overline{V}_d = \pi(V_d)$; then $\overline{V}_c$ is linearly compact and open, while $\overline{V}_d$ is discrete. Since $\overline{V}_c$ is open in $\overline{V}$, we have $\overline{V} = \overline{V}_c \oplus \overline{V}_d$.

Clearly, there exists a canonical isomorphism $\alpha\colon \overline{V}_d \to V_d/W_d$ of discrete vector spaces making the following diagram

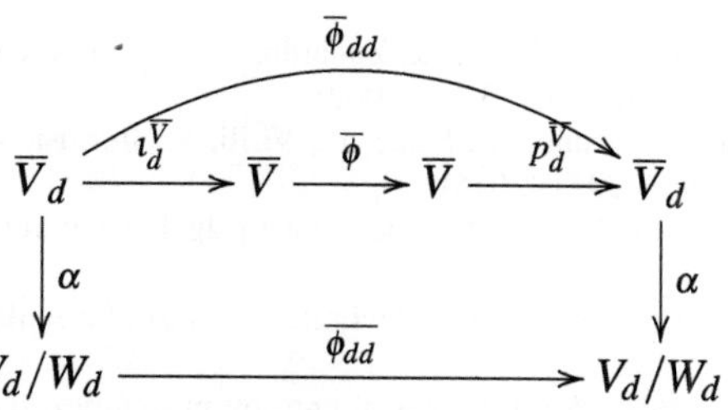

commute, where $\overline{\phi_{dd}}$ is the endomorphism induced by ϕ_{dd}. Then, by Propositions 13 and 7,

$$\mathrm{ent}(\phi) = \mathrm{ent}(\phi_{dd}), \quad \mathrm{ent}(\phi \restriction_W) = \mathrm{ent}(\phi_{dd} \restriction_{W_d}) \quad \text{and} \quad \mathrm{ent}(\overline{\phi}) = \mathrm{ent}(\overline{\phi_{dd}}).$$

Since $\mathrm{ent}(\phi_{dd}) = \mathrm{ent}(\phi_{dd} \restriction_{W_d}) + \mathrm{ent}(\overline{\phi_{dd}})$, in view of the Addition Theorem for $\mathrm{ent}_{\dim}$ (see [12, Theorem 5.1]) and Lemma 1, we can conclude that $\mathrm{ent}(\phi) = \mathrm{ent}(\phi \restriction_W) + \mathrm{ent}(\overline{\phi})$. □

Acknowledgements This work was supported by Programma SIR 2014 by MIUR (Project GADYGR, Number RBSI14V2LI, cup G22I15000160008), and partially also by the "National Group for Algebraic and Geometric Structures, and Their Applications" (GNSAGA—INdAM).

The first named author was partially supported by EPSRC Grant N007328/1 Soluble Groups and Cohomology.

It is a pleasure to thank our friends Dikran Dikranjan and Simone Virili for useful discussions on the topic of this paper.

The second named author gave a talk at the Conference "Recent Advances in Commutative Ring and Module Theory" (Brixen, June 13–17, 2016) that was partially based on the material presented in this paper, and so wants to thank all the participants of the conference, most of all the Scientific Committee.

References

1. Adler, R.L., Konheim, A.G., McAndrew, M.H.: Topological entropy. Trans. Am. Math. Soc. **114**, 309–319 (1965)
2. Belyaev, V.V.: Inert subgroups in infinite simple groups. Sib. Math. J. **34**, 218–232 (1993)
3. Bowen, R.: Entropy for group endomorphisms and homogeneous spaces. Trans. Am. Math. Soc. **153**, 401–414 (1971)
4. Castellano, I., Giordano Bruno, A.: Topological entropy for locally linearly compact vector spaces, Topology Appl., to appear (2018)
5. Dardano, U., Rinauro, S.: Inertial automorphisms of an abelian group. Rend. Sem. Mat. Univ. Padova **127**, 213–233 (2012)
6. Dikranjan, D., Giordano Bruno, A.: The connection between topological and algebraic entropy. Topology Appl. **159**, 2980–2989 (2012)
7. Dikranjan, D., Giordano Bruno, A.: Limit free computation of entropy. Rend. Istit. Mat. Univ. Trieste **44**, 297–312 (2012)
8. Dikranjan, D., Giordano Bruno, A.: The bridge theorem for totally disconnected LCA groups. Topology Appl. **169**, 21–32 (2014)
9. Dikranjan, D., Giordano Bruno, A.: Entropy on abelian groups. Adv. Math. **298**, 612–653 (2016)
10. Dikranjan, D., Goldsmith, B., Salce, L., Zanardo, P.: Algebraic entropy for abelian groups. Trans. Am. Math. Soc. **361**, 3401–3434 (2009)
11. Dikranjan, D., Giordano Bruno, A., Salce, L., Virili, S.: Intrinsic algebraic entropy. J. Pure Appl. Algebra **219**, 2933–2961 (2015)
12. Giordano Bruno, A., Salce, L.: A soft introduction to algebraic entropy. Arab. J. Math. **1**, 69–87 (2012)
13. Giordano Bruno, A., Virili, S.: On the Algebraic Yuzvinski Formula. Topol. Algebra Appl. **3**, 86–103 (2015)
14. Giordano Bruno, A., Virili, S.: Topological entropy in totally disconnected locally compact groups. Ergodic Theory Dyn. Syst. **37**, 2163–2186 (2017)
15. Hood, B.M.: Topological entropy and uniform spaces. J. Lond. Math. Soc. **8**, 633–641 (1974)
16. Köthe, G.: Topological Vector Spaces. Springer, Berlin/Heidelberg (1983)
17. Lefschetz, S.: Algebraic Topology, vol. 27. American Mathematical Society, Providence (1942)
18. Matlis, E.: Injective modules over Noetherian rings. Pac. J. Math. **8**, 511–528 (1958)
19. Peters, J.: Entropy on discrete abelian groups. Adv. Math. **33**, 1–13 (1979)

20. Peters, J.: Entropy of automorphisms on LCA groups. Pac. J. Math. **96**, 475–488 (1981)
21. Salce, L., Virili, S.: The Addition Theorem for algebraic entropies induced by non-discrete length functions. Forum Math. **28**, 1143–1157 (2016)
22. Salce, L., Zanardo, P.: A general notion of algebraic entropy and the rank-entropy. Forum Math. **21**, 579–599 (2009)
23. Salce, L., Vámos, P., Virili, S.: Length functions, multiplicities and algebraic entropy. Forum Math. **25**, 255–282 (2013)
24. Virili, S.: Algebraic and topological entropy of group actions, submitted
25. Virili, S.: Entropy for endomorphisms of LCA groups. Topology Appl. **159**, 2546–2556 (2012)
26. Weiss, M.D.: Algebraic and other entropies of group endomorphisms. Theory Comput. Syst. **8**: 243–248 (1974)
27. Willis, G.A.: The scale and tidy subgroups for endomorphisms of totally disconnected locally compact groups. Math. Ann. **361**, 403–442 (2015)
28. Yuzvinski, S.: Metric properties of endomorphisms of compact groups. Izv. Acad. Nauk SSSR Ser. Mat. **29**, 1295–1328 (1965) (in Russian). English Translation: Am. Math. Soc. Transl. **66**, 63–98 (1968)

Commutative Rings Whose Finitely Generated Ideals are Quasi-Flat

François Couchot

Abstract A definition of quasi-flat left module is proposed and it is shown that any left module which is either quasi-projective or flat is quasi-flat. A characterization of local commutative rings for which each ideal is quasi-flat (resp. quasi-projective) is given. It is also proven that each commutative ring R whose finitely generated ideals are quasi-flat is of λ-dimension ≤ 3, and this dimension ≤ 2 if R is local. This extends a former result about the class of arithmetical rings. Moreover, if R has a unique minimal prime ideal, then its finitely generated ideals are quasi-projective if they are quasi-flat.

Keywords Quasi-flat module • Chain ring • Arithmetical ring • fqf-ring • fqp-ring • λ-Dimension

2010 Mathematics Subject Classification: 13F05, 13B05, 13C13, 16D40, 16B50, 16D90

1 Introduction

In [1] Abuhlail, Jarrar and Kabbaj studied the class of commutative fqp-rings (**f**initely generated ideals are **q**uasi-**p**rojective). They proved that this class of rings strictly contains the one of the arithmetical rings and is strictly contained in the one of the Gaussian rings. It is also shown that the property for a commutative ring to be fqp is preserved by localization. It is known that a commutative ring R is arithmetical (resp. Gaussian) if and only if R_M is arithmetical (resp. Gaussian) for each maximal ideal M of R. But an example given in [7] shows that a commutative ring which is a locally fqp-ring is not necessarily a fqp-ring. So, in this cited paper the class of fqf-rings is introduced. Each local commutative fqf-ring is a fqp-ring, and a commutative ring is fqf if and only if it is locally fqf. These fqf-rings are defined in

F. Couchot (✉)
Normandie Université, F14032 Caen, France
e-mail: francois.couchot@unicaen.fr

M. Fontana et al. (eds.), *Rings, Polynomials, and Modules*,
https://doi.org/10.1007/978-3-319-65874-2_7

[7] without a definition of quasi-flat modules. Here we propose a definition of these modules and another definition of fqf-ring which is equivalent to the one given in [7]. We also introduce the module property of self-flatness. Each quasi-flat module is self-flat but we do not know if the converse holds. On the other hand, each flat module is quasi-flat and any finitely generated module is quasi-flat if and only if it is flat modulo its annihilator.

In Sect. 3 we give a complete characterization of local commutative rings for which each ideal is self-flat. These rings R are fqp and their nilradical N is the subset of zerodivisors of R. In the case where R is not a chain ring for which $N = N^2$ and R_N is not coherent every ideal is flat modulo its annihilator. Then in Sect. 4 we deduce that any ideal of a chain ring (valuation ring) R is quasi-projective if and only if it is almost maximal and each zerodivisor is nilpotent. This completes the results obtained by Hermann in [12] on valuation domains.

In Sect. 5 we show that each commutative fqf-ring is of λ-dimension ≤ 3. This extends the result about arithmetical rings obtained in [5]. Moreover it is shown that this λ-dimension is ≤ 2 in the local case. But an example of a local Gaussian ring R of λ-dimension ≥ 3 is given.

In this paper all rings are associative and commutative (except in the first section) with unity and all modules are unital.

2 Quasi-Flat Modules: Generalities

Let R be a ring, M a left R-module. A left R-module V is M-**projective** if the natural homomorphism $\mathrm{Hom}_R(V, M) \to \mathrm{Hom}_R(V, M/X)$ is surjective for every submodule X of M. We say that V is **quasi-projective** if V is V-projective. A ring R is said to be a left **fqp-ring** if every finitely generated left ideal of R is quasi-projective.

We say that V is M-**flat**[1] if for any epimorphism $p : M \to M'$, for any homomorphism $u : V \to M'$ and for any homomorphism $v : G \to V$, where M' is a left R-module and G a finitely presented left R-module, there exists a homomorphism $q : G \to M$ such that $pq = uv$. We call V **quasi-flat** (resp. **self-flat**) if V is V^n-flat for each integer $n > 0$ (resp. $n = 1$). Clearly each quasi-flat module is self-flat but we do not know if the converse holds.

An exact sequence $\mathscr{S}$ of left R-modules $0 \to F \to E \to G \to 0$ is **pure** if it remains exact when tensoring it with any right R-module. Then, we say that F is a **pure** submodule of E. Recall that $\mathscr{S}$ is pure if and only if $\mathrm{Hom}_R(M, \mathscr{S})$ is exact for each finitely presented left R-module M ([17, 34.5]). When E is flat, then G is flat if and only if $\mathscr{S}$ is pure ([17, 36.5]).

[1]The module property M-flat is generally used to define flat module.

Proposition 1 *Let R be a ring. Then:*

1. *each quasi-projective left R-module is quasi-flat;*
2. *each flat left R-module is quasi-flat.*

Proof 1. If V is a quasi-projective left R-module then, by Wisbauer [17, 18.2(2)], V is V^n-projective for each integer $n > 0$.

2. By Wisbauer [17, 36.8.3] a left R-module is flat if and only if it is M-flat for each left R-module M. □

Proposition 2 *Let R be a ring, $0 \to A \xrightarrow{t} B \to C \to 0$ an exact sequence of left R-modules and V a left module. If V is B-flat, then V is A-flat and C-flat.*

Proof Clearly V is C-flat. Let $p : A \to A'$ be an epimorphism of left R-modules. Consider the following pushout diagram of left R-modules:

$$\begin{array}{ccc} 0 & & 0 \\ \downarrow & & \downarrow \\ A & \xrightarrow{p} & A' \to 0 \\ t\downarrow & & t'\downarrow \\ B & \xrightarrow{p'} & B' \to 0 \end{array}$$

Let G be a finitely presented R-module and $V \xrightarrow{u} A'$ and $G \xrightarrow{v} V$ be homomorphisms. Since V is B-flat there exists a linear map $G \xrightarrow{d} B$ such that $t'uv = p'd$. By Wisbauer [17, 10.7] the above diagram is also a pullback diagram of left R-modules, so there exists a homomorphism $G \xrightarrow{q} A$ such that $pq = uv$. Hence V is A-flat. □

Corollary 1 *Let R be a ring, V a finitely generated left module and I its annihilator. Then V is flat over R/I if and only if V is quasi-flat.*

Proof If V is flat over R/I, then from Proposition 1 we deduce that it is quasi-flat. Conversely, if V is generated by n elements, then R/I is isomorphic to a submodule of V^n. It follows that $F = (R/I)^n$ is isomorphic to a submodule of V^{n^2}. By Proposition 2 V is F-flat. Since there exists an epimorphism $p : F \twoheadrightarrow V$, we get that $\ker(p)$ is a pure submodule of F. Hence V is flat over R/I. □

In Sect. 3 (Corollary 3 and Example 1) an example of a quasi-flat module (over a commutative ring) which is not flat modulo its annihilator is given.

We say that a ring R is a left **fqf-ring** if each finitely generated left ideal is quasi-flat. By Corollary 1 this definition is equivalent to the one given in [7, section 3].

3 Quasi-Flat Ideals Over Local fqp-Rings

In this section R is a commutative ring.

A module U is **uniserial** if its lattice of submodules is totally ordered by inclusion. A ring R is a **chain ring** (or a valuation ring) if it is a uniserial R-module. A chain ring which is an integral domain is a valuation domain. Recall that R is an **IF-ring** if each injective R-module is flat. When R is a chain ring, we denote by P its maximal ideal, by Z its subset of zerodivisors which is a prime ideal, by N its nilradical and by Q its quotient ring R_Z.

Lemma 1 *Let R be a chain ring and U an R-module. If U is quasi-flat (resp. quasi-projective) then aU is quasi-flat (resp. quasi-projective) too for each $a \in R$.*

Proof We consider the following homomorphisms: $p : (aU)^n \to U'$, $u : aU \to U'$ and $v : G \to aU$ where U' is an R-module, p is surjective, n an integer > 0 and G a finitely presented R-module. By Warfield [16, Theorem 1] G is a direct sum of cyclic submodules. It is easy to see that we may assume that G is cyclic. So $G = R/bR$ for some $b \in R$. If $x = v(1 + bR)$, then $bx = 0$ and there exists $y \in U$ such that $x = ay$. So, $bay = 0$. Let $v' : R/baR \to U$, $u' : U \to U'$ and $p' : U^n \to U'$ be the homomorphisms defined by $v'(r + baR) = ry$ for each $r \in R$, $u'(z) = u(az)$ for each $z \in U$ and $p'(w) = p(aw)$ for each $w \in U^n$.

The quasi-flatness of U implies that there exists a morphism $q' : R/baR \to U^n$ such that $p'q' = u'v'$. If we put $q(r + bR) = aq'(r + baR)$ for each $r \in R$, then the equalities $bq(1 + bR) = baq'(1 + baR) = 0$ imply that $q : G \to (aU)^n$ is a well-defined homomorphism, and we get $pq = uv$.

Now, suppose that $n = 1$ and U is quasi-projective. There exists $t' : U \to U$ such that $p't' = u'$. Let $t = t'|_{aU}$. Then $pt = u$. □

Let I be a non-zero proper ideal of a chain ring R. Then $I^\sharp = \{r \in R \mid rI \subset I\}$ is a prime ideal which is called the **top prime ideal** associated with I. It is easy to check that $I^\sharp = \{r \in R \mid I \subset (I : r)\}$. It follows that $I^\sharp/I$ is the inverse image of the set of zerodivisors of R/I by the natural map $R \to R/I$. So, $Z = 0^\sharp$.

Proposition 3 *Let R be a chain ring. Then each proper ideal I satisfying $Z \subset I^\sharp$ is flat modulo its annihilator.*

Proof First assume that $Z \subset I$. In this case I is a direct limit of free modules of rank one. So, it is flat. Now suppose that $I \subseteq Z$ and let $t \in I^\sharp \setminus Z$ and $a \in I \setminus tI$. Then $a = ts$ for some $s \in Z \setminus I$ and $t \in (I : s)$. So, $Z \subset (I : s)$. It is easy to check that $I = s(I : s)$, $I \cong (I : s)/(0 : s)$, $(0 : I) \supseteq (0 : s)$ and $(I : s)/(0 : s)$ strictly contains $Z/(0 : s)$ the subset of zerodivisors of $R/(0 : s)$ (see [4, Lemma 21]). □

Remark 1 If $P = Z$, then by Gill [11, Lemma 3] and Klatt and Levy[13, Proposition 1.3] we have $(0 : (0 : I)) = I$ for each ideal I which is not of the form Pt for some $t \in R$. In this case R is self-FP-injective and the converse holds. So, if A is a proper ideal such that $A^\sharp = P$, then R/A is self-FP-injective and it follows that $(A : (A : I)) = I$ for each ideal $I \supseteq A$ which is not of the form Pt for some $t \in R$.

Proposition 4 *Let R be a chain ring. Then any proper ideal I satisfying $I^\sharp \subset Z$ is not self-flat.*

Proof Let $s \in Z \setminus I^\sharp$. Since $s \notin s^2Q$, by applying the above remark to Q we get that there exists $a \in (0 : s^2) \setminus (0 : s)$. The multiplication by s in I induces an isomorphism $\sigma : I/(0 : s) \to I$. Let $u = \sigma^{-1}$, $p : I \to I/(0 : s)$ be the natural epimorphism and $v : R/sR \to I$ the homomorphism defined by $v(r + sR) = rsa$. Then $uv(1 + sR) = a + (0 : s)$ and $sb \neq 0$ for each $b \in a + (0 : s)$. So, there is no homomorphism $q : R/sR \to I$ such that $pq = uv$. □

Lemma 2 *Let R be a chain ring and I a non-zero proper ideal. Assume that $P = Z$ and $I \neq aP$ for each $a \in R$. Then I is FP-injective over R/A where $A = (0 : I)$.*

Proof By Remark 1 we have $I = (0 : A)$. Let $x \in I$ and $c \in R \setminus A$ such that $(A : c) \subseteq (0 : x)$. Then $(0 : c) \subseteq (0 : x)$. Since R is self-FP-injective there exists $y \in R$ such that $x = cy$. We have $(0 : y) = c(0 : x) \supseteq c(A : c) = A$ (the first equality holds by Couchot [4, Lemma 2]). Hence $y \in I$. □

Theorem 1 *Let R be a chain ring. Assume that either $Z \neq Z^2$ or Q is not coherent. Then the following conditions are equivalent:*

1. *$Z = N$;*
2. *each ideal I is flat over R/A where $A = (0 : I)$.*

Proof By Proposition 4 (2 ⇒ 1).

(1 ⇒ 2). Let I be an ideal and $A = (0 : I)$. By Proposition 3 it remains to examine the case where $I^\sharp = Z$. If $Z \neq Z^2$, then Z is principal over Q. It follows that Q is Artinian. Since I is a principal ideal of Q, then I is flat over Q/A and R/A. Now suppose that $Z = Z^2$ and Q is not coherent. By Couchot [4, Theorem 10] Z is flat, and we easily deduce that aZ is flat over $R/(0 : a)$ for each $a \in R$. Now suppose that I is neither principal over Q nor of the form aZ for each $a \in R$. By Lemma 2 I is FP-injective over R/A. From Q no coherent we deduce that $(0 : r)$ is not principal over Q for each $0 \neq r \in I$. By Couchot [8, Theorem 15(4)(c)] I is flat over R/A. □

Remark 2 If R is a chain ring such that either Z is principal over Q or Q is not coherent then each ideal I satisfying $Z \subseteq I^\sharp$ is flat modulo its annihilator.

Lemma 3 *Let R be a chain ring and M a finitely generated R-module. Then, for each proper ideal A which is not of the form rP for any $r \in P$, we have $AM = \cap_{s \in P \setminus A} sM$.*

Proof By Fuchs and Salce[9, Theorem 15] there is a finite sequence of pure submodules of M,

$$0 = M_0 \subset M_1 \subset \cdots \subset M_{n-1} \subset M_n = M.$$

such that M_k/M_{k-1} is cyclic for each $k = 1, \ldots, n$. We proceed by induction on n. When $n = 1$ M is cyclic and we use [4, Lemma 29] to conclude. Now suppose that $n > 1$. Let $x \in \cap_{s \in P \setminus A} sM$. We may assume that $x \notin M_{n-1}$. Since M/M_{n-1} is cyclic there exist $y \in M$ and $a \in A$ such that $(x - ay) \in M_{n-1}$. Moreover, by using the fact that M_{n-1} is a pure submodule of M we have that $(x-ay) \in \cap_{s \in P \setminus A} sM_{n-1}$. From the induction hypothesis we deduce that $x = ay + bz$ for some $z \in M_{n-1}$ and $b \in A$. □

Proposition 5 *Let R be a chain ring. Then, for each $a \in R$, aZ is quasi-flat.*

Proof We may assume that $Z = Z^2 \neq 0$. By Lemma 1 it is enough to study the case $a = 1$. First suppose that $Z = P$. We consider the following homomorphisms: $p : Z^n \to Z'$, $u : Z \to Z'$ and $v : G \to Z$ where Z' is an R-module, p is surjective, n an integer > 0 and $G = R/aR$ for some $a \in R$. If $r = v(1 + aR)$, then $ar = 0$. Let $s \in Z \setminus Rr$. Then $r = ss'$ for some $s' \in P$ and $u(r) = su(s')$. So, $u(r) \in \cap_{s \in Z \setminus Rr} sZ'$. Consider the following commutative pushout diagram:

$$\begin{array}{ccc} 0 & & 0 \\ \downarrow & & \downarrow \\ Z^n & \xrightarrow{p} & Z' \to 0 \\ t\downarrow & & t'\downarrow \\ R^n & \xrightarrow{p'} & R' \to 0 \end{array}$$

where t is the canonical inclusion. Clearly R' is finitely generated. So, by Lemma 3 $u(r) = rx'$ for some $x' \in R'$. Let $x \in R^n$ such that $p'(x) = x'$. Let $q : G \to Z^n$ be the homomorphism defined by $q(1 + aR) = rx$. Then $q : G \to Z^n$ is well defined because $ar = 0$ and we have $pq = uv$. Hence P is quasi-flat.

Now assume that $Z \neq P$ and Z is faithful. Let $a \in P$ and $t \in Z \setminus (0 : a)$. Let $K = \ker(p)$ and $G_1 = R/Rta$. Then $G \cong Rt/Rta \subseteq G_1$. Since Q is FP-injective v extends to $v_1 : G_1 \to Q$. But $v_1(1 + Rta) \in Z$ because it is annihilated by ta. There exists a homomorphism $q' : (G_1)_Z \to Z^n$ such that $p_Z q' = u_Z(v_1)_Z$. Let q_1 be the composition of the natural map $G_1 \to (G_1)_Z$ with q', $r = v_1(1 + atR)$ and $x = q_1(1+atR)$. Then $u(r) - p(x) \in K_Z/K$. Let $q = q_1|_G$. We have $v(t + taR) = tr$ and $q(t + taR) = tx$. So, $u(tr) - p(tx) = t(u(r) - p(x)) = 0$. Hence $uv = pq$. □

Proposition 6 *Let R be a chain ring. Then each ideal I satisfying $Z \subseteq I^\sharp$ is self-flat.*

Proof By Remark 2 we may assume that $0 \neq Z = Z^2$ and Q is coherent, and by Proposition 3 that $I^\sharp = Z$. We may suppose that I is neither principal over Q nor of the form aZ for any $a \in R$. We consider the following homomorphisms: $p : I \to I'$, $u : I \to I'$ and $v : G \to I$ where I' is an R-module, p is surjective and $G = R/aR$ for some $a \in P$. By Lemma 2 I is FP-injective over R/A, where $A = (0 : I)$ and by Couchot [8, Theorem 15(4)(c)] $Z \otimes_R I$ is flat over R/A because $(0 : r)$ is principal over Q for each $r \in I$. Since $I = ZI$ the canonical homomorphism $\phi : Z \otimes_R I \to I$ is surjective. Let $r = v(1 + aR)$. Then $r = \phi(s \otimes b)$ where $s \in Z$ and $b \in I$. Since $ar = 0$ then $a(s \otimes b) \in \ker(\phi) \cong \mathrm{Tor}_1^Q(Q/Z, I)$. So, $aZ \subseteq (0 : s \otimes b)$. Let $v' : R/taR \to Z \otimes_R I$ be the homomorphism defined by $v'(1 + taR) = s \otimes b$ where $t \in Z$. From the flatness of $Z \otimes_R I$ we deduce that there exists $q'_t : R/taR \to I$ such that $pq'_t = u\phi v'$. Let $x_t = q'_t(1+taR)$. Then $tax_t = 0$. Let t' be another element of Z. Thus $p(x_t) = p(x_{t'}) = u(r)$, whence $(x_{t'} - x_t) \in \ker(p) \subset Qx_t$ since I is a uniserial

Q-module. So, $Qx_t = Qx_{t'}$ and we can choose $x_t = x_{t'} = x$. Hence $aZ \subseteq (0:x)$. But $(0:x)$ is a principal ideal of Q, whence $ax = 0$. If we put $q(c + aR) = cx$ for each $c \in R$, then $pq = uv$. □

Theorem 2 *Let R be a chain ring. Then each ideal I is self-flat if and only if $Z = N$.*

Theorem 3 *[7, Theorem 4.1]. Let R be a local ring and N its nilradical. Then R is a fqp-ring if and only if either R is a chain ring or R/N is a valuation domain and N is a divisible torsionfree R/N-module.*

Proof Assume that R is fqp. By Singh Mohammad [14, Theorem 2] either R is chain ring or $N^2 = 0$. If R is not a chain ring, then N is the subset Z of all zerodivisors of R by Abuhlail et al. [1, Lemma 4.5]. Hence N is prime and it is a torsionfree R/N-module. Let $a \in R \setminus N$ and $b \in R$. By Abuhlail et al.[1, Lemma 3.8] the ideals Ra and Rb are comparable. So, if $b \in N$, then $b \in Ra$. We deduce that R/N is a valuation domain and N is divisible over R/N.

It is easy to show the converse. □

Corollary 2 *Let R be a local fqf-ring which is not a chain ring and N its nilradical. Then each ideal of R is flat modulo its annihilator.*

Proof Let I be an ideal. If $I \subseteq N$, then I is a torsionfree module over the valuation domain R/N. Hence it is a flat R/N-module. If $I \nsubseteq N$, then each finitely subideal of I is principal and free. So, I is flat. □

The following corollary and example allow us to see that there exist quasi-flat modules which are not flat modulo their annihilator.

Corollary 3 *Let R a chain ring. Assume that P is not principal and R is an IF-ring. Then, for each $a \in R$, aP is quasi-flat but it is not flat over $R/(0:aP)$.*

Proof Since R is coherent and P is not finitely generated we get that P is faithful. By Couchot [4, Theorem 10] P is not flat. Let $0 \neq a \in P$. There exists $b \in P$ such that $(0:a) = Rb$. So, $aP \cong P/Rb$, and $Rb = (0:aP)$ because P is faithful. By Couchot [4, Theorem 11] R/Rb is an IF-ring and consequently R/Rb satisfies the same conditions as R. Hence aP is quasi-flat but not flat over R/Rb. □

Example 1 Let $R = D/dD$, where D is a valuation domain with a non-principal maximal ideal and d a non-zero element of D which is not invertible. Then R satisfies the assumptions of Corollary 3.

4 Quasi-Projective Ideals Over Local fqp-Rings

An R-module M is said to be **linearly compact** if every finitely solvable set of congruences $x \equiv x_\alpha \pmod{M_\alpha}$ ($\alpha \in \Lambda$, $x_\alpha \in M$ and M_α is a submodule of M for each $\alpha \in \Lambda$) has a simultaneous solution in M. A chain ring R is **maximal** if it is linearly compact over itself and R is **almost maximal** if R/A is maximal for each non-zero ideal A.

Theorem 4 *Let R be a chain ring. The following conditions are equivalent:*

1. *R is almost maximal and $Z = N$;*
2. *each ideal is quasi-projective.*

Proof $(1 \Rightarrow 2)$. Let I be a non-zero proper ideal of R, $p : I \to I'$ an epimorphism, $K = \ker(p)$ and $u : I \to I'$ a homomorphism. First suppose that $I \subseteq Z$. By Theorem 2 I is self-flat. So, for each $r \in I$ and $b \in (0 : r)$ there exists $y_{r,b} \in I$ such that $by_{r,b} = 0$ and $u(r) = p(y_{r,b})$. Even if $K \neq 0$ we can take $y_{r,b} = y_{r,c} = y_r$ if c is another element of $(0 : r)$. So, $(0 : r) \subseteq (0 : y_r)$. Since Q is FP-injective then $y_r = rx_r$ where $x_r \in Q$. We put $R' = Q/K$, $p' : Q \to R'$ the canonical epimorphism and $x'_r = p'(x_r)$ for each $r \in I$. So, for each $r \in I$, $u(r) = rx'_r$. If $s \in I \setminus Rr$, then we easily check that $(x'_s - x'_r) \in R'[r] = \{y \in R' \mid ry = 0\}$. If R is almost maximal, then the family of cosets $(x'_r + R'[r])_{r \in I}$ has a non-empty intersection. Let x' be an element of this intersection. Then $u(r) = rx'$ for each $r \in I$. Let $x \in Q$ such that $p'(x) = x'$. For each $r \in I$, $rx \in rx_r + K \subseteq I$. If q is the multiplication by x in I, then $pq = u$. Hence I is quasi-projective. Now suppose that $Z \subset I$. Then for each $r \in I \setminus Z$ there exists $y_r \in I$ such that $u(r) = p(y_r)$. But $y_r = r(r^{-1}y_r) = rx_r$ where $x_r \in Q$. We do as above to show that I is quasi-projective. □

Proposition 7 *Let R a chain ring. Assume that $P = N$. Then R is almost maximal if each ideal I is quasi-projective.*

Proof If P is finitely generated, then R is Artinian. In this case R is maximal. Now assume that P is not finitely generated. Let $(a_\lambda + I_\lambda)_{\lambda \in \Lambda}$ be a totally ordered family of cosets such that $I = \cap_{\lambda \in \Lambda} I_\lambda \neq 0$. By Couchot [4, Lemma 29] $I \neq aP$ for each $a \in R$. Let $A = P(0 : I)$.

First we assume that I is different of the minimal non-zero ideal when it exists. So, $A \subset P$. We have $I = (0 : A) = \cap_{r \in A}(0 : r)$ (if $(0 : I)$ is not principal then $A = (0 : I)$). If not, either I is not principal and from $I = PI$ we deduce that $(0 : (0 : I)) = (0 : A)$, or I is principal which implies that P is faithful and $(0 : (0 : I)) = (0 : A))$. Let $r \in A$. We may assume that $I \subset I_\lambda$ for each $\lambda \in \Lambda$. Hence there exists $\lambda \in \Lambda$ such that $I_\lambda \subseteq (0 : r)$. We put $a(r) = a_\lambda r$. If $I_\mu \subset I_\lambda$, then $(a_\mu - a_\lambda) \in I_\lambda$, whence $a_\mu r = a_\lambda r$. So, in this manner, we define an endomorphism of A. Since $P = N$ there exists $c \in P \setminus A$ such that $c^2 \in A$. Let $B = (A : c)$. Then $A = cB$ and $c \in B$. Let $p : B \to A$ be the homomorphism defined by $p(r) = cr$ and $u : B \to A$ be the homomorphism defined by $u(r) = a(cr)$, for each $r \in B$. The quasi-projectivity of B implies there exists an endomorphism q of B such that $pq = u$. Since $(0 : c) \subseteq (0 : q(c))$ and R is self-FP-injective we deduce that $q(c) = ca'$ for some $a' \in R$ and $a(cr) = cq(r) = q(cr) = rq(c) = a'cr$ for each $r \in B$. Let $\lambda \in \Lambda$. We have $I = \cap_{r \in B}(0 : rc)$. Since $I \subset I_\lambda$ then $(0 : rc) \subseteq I_\lambda$ for some $r \in B$. From $I = \cap_{\mu \in \Lambda} I_\mu$ we deduce that there exists $\mu \in \Lambda$ such that $I_\mu \subseteq (0 : rc)$. It follows that $(a' - a_\mu) \in (0 : rc)$. But $(a_\mu - a_\lambda) \in I_\lambda$, so $a' \in (a_\lambda + I_\lambda)$ for each $\lambda \in \Lambda$.

Now we assume that I is the minimal non-zero ideal of R. In this case $A = P$. Let $s, t \in P$ such that $I = Rst$. There exists $\lambda_0 \in \Lambda$ such that $I_{\lambda_0} \subseteq Rt \subset P$. Let $\Lambda' = \{\lambda \in \Lambda \mid I_\lambda \subseteq I_{\lambda_0}\}$. Put $J_\lambda = (I_\lambda : t)$ and $J = \cap_{\lambda \in \Lambda'} J_\lambda$. Since $s \in J \setminus I$ then J

is not minimal. From $(a_\lambda - a_{\lambda_0}) \in I_{\lambda_0}$ we deduce that there exists $b_\lambda \in R$ such that $(a_\lambda - a_{\lambda_0}) = tb_\lambda$ for all $\lambda \in \Lambda'$. If $\lambda,\ \mu \in \Lambda'$ are such that $I_\mu \subseteq I_\lambda$, then we easily check that $b_\mu \in b_\lambda + J_\lambda$. From above it follows that there exists $b \in \cap_{\lambda\in\Lambda'} b_\lambda + J_\lambda$ and it is easy to see that $(a_{\lambda_0} + tb) \in \cap_{\lambda\in\Lambda} a_\lambda + I_\lambda$. Hence R is almost maximal. □

Proof (Proof of $(2 \Rightarrow 1)$ *in Theorem 4.)* Since each ideal I is self-flat we have $Z = N$ by Theorem 2. From the previous proposition we deduce that Q is almost maximal and we may assume that $P \neq Z$. When $Z = 0$ R is almost maximal by Hermann [12, Theorem 3.3]. Now suppose $Z \neq 0$. We shall prove that R/Z is maximal and we will conclude that R is almost maximal by using [6, Theorem 22]. Let $0 \neq x \in Z$ and I a proper ideal of R such that $Z \subset I$. Since I is quasi-projective then I is (Qx/Zx)-projective by Wisbauer [17, 18.2]. Let $q : I \to Q/Z$ be a homomorphism. If $z \in Z$ and $t \in I \setminus Z$, then $z = z't$ for some $z' \in Z$. So, $q(z) = z'q(t) = 0$, whence q factors through I/Z. It follows that I/Z is (Q/Z)-projective for each ideal I containing Z. By Hermann [12, Theorem 3.3] R/Z is almost maximal. Suppose that $Z^2 = Z$. We have that Qx is (Qx/Zx)-projective. Let $q : Q \to Q/Z$ be a homomorphism and $z \in Z$. There exist $z',\ t \in Z$ such that $z = z't$. So, $q(z) = z'q(t) = 0$ whence q factors through Q/Z. It follows that Q/Z is quasi-projective. If $Z \neq Z^2$, then Z is principal over Q and there exists $x \in Z$ such that $Z = (0 : x)$. Hence $Q/Z \cong Qx$ is quasi-projective. From R/Z almost maximal and [12, Theorem 3.4] we get that R/Z is maximal. □

Theorem 5 *Let R be a local fqp-ring which is not a chain ring and N its nilradical. Consider the following conditions:*

1. *R is a linearly compact ring;*
2. *each ideal is quasi-projective;*
3. *N is of finite rank over R/N.*

Then $(1 \Leftrightarrow ((2 \text{ and } 3))$.

Proof $(1 \Rightarrow (2 \text{ and } 3))$. Let $R' = R/N$ and Q' its quotient field. Then R' is a maximal valuation domain. Since N is a direct sum of modules isomorphic to Q' and a linearly compact module then N is of finite rank by Wisbauer [17, 29.8]. Let I be an ideal contained in N. By Hermann [12, Lemma 4.4] I is quasi-projective. Now suppose that $I \nsubseteq N$. In this case $N \subset I$ and since I/N is uniserial, by a similar proof as the one of Theorem 4 we show that I is quasi-projective.

$((2 \text{ and } 3) \Rightarrow 1)$. Since Q' is isomorphic to a direct summand of N, Q' and its submodules are quasi-projective. Hence, by Hermann [12, Theorems 3.3 and 3.4] R' is maximal. It follows that N is linearly compact, hence R is linearly compact by Wisbauer [17, 29.8]. □

Remark 3 In the previous theorem:

1. if N is the maximal ideal of R, then each ideal is quasi-projective even if N is not of finite rank over R/N;
2. if N is not the maximal ideal and if Q/N is countably generated over R/N, then $(1 \Leftrightarrow 2)$ because $(2 \Rightarrow 3)$ by Hermann [12, Lemma 4.3(b)].

5 λ-Dimension of Commutative fqf-Rings

In this section R is a commutative ring. We say that R is **arithmetical** if R_P is a chain ring for each maximal ideal P.

An R-module E is said to be of **finite n-presentation** if there exists an exact sequence:

$$F_n \to F_{n-1} \to \cdots F_1 \to F_0 \to E \to 0$$

with the F_i's free R-modules of finite rank. We write

$$\lambda_R(E) = \sup\{n \mid \text{there is a finite } n\text{-presentation of } E\}.$$

If E is not finitely generated we also put $\lambda_R(E) = -1$.

The **λ-dimension** of a ring R ($\lambda-\dim(R)$) is the least integer n (or ∞ if none such exists) such that $\lambda_R(E) \geq n$ implies $\lambda_R(E) = \infty$. See [15, chapter 8]. Recall that R is Noetherian if and only if $\lambda\text{-}dim(R) = 0$ and R is coherent if and only if $\lambda\text{-}dim(R) \leq 1$. The rings of λ-dimension $\leq n$ are also called **n-coherent** by some authors.

This notion of λ-dimension of a ring was formulated in [15, chapter 8] to study the rings of polynomials or power series over a coherent ring.

Theorem 6 *Let R be a local fqp-ring. Then* $\lambda-\dim(R) \leq 2$.

Proof By Couchot [5, Theorem II.11] $\lambda-\dim(R) \leq 2$ if R is a chain ring. Theorem 3 and the following proposition complete the proof. □

Proposition 8 *Let R be a local fqp-ring which is not a chain ring and N its nilradical. Then:*

1. *either R is Artinian or* $\lambda-\dim(R) = 2$ *if N is maximal;*
2. $\lambda-\dim(R) = 2$ *if N is not maximal.*

Moreover, if G is a finitely 2-presented module then:

3. *G is free if N is maximal and not finitely generated;*
4. *G is of projective dimension ≤ 1 if N is not maximal.*

Proof (1 and 3). If N is an R/N-vector space of finite dimension, then R is Artinian. Assume that N is not of finite dimension over R/N. Let G be an R-module of finite 2-presentation. So, there exists an exact sequence

$$F_2 \xrightarrow{u_2} F_1 \xrightarrow{u_1} F_0 \to G \to 0,$$

where F_i is free of finite rank for $i = 0, 1, 2$. Let G_i be the image of u_i for $i = 1, 2$. Since R is local we may assume that $G_i \subseteq NF_{i-1}$ for $i = 1, 2$ (see [2, p.222, 2. Local Rings and Projective Modules]). Then G_i is a module of finite length for $i = 1, 2$. It follows that F_1 is of finite length too. This is possible only if $F_1 = 0$. So, G is free. Hence $\lambda_R(G) = \infty$ and $\lambda\text{-dim}(R) \leq 2$. Let $0 \neq r \in N$. Since $(0 : r) = N$, $\lambda_R(R/rR) = 1$, whence $\lambda\text{-dim}(R) = 2$.

(2 and 4). Let P be the maximal ideal of R. Each $r \in P \setminus N$ is regular. So, Rr is free and since each finitely generated ideal which is not contained in N is principal, P is flat. Let G, G_1 and G_2 be as in 1. Since R is local we may assume that $G_i \subseteq PF_{i-1}$ for $i = 1, 2$. Then $\mathrm{Tor}_1^R(G_1, R/P) \cong \mathrm{Tor}_2^R(G, R/P) = 0$. So, the following sequence is exact:

$$0 \to G_2/PG_2 \to F_1/PF_1 \xrightarrow{v} G_1/PG_1 \to 0,$$

where v is induced by u_1. Since v is an isomorphism it follows that $G_2/PG_2 = 0$, and by Nakayama Lemma $G_2 = 0$. So, G_1 is free. Now, we do as in (1 and 3) to conclude (N is not finitely generated because it is divisible over R/N). □

Let A be a ring and E an A-module. The **trivial ring extension** of A by E (also called the idealization of E over A) is the ring $R := A \propto E$ whose underlying group is $A \times E$ with multiplication given by $(a, e)(a', e') = (aa', ae' + a'e)$. Let R be a ring. For a polynomial $f \in R[X]$, denote by $c(f)$ (the content of f) the ideal of R generated by the coefficients of f. We say that R is **Gaussian** if $c(fg) = c(f)c(g)$ for any two polynomials f and g in $R[X]$.

The following example shows that we cannot replace "fqp-ring" with "Gaussian ring" in Theorem 6.

Example 2 Let D be a non-almost maximal valuation domain and M its maximal ideal. Let $0 \neq d \in M$ such that D/Dd is not maximal and E the injective hull of D/Dd. Consider $R = D \propto E$ the trivial ring extension of D by E. Then R is a Gaussian local ring and $\lambda\text{-dim}(R) \geq 3$.

Proof By Couchot [4, Theorem 17] E is not uniserial. By Couchot [7, Corollary 4.3] R is Gaussian but not a fqp-ring because E is neither uniserial nor torsionfree. Let $e \in E$ such that $(0 : e) = Dd$. We put $a = (0, e)$ and $b = (d, 0)$. Then $(0 : a) = Rb$ and $(0 : b) = \{(0, x) \mid dx = 0\} = 0 \propto E[d]$, where $E[d] = \{x \in E \mid dx = 0\}$. If $D' = D/Dd$, then $E[d]$ is isomorphic to the injective hull of D' over D' and $D' \neq E[d]$ because D' is not maximal. By Couchot [4, Theorem 11] D' is an IF-ring and consequently $E[d]$ and $E[d]/D'$ are flat over D'. Then $E[d]$ is not finitely generated, else $E[d]/D'$ is a free D'-module and this contradicts that $E[d]$ is an essential extension of D'. So, $(0 : b)$ is not finitely generated, $\lambda_R(R/Ra) = 2$ and $\lambda\text{-dim}(R) \geq 3$. □

Proposition 9 *Let R be a fqf-ring with a unique minimal prime ideal N. The following assertions hold:*

1. *R_P is not a chain ring for each maximal ideal P if R is not arithmetical;*
2. *R is a fqp-ring.*

Proof 1. There exists a maximal ideal L such that R_L is a local fqp-ring which is not a chain ring. So, N_L is torsionfree and divisible over R_L/N_L. Moreover, since N_L is not uniserial over R_L, by Abuhlail et al. [1, Lemma 3.8] there exist $a, b \in N_L$

such that $aR_L \cap bR_L = 0$. It follows that $N_L = N_N$ and it is a vector space over R_N/N_N of dimension > 1. Let P be a maximal ideal. Then N_N is a localization of N_P. Consequently N_P is not uniserial. Hence, R_P is not a chain ring.

2. It follows that N is a torsionfree divisible module over R/N. So, if I is a finitely generated ideal contained in N, then I is a finitely generated flat module over the Prüfer domain R/N. So, I is projective over R/N. Now, if $I \not\subseteq N$, then I_P is a free R_P-module of rank 1. We conclude by Bourbaki [3, Chap.2, §5, 3, Théorème 2] that I is projective. □

Corollary 4 *Let R be a fqf-ring with a unique minimal prime ideal N. Assume that R is not arithmetical. Then either R is Artinian or $\lambda-\dim(R) = 2$.*

Proof When N is maximal we use Proposition 8. Now assume that N is not maximal. Let G be a R-module such that $\lambda_R(E) \geq 2$. We use the same notations as in the proof of Proposition 8. This proposition implies that G_1 is locally free. Since G_1 is a finitely presented flat module, we successively deduce that G_1 is projective, G_2 is projective and $\ker(u_2)$ is finitely generated. □

An integral domain D is said to be is **almost Dedekind** is D_P is a Noetherian valuation domain for each maximal ideal P.

The following example shows that we cannot remove the assumption "R is not arithmetical" in Corollary 4.

Example 3 Let D be an almost Dedekind domain which is not Noetherian (see [10, Example III.5.5]), Q its quotient field, P' a maximal ideal of D which is not finitely generated and $E = Q/D_{P'}$. Let $R = D \propto E$ and $N = \{(0, y) \mid y \in E\}$. Then R is an arithmetical ring, N is its unique minimal prime ideal and $\lambda-\dim R = 3$. Moreover, R_P is IF where P is the maximal ideal of R satisfying $P' = P/N$, and R_L is a valuation domain for each maximal ideal $L \neq P$.

Proof For each maximal ideal L of R let $L' = L/N$. Let $p \in P'$ such that $P'D_{P'} = pD_{P'}$, $x = 1/p + D_{P'}$, $a = (p, 0)$ and $b = (0, x)$. Since 0 is the sole prime ideal of D contained in $P' \cap L'$ for each maximal ideal $L' \neq P'$, then $E_{L'} = 0$. So, $R_L = D_{L'}$. Since E is uniserial and divisible over $D_{P'}$, R_P is a chain ring by Couchot [7, Proposition 1.1]. So, R is arithmetical. We have $(0 :_{R_P} b) = aR_P = PR_P$. By Couchot [4, Theorem 10] R_P is IF. Clearly Dx is the minimal submodule of E. So, $P' = (0 : x)$ and $D_{P'}x = Dx$. If $q \in Q \setminus D_{P'}$, then $q = sp^n/t$ where $s, t \in D \setminus P'$ and n an integer > 0. So, $pq \in D_{P'}$ if and only if $n = 1$. It follows that $Dx = \{y \in E \mid py = 0\}$. Let $\hat{a}$ and $\hat{b}$ be the respective multiplications in R by a and b. Then $\ker(\hat{a}) = Rb$ and $\ker(\hat{b}) = P$ which is not finitely generated. So, $\lambda_R(R/Ra) = 2$ and $\lambda-\dim(R) = 3$ by Couchot [5, Theorem II.1]. □

Theorem 7 *Let R be a fqf-ring. Then $\lambda-\dim(R) \leq 3$.*

Proof Let G be an R-module of finite 3-presentation. So, there exists an exact sequence

$$F_3 \xrightarrow{u_3} F_2 \xrightarrow{u_2} F_1 \xrightarrow{u_1} F_0 \rightarrow G \rightarrow 0,$$

where F_i is free of finite rank for $i = 0, 1, 2, 3$. Let G_i be the image of u_i for $i = 1, 2, 3$.

We do as in the proof of [5, Theorem II.1]. For each maximal ideal P we shall prove that there exist $t_P \in R \setminus P$ such that $\lambda_{R_{t_P}}(G_{t_P}) \geq 4$. We end as in the proof of [5, Theorem II.1] to show that $\ker(u_3)$ is finitely generated, by using the fact that Max R is a quasi-compact topological space.

Let P be a maximal ideal. First assume that R_P is a chain ring. As in the proof of [5, Theorem II.1] we show there exists $t_P \in R \setminus P$ such that $\lambda_{R_{t_P}}(G_{t_P}) \geq 4$.

Now assume that R_P is not a chain ring. We suppose that either P is not minimal or P is minimal but PR_P is not finitely generated over R_P. In this case $(G_1)_P$ is free over R_P by Proposition 8. Since G_1 is finitely presented, there exists $t_P \in R \setminus P$ such that $(G_1)_{t_P}$ is free over R_{t_P} by Bourbaki [3, Chapitre 2, §5, 1, Corollaire de la proposition 2]. It follows that $(G_2)_{t_P}$ and $(G_3)_{t_P}$ are projective. So, $\ker((u_3)_{t_P})$ is finitely generated.

Finally assume that R_P is not a chain ring, P is minimal and PR_P is finitely generated over R_P. We have $P^2R_P = 0$. Since $P^2R_L = R_L$ for each maximal ideal $L \neq P$, P^2 is a pure ideal of R. It follows that R/P^2 is flat. Clearly R/P^2 is local. So, $R_P = R/P^2$. If P^2 is finitely generated, then $P^2 = Re$ where e is an idempotent of R. So, if $t_P = 1 - e$, then $D(t_P) = \{P\}$, $R_{t_P} = R_P$ and $\ker((u_3)_{t_P})$ is finitely generated. If P^2 is not finitely generated, then $P = I + P^2$ where I is finitely generated but not principal because so is P/P^2. Since I^2 is a finitely generated subideal of the pure ideal P^2 there exists $a \in P^2$ such that $r = ar$ for each $r \in I^2$. It follows that $(1 - a)I^2 = 0$. Hence $I^2R_t = 0$ where $t = (1 - a)$ and $IR_t \neq 0$ because for each $s \in I\setminus P^2$, $s \neq sa$. Since G_t is finitely generated, after possibly multiplying t with an element in $R\setminus P$, we may assume that G_t has a generating system $\{g_1, \dots, g_p\}$ whose image in $(G_t)_P$ is a minimal generating system of $(G_t)_P$ containing p elements. Let F'_0 be a free R_t-module with basis $\{e_1, \dots, e_p\}$, $\pi : F'_0 \to G_t$ be the homomorphism defined by $\pi(e_k) = g_k$ for $k = 1, \dots, p$ and $G'_1 = \ker(\pi)$. We get the following commutative diagram with exact rows and columns:

$$\begin{array}{ccccccccc}
 & & 0 & & 0 & & 0 & & \\
 & & \downarrow & & \downarrow & & \downarrow & & \\
0 & \to & P^2G'_1 & \to & P^2F'_0 & \xrightarrow{\pi} & P^2G_t & \to & 0 \\
 & & \downarrow & & \downarrow & & \downarrow & & \\
0 & \to & G'_1 & \to & F'_0 & \xrightarrow{\pi} & G_t & \to & 0 \\
 & & \downarrow & & \downarrow & & \downarrow & & \\
0 & \to & (G'_1)_P & \to & (F'_0)_P & \xrightarrow{\pi_P} & (G_t)_P & \to & 0 \\
 & & \downarrow & & \downarrow & & \downarrow & & \\
 & & 0 & & 0 & & 0 & &
\end{array}$$

Since $(G'_1)_P \subseteq P(F'_0)_P$, we have that $G'_1 \subseteq PF'_0$. But, since G'_1 is finitely generated, after possibly multiplying t with and element of $R\setminus P$, we may assume that $G'_1 \subseteq IF'_0$ and that G'_1 has a generating system $\{g'_1, \dots, g'_q\}$ whose image in $(G'_1)_P$ is a minimal generating system of $(G'_1)_P$ with q elements. Let F'_1 be a free R_t-module with basis $\{e'_1, \dots, e'_q\}$, $u'_1 : F'_1 \to F'_0$ defined by $u'_1(e'_k) = g'_k$ for $k = 1, \dots, q$ and $G'_2 = \ker(u'_1)$. Again, for a suitable $t \in R \setminus P$ we may assume that G'_2 is contained in IF'_1 and has a generating system whose image in $(G'_2)_P$ is a minimal generating system with the same cardinal. Since $I_t^2 = 0$, it follows that $G'_2 = IF'_1$. Consequently, if I_t is generated by $\{r_1, \dots, r_n\}$, then G'_2 is generated $\{r_i e'_k \mid 1 \leq i \leq n,\ 1 \leq k \leq q\}$. Let F'_2 be a free R_t-module with basis $\{\epsilon_{i,k} \mid 1 \leq i \leq n,\ 1 \leq k \leq q\}$, $u'_2 : F'_2 \to F'_1$ be the homomorphism defined by $u'_2(\epsilon_{i,k}) = r_i e'_k$ for $i = 1, \dots, n$ and $k = 1, \dots, q$ and $G'_3 = \ker(u'_2)$. Since G'_3 is finitely generated, as above, for a suitable $t \in R \setminus P$, we get $G'_3 = I_t F'_2$. We easily deduce that $I_t = (0 :_{R_t} r_i)$ for each $i = 1, \dots, n$. Now, let F'_3 be a free R_t-module of rank qn^2 and $u'_3 : F'_3 \to F'_2$ be the homomorphism defined like u'_2. Then we get $\ker(u'_3) = IF'_3$, hence it is finitely generated. So, for a suitable $t_P \in R \setminus P$ we have $\lambda_{R_{t_P}}(G_{t_P}) \geq 4$. □

With a similar proof as the one of [5, Corollary II.13], and by using Proposition 8 we get the following theorem.

Theorem 8 *Let R be a fqf-ring. Assume that R_P is either an integral domain or a non-coherent ring for each maximal ideal P which is not an isolated point of* Max R. *Then $\lambda-\dim(R) \leq 2$.*

Proof Let G be an R-module of finite 2-presentation. So, there exists an exact sequence

$$F_2 \xrightarrow{u_2} F_1 \xrightarrow{u_1} F_0 \to G \to 0,$$

where F_i is free of finite rank for $i = 0, 1, 2$. Let G_i be the image of u_i for $i = 1, 2$.

We do as in the proof of the previous theorem. First suppose that P is a non-isolated point of Max R. In this case $(G_1)_P$ is free over R_P by Proposition 8. Since G_1 is finitely presented, there exists $t_P \in R \setminus P$ such that $(G_1)_{t_P}$ is free over R_{t_P} by Bourbaki [3, Chapitre 2, §5, 1, Corollaire de la proposition 2]. It follows that $(G_2)_{t_P}$ is projective. So, $\ker((u_2)_{t_P})$ is finitely generated. Now assume that P is isolated. There exists $t_P \in R \setminus P$ such that $R_P \cong R_{t_P}$. By Theorem 6 $\ker((u_2)_{t_P})$ is finitely generated. □

Example 3 and the following show that the assumption "R_P is a non-coherent ring" cannot be removed in Theorem 8.

Example 4 Let A be a von Neumann regular ring which is not self-injective, H the injective hull of A, $x \in H \setminus A$, $E = A + Ax$ and $R = A \propto E$. Then:

1. R is a fqf-ring which is not a fqp-ring;
2. for each maximal ideal P, R_P is Artinian;
3. $\lambda-\dim(R) = 3$.

Proof Let $N = \{(0, y) \mid y \in E\}$, $a = (0, 1)$ and $b = (0, x)$.

1. See [7, Example 4.6].
2. If P is a maximal ideal of R, then R_P is the trivial ring extension of the field $A_{P'}$ by the finite dimensional vector space $E_{P'}$ where $P' = P/N$. Hence R_P is Artinian.
3. Consider the following free resolution of R/aR:

$$R^2 \xrightarrow{u_2} R \xrightarrow{u_1} R \to R/Ra \to 0$$

where $u_2((r,s)) = ra + sb$ for each $(r,s) \in R^2$ and $u_1(r) = ra$ for each $r \in R$. We easily check that this sequence of R-modules is exact. The A-module E is not finitely presented, otherwise E/A is finitely presented and, since each exact sequence of A-modules is pure, A is a direct summand of E which contradicts that A is essential in E. Consequently, N, which is the image of u_2, is not finitely presented. So, $\lambda_R(R/Ra) = 2$ and $\lambda{-}\dim(R) = 3$. □

Acknowledgements At the Conference on Rings and Polynomials held in Graz (Austria), 3–8 July 2016, I did a talk entitled "Finitely generated powers of prime ideals". An article with this title is appeared in Palest. J. Math., vol. 6(2). I thank again the organizers of this conference.

References

1. Abuhlail, J., Jarrar, V., Kabbaj, S.: Commutative rings in which every finitely generated ideal is quasi-projective. J. Pure Appl. Algebra **215**, 2504–2511 (2011)
2. Bland, P.E.: Rings and Their Modules. De Gruyter, Berlin (2011)
3. Bourbaki, N.: Algèbre Commutative, Chapitres 1 et 2. Hermann, Paris (1961)
4. Couchot, F.: Injective modules and fp-injective modules over valuation rings. J. Algebra **267**, 359–376 (2003)
5. Couchot, F.: The λ-dimension of commutative arithmetic rings. Commun. Algebra **31**(7), 3143–3158 (2003)
6. Couchot, F.: Pure-injective hulls of modules over valuation rings. J. Pure Appl. Algebra **207**, 63–76 (2006)
7. Couchot, F.: Trivial ring extensions of Gaussian rings and fqp-rings. Commun. Algebra **43**(7), 2863–2874 (2015)
8. Couchot, F.: Weak dimension of fp-injective modules over chain rings. Commun. Algebra **44**, 381–389 (2016)
9. Fuchs, L., Salce, L.: In: Modules over Valuation Domains. Lecture Notes in Pure and Applied Mathematics, vol. 97. Marcel Dekker, New York (1985)
10. Fuchs, L., Salce, L.: In: Modules over Non-Noetherian Domains. Mathematical Surveys and Monographs, vol. 84. American Mathematical Society, Providence (2001)
11. Gill, D.T.: Almost maximal valuation rings. J. Lond. Math. Soc. **4**, 140–146 (1971)
12. Hermann, P.: Self-projective modules over valuation rings. Arch. Math. **43**, 332–339 (1984)
13. Klatt, G.B., Levy, L.S.: Pre-self injectives rings. Trans. Am. Math. Soc. **137**, 407–419 (1969)
14. Singh, S., Mohammad, A.: Rings in which every finitely generated left ideal is quasi-projective. J. Indian Math. Soc **40**(1–4), 195–205 (1976)
15. Vasconcelos, W.V.: The Rings of Dimension Two. Lecture Notes in Pure and Applied Mathematics, vol. 22. Marcel Dekker, New York (1976)
16. Warfield, R.: Decomposability of finitely presented modules. Proc. Am. Math. Soc. **25**, 167–172 (1970)
17. Wisbauer, R.: Foundations of Module and Ring Theory. Gordon and Breach Science, Publishers Reading (1991)

Commutative Rings with a Prescribed Number of Isomorphism Classes of Minimal Ring Extensions

David E. Dobbs

Abstract Let κ be a cardinal number. If $\kappa \geq 2$, then there exists a (commutative unital) ring A such that the set of A-algebra isomorphism classes of minimal ring extensions of A has cardinality κ. The preceding statement fails for $\kappa = 1$ and, if A must be nonzero, it also fails for $\kappa = 0$. If $\kappa \leq \aleph_0$, then there exists a ring whose set of maximal (unital) subrings has cardinality κ. If an infinite cardinal number κ is of the form $\kappa = 2^{\lambda}$ for some (infinite) cardinal number λ, then there exists a field whose set of maximal subrings has cardinality κ.

Keywords Commutative ring • Minimal ring extension • Cardinal number • Polynomial ring • Prime ideal • Valuation domain • Maximal subring • Idealization • Ordinal number

Subject Classifications [2010]: Primary: 13B99; Secondary: 13A15

1 Introduction

All rings and algebras considered in this note are commutative with identity; all subrings, inclusions of rings, ring extensions, ring homomorphisms, and modules are unital. As usual, we say (cf. [12]) that a ring extension $A \subset B$ is a *minimal ring extension* (or that B is a *minimal ring extension of* A) if there does not exist a ring properly contained between A and B. Similarly, given a ring extension $A \subset B$, one says that A is a *maximal subring of* B if $A \subset B$ is a minimal ring extension. Given a cardinal number κ, our interests here are twofold: to determine if there exists a ring A such that the cardinal number of the collection of A-algebra isomorphism classes of minimal ring extensions of A is κ; and to determine if there exists a ring B such that the cardinal number of the set of maximal subrings of B is κ. (For the first of these issues, the focus on A-algebra isomorphism classes, rather than on A-algebras,

D.E. Dobbs (✉)
Department of Mathematics, University of Tennessee, Knoxville, TN 37996-1320, USA
e-mail: dedobbs@comporium.net

M. Fontana et al. (eds.), *Rings, Polynomials, and Modules*,
https://doi.org/10.1007/978-3-319-65874-2_8

is necessary in order to avoid set-theoretic paradoxes, as the class of minimal ring extensions of a nonzero ring A is not a set.) Our results will answer the question about isomorphism classes of minimal ring extensions for any κ; we shall answer the corresponding question about maximal subrings if $\kappa \leq \aleph_0$ and also for certain greater values of κ. The next paragraph will give some background on minimal ring extensions and then summarize our results concerning that concept. The subsequent paragraph will do the analogue for the concept of a maximal subring.

Many of the minimal ring extensions considered below will be integral ring extensions. Let $A \subset B$ be an integral ring extension, with the conductor $M := (A : B)$. Then (cf. [12, Lemme 1.2 and Proposition 4.1], [11, Lemma II.3], [20, Theorem 3.3]), $A \subset B$ is a (an integral) minimal ring extension if and only if M is a maximal ideal of A and (exactly) one of the following three conditions holds: $A \subset B$ is said to be, respectively, *inert*, *ramified*, or *decomposed* if B/MB ($= B/M$) is isomorphic, as an algebra over the field $F := A/M$, to a minimal field extension of F, $F[X]/(X^2)$, or $F \times F$. (As usual, X denotes an indeterminate over the ambient base ring.) As above, let κ be a cardinal number. We show (see Lemma 1 and Theorem 1) that 1 is the only cardinal number which *cannot* be realized for some ring A as the cardinal number of the collection of A-algebra isomorphism classes of minimal ring extensions of A. If one is interested only in nonzero rings, then 0 is the only other cardinal number which cannot be realized in this manner. This observation is related to the fact that a ring A is a/the zero ring if and only if A does not have any (unital) proper ring extensions (in which case, A also does not have any proper subrings).

Recently, the dual concept of a "maximal subring" has been investigated by A. Azarang, often in collaboration with O. A. S. Karamzadeh, in a series of papers (four of which we discuss below in relation to our work). Suffice it to recall here that, while many rings have maximal subrings (cf. [1, Corollary 2.8]), some (nonzero non-prime) rings do not have maximal subrings [2, Example 2.6]. Our main results on maximal subrings may be summarized as follows. If κ is a cardinal number such that either $\kappa \leq \aleph_0$ or $\kappa = 2^\lambda$ for some infinite cardinal number λ, then there exists a ring whose set of maximal subrings has cardinality κ (see Proposition 1 (b), (c) and Corollary 1). Given the above mention of 2^λ, we wish to emphasize that we assume only the usual logical foundation of ZFC (Zermelo–Fraenkel set theory, together with the Axiom of Choice), as that foundation will allow us to use the well-known rules for the arithmetic of infinite cardinal numbers (as in, for instance, [14, pages 94–99]).

Proofs in both parts of this work (i.e., on minimal ring extensions and on maximal subrings) will need the definition of an idealization. (See the proofs of Lemma 1, Theorem 1, and Proposition 1 (a).) Recall that if A is a ring and E is an A-module, then the *idealization* $A(+)E$ is the ring whose additive structure is given by the group $A \oplus E$ and whose multiplication is given by the following: if $a_1, a_2 \in A$ and $e_1, e_2 \in E$, then $(a_1, e_1)(a_2, e_2) := (a_1a_2, a_1e_2 + a_2e_1)$. We recommend [15] as a good reference for the basic facts about idealizations.

If A is a ring, then Spec(A) (resp., Max(A)) denotes the set of prime (resp., maximal) ideals of A. By the "dimension" of a ring, we mean its Krull dimension. If F is a field, $\overline{F}$ denotes a/the algebraic closure of F. If $\mathscr{S}$ is a set, then $|\mathscr{S}|$ denotes the cardinal number of $\mathscr{S}$. Any unexplained material is standard, as in [13, 18].

2 Results

We begin by showing in Lemma 1 (a) that the minimal ring extensions of a ring A fill up a number of A-algebra isomorphism classes. Parts (b) and (c) of Lemma 1 then determine when that (cardinal) number is 0 or 1.

Lemma 1

(a) Let $A \subset B$ and $A \subset C$ be rings such that B and C are isomorphic as A-algebras. Then $A \subset B$ is a minimal ring extension if and only if $A \subset C$ is a minimal ring extension.

(b) Let A be a ring. Then the cardinal number of the set of A-algebra isomorphism classes of minimal ring extensions of A is 0 if and only if A is a/the zero ring.

(c) There does not exist a ring A such that the cardinal number of the set of A-algebra isomorphism classes of minimal ring extensions of A is 1.

Proof (a) It suffices to prove that if $A \subset B$ is a minimal ring extension, then $A \subset C$ is a minimal ring extension. We shall prove the contrapositive. Assume, then, that there exists a ring D such that $A \subset D \subset C$. Let $f : C \to B$ be an A-algebra isomorphism. Then $f(D)$ is a ring such that $A \subset f(D) \subset B$, as desired.

(b), (c): It was noted in the Introduction that if A is a/the zero ring, then A has no proper ring extension, and so no such A can have a minimal ring extension. Next, suppose that A is a nonzero ring. Then there exists $M \in \text{Max}(A)$. It was proved in [7] that $A(+)A/M$ is (A-algebra isomorphic to) a minimal ring extension of A. Moreover, it follows from case (b) of [12, Lemme 1.5] that $A \times A/M$ is (A-algebra isomorphic to) a minimal ring extension of A. Therefore, it suffices to prove that $A(+)A/M$ is not A-algebra isomorphic to $A \times A/M$. This was shown in [10, Corollary 2.5] for the special case of a reduced nonzero ring A, but we prove it next without the "reduced" restriction.

It was shown in [21, Lemma 2.1] that the canonical ring extension $A \hookrightarrow E$ is subintegral for each A-module E. Therefore, the minimal ring extension $A \hookrightarrow A(+)A/M$ is ramified. In particular, only one prime ideal of $A(+)A/M$ lies over M (cf. [11, Corollary II.2], [20, Theorem 3.3]). (That prime ideal is $M(+)A/M$; this fact can also be found in [15].) On the other hand, the minimal ring extension $A \hookrightarrow A \times A/M$ is decomposed, and so two distinct prime ideals of $A \times A/M$ lie over M. They are, of course, $Q_1 := M \times A/M$ and $Q_2 := A \times \{0\}$. Consequently, there is no A-algebra isomorphism g from $A \times A/M$ onto $A(+)A/M$ (for otherwise, $g(Q_1)$ and $g(Q_2)$ would be distinct prime ideals of $A(+)A/M$ which lie over M, a contradiction).

Theorem 1 presents our main realizability results concerning minimal ring extensions.

Theorem 1 *For each cardinal number $\kappa \geq 2$, there exists a ring A such that the cardinal number of the set of A-algebra isomorphism classes of minimal ring extensions of A is κ.*

Before proving Theorem 1, we shall give two lemmas which will be used in its proof.

The inert/ramified/decomposed trichotomy that was recalled in the Introduction is ultimately derived from Ferrand and Olivier's classification of the minimal ring extensions of a field. The substance of that result is given next for reference purposes. The statement of Lemma 2 replaces $F[X]/(X^2)$ from the statement of [12, Lemme 1.2] with the idealization $F(+)F$ for two reasons: $F(+)F \cong F[X]/(X^2)$ as F-algebras; and we wish to avoid confusing the symbol X with any of the indeterminates that will appear in the proof of Theorem 1.

Lemma 2 (*Ferrand-Olivier [12, Lemme 1.2]*) *Let F be a field. Then an F-algebra B can be viewed as a minimal ring extension of F if and only if B is F-algebra isomorphic to either a minimal field extension of F, the idealization $F(+)F$, or $F\times F$.*

The final lemma before the proof of Theorem 1 concerns real closed fields. Relevant background on this concept can be found in, for instance, [17], which will be cited as needed in the proof of Lemma 3.

Lemma 3 *Let s and t be nonnegative integers, at least one of which is positive. Then there exist real closed fields $R_1, \ldots, R_s$ and algebraically closed fields $C_1, \ldots, C_t$ such that*

$$|R_1| < \ldots < |R_s| < |C_1| < \ldots < |C_t|.$$

Proof We claim that there exist real closed fields of arbitrarily large cardinality. To see this, let κ be an infinite cardinal number and start with any real closed field, F (for instance, $\mathbb{R}$). Of course, F is infinite since it is of characteristic 0. By replacing κ with $2^{\max(\kappa,|F|)}$, we can assume that $\kappa > |F|$. Next, let $\{X_i\}$ be a set of algebraically independent indeterminates over F such that $|\{X_i\}| = \kappa$. Using the standard facts about arithmetic for infinite cardinal numbers (cf. [14, pages 94–98]), one verifies that the field $L := F(\{X_i\})$ has cardinality κ. As L is a purely transcendental field extension of F, it follows that L is a formally real field (cf. [17, Exercise 4, page 272]). It follows that L is an ordered field [17, Corollary 2, page 274]. Hence, by Jacobson [17, Theorem 8, page 285], L has a real closure, say Ω. As Ω is an algebraic extension of the (infinite) field L, we have $|\Omega| = |L| = \kappa > |F|$ (cf. [17, Lemma, page 143]). This proves the above claim.

By iterating what was just proved, we get the following consequence. For each positive integer s, there exist s real closed fields $R_1 \subset \ldots \subset R_s$ such that $|R_1| < \ldots < |R_s|$.

We next show that for each positive integer t, there exist algebraically closed fields $C_1 \subset \ldots \subset C_t$ such that $|C_1| < \ldots < |C_t|$. To see this, start by taking

C_1 to be any algebraically closed field (for instance, $\mathbb{C}$). For the inductive step, given a suitable C_i with $1 \leq i < t$, construct C_{i+1} as follows. Let $\aleph$ be a cardinal number exceeding $|C_i|$ (for instance, $2^{|C_i|}$). Let E be the field obtained by adjoining to C_i a set, with cardinality $\aleph$, consisting of indeterminates which are algebraically independent over C_i. Then it suffices to let C_{i+1} be a/the algebraic closure of E.

Suppose that we first construct the s real closed fields for some $s > 0$ and then construct the t algebraically closed fields as above. To complete the proof, it suffices to ensure that $R_s \subset C_1$ and $|R_s| < |C_1|$. This, in turn, can be done by modifying the above construction of C_1, as follows. Let λ be a cardinal number exceeding $|R_s|$ (for instance, $2^{|R_s|}$). Let N be the field obtained by adjoining to R_s a set, with cardinality λ, consisting of indeterminates which are algebraically independent over R_s. Then it suffices to let C_1 be a/the algebraic closure of N.

Proof of Theorem 1 Suppose first that κ is an integer. As $\kappa \geq 2$, there exist nonnegative integers s and t such that $\kappa = 3s + 2t$. Next, choose real closed fields $R_1, \ldots, R_s$ and algebraically closed fields $C_1, \ldots, C_t$ as in the statement of Lemma 3. By Lemma 2, if $1 \leq i \leq s$, there are exactly three R_i-algebra isomorphism classes of minimal ring extensions of R_i and a set of representatives for these classes is $\{\overline{R_i}, R_i(+)R_i, R_i \times R_i\}$. (Indeed, $\overline{R_i}$ is, up to isomorphism, the only minimal field extension of R_i because the algebraic closure of any real closed field $\mathfrak{R}$ is a two-dimensional field extension of $\mathfrak{R}$ [17, Corollary, page 276].) It follows that all the minimal ring extensions of R_i have the same cardinality as R_i. Similarly, by Lemma 2, if $1 \leq j \leq t$, there are exactly two C_j-algebra isomorphism classes of minimal ring extensions of C_j and a set of representatives for these classes is $\{C_j(+)C_j, C_j \times C_j\}$; it follows that all the minimal ring extensions of C_j have the same cardinality as C_j.

Consider a direct product $A = D_1 \times \cdots \times D_{s+t}$, where $D_1, \ldots, D_{s+t}$ is a (finite) list of nonzero rings. It is well known (cf. [11, Lemma III.3 (d)]) that a ring extension, say $\mathscr{E}$, of A can be expressed (up to isomorphism) as $\prod_{i=1}^{s+t} E_i$ where E_i is a ring extension of D_i for each $i = 1, \ldots, s+t$. Moreover, by (an easy inductive application of) [8, Lemma 2.14], $\mathscr{E}$ is a minimal ring extension of A if and only if there exists (a necessarily unique) $j \in \{1, \ldots, s+t\}$ such that $D_i = E_i$ if $i \neq j$ and $D_j \subset E_j$ is a minimal ring extension.

Apply the preceding paragraph, with $D_i := R_i$ if $1 \leq i \leq s$ and $D_{s+j} := C_j$ if $1 \leq j \leq t$. In other words,

$$A = R_1 \times \cdots \times R_s \times C_1 \times \cdots \times C_t.$$

The recipe in that paragraph allows us to build (A-algebra) isomorphic copies of all the minimal ring extensions of A. When this is done by using the information from the first paragraph, the number of minimal ring extensions of A that are built is

$$3 + \cdots + 3 + 2 + \cdots + 2 = 3s + 2t = \kappa.$$

We need only to show that no two of those minimal ring extensions, say $\mathcal{E}_1$ and $\mathcal{E}_2$, are isomorphic as A-algebras.

Suppose, on the contrary, that $\mathcal{E}_1 \cong \mathcal{E}_2$ as A-algebras. It is well known that when finitely many nonzero Artinian rings are used as factors to build a (finite) direct product of rings, those factors are determined up to isomorphism by that direct product (cf. [5, Theorem 8.7]). Thus, by the cardinality restrictions in Lemma 3 and the observations about cardinality in the first paragraph, either $\mathcal{E}_1$ and $\mathcal{E}_2$ were each built using a minimal ring extension of the same R_i or $\mathcal{E}_1$ and $\mathcal{E}_2$ were each built using a minimal ring extension of the same C_j. The arguments (leading to a contradiction) are the same for each of these possibilities and so, to ease the notation, we will assume that $\mathcal{E}_1$ and $\mathcal{E}_2$ were built, respectively, using the minimal ring extensions E_1 and E_2 of R_1. In other words,

$$\mathcal{E}_1 = E_1 \times R_2 \times \cdots \times R_s \times C_1 \times \cdots \times C_t$$

and $\mathcal{E}_2 = E_2 \times R_2 \times \cdots \times R_s \times C_1 \times \cdots \times C_t$, where we ignore the factors $R_2, \ldots, R_s$ if $s = 1$.

By hypothesis, there is an A-algebra isomorphism $\varphi : \mathcal{E}_1 \to \mathcal{E}_2$. To obtain the desired contradiction, it suffices to know that φ can be built coordinate-wise by using a suitable R_1-algebra homomorphism $E_1 \to E_2$ (which would necessarily be an isomorphism), along with the identity maps $R_i \to R_i$ (if $2 \leq i \leq s$) and $C_j \to C_j$ (if $1 \leq j \leq t$). This is known: the more general piece of folklore describing A-algebra homomorphisms when A is any finite direct product of rings can be proved by reasoning as in [6, pages 7–8]. This completes the proof in case κ is an integer.

Next, suppose that $\kappa = \aleph_0$. Let p be a prime number. We shall prove that $A := \mathbb{F}_p$ has the asserted behavior. In view of Lemma 2 (and the fact that $2 + \aleph_0 = \aleph_0$), it suffices to show that the cardinality of the set of $\mathbb{F}_p$-algebra isomorphism classes of minimal field extensions of $\mathbb{F}_p$ is $\aleph_0$. We can find representatives of these classes by working inside a fixed algebraic closure of $\mathbb{F}_p$. There, using the classical Galois theory of finite fields, we see that the minimal field extensions of $\mathbb{F}_p$ are the fields $\mathbb{F}_{p^q}$, where q runs over the set of all prime numbers. Different values of q lead to non-isomorphic minimal field extensions of $\mathbb{F}_p$, since $[\mathbb{F}_{p^q} : \mathbb{F}_p] = q$. Thus, the cardinal number of the set of A-algebra isomorphism classes of the minimal ring extensions of A is the cardinal number of the set of prime numbers, namely $\aleph_0 = \kappa$, as desired.

Finally, suppose that $\kappa > \aleph_0$. (Actually, the following argument also applies if $\kappa = \aleph_0$, but we thought it appropriate to include the direct proof for that case in the preceding paragraph.) Let I be a set with cardinal number κ and let $\{X_i \mid i \in I\}$ be a set of algebraically independent indeterminates over $\mathbb{Q}$, with $X_i \neq X_j$ if $i \neq j$ in I. (One could replace $\mathbb{Q}$ here with any other countable field of characteristic 0.) Let $K := \mathbb{Q}(\{X_i^2 \mid i \in I\})$, the quotient field of the polynomial ring $\mathbb{Q}[\{X_i^2 \mid i \in I\}]$. We shall prove that $A := K$ has the asserted behavior.

In view of Lemma 2 (and the fact that $2 + \kappa = \kappa$), it suffices to show that the cardinality of the set $\gimel$ of K-algebra isomorphism classes of minimal field extensions of K is κ. We can find representatives of these classes by working inside $\overline{K}$, a fixed algebraic closure of K. Since the usual rules of arithmetic for infinite cardinal numbers ensure that $|K| = \kappa$, we also have $|\overline{K}| = \kappa$ (cf. [17, Lemma, page 143]). Hence, there are at most κ K-subalgebras of $\overline{K}$ of the form $K[u]$ $(= K(u))$ for some $u \in \overline{K}$. As K has characteristic 0, every finite-dimensional algebraic field extension of K is separable (cf. [17, page 39], [16, Remarks (i), page 261]) and hence, by the Primitive Element Theorem (cf. [17, Theorem 14, page 54], [16, Proposition 6.15 (c), pages 287–288]) is of the form $K(u)$ for some $u \in \overline{K}$. Moreover every minimal field extension of K is an (a finite-dimensional) algebraic field extension of K. The upshot is that $\beth \leq \kappa$. It therefore will suffice to find a set $\{\xi_i \in \overline{K} \mid i \in I\}$ such that $K \subset K(\xi_i)$ is a minimal field extension for each $i \in I$ and $K(\xi_i)$ is not K-algebra isomorphic to $K(\xi_j)$ whenever $i \neq j$ in I.

Let $i \in I$. Note that X_i is algebraic over K since $X_i^2 \in K$. Thus $[K(X_i) : K] \leq 2$. We claim that $[K(X_i) : K] = 2$. This fact was essentially proved in [9, page 4497], but for the sake of completeness, we include its proof. Suppose the claim is false. Then there exist $f, g \in \mathbb{Q}[\{X_j^2 \mid j \in I\}]$ such that $X_i = f/g$ in K, and so $X_i g = f$ in the polynomial ring $\mathbb{Q}[\{X_j \mid j \in I\}]$. Let δ_1 (resp., δ_2) denote the degree, as a polynomial in X_i with coefficients in $\mathbb{Q}[\{X_j^2 \mid j \in I \setminus \{i\}\}]$, of g (resp., of f). Both δ_1 and δ_2 are even integers, since f and g are elements of $\mathbb{Q}[\{X_j^2 \mid j \in I\}]$. However, calculating degrees in the variable X_i leads to

$$1 + \delta_1 = \deg_{X_i}(X_i) + \deg_{X_i}(g) = \deg_{X_i}(X_i g) = \deg_{X_i}(f) = \delta_2,$$

so that $1 = \delta_2 - \delta_1$ is the difference of even integers, whence $1/2 \in \mathbb{Z}$, the desired contradiction. This proves the above claim. It follows that $K \subset K(X_i)$ is a minimal field extension. We can now revisit the above choice of the algebraic closure $\overline{K}$ and arrange that $\{X_i \mid i \in I\} \subseteq \overline{K}$. It now makes sense, for each $i \in I$, to put $\xi_i := X_i$. The proof will be complete if we show that $K(\xi_i)$ is not K-algebra isomorphic to $K(\xi_j)$ whenever $i \neq j$ in I.

Suppose, on the contrary, that there exists a K-algebra isomorphism $\psi : K(\xi_i) \to K(\xi_j)$ for some $i \neq j$ in I. As $\eta_i := \psi(\xi_i) \in \overline{K}$ satisfies

$$\eta_i^2 = \psi(\xi_i^2) = \psi(X_i^2) = X_i^2,$$

we have $\eta_i = \pm X_i$. Thus $X_i = \pm\eta_i \in K(\xi_j) = K + K\xi_j = K + KX_j$, and so $X_i = a + bX_j$ for some $a, b \in K$. Recall from the preceding paragraph that $X_i \notin K$ (and, similarly, $X_j \notin K$). Therefore $b \neq 0$. Also, since $X_i^2 = a^2 + 2abX_j + b^2X_j^2$,

$$2abX_j = X_i^2 - a^2 - b^2X_j^2 \in K + K + K = K.$$

Since $X_j \notin K$ and $2b \neq 0$, it follows that $a = 0$, and so $X_i = bX_j$.

Since $T := \mathbb{Q}[\{X_\nu^2 \mid \nu \in I\}]$ is a domain whose quotient field is K, we can write $b = f_1/g_1$, where f_1 and g_1 are nonzero elements of T. Let $\delta_3 := \deg_{X_i^2}(f_1)$ and $\delta_4 := \deg_{X_i^2}(g_1)$. We have $X_i/X_j = b = f_1/g_1$. Square the left- and right-hand sides and cross-multiply. The result is $X_i^2 g_1^2 = X_j^2 f_1^2$. Then

$$1 + 2\delta_4 = \deg_{X_i^2}(X_i^2 g_1^2) = \deg_{X_i^2}(X_j^2 f_1^2) = 0 + \deg_{X_i^2}(f_1^2) = 2\delta_3,$$

so that $1 = 2\delta_3 - 2\delta_4$ is an even integer, the desired contradiction, thus completing the proof. □

Remark 1 In view of the approach in Theorem 1 to the case of an integer $\kappa \geq 2$, one may expect a similar use of finite direct products when pursuing "dual" considerations for maximal subrings. However, that study will not be "dual" to the above work on minimal ring extensions. For instance, if B_1, B_2 are nonzero rings and $B := B_1 \times B_2$, then a subring A of B need not be of the form $A_1 \times A_2$ for suitable rings $A_1 \subseteq B_1$ and $A_2 \subseteq B_2$. The simplest instance of this phenomenon was noted by Werner [23, Remark 2.3]: if p is a prime number and $B := \mathbb{F}_p \times \mathbb{F}_p$, then the only proper subring of B is $\Delta := \{(a,a) \in B \mid a \in \mathbb{F}_p\}$. In this example, Δ is not of the form $A_1 \times A_2$ for suitable rings $A_1, A_2 \subseteq \mathbb{F}_p$. (Indeed, the only ring having that form is B itself.) It is interesting that Δ is isomorphic to B_1 $(= \mathbb{F}_p)$ and so, also isomorphic to B_2; cf. also Lemma 2. However, it would take us far afield if we were to pursue an approach emphasizing such isomorphisms in studying maximal subrings, and so that study below will mostly proceed via other means, with a notable exception in Proposition 1 (d).

We next examine the extent to which the concept of "maximal subring" admits analogues of the above results on the concept of "minimal ring extension." Proposition 1 will collect our initial results on maximal subrings. In case the cardinal number κ is uncountable, our results will not be as comprehensive as the conclusions in Theorem 1. After stating Proposition 1, we will comment on how its various assertions relate to the literature before proceeding with its proof.

Proposition 1

(a) There exists a ring B such that B has infinitely many subrings but B has no maximal subring.

(b) Let n be a positive integer. Then there exists a ring B such that B has exactly n maximal subrings.

(c) There exists a ring B such that the set of maximal subrings of B has cardinality $\aleph_0$.

(d) For each cardinal number $\kappa > \aleph_0$, there exists a ring B such that the set of maximal subrings of B has cardinality at least κ.

Azarang and Karamzadeh have constructed a Noetherian ring that has no maximal subring [2, Example 2.6]. For the sake of completeness, Proposition 1 (a) also constructs such a ring. Note also that much of the recent literature has provided several examples of rings that *do* have maximal subrings. For instance, it was shown in [1, Corollary 2.8] that every finite-type algebra B over a ring Λ has a maximal subring. (In reviewing [1] for Mathematical Reviews, we noted that one should make explicit in the preceding statement that B and Λ are nonzero. We would like to add here that one should also make explicit that B is a commutative unital Λ-algebra and that B is not its own prime subring; i.e., B is not isomorphic to $\mathbb{Z}$ or $\mathbb{Z}/n\mathbb{Z}$ for any integer $n \geq 2$.) One could view the rings B in [1, Corollary 2.8] as being somewhat "small." However, some "large" (uncountable) rings B also have maximal subrings, as Azarang and Karamzadeh have shown this to be the case whenever B is a ring such that $|\text{Max}(B)| > 2^{\aleph_0}$ [3]. In light of these results, it is natural to ask exactly how many maximal subrings a ring can have. In the spirit of Theorem 1, we show in Proposition 1 (b), (c) that any positive integer, as well as $\aleph_0$, can be realized as the cardinal number of the set of maximal subrings of a suitable ring.

It does not seem to be known whether every cardinal number $\kappa > \aleph_0$ can be realized as the cardinal number of the set of maximal subrings of some ring. Somewhat along these lines, Azarang and Karamzadeh have shown in [4] that if B is a Noetherian ring such that $|B| > 2^{\aleph_0}$, then the cardinal number of the set of maximal subrings of B is at least $|\text{Max}(R)|$. If $\kappa > \aleph_0$, we show in Proposition 1 (d) that some ring has at least κ maximal subrings. In addition, Corollary 1 establishes that if $\kappa = 2^\lambda$ for some infinite cardinal number λ, then there exists a field whose set of maximal subrings has cardinality κ. However, for other kinds of transfinite cardinal numbers κ (for instance, $\kappa = \aleph_\alpha$ for a limit ordinal α), we leave open the question whether there exists a ring having exactly κ maximal subrings.

Proof of Proposition 1 (a) Of course, the zero ring has no proper subring; nor do $\mathbb{Z}$ or $\mathbb{Z}/n\mathbb{Z}$ for any integer $n \geq 2$. Thus $\{0\}$, $\mathbb{Z}$ and $\mathbb{Z}/n\mathbb{Z}$ (for $n \geq 2$) are all examples of (Noetherian) rings whose set of maximal subrings has cardinality 0. These are, admittedly, somewhat trivial examples, as they each have no proper subrings. We next give an example where B has infinitely many subrings, none of which is a maximal subring.

Take E to be any nonzero abelian group that has infinitely many subgroups but no maximal (proper) subgroup. (Perhaps the best-known such E is the group

$$\mathbb{Z}_{p^\infty} := \{v \in \mathbb{Q}/\mathbb{Z} \mid p^n v = 0 \text{ for some positive integer } n\},$$

where p is any prime number: cf. [16, Example 7 (c), (d), page 37].) Put $B := \mathbb{Z}(+)E$. View $\mathbb{Z}$ as a subring of B via the (unital) ring homomorphism $\mathbb{Z} \to B$, $m \mapsto (m, 0)$. By Dobbs [7, Remark 2.9], the set of rings T such that $\mathbb{Z} \subseteq T \subseteq B$ (that is, the set of subrings of B) is the set of idealizations $\mathscr{S} := \{\mathbb{Z}(+)G \mid G \text{ is a subgroup of } E\}$. It follows, because of the conditions on E, that $\mathscr{S}$ is infinite; and, for all $\mathbb{Z}(+)G \in \mathscr{S}$ with G a proper subgroup of E, there exists a proper subgroup H of E which properly contains G, so that $\mathbb{Z}(+)G \subset \mathbb{Z}(+)H \subset \mathbb{Z}(+)E = B$ in $\mathscr{S}$. So B has no maximal subring, which completes the proof of (a).

(b) Since $\mathbb{Z}$ is a Bézout domain, it follows that each overring T of $\mathbb{Z}$ (that is, each subring T of $\mathbb{Q}$) is of the form $T = \mathbb{Z}_S$ for some saturated multiplicatively closed subset S of $\mathbb{Z}$ [13, Exercise 10 (b), page 250]. As is well known (cf. [5, Exercises 7 and 8, page 44]), such S are the sets of the form $\mathbb{Z} \setminus \cup_{i \in I} P_i$, where $\{P_i \mid i \in I\} \subseteq \text{Spec}(\mathbb{Z})$. (In this proof, we shall always assume that indexing is "efficient"; in the present situation, "efficient" means that $P_{i_1} \neq P_{i_2}$ whenever $i_1 \neq i_2$ in I.) Now, let $\{Q_j \mid j \in J\}$ and $\{P_i \mid i \in I\}$ be subsets of $\text{Spec}(\mathbb{Z})$. Put

$$A := \mathbb{Z}_{\mathbb{Z} \setminus \cup_{j \in J} Q_j} \text{ and } B := \mathbb{Z}_{\mathbb{Z} \setminus \cup_{i \in I} P_i}.$$

As $\mathbb{Z}$ is a (principal ideal domain and hence a) unique factorization domain, it is easy to see that $A \subseteq B$ if and only if $\cup_{i\in I}P_i \subseteq \cup_{j\in J}Q_j$.

Next, let $z_1, \ldots, z_n$ be n pairwise-distinct prime numbers, and write

$$\mathrm{Spec}(\mathbb{Z}) \setminus \{z_1\mathbb{Z}, \ldots, z_n\mathbb{Z}\} \text{ as } \{P_i \mid i \in I\}.$$

Take $A \subseteq B$ as above. If possible, do not allow a P_i or a Q_j to be $\{0\}$. Then it follows from unique factorization that $A = B$ if and only if there exists a bijection $\sigma : I \to J$ so that $P_i = Q_{\sigma(i)}$ for all $i \in I$. Suppose that $A \subset B$; that is,

$$\mathrm{Spec}(\mathbb{Z}) \setminus \cup_{j\in J}Q_j \subset \mathrm{Spec}(\mathbb{Z}) \setminus \cup_{i\in I}P_i = \{z_1\mathbb{Z}, \ldots, z_n\mathbb{Z}\}.$$

Then $|\mathrm{Spec}(\mathbb{Z}) \setminus \cup_{j\in J}Q_j| < n$. Moreover, A is a maximal subring of B if and only if $|\mathrm{Spec}(\mathbb{Z}) \setminus \cup_{j\in J}Q_j| = n - 1$; indeed, if

$$m := |\mathrm{Spec}(\mathbb{Z}) \setminus \cup_{j\in J}Q_j| < n - 1,$$

then there exists $J' \subset J$ with $|\mathrm{Spec}(\mathbb{Z}) \setminus \cup_{j\in J'}Q_j| = m + 1$, so that

$$A \subset \mathbb{Z}_{\mathbb{Z}\setminus\cup_{j\in J'}Q_j} \subset B.$$

There are exactly n subsets of $\{z_1\mathbb{Z}, \ldots, z_n\mathbb{Z}\}$ which have cardinality $n - 1$. Each of those subsets takes the form $\mathrm{Spec}(\mathbb{Z}) \setminus \cup_{j\in J}Q_j$ for a unique set $\{Q_j \mid j \in J\} \subseteq \mathrm{Spec}(\mathbb{Z})$ and hence determines a unique maximal subring $A = \mathbb{Z}_{\mathbb{Z}\setminus\cup_{j\in J}Q_j}$ of B. The upshot is that B has exactly n maximal subrings, as asserted.

(c) Put $B := \mathbb{Q}$. Since B is a field, the maximal subrings of B are the one-dimensional valuation domains having B as their quotient field (cf. [18, Exercise 29, page 43]), that is, the one-dimensional valuation overrings of $\mathbb{Z}$. As $\mathbb{Z}$ is a Prüfer domain, the valuation overrings of $\mathbb{Z}$ are (apart from $\mathbb{Q}$) the domains of the form $\mathbb{Z}_{p\mathbb{Z}}$ where p is any prime number (cf. [18, Theorems 64 and 65]). Each such $\mathbb{Z}_{p\mathbb{Z}}$ is one-dimensional. Moreover, the assignment $p \mapsto \mathbb{Z}_{p\mathbb{Z}}$ is an injection, since the maximal ideal of $\mathbb{Z}_{p\mathbb{Z}}$ meets $\mathbb{Z}$ in $p\mathbb{Z}$. Therefore, the cardinality of the set of maximal subrings of B is the same as the cardinality of the set of prime numbers, namely $\aleph_0$.

(d) Let I be a set with cardinality κ. Let B be the direct product, $B := \prod_I \mathbb{F}_2$, of a family of copies of the finite field $\mathbb{F}_2 = \{0, 1\}$ indexed by I. View elements of B as tuples of the form (a_i) as i runs over I. Well-order I. There is no harm in viewing I as the ordinal number κ. It is well known that since κ is a transfinite cardinal number, κ must be a limit ordinal (cf. [14, Exercise, page 100]). Thus, if $j \in I$, then $j + 1 \in I$. Hence, for each $j \in I$, the following set is well-defined: $T_j := \{(a_i) \in B \mid a_j = a_{j+1}\}$. It is clear that T_j is a proper subring of B, for each $j \in I$; and that $A_{j_1} \not\subseteq A_{j_2}$ if $j_1 \neq j_2$ in I. (The last assertion should be verified for the two natural cases, namely when $j_1 < j_2$ and when $j_2 < j_1$.) Thus, the assignment $j \mapsto T_j$ gives an injection from I to the set of proper subrings of B. To complete the proof, it suffices to show that if $j \in I$, then T_j is a maximal (proper) additive subgroup (and hence a maximal subring) of B. This would, in turn, follow from Lagrange's Theorem if the index $[B : T_j]$ is a prime number. In fact, $[B : T_j] = 2$. To see this, define $c := (c_i) \in B$ by $c_j := 1$ and $c_i := 0$ for all $i \in J \setminus \{j\}$, and note that $B = T_j \cup (T_j + c)$. The proof is complete. □

Remark 2 (a) The following is a more basic rendering of the proof of Proposition 1 (b), as it does not explicitly mention rings of fractions. Let $z_1, \ldots, z_n$ be n pairwise-distinct prime numbers. Let the set B consist of 0 and all the nonzero rational numbers which, when written in "lowest terms," have denominators divisible (in $\mathbb{Z}$) by no prime numbers other than $z_1, \ldots, z_n$. Using the usual rules for addition and multiplication, we see that B is a subring of $\mathbb{Q}$. Next, let $j \in \{1, \ldots, n\}$, and let A_j be the analogue of B where the above role of $\{z_1, \ldots, z_n\}$ is played by $\{z_1, \ldots, z_n\} \setminus \{z_j\}$. (If $n = 1$, this means that $A_j = \mathbb{Z}$.) Note that $1/z_j \in B \setminus A_j$. It follows that A_j is a proper subring of B. Similarly, $A_{j_1} \neq A_{j_2}$ if $1 \leq j_1 < j_2 \leq n$. Therefore, to show that B has exactly n maximal subrings (namely, the A_j), it suffices to prove the following. If D is a subring of B such that, for all $j \in \{1, \ldots, n\}$, there exists a nonzero element $\xi_j \in D$ whose expression in lowest terms has a denominator divisible by z_j, then $D = B$ (cf. [13, Lemma 27.2]). We next prove this statement.

It suffices to prove that if $j \in \{1, \ldots, n\}$, then $1/z_j \in D$. By hypothesis, some nonzero (integral) multiple of ξ_j can be expressed as a_j/z_j for some integer a_j which is relatively prime to z_j. There is no harm in replacing ξ_j with a_j/z_j. As the greatest common divisor of a_j and z_j (in $\mathbb{Z}$) is 1, there exist $\lambda, \mu \in \mathbb{Z}$ such that

$$\lambda a_j + \mu z_j = 1.$$

Dividing through by z_j gives $1/z_j = \lambda \xi_j + \mu \in D + \mathbb{Z} = D$, as required.

The referee has kindly advised the inclusion of another way to express the perspective on the proof of Proposition 1 (b) that was suggested in the preceding two paragraphs. Let $\Lambda := \{z_1, \ldots, z_n\}$ and B be as in the above proof of Proposition 1 (b). Since $\mathbb{Z}$ is a UFD, the elements of B are the rational numbers of the form a/b where $a, b \in \mathbb{Z}$, $b \neq 0$ and each prime number which divides b (in $\mathbb{Z}$) must belong to Λ. Next, observe as above that if a and b are relatively prime integers, then $\mathbb{Z}[a/b] = \mathbb{Z}[1/b]$. It follows easily that $B = \mathbb{Z}[1/z_1, \ldots, 1/z_n]$ and that B has exactly n maximal subrings $A_i = \mathbb{Z}[\Lambda_i]$, where $\Lambda_i := \Lambda \setminus \{1/z_i\}$, for $i = 1, \ldots, n$.

(b) From the point of view of multiplicative ideal theory, the following alternate proof of Proposition 1 (d) may be of interest, as it produces a field with the asserted behavior. Let κ be an infinite cardinal number. Let I be a set with cardinal number κ and let F be a field. Let $\{X_i \mid i \in I\}$ be a set of algebraically independent indeterminates over F (with $X_i \neq X_j$ if $i \neq j$ in I). Then the polynomial ring $R := F[\{X_i \mid i \in I\}]$ is an infinite-dimensional unique factorization domain. Let K denote the quotient field of R. Fix $j \in I$. As X_j is irreducible in R (by a degree argument) and R is a UFD, it follows that X_j is a prime element of R. Thus, since R is a UFD, $P_j := X_jR$ is a height one prime ideal of R (cf. [18, Theorem 5]). Next, by Gilmer [13, Theorem 19.6], there exists a valuation overring V_j of R (contained in

K) whose maximal ideal meets R at P_j. As we are working inside the quotient field of R and $\text{ht}_R(P_j) = 1$, it follows that each nonzero prime ideal $\mathfrak{P}$ of V_j meets R in P_j (cf. [13, Lemma 11.1]). Hence, $\mathfrak{Q} := \cap\mathfrak{P}$ (the so-called *pseudo-radical* of R) also satisfies $\mathfrak{Q} \cap R = P_j$. As the prime ideals of (the valuation domain) V_j are linearly ordered by inclusion, $\mathfrak{Q} \in \text{Spec}(V_j)$ and $\text{ht}_{V_j}(\mathfrak{Q}) = 1$. Therefore, $W_j := (V_j)_{\mathfrak{Q}}$ is a one-dimensional valuation overring of R whose maximal ideal, $\mathfrak{Q}$, lies over P_j. By Kaplansky [18, Exercise 29, page 43], W_j is a maximal subring of K. As the assignment $j \mapsto W_j$ gives an injection from I into the set of maximal subrings of K, the alternate proof of Proposition 1 (d) can be completed by taking $B := K$.

(c) It should be noted that the proof in (b) actually produces a ring B that has more than κ maximal subrings, in fact, at least 2^κ maximal subrings. To see this, note first the construction of W_j in (b) ensured that X_j is a nonunit of W_j and X_i is a unit of W_j for all $i \in I \setminus \{j\}$. Next, with $\epsilon_i \in \{1, -1\}$ for each $i \in I$, repeat the argument in (b), replacing R with the unique factorization domain $F[\{X_i^{\epsilon_i} \mid i \in I\}]$. For each $j \in I$, we have that in the resulting maximal subring of K, $X_j^{\epsilon_j}$ is a nonunit and $X_i^{\epsilon_i}$ is a unit for all $i \in I \setminus \{j\}$. Thus, as all of the 2^κ possible choices of the list $(\epsilon_i)_{i \in I}$ are considered, we end up constructing $2^\kappa \cdot \kappa = 2^\kappa$ (pairwise distinct) maximal subrings of K, as asserted.

We next give what may be our deepest realizability result for infinite cardinal numbers.

Corollary 1 *Let κ be an infinite cardinal number of the form $\kappa = 2^\lambda$ for some (infinite) cardinal number λ. Then there exists a field K whose set of maximal subrings has cardinality κ.*

Proof Rework parts (b) and (c) of Remark 2, with λ playing the earlier role of κ. The upshot is that the cardinal number of the set of maximal subrings of K is at least $2^\lambda = \kappa$. Now, specialize the earlier construction by taking the field F in Remark 2 (b) to be finite. As it suffices to show that the cardinal number of the set of maximal subrings of K is at most 2^λ, we need only prove that $|K| \leq \lambda$. This, in turn, follows from the usual rules of arithmetic for infinite cardinal numbers, since $|F| < \lambda$ leads to $|R| = \lambda$.

Remark 3 (a) Despite Remark 2 (c), it is possible to prove the following. For each nonzero cardinal number κ, there exists an integral domain D, with quotient field K, such that the set of maximal overrings of D (inside K) has cardinal number κ; that is, such that κ is the cardinal number of the set of maximal subrings of K which contain D. We next prove this statement.

Let $\mathcal{M}$ be a set of cardinality κ and let θ be an element such that $\theta \notin \mathcal{M}$. Consider the set $\mathcal{T} := \mathcal{M} \cup \{\theta\}$. Give $\mathcal{T}$ the structure of a partially ordered set by decreeing that ($x \leq x$ for each $x \in \mathcal{T}$ and) $\theta \leq m$ for each $m \in \mathcal{M}$ (with no other instances where the relation $\leq$ holds). Note that $\mathcal{T}$ is a tree. (Recall that if $x \in \mathcal{T}$, then $x^\downarrow := \{y \in \mathcal{T} \mid y \leq x\}$. The conclusion that $\mathcal{T}$ is a tree holds since $x^\downarrow$ is linearly ordered for each $x \in \mathcal{T}$. Indeed, if $m \in \mathcal{M}$, then $m^\downarrow = \{\theta, m\}$; and $\theta^\downarrow = \{\theta\}$.) Moreover, θ is the unique minimal element of $\mathcal{T}$. In addition, since the only nontrivial chains in $\mathcal{T}$ are of the form $\theta \leq m$ (with $m \in \mathcal{M}$), it is clear that $\mathscr{T}$ satisfies Kaplansky's conditions (K_1) and (K_2). (In other words, each chain in $\mathscr{T}$ has an infimum and a supremum; and whenever $x < y$ in $\mathscr{T}$, there exist "immediate neighbors" $x_1 < y_1$ in $\mathscr{T}$ such that $x \leq x_1 < y_1 \leq y$.) Therefore, by Lewis [19, Theorem 3.1], there exists a Prüfer domain D such that Spec(D), as a partially ordered set under inclusion, is order-isomorphic to $\mathscr{T}$. Under any such order-isomorphism, θ corresponds to the prime ideal $\{0\}$ of D and each $m \in \mathscr{M}$ corresponds to a maximal ideal of D. Thus, R is one-dimensional and $|\text{Max}(D)| = |\mathscr{M}| = \kappa$. As in the proof of Proposition 1 (c), it follows from [18, Exercise 29, page 43 and Theorems 64 and 65] that the assignment $P \mapsto D_P$ defines a bijection from Max(D) to the set of maximal subrings of K which contain D. Therefore, the cardinal number of this set is $|\text{Max}(D)| = |\mathscr{M}| = \kappa$, as asserted.

(b) In view of Corollary 1, it is natural to ask if the field K in (a) has exactly 2^κ maximal subrings. We do not know the answer in general. But we shall next show that if the field used in proving Jaffard's Theorem is a finite field L, then K has at most 2^κ maximal subrings. (Recall from (a) that K has at least κ maximal subrings. Of course, given the result that we just announced, the answer to the question of exactly how many maximal subrings K has will then be one of the cardinal numbers between κ and 2^κ, and the nature of those cardinal numbers depends on the model of set theory being used.)

Recall from the proof of (a) that K is the quotient field of a certain integral domain D, and so it suffices to prove that $|D| = \kappa$ (for then $|K| = \kappa$). In the proof of (a), D was found using the proof of [19, Theorem 3.1]. That proof involved the lattice ordered abelian group $A := \{f : \mathscr{T} \to \mathbb{Z} \mid f(x) = 0$ for all but at most finitely many $x \in \mathscr{T}\}$. Since $|\mathscr{T}| = \kappa$ is infinite, it follows from the usual rules for arithmetic of infinite cardinal numbers that $|A| = \kappa$. The proof of [19, Theorem 3.1] went on to cite the proof of Jaffard's Theorem in [13, Theorem 18.6]. In that last-mentioned proof, the domain D is defined as the group ring LA. As L is finite and $|A| = \kappa$, a final appeal to the laws of arithmetic for infinite cardinal numbers gives $|D| = \kappa$, thus completing the proof.

(c) Suppose that the field L that was mentioned in (b) is finite. Then, combining the results in (a) and (b), we see that the cardinal number $\beth$ of the set of maximal subrings of the field K satisfies $\kappa \leq \beth \leq 2^\kappa$. If κ is infinite, the nature of the cardinal numbers between κ and 2^κ can depend on the model of set theory being used. Perhaps it would be prudent, in closing, to note that similar questions occasionally do not admit simple answers. See, for instance, [22]: it is shown there that if R is a countable ring and E is an R-module with $|E| = \aleph_1$, the answer to the question of whether $2^{\aleph_1}$ is the cardinal number of the set of (not necessarily maximal) submodules of E is independent of ZFC. Thus, by Dobbs [7, Remark 2.9], one cannot use ZFC alone to determine the cardinal number of the set of (not necessarily maximal) R-subalgebras of $R(+)E$.

References

1. Azarang, A.: On maximal subrings. Far East J. Math. Sci. **32**, 107–118 (2009)
2. Azarang, A., Karamzadeh, O.A.S.: On the existence of maximal subrings in commutative Artinian rings. J. Algebra Appl. **9**, 771–778 (2010)
3. Azarang, A., Karamzadeh, O.A.S.: Which fields have no maximal subrings? Rend. Sem. Mat. Univ. Padova **126**, 213–228 (2011)
4. Azarang, A., Karamzadeh, O.A.S.: On maximal subrings of commutative rings. Algebra Colloq. **19**, 1125–1138 (2012)
5. Atiyah, M.F., Macdonald, I.G.: Introduction to Commutative Algebra. Addison-Wesley, Reading, MA (1969)
6. Dobbs, D.E.: Cech cohomology and a dimension theory for commutative rings. Ph.D. thesis, Cornell University, Ithaca, New York (1969)
7. Dobbs, D.E.: Every commutative ring has a minimal ring extension. Commun. Algebra **34**, 3875–3881 (2006)
8. Dobbs, D.E.: On the commutative rings with at most two proper subrings. Int. J. Math. Math. Sci. **2016**, 1–13 (2016). Article ID 6912360. doi:10.1155/2016/6912360
9. Dobbs, D.E., Mullins, B.: On the lengths of maximal chains of intermediate fields in a field extension. Commun. Algebra **29**, 4487–4507 (2001)
10. Dobbs, D.E., Shapiro, J.: A classification of the minimal ring extensions of certain commutative rings. J. Algebra **308**, 800–821 (2007)
11. Dobbs, D.E., Mullins, B., Picavet, G., Picavet-L'Hermitte, M.: On the FIP property for extensions of commutative rings. Commun. Algebra **33**, 3091–3119 (2005)
12. Ferrand, D., Olivier, J.-P.: Homomorphismes minimaux d'anneaux. J. Algebra **16**, 461–471 (1970)
13. Gilmer, R.: Multiplicative Ideal Theory. Dekker, New York (1972)
14. Halmos, P.R.: Naive Set Theory. Van Nostrand, Princeton (1960)
15. Huckaba, J.A.: Commutative Rings with Zero Divisors. Dekker, New York (1988)
16. Hungerford, T.W.: Algebra. Springer, New York (1974)
17. Jacobson, N.: Lectures in Abstract Algebra, Volume III - Theory of Fields and Galois Theory. Van Nostrand, Princeton (1964)
18. Kaplansky, I.: Commutative Rings, rev. ed. University of Chicago Press, Chicago (1974)
19. Lewis, W.J.: The spectrum of a ring as a partially ordered set. J. Algebra **25**, 419–434 (1973)
20. Picavet, G., Picavet-L'Hermitte, M.: About minimal morphisms. In: Multiplicative Ideal Theory in Commutative Algebra, pp. 369–386. Springer, New York (2006)
21. Picavet, G., Picavet-L'Hermitte, M.: Modules with finitely many submodules. Int. Electron. J. Algebra **19**, 119–131 (2016)
22. Stepräns, J.: The number of submodules. Proc. Lond. Math. Soc. **49**, 183–192 (1984)
23. Werner, N.J.: Covering numbers of finite rings. Am. Math. Mon. **122**, 552–566 (2015)

Applications of Multisymmetric Syzygies in Invariant Theory

M. Domokos

Abstract A presentation by generators and relations of the nth symmetric power B of a commutative algebra A over a field of characteristic zero or greater than n is given. This is applied to get information on a minimal homogeneous generating system of B (in the graded case). The known result that in characteristic zero the algebra B is isomorphic to the coordinate ring of the scheme of n-dimensional semisimple representations of A is also recovered. The special case when A is the two-variable polynomial algebra and $n = 3$ is applied to find generators and relations of an algebra of invariants of the symmetric group of degree four that was studied in connection with the problem of classifying sets of four unit vectors in the Euclidean space.

Keywords Symmetric product • Generators and relations • Multisymmetric polynomials • Trace identities • Cayley–Hamilton theorem

MSC: 13A50, 16R30

1 Introduction

Let n be a positive integer and let A be a commutative K-algebra (with identity 1), where K is a field of characteristic zero or $\mathrm{char}(K) > n$. Denote by $T^n(A)$ the nth tensor power of A. This is a commutative K-algebra in the standard way. The symmetric group S_n acts on $T^n(A)$ by permuting the tensor factors. This is an action via K-algebra automorphisms. Write $T^n(A)^{S_n}$ for the subalgebra of S_n-invariants. First we give a presentation of this commutative K-algebra in terms of generators and relations. Note that even if this algebra is finitely generated, we have to take a redundant (typically infinite) generating system that allows a simple description of the relations.

M. Domokos (✉)
MTA Alfréd Rényi Institute of Mathematics, Reáltanoda utca 13–15, 1053 Budapest, Hungary
e-mail: domokos.matyas@renyi.mta.hu

M. Fontana et al. (eds.), *Rings, Polynomials, and Modules*,
https://doi.org/10.1007/978-3-319-65874-2_9

The key step is to pay attention to a K-vector space basis of $T^n(A)^{S_n}$ which comes together with a rule to rewrite products of the generators of $T^n(A)^{S_n}$ into normal form. The rewriting algorithm is furnished by relations that come uniformly from one master relation. This yields the desired presentation of $T^n(A)^{S_n}$, see Theorem 1. As an application we deduce information on a minimal generating system of $T^n(A)^{S_n}$ in the case when A is graded, see Corollary 1 and Corollary 2. This has relevance, for instance, for the study of polynomial invariants associated with representations of wreath products, see Sect. 3.

Since the master relation mentioned above comes from the Cayley–Hamilton identity of matrices, as a corollary of Theorem 1 we obtain Corollary 4 asserting that $T^n(A)^{S_n}$ is isomorphic to the subalgebra $\mathcal{O}(\mathrm{rep}(A,n))^{GL(n,K)}$ of $GL(n,K)$-invariants in the coordinate ring of the scheme of n-dimensional representations of A (where A is a finitely generated K-algebra and $\mathrm{char}(K) = 0$). This latter result is due to Vaccarino [30, Theorem 4.1.3], who proved it by a different approach.

In Sect. 5 we turn to a very concrete application of Theorem 1. Its special case when $A = K[x_1, x_2]$ is a two-variable polynomial ring was used in [10, 11] to derive generators of the ideal of relations among a minimal generating system of $T^3(A)^{S_3}$. This is applied here to give a minimal presentation of the ring of invariants R^{S_4} of the permutation representation of the symmetric group S_4 associated with the action of S_4 on the set of two-element subsets of the set $\{1, 2, 3, 4\}$. This ring of invariants was studied before by Aslaksen et al. [2] because of its relevance for classifying sets of four unit vectors in an Euclidean space. A Hironaka decomposition and a minimal generating system of R^{S_4} was computed in [2]. Here we get simultaneously the generators and relations essentially as a consequence of the minimal presentation of $T^3(K[x_1, x_2])^{S_3}$ mentioned above. We note that the ring of invariants R^{S_4} fits into a series that has relevance for the graph isomorphism problem, and has been studied, for example, in [28].

2 Generators and Relations for Symmetric Tensor Powers of a Commutative Algebra

Choose a subset $\mathcal{M} \subset A$ such that $1 \notin \mathcal{M}$ and $\{1\} \cup \mathcal{M}$ is a K-vector space basis in A. For $w \in \mathcal{M}$ set

$$[w] := w \otimes 1 \otimes \cdots \otimes 1 + 1 \otimes w \otimes 1 \otimes \cdots \otimes 1 + \cdots + 1 \otimes \cdots \otimes 1 \otimes w. \qquad (1)$$

Proposition 1 *The products $[w_1] \cdots [w_r]$ with $r \leq n$, $w_i \in \mathcal{M}$ constitute a K-vector space basis of $T^n(A)^{S_n}$.*

Proof The elements $w_1 \otimes \cdots \otimes w_n$ with $w_i \in \{1\} \cup \mathcal{M}$ constitute a basis in $T^n(A)$, and S_n permutes these basis vectors. For a multiset $\{w_1, \ldots, w_r\}$ with $r \leq n$, $w_i \in \mathcal{M}$, denote by $O_{\{w_1,\ldots,w_r\}}$ the S_n-orbit sum of $w_1 \otimes \cdots \otimes w_r \otimes 1 \otimes \cdots \otimes 1$, and call r its *height*. Clearly these elements constitute a basis in $T^n(A)^{S_n}$. Assume that the multiset $\{w_1, \ldots, w_r\}$ contains d distinct elements with multiplicities $r_1, \ldots, r_d$ (so $r_1 + \cdots + r_d = r$), then expanding $[w_1] \cdots [w_r]$ as a linear combination of the above basis elements, the coefficient of $O_{\{w_1,\ldots,w_r\}}$ is $r_1! \cdots r_d!$, and all other basis elements contributing have strictly smaller height. This clearly shows the claim. □

Remark 1 The special case when $A = K[x_1, \ldots, x_m]$ is a polynomial ring and $\mathcal{M}$ is the set of monomials appears in [3, 15, Proposition 2.5'] (see [4] for the interpretation needed here), and [10]. A formally different but related statement is Lemma 3.2 in [30].

Take commuting indeterminates T_w ($w \in \mathcal{M}$), and write $\mathcal{F}$ for the commutative polynomial algebra $\mathcal{F} := K[T_w \mid w \in \mathcal{M}]$. Write

$$\varphi : \mathcal{F} \to T^n(A)^{S_n} \tag{2}$$

for the K-algebra homomorphism given by $T_w \mapsto [w]$ for all $w \in \mathcal{M}$.

To a multiset $\{w_1, \ldots, w_{n+1}\}$ of $n+1$ elements from $\mathcal{M}$ we associate an element $\Psi_{\{w_1\ldots,w_{n+1}\}} \in \mathcal{F}$ as follows. Write $\mathcal{P}_{n+1}$ for the set of partitions $\lambda = \lambda_1 \bigcup \cdots \bigcup \lambda_h$ of the set $\{1, \ldots, n+1\}$ into the disjoint union of non-empty subsets λ_i, and denote $h(\lambda) = h$ the number of parts of the partition λ. Set

$$\Psi_{\{w_1,\ldots,w_{n+1}\}} = \sum_{\lambda \in \mathcal{P}_{n+1}} (-1)^{h(\lambda)} \prod_{i=1}^{h(\lambda)} \left((|\lambda_i| - 1)! \cdot T_{\prod_{s \in \lambda_i} w_s} \right),$$

where for a general element $a \in A$ (say for $a = \prod_{s \in \lambda_i} w_s$) we write

$$T_a := c_0 n + \sum_{w \in \mathcal{M}} c_w T_w$$

provided that $a = c_0 + \sum_{w \in \mathcal{M}} c_w w$, with $c_0, c_w \in K$.

Theorem 1 *The K-algebra homomorphism φ is surjective onto $T^n(A)^{S_n}$, and its kernel is the ideal generated by the $\Psi_{\{w_1,\ldots,w_{n+1}\}}$, where $\{w_1, \ldots, w_{n+1}\}$ ranges over all multisets of $n+1$ elements in $\mathcal{M}$.*

Proof Surjectivity of φ follows from Proposition 1. The fact that $\Psi_{\{w_1,\ldots,w_{n+1}\}}$ belongs to $\ker(\varphi)$ follows from the Cayley–Hamilton identity. Let $Y(1), \ldots, Y(n+1)$ be $n \times n$ matrices over an arbitrary commutative ring. For a permutation $\pi \in S_{n+1}$ with cycle decomposition

$$\pi = (i_1 \cdots i_d) \cdots (j_1 \ldots j_e)$$

set

$$\mathrm{Tr}^\pi = \mathrm{Tr}(Y(i_1)\cdots Y(i_d))\cdots \mathrm{Tr}(Y(j_1)\cdots Y(j_e)).$$

Then we have the equality

$$\sum_{\pi \in S_{n+1}} \mathrm{sign}(\pi)\mathrm{Tr}^\pi = 0 \tag{3}$$

called the *fundamental trace identity* of $n \times n$ matrices. This can be proved by multilinearizing the Cayley–Hamilton identity to get an identity multilinear in the n matrix variables $Y(1), \dots, Y(n)$, and then multiplying by $Y(n+1)$ and taking the trace; see, for example, [13, 21, 23] for details. For $a \in A$ denote by $\tilde{a}$ the diagonal $n \times n$ matrix whose ith diagonal entry is $1 \otimes \cdots \otimes 1 \otimes a \otimes 1 \otimes \cdots \otimes 1$ (the ith tensor factor is a). The substitution $Y(i) \mapsto \tilde{w}_i$ $(i = 1, \dots, n+1)$ in (3) yields that $\varphi(\Psi_{\{w_1,\dots,w_{n+1}\}}) = 0$.

The coefficient in $\Psi_{\{w_1,\dots,w_{n+1}\}}$ of the term $T_{w_1}\cdots T_{w_{n+1}}$ is $(-1)^{n+1}$, and all other terms are products of at most n variables T_u. Therefore the relation $\varphi(\Psi_{\{w_1,\dots,w_{n+1}\}}) = 0$ can be used to rewrite $[w_1]\cdots[w_{n+1}]$ as a linear combination of products of at most n invariants of the form $[u]$ (where $u \in \mathcal{M}$). So these relations are sufficient to rewrite an arbitrary product of the generators $[w]$ in terms of the basis given by Proposition 1. This implies our statement about the kernel of φ. □

Remark 2 (i) The ideal of relations among the generators of the algebra of multi-symmetric polynomials (i.e. the special case $T^n(A)^{S_n}$ with $A = K[x_1, \dots, x_m]$) has been studied by several authors, see [5–7, 16–18, 29]. The case of Theorem 1 when A is the qth Veronese subalgebra of the m-variable polynomial algebra $K[x_1, \dots, x_m]$ was given in Theorem 2.5 of [10]. The approach of Theorem 2.5 in [10] is close to an argument for Theorem 2.1 in [3]. The proof of Theorem 1 is a generalization to arbitrary commutative A of the proof of Theorem 2.5 in [10].

(ii) Although the presentation of $T^n(A)^{S_n}$ given in Theorem 1 is infinite, in certain cases an a priori upper bound for the degrees of relations in a minimal presentation is available, and a finite presentation can be obtained from the infinite presentation above (see, for instance, [10, Theorem 3.2], building on [8]). Based on this procedure even a minimal presentation is worked out in [11] for $T^3(K[x_1, \dots, x_m])^{S_3}$.

Suppose that A is generated as a K-algebra by the elements $a_1, \dots, a_m$. Take the m-variable polynomial algebra $K[x_1, \dots, x_m]$ and denote by $\rho : K[x_1, \dots, x_m] \to A$ the K-algebra surjection given by $\rho(x_i) = a_i$ for $i = 1, \dots, m$. This induces a K-algebra surjection $T^n(\rho) : T^n(K[x_1, \dots, x_m]) \to T^n(A)$ in the obvious way. Since $T^n(\rho)$ is S_n-equivariant and char(K) does not divide $|S_n|$, we deduce that it restricts to a K-algebra surjection $T^n(\rho) : T^n(K[x_1, \dots, x_m])^{S_n} \to T^n(A)^{S_n}$. Since the algebra $T^n(K[x_1, \dots, x_m])^{S_n}$ is classically known to be generated by $[x_{i_1} \dots x_{i_d}]$ where $d \le n$, $1 \le i_1 \le \cdots \le i_d \le m$ (see [19, 20, 25, 27] or [31, Chapter II.3], and for a characteristic free statement see [12, 24] or [30, Corollary 3.14]), we obtain the following known fact:

Proposition 2 *The K-algebra $T^n(A)^{S_n}$ is generated by*

$$\{[a_{i_1}\cdots a_{i_d}] \mid \quad d \le n, \quad 1 \le i_1 \le \cdots \le i_d \le m\}.$$

Remark 3 Proposition 2 follows also directly from Theorem 1, since the relation $\varphi(\Psi_{\{w_1,\dots,w_{n+1}\}}) = 0$ can be used to rewrite $[w_1 \cdots w_{n+1}]$ as a linear combination of products of invariants of the form $[u]$ where u is a proper subproduct of $w_1 \cdots w_{n+1}$.

Moreover, when $A = K[x_1, \dots, x_m]$ is a polynomial ring, the above generating set is minimal. For a general commutative K-algebra A the generating set in Proposition 2 may not be minimal:

Example 1 Let $A = K[x^2, x^3]$ be the K-subalgebra of the polynomial ring $K[x]$ generated by $s = x^2$ and $t = x^3$. Then $\{s^i t^j \mid i = 0, 1, 2; \quad j = 0, 1, \dots\}$ is a K-vector space basis of A. The K-algebra generating system of $T^2(A)^{S_2}$ given by Proposition 2 is $\{[s], [t], [s^2], [st], [t^2]\}$. However, this generating system is not minimal, because $t^2 = s^3$ implies that

$$[t^2] = [s^3] = \frac{3}{2}[s^2][s] - \frac{1}{2}[s]^3,$$

thus the algebra $T^2(A)^{S_2}$ is generated by $\{[s], [t], [s^2], [st]\}$.

As an application of Theorem 1 we shall derive some information on a minimal homogeneous generating system of $T^n(A)^{S_n}$ when A is graded. We shall work with graded algebras $R = \bigoplus_{d=0}^{\infty} R_d$, where $R_0 = K$. Set $R_+ := \bigoplus_{d>0} R_d$. We say that a homogeneous $r \in R_+$ is *indecomposable* if $r \notin (R_+)^2$; that is, r is not contained in the subalgebra generated by lower degree elements.

Corollary 1 *Let A be a graded algebra whose minimal positive degree homogeneous component has degree q.*

(i) Suppose that b is a non-zero homogeneous element in A with $\deg(b) < (n+1)q$*. Then $[b]$ is an indecomposable element in the graded algebra $T^n(A)^{S_n}$ (whose grading is induced by the grading on A).*

(ii) Suppose that $\mathscr{B}$ is a K-linearly independent subset of A consisting of homogeneous elements of positive degree strictly less than $(n+1)q$. Then the set $\{[b] \mid b \in \mathscr{B}\}$ is part of a minimal homogeneous K-algebra generating system of $T^n(A)^{S_n}$.

Proof Obviously (i) is a special case of (ii), so we shall prove (ii). Extend the set $\mathscr{B}$ to a subset $\mathscr{M} \subset A$ such that

1. $\mathscr{M}$ consists of homogeneous elements of positive degree;
2. $\{1\} \cup \mathscr{M}$ is a K-vector space basis of A.

This is possible by the assumptions. Consider the corresponding (infinite) presentation of $T^n(A)^{S_n}$ by generators and relations given in Theorem 1. Endow $\mathscr{F}$ with a grading such that $\deg(T_w) = \deg(w)$ for all $w \in \mathscr{M}$. According to Theorem 1 the ideal $\ker(\varphi)$ is generated by $\Psi_{\{w_1,\dots,w_{n+1}\}}$, where $\{w_1, \dots, w_{n+1}\}$ ranges over all multisets of $n+1$ elements in $\mathscr{M}$. Clearly $\Psi_{\{w_1,\dots,w_{n+1}\}} \in \mathscr{F}$ is homogeneous of degree $\deg(w_1)+\dots+\deg(w_{n+1})$. For each w_i we have $\deg(w_i) \geq q$, implying

$$\deg(w_1)+\dots+\deg(w_{n+1}) \geq (n+1)q > \deg(b) \text{ for all } b \in \mathscr{B}.$$

Consequently $\ker(\varphi)$ is a homogeneous ideal generated by elements of degree strictly greater than $\deg(b)$ for any $b \in \mathscr{B}$. By the graded Nakayama Lemma (see, for example, [9, Lemma 3.5.1]) it is sufficient to show that setting $R = T^n(A)^{S_n}$, the cosets $\{[b]+(R_+)^2 \mid b \in \mathscr{B}\}$ are K-linearly independent in the factor space $R/(R_+)^2$. This is equivalent to the condition that if

$$h = \sum_{b\in\mathscr{B}} c_b T_b \in \ker(\varphi) + (\mathscr{F}_+)^2 \text{ for some } c_b \in K, \tag{4}$$

then all coefficients c_b are zero, that is, $h = 0$. The ideals $\ker(\varphi)$ and $(\mathscr{F}_+)^2$ are homogeneous, and since all non-zero homogeneous components of $\ker(\varphi)$ have degree strictly grater than $\deg(b)$ for all $b \in \mathscr{B}$, we deduce from (4) that $h \in (\mathscr{F}_+)^2$. Now h is a linear combination of the variables in the polynomial ring $\mathscr{F}$ (in possibly infinitely many variables), so this leads to the conclusion $h = 0$. □

Proposition 2 and Corollary 1 have the following immediate corollary:

Corollary 2 *Suppose that A is a graded algebra generated by homogeneous elements of the same positive degree q. Let $\mathscr{B}$ be the union of K-vector space bases of the positive degree homogeneous components of A of degree at most nq (for example, $\mathscr{B}$ can be chosen to be a set of products of length at most n of the generators of A). Then $\{[b] \mid b \in \mathscr{B}\}$ is a minimal homogeneous K-algebra generating system of the algebra $T^n(A)^{S_n}$.*

3 Wreath Products

Corollary 1 and Corollary 2 can be applied in invariant theory, one of whose basic targets is to find a minimal homogeneous K-algebra generating system in an *algebra of polynomial invariants* $S(V)^G$. Here G is a group acting on a finite dimensional vector space V via linear transformations, and $S(V)$ is the symmetric tensor algebra of V (i.e. a polynomial algebra with a basis of V as the variables) endowed with the induced G-action via K-algebra automorphisms, and $S(V)^G$ is the subalgebra consisting of the elements fixed by G.

For a group G and a positive integer n the *wreath product* $G \wr S_n$ is defined as the semidirect product $H \rtimes S_n$, where $H = G \times \dots \times G$ is the direct product of n copies of G, and conjugation by $\sigma \in S_n$ yields the corresponding permutation of the direct factors of H. Given a G-module V there is a natural $G \wr S_n$-module structure on $V^n = V \oplus \dots \oplus V$ (n direct summands) given by

$$(g_1, \ldots, g_n, \sigma) \cdot (v_1, \ldots, v_n) = (g_1 \cdot v_{\sigma^{-1}(1)}, \ldots, g_n \cdot v_{\sigma^{-1}(n)}). \tag{5}$$

Consider the corresponding *algebra* $S(V^n)^{G \wr S_n}$ *of polynomial invariants*. Clearly we have $S(V^n)^{G \wr S_n} \subseteq S(V^n)^H \subseteq S(V^n)$. Since H is normal in $G \wr S_n$, the subalgebra $S(V^n)^H \subseteq S(V^n)$ is S_n-stable, and $S(V^n)^{G \wr S_n} = (S(V^n)^H)^{S_n}$. With the usual identification $S(V^n) = T^n(S(V))$ we obtain

$$S(V^n)^H = S(V^n)^{G \times \cdots \times G} = T^n(S(V)^G),$$

and formula (5) shows that the action of S_n on $S(V^n)$ corresponds to the action on $T^n(S(V))$ via permutation of the tensor factors. We conclude the identification

$$S(V^n)^{G \wr S_n} = T^n(S(V)^G)^{S_n}.$$

Therefore Corollary 1 and Corollary 2 have the following consequence:

Corollary 3 *Let q denote the minimal positive degree of a homogeneous element in $S(V)^G$, and let $\mathscr{B}$ be the union of K-vector space bases of the homogeneous components of $S(V)^G$ of positive degree strictly less than $(n+1)q$.*

(i) Then $\{[b] \mid b \in \mathscr{B}\}$ is part of a minimal homogeneous K-algebra generating system of $S(V^n)^{G \wr S_n}$.

(ii) Assume in addition that $S(V)^G$ is generated by its homogeneous component of degree q. Then $\{[b] \mid b \in \mathscr{B}\}$ is a minimal homogeneous K-algebra generating system of $S(V^n)^{G \wr S_n}$.

Example 2 (i) Let G be the special linear group $SL_q(K)$ acting by left multiplication on the space $V = K^{q \times r}$ of $q \times r$ matrices. Then by classical invariant theory (see [31]) we know that $S(V)^G$ is generated by the determinants of $q \times q$ minors, all having degree q, so Corollary 3 (ii) applies for $S(V^n)^{G \wr S_n}$.

(ii) Let G be the cyclic group of order q acting by multiplication by a primitive qth root of 1 on $V = K^m$. In this case the ring $S(V^n)^{G \wr S_n}$ can be interpreted as the ring of vector invariants of some pseudo-reflection group. Note that [10, Theorem 2.5 (ii)] is a special case of Corollary 3 (ii).

4 The Scheme of Semisimple Representations of A

In this section we assume that $\mathrm{char}(K) = 0$, and A is a finitely generated commutative K-algebra. We shall use the following notation: given a homomorphism $\mu : C \to D$ of commutative K-algebras we write $\mu^{n \times n} : C^{n \times n} \to D^{n \times n}$ for the homomorphism induced in the obvious way between the corresponding algebras of $n \times n$ matrices.

Now choose K-algebra generators $a_1, \ldots, a_m$ of A, and consider the K-algebra surjection $\pi : K\langle x_1, \ldots, x_m \rangle \to A$, $x_i \mapsto a_i$ from the free associative K-algebra.

Take m generic $n \times n$ matrices $X(1), \dots, X(m)$ (their mn^2 entries are indeterminates in an mn^2-variable polynomial algebra $K[x_{ij}^{(r)} \mid 1 \leq i,j \leq n,\ r = 1,\dots,m]$). Take the factor of this polynomial algebra by the ideal generated by all entries of $f(X(1),\dots,X(m))$, where f ranges over $\ker(\pi)$ (in particular, the entries of $X(r)X(s) - X(s)X(r)$ are among the generators of this ideal). This algebra is the *coordinate ring* $\mathcal{O}(\mathrm{rep}(A,n))$ of the *scheme* $\mathrm{rep}(A,n)$ *of n-dimensional representations of A* (by definition of this scheme). We write $\eta : K[x_{ij}^{(r)} \mid 1 \leq i,j \leq n,\ r = 1,\dots,m] \to \mathcal{O}(\mathrm{rep}(A,n))$ for the natural surjection. Denote by $Y(1),\dots,Y(m) \in \mathcal{O}(\mathrm{rep}(A,n))^{n\times n}$ the images of the generic matrices $X(1),\dots,X(m)$ under $\eta^{n\times n}$. Then the K-algebra homomorphism $\rho : K\langle x_1,\dots,x_m\rangle \to \mathcal{O}(\mathrm{rep}(A,n))^{n\times n}$ given by $x_i \mapsto Y(i)$ factors through a K-algebra homomorphism $\overline{\rho} : A \to \mathcal{O}(\mathrm{rep}(A,n))^{n\times n}$. The conjugation action of the general linear group $GL(n,K)$ on $n \times n$ matrices induces an action (via K-algebra automorphisms) on $\mathcal{O}(\mathrm{rep}(A,n))$. Namely $g \in GL(n,K)$ maps the (i,j)-entry of $Y(l)$ to the (i,j)-entry of $gY(l)g^{-1}$. Consider the subalgebra $\mathcal{O}(\mathrm{rep}(A,n))^{GL(n,K)}$ of $GL(n,K)$-invariants. Motivated by Artin [1] we call it the *coordinate ring of the scheme of semisimple n-dimensional representations of A*.

Define the K-algebra homomorphism

$$\gamma : \mathcal{F} \to \mathcal{O}(\mathrm{rep}(A,n)) \text{ given by } T_w \mapsto \mathrm{Tr}(\overline{\rho}(w))$$

where $\mathcal{F}$ stands for the same polynomial algebra (possibly in infinitely many variables) as in Sect. 2.

Corollary 4 *The K-algebra homomorphism γ factors through an isomorphism*

$$\overline{\gamma} : T^n(A)^{S_n} \xrightarrow{\cong} \mathcal{O}(\mathrm{rep}(A,n))^{GL(n,K)}$$

(so $\gamma = \overline{\gamma} \circ \varphi$, where $\varphi : \mathcal{F} \to T^n(A)^{S_n}$, $T_w \mapsto [w]$ is defined in (2)).

Proof Since the fundamental trace identity holds for matrices over the commutative ring $\mathcal{O}(\mathrm{rep}(A,n))$, we conclude that all the $\Psi_{\{w_1,\dots,w_{n+1}\}}$ belong to $\ker(\gamma)$. By Theorem 1 these elements generate the ideal $\ker(\varphi)$, hence $\ker(\gamma) \supseteq \ker(\varphi)$, implying the existence of a (unique) K-algebra homomorphism $\overline{\gamma}$ with $\gamma = \overline{\gamma} \circ \varphi$.

The homomorphism γ (and hence $\overline{\gamma}$) is surjective onto $\mathcal{O}(\mathrm{rep}(A,n))^{GL(n,K)}$ because the algebra of $GL(n,K)$-invariants is generated by traces of monomials in the generic matrices $Y(1),\dots,Y(m)$ (this follows from [26] since the characteristic of K is zero, hence $GL(n,K)$ is linearly reductive).

Define $\beta : K[x_{ij}^{(r)} \mid 1 \leq i,j \leq n,\quad r = 1,\dots,m] \to T^n(A)$ by $\beta^{n\times n} : X(r) \mapsto \tilde{a}_r$ where we use the notation of the proof of Theorem 1, so $\tilde{a}_r$ is a diagonal $n\times n$ matrix over $T^n(A)$ whose jth diagonal entry is $1 \otimes \cdots \otimes 1 \otimes a_r \otimes 1 \otimes \cdots \otimes 1$ (the jth tensor factor is a_r). Since $f(\tilde{a}_1,\dots,\tilde{a}_m) = 0$ for any $f \in \ker(\pi)$, we conclude that β factors through a homomorphism $\overline{\beta} : \mathcal{O}(\mathrm{rep}(A,n)) \to T^n(A)$. We have $\overline{\beta}^{n\times n} : Y(i) \mapsto \tilde{a}_i$ and more generally, $\overline{\beta}^{n\times n} : \overline{\rho}(w) \mapsto \tilde{w}$. It follows that

$$\overline{\beta}(\overline{\gamma}([w])) = \overline{\beta}(\gamma(T_w)) = \overline{\beta}(\mathrm{Tr}(\overline{\rho}(w))) = \mathrm{Tr}(\overline{\beta}^{n\times n}(\overline{\rho}(w))) = \mathrm{Tr}(\tilde{w}) = [w],$$

hence $\overline{\beta} \circ \overline{\gamma}$ is the identity map on $T^n(A)^{S_n}$. This shows that $\overline{\gamma}$ is injective as well, and hence it is an isomorphism. □

Remark 4 The isomorphism $T^n(A)^{S_n} \cong \mathcal{O}(\mathrm{rep}(A,n))^{GL(n,K)}$ is a result of Vaccarino [30, Theorem 4.1.3]. The proof given above is different, and it is an adaptation of the proof of [10, Theorem 4.1]. The special case when A is a polynomial ring is discussed also in [14] and [22].

5 The Symmetric Group Acting on Pairs

Write $\binom{[n]}{2}$ for the set of two-element subsets of $\{1,\dots,n\}$. The symmetric group S_n acts on the $\binom{n}{2}$-variable polynomial algebra

$$R_n = K[x_{\{i,j\}} \mid \{i,j\} \in \binom{[n]}{2}]$$

as

$$\sigma \cdot x_{\{i,j\}} = x_{\{\sigma(i),\sigma(j)\}} \qquad \text{for } \sigma \in S_n.$$

The subalgebra $R_n^{S_n}$ was studied for two reasons, the first comes from graph theory. Given a simple graph Γ with vertex set $\{1,\dots,n\}$ and a polynomial $f \in R_n$ denote by $f(\Gamma)$ the value of f under the substitution

$$x_{\{i,j\}} \mapsto \begin{cases} 1 \text{ if } \{i,j\} \text{ is an edge in } \Gamma \\ 0 \text{ otherwise.} \end{cases}$$

Suppose that $f_1,\dots,f_r$ generate the K-algebra $R_n^{S_n}$. The following statement is well known (see, for example, [9, Lemma 5.5.1]):

Proposition 3 *The graphs Γ and Γ' on the vertex set $\{1,\dots,n\}$ are isomorphic if and only if $f_i(\Gamma) = f_i(\Gamma')$ for all $i = 1,\dots,r$.*
The second motivation to study $R_n^{S_n}$ comes from the problem of classifying orbits of n-element sets of unit vectors in the Euclidean space $\mathbb{R}^m$ under the action of the orthogonal group $O(\mathbb{R}^m)$. We refer to [2] for the details (see also [9, section 5.10.2]).

From now on we focus on the case $n = 4$ and set $R := R_4$. For our purposes a more convenient generating system of R is $x_1, x_2, x_3, z_1, z_2, z_3$ where

$$x_1 = x_{\{1,2\}} + x_{\{3,4\}}, \quad x_2 = x_{\{1,3\}} + x_{\{2,4\}}, \quad x_3 = x_{\{1,4\}} + x_{\{2,3\}}$$

$$z_1 = x_{\{1,2\}} - x_{\{3,4\}}, \quad z_2 = x_{\{1,3\}} - x_{\{2,4\}}, \quad z_3 = x_{\{1,4\}} - x_{\{2,3\}}.$$

We shall use the notation of Sect. 2. Identify R with the third tensor power $T^3(K[x,z])$ of the two-variable polynomial algebra $K[x,z]$ as follows:

$$x_1 = x \otimes 1 \otimes 1, \quad x_2 = 1 \otimes x \otimes 1, \quad x_3 = 1 \otimes 1 \otimes x$$

$$z_1 = z \otimes 1 \otimes 1, \quad z_2 = 1 \otimes z \otimes 1, \quad z_3 = 1 \otimes 1 \otimes z.$$

Consequently we have

$$x_1 + x_2 + x_3 = [x], \quad z_1^2 + z_2^2 + z_3^2 = [z^2], \quad x_1 z_1^2 + x_2 z_2^2 + x_3 z_3^2 = [xz^2],$$
$$\text{and} \quad x_1^i z_1^j + x_2^i z_2^j + x_3^i z_3^j = [x^i z^j] \quad \text{in general.}$$

The symmetric group S_3 acts on $T^3(K[x,z])$ by permutation of the tensor factors (as in Sect. 2), and $T^3(K[x,z^2])$ is an S_3-stable subalgebra of $T^3(K[x,z])$.

Proposition 4 *The algebra R^{S_4} is a rank two free module over a subalgebra isomorphic to the third symmetric power of the two-variable polynomial ring. More concretely, under the identification $R = T^3(K[x,z])$ we have*

$$R^{S_4} = T^3(K[x,z^2])^{S_3} \oplus z_1 z_2 z_3 T^3(K[x,z^2])^{S_3}, \tag{6}$$

and

$$(z_1 z_2 z_3)^2 = \frac{1}{3}[z^6] - \frac{1}{2}[z^4][z^2] + \frac{1}{6}[z^2]^3 \in T^3(K[x,z^2])^{S_3}. \tag{7}$$

Proof Relation (7) is just the Newton–Girard formula expressing the third elementary symmetric polynomial of z_1^2, z_2^2, z_3^2 in terms of their power sum symmetric functions. Denote by H the abelian normal subgroup of S_4 consisting of the three double transpositions and the identity. The variables x_1, x_2, x_3 are H-invariant whereas z_1, z_2, z_3 span H-invariant subspaces on which H acts via its three non-trivial characters. It follows that the subalgebra R^H is generated by

$$x_1, x_2, x_3, z_1^2, z_2^2, z_3^2, z_1 z_2 z_3.$$

Denote by P the subalgebra of R^H generated by the six algebraically independent elements $x_1, x_2, x_3, z_1^2, z_2^2, z_3^2$. Thus under the identification $R = T^3(K[x,z])$ we have $P = T^3(K[x,z^2])$. The square of the remaining generator $z_1 z_2 z_3$ belongs to P by (7), hence

$$R^H = P \oplus P z_1 z_2 z_3 \tag{8}$$

is a rank two free P-module. Since H is normal, R^H is an S_4-stable subalgebra of R, and the action of S_4 on R^H factors through an action of $S_4/H \cong S_3$. More concretely, S_4 permutes the elements x_1, x_2, x_3 and it permutes the elements z_1, z_2, z_3 up to sign.

In fact $z_1z_2z_3$ is S_4-invariant (as one can easily check) and there exists a surjective group homomorphism $\pi : S_4 \to S_3$ (with kernel H) such that for any $g \in S_4$ we have

$$g \cdot x_i = x_{\pi(g)(i)}, \quad g \cdot z_i^2 = z_{\pi(g)(i)}^2 \quad (i = 1, 2, 3).$$

This shows in particular that P is an $S_3 = S_4/H$-stable subalgebra of R^H, and under the identification $P = T^3(K[x, z^2])$ the $S_3 = S_4/H$-action on P is identified with the S_3-action on $T^3(K[x, z^2])$ via permutation of the tensor factors. Since $z_1z_2z_3$ is S_3-invariant, we deduce from (8) that $R^{S_4} = (R^H)^{S_3} = P^{S_3} \oplus P^{S_3}z_1z_2z_3$ is a rank two free P^{S_3}-module, and in fact (6) holds. □

Theorem 2 *Identify R and $T^3(K[x, z])$ as above.*

(i) The algebra R^{S_4} is generated by the ten elements

$$[x],\ [x^2],\ [x^3],\ [z^2],\ [z^4],\ [z^6],\ [xz^2],\ [x^2z^2],\ [xz^4],\ z_1z_2z_3.$$

(ii) Consider the surjective K-algebra homomorphism ϕ from the ten-variable polynomial algebra

$$\mathcal{F} = K[T_w, S \mid w \in \{x, x^2, x^3, y, y^2, y^3, xy, x^2y, xy^2\}]$$

onto R^{S_4} given by

$$\phi(S) = z_1z_2z_3 \quad \text{and} \quad \phi(T_w) = [\psi(w)],$$

where $\psi : K[x, y] \to K[x, z]$ is the K-algebra homomorphism mapping $x \mapsto x$ and $y \mapsto z^2$. The kernel of ϕ is minimally generated (as an ideal) by the element

$$S^2 - \frac{1}{3}T_{y^3} + \frac{1}{2}T_{y^2}T_y - \frac{1}{6}T_y^3 \tag{9}$$

and the five elements (given explicitly in the proof)

$$J_{3,2},\quad J_{2,3},\quad J_{4,2},\quad J_{3,3},\quad J_{2,4}. \tag{10}$$

Proof We know from Proposition 2 that the subalgebra $T^3(K[x, z^2])^{S_3}$ of R^{S_4} is generated by $\{[x], [x^2], [x^3], [z^2], [z^4], [z^6], [xz^2], [x^2z^2], [xz^4]\}$, hence statement (i) follows from (6) in Proposition 4.

Next we turn to the proof of (ii). Denote by ϕ' the restriction of ϕ to the nine-variable polynomial algebra

$$\mathcal{F}' = K[T_w \mid w \in \{x, x^2, x^3, y, y^2, y^3, xy, x^2y, xy^2\}].$$

Then $\phi' : \mathcal{F}' \to T^3(K[x,z^2])^{S_3}$ is a surjective K-algebra homomorphism (as we pointed out above). The element (9) belongs to $\ker(\phi)$ by (7). In particular, relation (9) allows to rewrite the square $(z_1z_2z_3)^2$ of the 10th generator of R^{S_4} given in (i) as a polynomial in the first 9 generators. It follows from (6) in Proposition 4 that relation (9) together with $\ker(\phi')$ generates the ideal $\ker(\phi)$. Moreover, since the indeterminate $S \in \mathcal{F}$ occurs only in the relation (9) and does not occur in $\ker(\phi')$, a minimal generating system of the ideal $\ker(\varphi')$ together with the element (9) constitutes a minimal homogeneous generating system of $\ker(\phi)$.

Therefore it is sufficient to prove that $\ker(\phi')$ is minimally generated by the elements (10). The K-algebra homomorphism $\psi : K[x,y] \to K[x,z]$ induces an isomorphism

$$\tilde{\psi} : T^3(K[x,y]) \to T^3(K[x,z^2]), \quad x_i \mapsto x_i, \quad y_i \mapsto z_i^2, \quad i = 1,2,3$$

(where we identify $T^3(K[x,y])$ with $K[x_1,x_2,x_3,y_1,y_2,y_3]$ similarly to the identification of $T^3(K[x,z])$ and R). Thus we have

$$\phi' = \tilde{\psi} \circ \mu \tag{11}$$

where μ stands for the K-algebra surjection

$$\mu : \mathcal{F}' \to T^3(K[x,y])^{S_3}, \quad T_w \mapsto [w] \text{ for } w \in \{x, x^2, x^3, y, y^2, y^3, xy, x^2y, xy^2\}$$

studied in detail in [10] and [11]. Moreover, by (11) we have

$$\ker(\phi') = \ker(\mu),$$

since $\tilde{\psi}$ is injective. It was explained first in section 6.2 of [10] how a minimal generating system of $\ker(\mu)$ can be deduced from a special case of Theorem 1. Later in [11] a natural action of the general linear group $GL_2(K)$ was also taken into account. Namely $GL_2(K)$ acts on $K[x,y]$ via linear substitutions of the variables. That is, $\begin{pmatrix} a & b \\ c & d \end{pmatrix} \in GL_2(K)$ maps $f(x,y) \in K[x,y]$ to $f(ax+cy, bx+dy)$. Take the third tensor power of this $GL_2(K)$-representation. We get an action of $GL_2(K)$ on $T^3(K[x,y])$ which obviously commutes with the S_3-action we considered. Therefore $T^3(K[x,y])^{S_3}$ is a $GL_2(K)$-stable subalgebra. There is an obviously defined $GL_2(K)$-action on $\mathcal{F}'$ such that μ is $GL_2(K)$-equivariant, hence $\ker(\mu)$ is a $GL_2(K)$-stable ideal. Theorem 3.1 in [11] asserts in this special case that $\ker(\mu)$ is minimally generated as a $GL_2(K)$-stable ideal by the elements $J_{3,2}$ and $J_{4,2}$ given as follows:

$$\begin{aligned} J_{3,2} = {} & 6T_{x^2y}T_{xy} - 3T_{xy^2}T_{x^2} - 2T_{x^2y}T_xT_y + T_{xy^2}T_x^2 - 4T_{xy}^2T_x \\ & + 2T_{xy}T_x^2T_y - 3T_{x^3}T_{y^2} + 4T_{x^2}T_xT_{y^2} - T_x^3T_{y^2} + T_{x^3}T_y^2 - T_{x^2}T_xT_y^2 \end{aligned}$$

$$
\begin{aligned}
J_{4,2} = {} & 6T_{x^2y}^2 + T_{xy}^2 T_{x^2} - 3T_{xy}^2 T_x^2 - 6T_{x^3} T_{xy^2} + 2T_{x^2} T_{xy^2} T_x \\
& + 4T_{x^3} T_{xy} T_y - 2T_{x^2} T_{xy} T_x T_y + 2T_{xy} T_x^3 T_y - 4T_{x^2y} T_{x^2} T_y - T_{x^2}^2 T_{y^2} \\
& + T_{x^2}^2 T_y^2 + 4T_{x^2} T_x^2 T_{y^2} - T_{x^2} T_x^2 T_y^2 - T_x^4 T_{y^2} - 2T_{x^3} T_x T_{y^2}
\end{aligned}
$$

The $GL_2(K)$-submodule of $\ker(\mu)$ generated by $J_{3,2}$ is two-dimensional with K-basis $J_{3,2}, J_{2,3}$ where

$$
\begin{aligned}
J_{2,3} = {} & 6T_{xy^2} T_{xy} - 3T_{x^2y} T_{y^2} - 2T_{xy^2} T_x T_y + T_{x^2y} T_y^2 - 4T_{xy}^2 T_y \\
& + 2T_{xy} T_y^2 T_x - 3T_{y^3} T_{x^2} + 4T_{y^2} T_y T_{x^2} - T_y^3 T_{x^2} + T_{y^3} T_x^2 - T_{y^2} T_y T_x^2 .
\end{aligned}
$$

The $GL_2(K)$-submodule of $\ker(\mu)$ generated by $J_{4,2}$ is three-dimensional with K-basis $J_{4,2}, J_{3,3}, J_{2,4}$ where

$$
\begin{aligned}
J_{3,3} = {} & 3T_{x^2y} T_{xy^2} - T_{xy} T_{x^2} T_{y^2} + T_{xy}^3 + T_{xy} T_x^2 T_{y^2} - 5T_{xy}^2 T_x T_y - 3T_{x^3} T_{y^3} \\
& + 2T_{xy} T_{xy^2} T_x + T_{x^2} T_x T_{y^3} - 3T_{x^2} T_{xy^2} T_y + 2T_{x^2y} T_{xy} T_y + 3T_{x^2} T_x T_{y^2} T_y \\
& + T_{x^3} T_{y^2} T_y + T_{x^2} T_{xy} T_y^2 - T_x^3 T_{y^2} T_y + 2T_x^2 T_{xy} T_y^2 - T_{x^2} T_x T_y^3 - 3T_x T_{x^2y} T_{y^2}
\end{aligned}
$$

$$
\begin{aligned}
J_{2,4} = {} & 6T_{xy^2}^2 + T_{xy}^2 T_{y^2} - 3T_{xy}^2 T_y^2 - 6T_{y^3} T_{x^2y} + 2T_{y^2} T_{x^2y} T_y \\
& + 4T_{y^3} T_{xy} T_x - 2T_{y^2} T_{xy} T_x T_y + 2T_{xy} T_y^3 T_x - 4T_{xy^2} T_{y^2} T_x - T_{y^2}^2 T_{x^2} \\
& + T_{y^2}^2 T_x^2 + 4T_{y^2} T_y^2 T_{x^2} - T_{y^2} T_y^2 T_x^2 - T_y^4 T_{x^2} - 2T_{y^3} T_y T_{x^2} .
\end{aligned}
$$

□

Relation (9) shows that the generator $\phi(T_{y^3}) = [z^6]$ of R^{S_4} is redundant. Consider the subalgebra

$$
\mathscr{F}_1 = K[T_w, S \mid w \in \{x, x^2, x^3, y, y^2, xy, x^2y, xy^2\}]
$$

of $\mathscr{F}$ in Theorem 2. A minimal presentation of R^{S_4} in terms of generators and relations is as follows:

Corollary 5

(i) The algebra R^{S_4} is minimally generated by the nine elements

$$
[x],\ [x^2],\ [x^3],\ [z^2],\ [z^4],\ [xz^2],\ [x^2z^2],\ [xz^4],\ z_1z_2z_3 .
$$

(ii) The kernel of the surjective K-algebra homomorphism

$$
\phi|_{\mathscr{F}_1} : \mathscr{F}_1 \to R^{S_4}
$$

is minimally generated (as an ideal) by the five elements

$$\tilde{J}_{3,2}, \quad \tilde{J}_{2,3}, \quad \tilde{J}_{4,2}, \quad \tilde{J}_{3,3}, \quad \tilde{J}_{2,4}$$

obtained from the elements (10) *via the substitution*

$$T_{y^3} \mapsto 3S^2 + \frac{3}{2}T_{y^2}T_y - \frac{1}{2}T_y^3.$$

Proof It is well known (and follows, for example, from Corollary 2) that $T^3(K[x,y])^{S_3}$ is minimally generated by $[x], [x^2], [x^3], [y], [y^2], [y^3], [xy], [x^2y], [xy^2]$. Consequently $T^3(K[x,z^2])^{S_3}$ is minimally generated by $[x], [x^2], [x^3], [z^2], [z^4]$, $(z_1z_2z_3)^2, [xz^2], [x^2z^2], [xz^4]$. It is now easy to deduce from Proposition 4 that the generating set of R^{S_4} given in the statement is *minimal*. The statement about $\ker(\phi|_{\mathscr{F}_1})$ follows from Theorem 2 (ii). □

Remark 5

(i) A minimal generating system of R^{S_4} was given by Aslaksen et al. [2, Theorem 4] by different methods. The authors also mention that they found the basic syzygies (relations) among the generators with the aid of computer, but the relations turned out to be quiet complicated, and so they left it out from their paper.

(ii) The approach of Aslaksen et al. in [2, Theorem 4] is based on finding a Hironaka decomposition of R^{S_4} (i.e. a presentation of R^{S_4} as a finite rank free module over a polynomial subalgebra). We note that a Hironaka decomposition of R^{S_4} can be derived also from Proposition 4 and the Hironaka decomposition of $T^3(K[x,y])^{S_3}$ given in [10, section 6.2]. Setting $Q = K[[x],[x^2],[x^3],[z^2],[z^4],[z^6]]$ we have that Q is a 6-variable polynomial algebra and

$$P = T^3(K[x,z^2])^{S_3} = Q \oplus [xz^2]Q \oplus [x^2z^2]Q \oplus [xz^4]Q \oplus [xz^2]^2Q \oplus [x^2z^2][xz^4]Q \tag{12}$$

is a free Q-module of rank 6. Note that

$$S = Q \oplus z_1z_2z_3Q = K[[x],[x^2],[x^3],[z^2],[z^4],z_1z_2z_3]$$

is a 6-variable polynomial algebra as well, and by (12) and by (6) we obtain

$$R^{S_4} = S \oplus [xz^2]S \oplus [x^2z^2]S \oplus [xz^4]S \oplus [xz^2]^2S \oplus [x^2z^2][xz^4]S$$

is a free S-module of rank 6. This is compatible with the result of [2, Theorem 4].

Acknowledgements I thank Wai-kit Yeung for a question about the topic of Sect. 4 that inspired me to formulate Theorem 1. This research was supported by National Research, Development and Innovation Office, NKFIH K 119934.

References

1. Artin, M.: On Azumaya algebras and finite dimensional representations of rings. J. Algebra **11**, 532–563 (1969)
2. Aslaksen, H., Chan, S.-P., Gulliksen, T.: Invatiants of S_n and the shape of sets of vectors. Appl. Algebra Eng. Commun. Comput. **7**, 53–57 (1996)
3. Berele, A.: Trace identities for diagonal and upper triangular matrices. Int. J. Algebra Comput. **6**, 645–654 (1996)
4. Berele, A., Regev, A.: Some results on trace cocharacters. J. Algebra **176**, 1013–1024 (1995)
5. Briand, E.: Polynômes multisymétriques. Ph.D. dissertation, Université de Rennes I and Universidad de Cantabria (October 2002)
6. Bukhshtaber, V.M., Rees, E.G.: Rings of continuous functions, symmetric products, and Frobenius algebras, (in Russian). Usp. Mat. Nauk **59**, 125–144 (2004)
7. Dalbec, J.: Multisymmetric functions. Beiträge Algebra Geom. **40**, 27–51 (1999)
8. Derksen, H.: Degree bounds for syzygies of invariants. Adv. Math. **185**, 207–214 (2004)
9. Derksen, H., Kemper, G.: Computational Invariant Theory, 1st edn. Springer, Berlin (2002)
10. Domokos, M.: Vector invariants of a class of pseudo-reflection groups and multisymmetric syzygies. J. Lie Theory **19**, 507–525 (2009)
11. Domokos, M., Puskás, A.: Multisymmetric polynomials in dimension three. J. Algebra **356**, 283–303 (2012)
12. Fleischmann, P.: A new degree bound for vector invariants of symmetric groups. Trans. Am. Math. Soc. **350**, 1703–1712 (1998)
13. Formanek, E.: The polynomial identities and invariants of $n \times n$ matrices. In: Regional Conference Series in Mathematics, vol. 78. American Mathematical Society, Providence (1991)
14. Gan, W.L., Ginzburg, V.: Almost-commuting variety, D-modules, and Cherednik algebras. Int. Math. Res. Pap. **2**, 1–54 (2006)
15. Gelfand, I.M., Kapranov, M.K., Zelevinsky, A.V.: Discriminants, Resultants, and Multidimensional Determinants. Birkhauser, Boston (1994)
16. Junker, Fr.: Die Relationen, welche zwischen den elementaren symmetrischen Functionen bestehen. Math. Ann. **38**, 91–114 (1891)
17. Junker, Fr.: Ueber symmetrische Functionen von mehreren Reihen von Veränderlichen. Math. Ann. **43**, 225–270 (1893)
18. Junker, Fr.: Die symmetrischen Functionen und die Relationen zwischen den Elementarfunctionen derselben. Math. Ann. **45**, 1–84 (1894)
19. MacMahon, P.A.: Combinatory Analysis, vol. 2. Cambridge University Press, Cambridge (1916)
20. Noether, E.: Der Endlichkeitssatz der Invarianten endlicher Gruppen. Math. Ann. **77**, 89–92 (1915)
21. Procesi, C.: The invariant theory of $n \times n$ matrices. Adv. Math. **19**, 306–381 (1976)
22. Procesi, C.: Invariants of commuting matrices. arXiv:1501.05190
23. Razmyslov, Y.P.: Trace identities of full matrix algebras over a field of characteristic zero, (in Russian). Izv. Akad. Nauk. SSSR Ser. Mat. **38**, 723–756 (1974)
24. Rydh, D.: A minimal set of generators for the ring of multisymmetric functions. Ann. Inst. Fourier (Grenoble) **57**, 1741–1769 (2007)
25. Schläfli, L.: Über die Resultante eines Systemes mehrerer algebraischer Gleichungen. Vienna Acad. Denkschriften **4**, 1–74 (1852)

26. Sibirskiĭ, K.S.: Algebraic invariants for a set ofmatrices, (in Russian). Sib. Math. Zhurnal **9**, 152–164 (1968)
27. Stepanov, S.A.: Polynomial invariants of finite groups over fields of prime characteristic, (in Russian). Diskret. Mat. **11**, 3–14 (1999)
28. Thiéry, N.M.: Algebraic invariants of graphs; a study based on computer exploration. SIGSAM Bull. **34**, 9–20 (2000)
29. Vaccarino, F.: The ring of multisymmetric functions. Ann. Inst. Fourier (Grenoble) **55**, 717–731 (2005)
30. Vaccarino, F.: Moduli of linear representations, symmetric products and the non commutative Hilbert scheme, Geometric methods in representation theory. II, 437–458, Séminaires et Congrès, 24-II, Société Mathématique, France, Paris, 2012
31. Weyl, H.: The Classical Groups. Princeton University Press, Princeton (1939)

Functorial Properties of Star Operations: New Developments

Jesse Elliott

Abstract We generalize the definition of a star operation and related notions to rings with zerodivisors and investigate, for any universal star operation $*$, various generalizations of the $*$-linked extensions that yield functoriality of $*$-class groups. These include a generalization of the PDE extensions of a Krull domain that also yields functoriality of divisor class groups.

Keywords Commutative ring • Star operation • Krull domain • Class group • t-class group • t-linked extension

Mathematics Subject Classification (2010) 13A15, 13F05, 13B02

1 Introduction

The divisor class group $\mathrm{Cl}(A)$ of a Krull domain A is a measure of the extent to which unique factorization fails in A. For flat extensions, or more generally for extensions that satisfy a condition known as PDE [6, Section 6], it is known that the divisor class group $\mathrm{Cl}(A)$ is functorial in A: if an extension $B \supseteq A$ of Krull domains satisfies the particular condition PDE, then the function $\mathrm{Cl}(A) \longrightarrow \mathrm{Cl}(B)$ acting by $[I] \longmapsto [(IB)^v]$ is a group homomorphism that is functorial in towers, where $v : I \longmapsto I^v = (I^{-1})^{-1}$ is the operation of divisorial closure on fractional ideals. This fact is useful for relating the properties of a Krull domain with those of its PDE extensions. However, the divisor class group is not functorial in A for all extensions of Krull domains.

The notion of a *star operation* on an integral domain, introduced by Krull in the guise of his *′-operations* [7, 8], allows one to generalize the theories of Dedekind domains and Krull domains to far more generally classes of domains. The notion can be extended naturally to all commutative rings as follows. All rings

J. Elliott (✉)
California State University, Channel Islands, One University Drive, Camarillo, CA 93012, USA
e-mail: jesse.elliott@csuci.edu

M. Fontana et al. (eds.), *Rings, Polynomials, and Modules*,
https://doi.org/10.1007/978-3-319-65874-2_10

and algebras are assumed commutative with identity. Let A be a ring with total quotient ring K. An element a of A is said to be *regular* if it is a nonzerodivisor. We let A^{reg} denote the monoid of all regular elements of A. We let $\mathbf{K}(A)$ denote the lattice of all A-submodules of K. The lattice $\mathbf{K}(A)$ is an ordered monoid under the operation $(I, J) \longmapsto IJ$, where IJ denotes the A-submodule of K generated by the set $\{ab : a \in I, b \in J\}$. An A-submodule of K is said to be *regular* if it contains a regular element of A. A *fractional ideal* of A is an A-submodule I of K such that $I^{-1} = (A :_K I)$ is regular, or equivalently such that $aI \subseteq A$ for some regular $a \in A$. We let $\mathbf{F}(A)$ denote the ordered monoid of all regular fractional ideals of A. A *star operation on* A is a nucleus $*$ on the ordered monoid $\mathbf{F}(A)$ such that $A^* = A$, that is, it is a closure operation $* : I \longmapsto I^*$ on the lattice $\mathbf{F}(A)$ such that $I^*J^* \subseteq (IJ)^*$ for all $I, J \in \mathbf{F}(A)$ and $A^* = A$ [4, 5].

Among the most important star operations are the operation

$$v : I \longmapsto I^v = (I^{-1})^{-1}$$

of *divisorial closure*, or *v-closure*, the operation

$$t : I \longmapsto I^t = \bigcup\{J^v : J \in \mathbf{F}(A),\ J \subseteq I \text{ is fin.gen.}\}$$

of *t-closure* (equal to v if A is a Krull domain), the operation

$$w : I \longmapsto I^w = \bigcup\{(I :_K J) : J \in \mathbf{F}(A) \text{ is fin.gen.},\ J^t = A\}$$

of *w-closure* (equal to t if A is a PVMD), and the identity operation $d : I \longmapsto I$ (equal to v if A is a Dedekind domain). Using these star operations, one can generalize many well-known results about Dedekind domains, UFDs, Krull domains, Prüfer domains, and PVMDs to rings with zerodivisors [5].

The set of all star operations on A is a complete lattice, where $* \leq *'$ if $I^* \subseteq I^{*'}$ for all $I \in \mathbf{F}(A)$. The divisorial closure star operation v is the largest star operation on A, and the identity star operation d is the smallest. A star operation $*$ on A is said to be of *finite type* if $I^* = \bigcup\{J^* : J \in \mathbf{F}(A),\ J \subseteq I \text{ is fin.gen.}\}$ for all $I \in \mathbf{F}(A)$, or equivalently if $*$ is Scott continuous, that is, if $* : \mathbf{F}(A) \longrightarrow \mathbf{F}(A)$ is continuous when $\mathbf{F}(A)$ is endowed with the Scott topology. The star operation t is the largest finite type star operation on A. More generally, if $*$ is any star operation on A, then the map $*_t : I \longmapsto I^{*_t} = \bigcup\{J^* : J \in \mathbf{F}(A),\ J \subseteq I \text{ is fin.gen.}\}$ is the largest finite type star operation on A that is less than or equal to $*$.

Let $*$ be a star operation on a ring A. The operation $(I, J) \longmapsto (IJ)^*$ on $\mathbf{F}(A)$ is called *$*$-multiplication*. A regular fractional ideal I of A has an inverse J with respect to $*$-multiplication if and only if $(II^{-1})^* = A$, in which case $J^* = I^{-1}$; such a regular fractional ideal I is said to be *$*$-invertible*. We let $*$-$\mathrm{Inv}(A)$ denote the submonoid of $\mathbf{F}(A)$ consisting of the $*$-invertible fractional ideals of A under multiplication. The *$*$-class semigroup* $\mathbf{S}^*(A)$ of a ring A is the monoid of all $*$-closed regular fractional ideals of A under $*$-multiplication modulo the subgroup of principal regular fractional ideals of A. The $*$-*class group* $\mathrm{Cl}^*(A)$ of a ring A is the group of invertible elements of the monoid $\mathbf{S}^*(A)$; equivalently, it is the group of $*$-invertible $*$-closed fractional ideals under $*$-multiplication modulo the subgroup of principal fractional ideals.

The t-closure operation, above nearly all other star operations, is a powerful tool well known among specialists in multiplicative ideal theory. It can be used, for example, to uniquely characterize the UFDs, the Krull domains, the PVMDs, and the integrally closed domains. For example, a domain A is a UFD if and only if the t-closure of every nonzero ideal of A is principal. The t-closure operation also yields a measure of the extent to which unique factorization fails in a given domain, since the t-class semigroup $\mathbf{S}^t(A)$ of A is trivial if and only if A is a UFD. A domain A is a Krull domain if and only if $\mathbf{S}^t(A)$ is a group, in which case it is the usual class group of A, generated by the classes of the height one primes, which are equivalently the t-maximal ideals of A. The t-class group $\mathrm{Cl}^t(A)$ of a domain A contains the Picard group $\mathrm{Pic}(A) = \mathrm{Cl}^d(A)$ but typically carries far more information.

An A-algebra B is A-torsion-free (as an A-module) if and only if $A^{\mathrm{reg}} \subseteq B^{\mathrm{reg}}$ (or $A^{\mathrm{reg}}1_B \subseteq B^{\mathrm{reg}}$, where 1_B is the identity of B). If B is A-torsion-free, then the homomorphism $A \longrightarrow B$ extends uniquely to a homomorphism $K \longrightarrow L$ of the respective total quotient rings of A and B, and the map $\mathbf{F}(A) \longrightarrow \mathbf{F}(B)$ given by $I \longmapsto IB$ is a well-defined homomorphism of ordered monoids.

In [3], to study the functoriality of star operations and $*$-class groups and semigroups, or lack thereof, we defined a *universal star operation* to be an association $* : A \longmapsto *_A$ of a star operation $* = *_A$ on A to every integral domain A [3]. We expand this definition here as follows: we define a *universal star operation* to be an association $* : A \longmapsto *_A$ of a star operation $* = *_A$ on A to every ring A. If $*$ is a universal star operation and A is a ring, then $\mathrm{Cl}^*(A)$ (resp., $\mathbf{S}^*(A)$, $*$-$\mathrm{Inv}(A)$) denotes $\mathrm{Cl}^{*_A}(A)$ (resp., $\mathbf{S}^{*_A}(A)$, $*_A$-$\mathrm{Inv}(A)$). We often use $*$ to denote $*_A$ when the ring A is understood. The star operations v, t, and d are very natural examples of universal star operations, yet v and t, Cl^v and Cl^t, etc., are not functorial in A.

Nevertheless, there are various conditions one can impose, such as flatness, that guarantee functoriality. Let $*$ be a universal star operation and A a ring. We say that an A-torsion-free A-algebra B is $*$-*linked* if $(IB)^{*_B} = (I^{*_A}B)^{*_B}$ for all $I \in *$-$\mathrm{Inv}(A)$, and the extension is said to be $*$-*compatible* if the same equation holds for all $I \in \mathbf{F}(A)$. It is well known that flat $\Longrightarrow$ t-compatible $\Longrightarrow$ t-linked for integral domains, and the same is true for arbitrary ring extensions. For extensions of Krull domains the conditions of PDE, t-compatibility, and t-linkedness are equivalent [1, Theorem 3.2].

The following condition is more general than $*$-linkedness. Following [3], we say that an A-torsion-free A-algebra B is $*$-*ideal class linked*, or $*$*ICL*, if there is a group homomorphism $\mathrm{Cl}^*(A) \longrightarrow \mathrm{Cl}^*(B)$ of $*$-class groups induced by the (well defined) map $I \longmapsto (I^{*_A}B)^{*_B}$ from $\mathbf{F}(A)$ to $\mathbf{F}(B)$. Equivalently this means that for all $I, J \in$ $*$-$\mathrm{Inv}(A)$ one has $a((IJ)^{*_A}B)^{*_B} = b(I^{*_A}J^{*_A}B)^{*_B}$ for some regular elements a and b of B. Any $*$-linked extension is $*$ICL; thus any PDE (or t-linked) extension of Krull domains is tICL. Theorem 1 provides an example of an extension of Krull domains that is not tICL. This might seem to imply that the tICL condition is more useful than the t-linkedness condition, being more general and yet apparently sufficient for functoriality of t-class groups. However, although the induced homomorphisms of t-class groups are functorial for t-linked extensions, they are not functorial for tICL extensions. In fact, perhaps surprisingly, the condition tICL is not stable under towers: in Theorem 1 we construct a tower $C \supseteq B \supseteq A$ of Krull domains such that the extensions $C \supseteq B$ and $B \supseteq A$ are tICL but the extension $C \supseteq A$ is not. This provides in particular an example of an extension $C \supseteq A$ of Krull domains such that the (well-defined) map $\mathrm{Cl}(A) \longrightarrow \mathrm{Cl}(C)$ acting by $[I] \longmapsto [(IC)^v]$ is not a group homomorphism.

Nevertheless, the following condition, which is in a precise sense "halfway" between $*$-linkedness and $*$ICL, *is* stable under towers and yields functoriality of $*$-class groups: an A-torsion-free A-algebra B is *$*$-class linked*, or *$*$CL*, if for all $I \in *\text{-Inv}(A)$ one has $a(IB)^{*_B} = b(I^{*_A}B)^{*_B}$ for some regular elements a and b of B. All of this suggests that t-class linkedness is likely to be the most general condition on torsion-free algebras that yields functoriality of t-class groups. Moreover, the proof of Proposition 4 provides an example of an extension of Krull domains that is t-class linked but not t-linked, i.e., not PDE. This shows that functoriality of class groups for Krull domains holds for a larger class of extensions than the PDE extensions.

2 Functorial Properties of Star Class Groups

Let $*$ be a universal star operation and A a ring. Recall that $\mathbf{F}(A)$ denotes the ordered monoid of all regular fractional ideals of A and $*\text{-Inv}(A)$ denotes the submonoid of $\mathbf{F}(A)$ consisting of the $*$-invertible fractional ideals of A.

In [3] we defined eight classes of extensions $B \supseteq A$, namely, those that are *$*$-compatible*, *$*$-linked*, *$*$-class compatible*, *$*$-class linked*, *$*$-ideal compatible*, *$*$-ideal linked*, *$*$-ideal class compatible*, and *$*$-ideal class linked*, respectively. Here will use the abbreviations $*$C, $*$L, $*$CC, $*$CL, $*$IC, $*$IL, $*$ICC, and $*$ICL, respectively, for these properties. Their definitions, more generally, for any ring A and any A-torsion-free A-algebra B are summarized in Table 1. In particular, for example, we say that an A-torsion-free A-algebra B is *$*$-ideal class linked*, or *$*$ICL*, if for all $I, J \in *\text{-Inv}(A)$ one has $a((IJ)^*B)^* = b(I^*J^*B)^*$ for some regular elements a and b of B. Equivalently this means that there is a group homomorphism $\mathrm{Cl}^*(A) \longrightarrow \mathrm{Cl}^*(B)$ of $*$-class groups induced by the map $I \longmapsto (I^*B)^*$ from $\mathbf{F}(A)$ to $\mathbf{F}(B)$.

The eight properties are connected by the implications given in Figure 1. Notice that there is a "duality" among the eight properties: $*$C and $*$ICL are dual, $*$L and $*$ICC are dual, $*$CC and $*$IL are dual, and $*$CL and $*$IC are dual. Duality here just means that the implications in Figure 1 holding among the various properties also hold among the negations of their dual properties.

Table 1 Definitions of eight classes of A-torsion-free A-algebras B

B/A is			
$*$C	$\forall I \in \mathbf{F}(A)$		$(IB)^* = (I^*B)^*$
$*$L	$\forall I \in *\text{-Inv}(A)$		$(IB)^* = (I^*B)^*$
$*$CC	$\forall I \in \mathbf{F}(A)$	$\exists a, b \in B^{\text{reg}}$	$a(IB)^* = b(I^*B)^*.$
$*$CL	$\forall I \in *\text{-Inv}(A)$	$\exists a, b \in B^{\text{reg}}$	$a(IB)^* = b(I^*B)^*.$
$*$IC	$\forall I, J \in \mathbf{F}(A)$		$((IJ)^*B)^* = (I^*J^*B)^*$
$*$IL	$\forall I, J \in *\text{-Inv}(A)$		$((IJ)^*B)^* = (I^*J^*B)^*$
$*$ICC	$\forall I, J \in \mathbf{F}(A)$	$\exists a, b \in B^{\text{reg}}$	$a((IJ)^*B)^* = b(I^*J^*B)^*$
$*$ICL	$\forall I, J \in *\text{-Inv}(A)$	$\exists a, b \in B^{\text{reg}}$	$a((IJ)^*B)^* = b(I^*J^*B)^*$

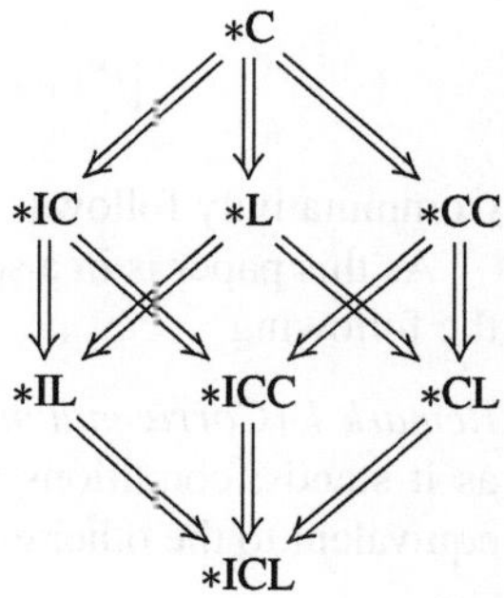

Fig. 1 Implications among eight classes of A-torsion-free A-algebras

The following extension of [3, Proposition 5.9] shows in particular that the properties $*$C, $*$L, $*$CC, and $*$CL are stable in towers and yield functoriality of $*$-class groups. In marked contrast, however, we will show in Sect. 6 that the conditions tIC, tIL, tICC, and tICL are not stable under towers, even among towers of Krull domains.

Proposition 1 *Let $*$ be a universal star operation, let A be a ring, B an A-torsion-free A-algebra, and C a B-torsion-free B-algebra. One has the following:*

1. *If C/B is $*$C and B/A is $*$C, then C/A is $*$C.*
2. *If C/B is $*$C and B/A is $*$IC, then C/A is $*$IC.*
3. *If C/B is $*$L and B/A is $*$L, then C/A is $*$L.*
4. *If C/B is $*$L and B/A is $*$IL, then C/A is $*$IL.*
5. *If C/B is $*$CC and B/A is $*$CC, then C/A is $*$CC.*
6. *If C/B is $*$CC and B/A is $*$ICC, then C/A is $*$ICC.*
7. *If C/B is $*$CL and B/A is $*$CL, then C/A is $*$CL.*
8. *If C/B is $*$CL and B/A is $*$ICL, then C/A is $*$ICL.*

Moreover, in any of the above cases the diagram

$$\begin{array}{ccc} \mathrm{Cl}^*(A) & \longrightarrow & \mathrm{Cl}^*(B) \\ & \searrow & \downarrow \\ & & \mathrm{Cl}^*(C) \end{array}$$

of induced group homomorphisms is commutative.

Proof Statements 1 through 8 are [3, Proposition 5.9] generalized to rings with zero-divisors. To prove the given diagram commutative we may assume the hypotheses of statement 8. Let $I \in *\text{-Inv}(A)$. Then under the composition in the given diagram one has

$$[I^*] \longmapsto [((I^*B)^*C)^*] = [(I^*BC)^*] = [(I^*C)^*].$$

Commutativity follows.

As this paper is in a sense a sequel to the paper [3], we feel compelled to remark the following.

Remark 1 (Corrigendum to [3]) We remark that [3, Proposition 3.11] is not correct as it stands: conditions (5) and (7) of the proposition are implied by but are not equivalent to the other eight conditions of the proposition.

3 Classifying Extensions via Commutative Diagrams

Let $*$ be a universal star operation and A a ring. Let $\mathbf{F}^*(A)$ denote the monoid of $*$-closed fractional ideals of A under the operation $(I, J) \longmapsto (IJ)^*$ of $*$-multiplication. Thus $\mathbf{F}(A) = \mathbf{F}^d(A)$. The map $I \longmapsto I^*$ defines a surjective monoid homomorphism $\mathbf{F}(A) \longrightarrow \mathbf{F}^*(A)$. The group $\mathrm{Inv}^*(A)$ of $*$-invertible $*$-closed fractional ideals of A is the group of units of the monoid $\mathbf{F}^*(A)$. The monoid $*\text{-Inv}(A)$ of $*$-invertible fractional ideals of A under ordinary multiplication is the pullback of $\mathrm{Inv}^*(A)$ along the monoid homomorphism $\mathbf{F}(A) \longrightarrow \mathbf{F}^*(A)$. The $*$-*class semigroup* $\mathbf{S}^*(A)$ of A is defined to be the monoid $\mathbf{F}^*(A)$ modulo the subgroup $\mathrm{Prin}(A)$ of principal fractional ideals of A. We let $\mathbf{S}(A) = \mathbf{S}^d(A)$ denote the *ideal class semigroup* of A. The $*$-*class group* $\mathrm{Cl}^*(A)$ of A is the group of units of $\mathbf{S}^*(A)$; alternatively it is the group $\mathrm{Inv}^*(A)/\mathrm{Prin}(A)$. We also let $*$-$\mathrm{Cl}(A)$ denote the monoid $*$-$\mathrm{Inv}(A)/\mathrm{Prin}(A)$, which is the pullback of $\mathrm{Cl}^*(A)$ along the monoid homomorphism $\mathbf{S}(A) \longrightarrow \mathbf{S}^*(A)$. Given these definitions, one has the following commutative diagram of monoid homomorphisms, in which all rising arrows are injective and all falling arrows are surjective.

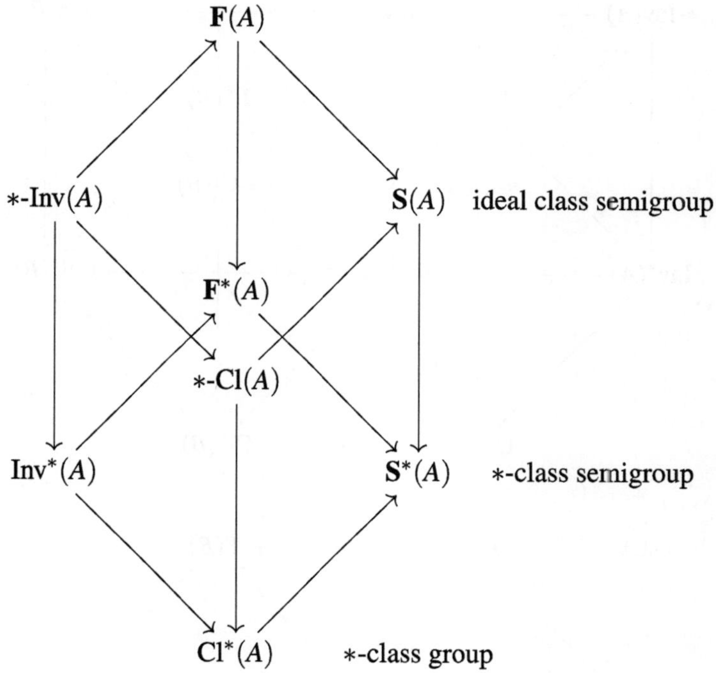

As we have seen, of the eight classes of torsion-free extensions that we have considered, four are stable in towers: $*$C, $*$L, $*$CC, and $*$CL. It is also useful to note that an A-torsion-free A-algebra B is $*$C (resp., $*$L, $*$CC, $*$CL) if and only if there is a commutative diagram ($*$C) (resp., ($*$L), ($*$CC), ($*$CL)) of monoid homomorphisms as shown below. (The dotted arrows represent maps for which the resulting diagram is commutative that exist if and only if the extension B/A has the respective property.)

(∗C)

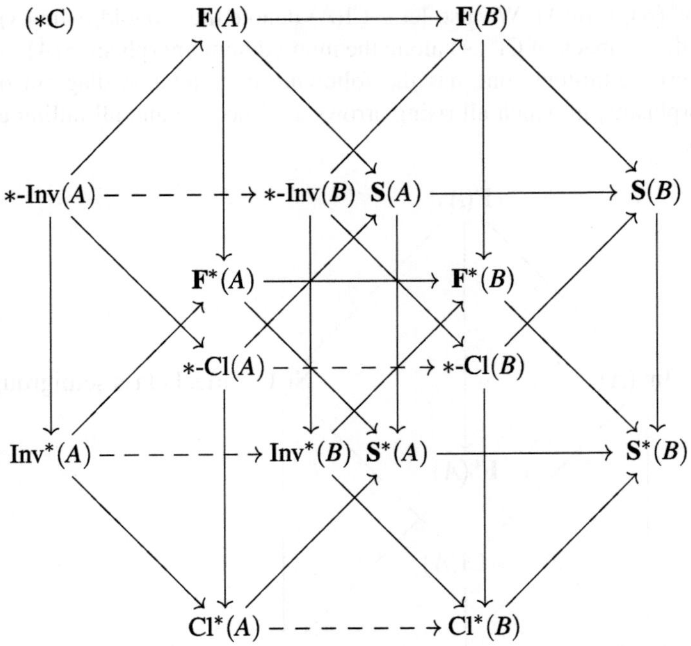
F(A)
F(B)
∗-Inv(A)
∗-Inv(B)
S(A)
S(B)
F*(A)
F*(B)
∗-Cl(A)
∗-Cl(B)
Inv*(A)
Inv*(B)
S*(A)
S*(B)
Cl*(A)
Cl*(B)

(∗L)

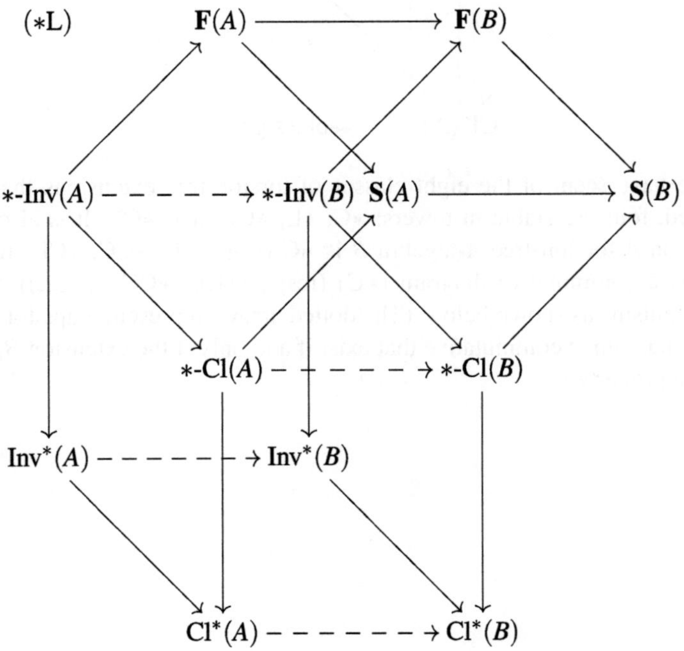
F(A)
F(B)
∗-Inv(A)
∗-Inv(B)
S(A)
S(B)
∗-Cl(A)
∗-Cl(B)
Inv*(A)
Inv*(B)
Cl*(A)
Cl*(B)

(∗CC)

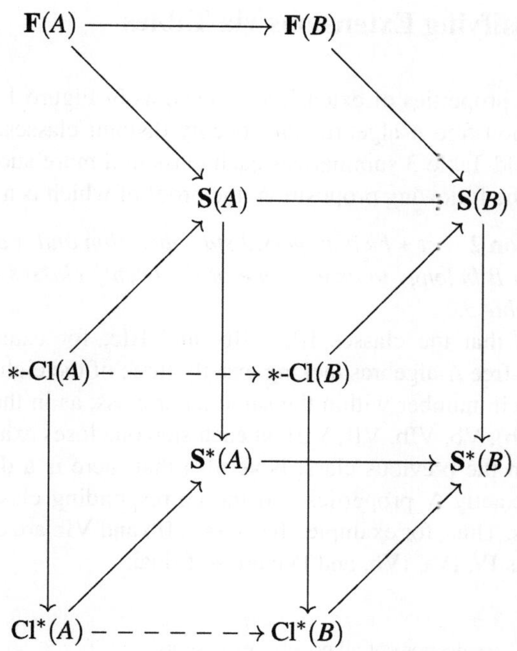

(∗CL)

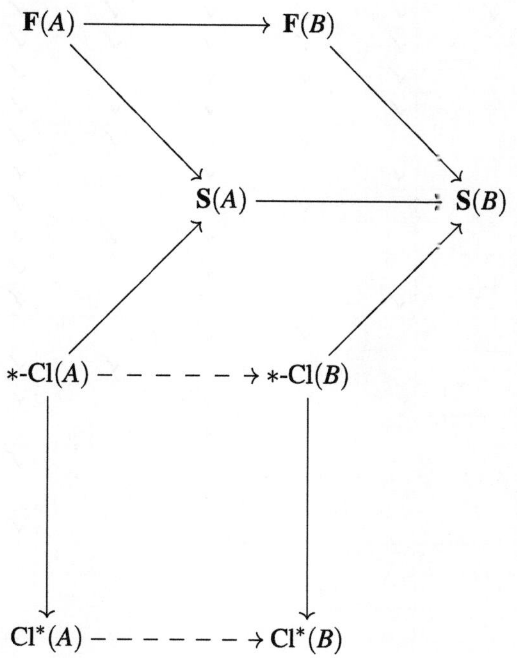

These equivalences follow from [3, Theorem 1.1 and Propositions 5.3 and 5.4] generalized to rings with zerodivisors.

4 Classifying Extensions via Tables

The eight properties of extensions, related as in Figure 1, can be used to partition all A-torsion-free A-algebras into twenty disjoint classes. Table 2 lists all twenty classes, and Table 3 summarizes each class in a more succinct form. In particular, we have the following proposition, the proof of which is a simple logic exercise.

Proposition 2 *Let $*$ be a universal star operation and A a ring. Any A-torsion-free A-algebra B belongs to exactly one of the twenty classes characterized by Table 2 and by Table 3.*

Notice that the classes IIIa, IIIb, and IIIc, for example, are together those A-torsion-free A-algebras lacking exactly three of the eight properties. Also, as one moves up in number within the same letter class, as in the sequence 0, I, IIb, IIIb, IV (or IVb), Vb, VIb, VII, VIII, at each step one loses exactly one of the properties holding in the previous class. Note also that there is a duality between any class lacking exactly N properties and the corresponding class lacking exactly $8 - N$ properties. Thus, for example, the classes IIb and VIb are dual to one another, while the classes IV, IVa, IVb, and IVc are self-dual.

Table 2 Classifications of all possible extensions

Class	$*$C	$*$L	$*$CC	$*$CL	$*$IC	$*$IL	$*$ICC	$*$ICL
0	✓	✓	✓	✓	✓	✓	✓	✓
I		✓	✓	✓	✓	✓	✓	✓
IIa		✓	✓	✓		✓	✓	✓
IIb		✓		✓	✓	✓	✓	✓
IIc			✓	✓	✓	✓	✓	✓
IIIa		✓		✓		✓	✓	✓
IIIb			✓	✓		✓	✓	✓
IIIc				✓	✓	✓	✓	✓
IV				✓		✓	✓	✓
IVa		✓		✓		✓		✓
IVb			✓	✓			✓	✓
IVc					✓	✓	✓	✓
Va				✓		✓		✓
Vb				✓			✓	✓
Vc						✓	✓	✓
VIa				✓				✓
VIb						✓		✓
VIc							✓	✓
VII								✓
VIII								

Table 3 Classification summary

Class	is	is not	Class	is	is not
0	∗C		VIII		∗ICL
I	∗L	∗C	VII	∗ICL	∗ICC
	∗CC				∗IL
	∗IC				∗CL
IIa	∗L	∗IC	VIa	∗CL	∗ICC
	∗CC				∗IL
IIb	∗L	∗CC	VIb	∗IL	∗ICC
	∗IC				∗CL
IIc	∗CC	∗L	VIc	∗ICC	∗IL
	∗IC				∗CL
IIIa	∗L	∗CC	Va	∗IL	∗ICC
	∗ICC	∗IC		∗CL	∗L
IIIb	∗CC	∗L	Vb	∗ICC	∗IL
	∗IL	∗IC		∗CL	∗CC
IIIc	∗CL	∗L	Vc	∗ICC	∗IC
	∗IC	∗CC		∗IL	∗CL
IV	∗ICC	∗L	IVa	∗L	∗ICC
	∗IL	∗CC	IVb	∗CC	∗IL
	∗CL	∗IC	IVc	∗IC	∗CL

5 *t*-Closure

If A or B is restricted to lie in a certain class of rings, then B/A may consequently be restricted from lying in some of the twenty classes of extensions in Table 2. In the extreme case, for example, if every ideal of A is a $*$-ideal, then all extensions of A lie in class 0. This happens for $* = t$, for example, if A is a Prüfer domain. Similarly, as the following proposition shows, any extension of a Krull domain must lie in exactly one of the classes 0, IIc, IVb, IVc, VIc, and VIII.

Proposition 3 *Let A be a Krull domain and B any A-torsion-free A-algebra. Then for B/A one has tC* $\Longleftrightarrow$ *tL* $\Longleftrightarrow$ *PDE, tCC* $\Longleftrightarrow$ *tCL, tIC* $\Longleftrightarrow$ *tIL, and tICC* $\Longleftrightarrow$ *tICL. Moreover, B must lie in exactly one of the classes 0, IIc, IVb, IVc, VIc, and VIII.*

Proof It is well known that a Krull domain is equivalently a domain A in which every fractional ideal is t-invertible, that is, in which $\mathbf{F}(A) = t\text{-Inv}(A)$. The proposition therefore follows immediately from the definitions in Table 1.

The following is a partial converse to the result above.

Proposition 4 *For* $* = t$*, the classes of extensions 0, IIc, IVb, IVc, and VIII listed in Table 2 each contain some extensions of the form* $B \supseteq A$ *where both A and B are Krull domains.*

Proof If $B = A$, then the extension is in class 0. If A and B are UFDs and $B \supseteq A$ is not t-linked, then the extension is in class IIc; thus, for example, the extension $\mathbb{Z}[X/2] \supseteq \mathbb{Z}[X]$ is in class IIc [3, Example 6.10]. If A is any Noetherian integrally closed domain that is not locally factorial, then by [3, Proposition 6.6 and Example 6.7] there is a DVR overring B of A such that $B \supseteq A$ is in class IVb. Finally, Theorem 1 of the next section shows that the classes IVc and VIII also contain extensions of Krull domains.

Conjecture 1 For $* = t$, the class VIc also contains some extensions of the form $B \supseteq A$ where both A and B are Krull domains. Equivalently, there exists an extension $B \supseteq A$ of Krull domains such that (1) the map

$$\mathbf{F}^t(A) \longrightarrow \mathbf{F}^t(B)$$
$$I \longmapsto (IB)^t$$

is not a monoid homomorphism, (2) the induced map $\mathbf{S}^t(A) \longrightarrow \mathbf{S}^t(B)$ is a monoid homomorphism, and (3) the square

$$\begin{array}{ccc} \mathbf{S}(A) & \longrightarrow & \mathbf{S}(B) \\ \downarrow & & \downarrow \\ \mathbf{S}^t(A) & \longrightarrow & \mathbf{S}^t(B) \end{array}$$

is not commutative.

By Propositions 3 and 4, a proof or disproof of the above conjecture would complete a characterization of the possible extensions of a Krull domain according to the classification scheme of Table 2.

For a "random" universal star operation $*$, the twenty classes of extensions defined in Table 2 will almost certainly be nonempty. Regarding the t-operation, however, the situation is not so clear.

Conjecture 2 For $* = t$, all twenty classes of extensions listed in Table 2 are nonempty.

Conjecture 3 For $* = t$, all twenty classes of extensions listed in Table 2 contain some extensions of the form $B \supseteq A$ where both A and B are PVMDs.

It is known also that the class I for $* = t$ contains some extensions of PVMDs [3, Examples 4.12 and 6.9]. Together with Proposition 4, this verifies as nonempty only six of the twenty classes. Assuming Conjecture 2 is true, Table 3 for $* = t$ yields the shortest possible path towards determining whether or not an A-algebra lies in a given class. Note that class IV is the most difficult to verify. Constructing such an example seems difficult since one has to control simultaneously for six of the eight properties. Indeed, class IV is empty if and only if

$$t\mathrm{IL} \wedge t\mathrm{ICC} \wedge t\mathrm{CL} \Longrightarrow t\mathrm{IC} \vee t\mathrm{L} \vee t\mathrm{CC}.$$

It would be very surprising if the given conjunction implied the disjunction while not implying any of its disjuncts. Thus, if class IV were empty but the given conjunction did not imply any of the disjuncts, then an examination of Table 2 and the fact that IIc is nonempty shows that at least three of the other classes would also have to be empty, namely, IIIa and either IIa and IIIb or IIb and IIIc.

We end this section by proving that all flat algebras are t-compatible. This fact is well known for integral domains. Recall that $\mathbf{K}(A)$ denotes the ordered monoid of all A-submodules of the total quotient ring of A.

Proposition 5 *Let A be a ring with total quotient ring K and B a flat A-algebra with total quotient ring L, so L is a K-algebra. Then $IB \cap JB = (I \cap J)B$ in $\mathbf{K}(B)$ for all $I, J \in \mathbf{K}(A)$.*

Proof Tensoring the exact sequence

$$0 \longrightarrow I \cap J \longrightarrow K \longrightarrow K/I \oplus K/J$$

of A-modules with the flat A-module B, we get the exact sequence

$$0 \longrightarrow (I \cap J) \otimes_A B \longrightarrow K \otimes_A B \longrightarrow (K/I \oplus K/J) \otimes_A B,$$

which since $K \otimes_A B$ is isomorphic to the compositum KB of K and B in L is equivalent to the exact sequence

$$0 \longrightarrow (I \cap J)B \longrightarrow KB \longrightarrow (KB/IB) \oplus (KB/JB).$$

The desired equality follows.

Proposition 6 *Let A be a ring with total quotient ring K and B a flat A-algebra with total quotient ring L, so L is a K-algebra. One has the following:*

1. $(I :_K J)B = (IB :_{KB} JB)$ *for all $I, J \in \mathbf{K}(A)$ with J finitely generated.*
2. $(I :_K J)B = (IB :_L JB)$ *for all $I, J \in \mathbf{K}(A)$ with J finitely generated and regular.*
3. $(IB)^{-1} = I^{-1}B$ *in L for all finitely generated $I \in \mathbf{F}(A)$.*

Proof Let $I \in \mathbf{K}(A)$ and $a \in K$. Consider the multiplication by a map $K \longrightarrow K$ of A-modules followed by the projection $K \longrightarrow K/I$. The kernel of the composed map $K \longrightarrow K/I$ is the A-module $(I :_K aA)$, so we have an exact sequence

$$0 \longrightarrow (I :_K aA) \longrightarrow K \longrightarrow K/I$$

of A-modules. Since B is flat over A, tensoring with B yields the exact sequence

$$0 \longrightarrow (I :_K aA) \otimes_A B \longrightarrow K \otimes_A B \longrightarrow (K/I) \otimes_A B,$$

which is equivalent to the exact sequence

$$0 \longrightarrow (I :_K aA)B \longrightarrow KB \longrightarrow KB/IB.$$

Therefore the kernel $(IB :_{KB} aB)$ of the multiplication by a map $KB \longrightarrow KB/IB$ is equal to $(I :_K aA)B$. This proves statement 1 when $J = aA$ is principal. Now suppose that $J \in \mathbf{K}(A)$ is finitely generated, say, $J = (a_1, \ldots, a_n)$, where $a_i \in K$ for all i. Then, using also Proposition 5, we see that $(I :_K J)B = ((I :_K a_1) \cap \cdots \cap (I :_K a_n))B = (I :_K a_1)B \cap \cdots \cap (I :_K a_n)B = (IB :_{KB} a_1B) \cap \cdots \cap (IB :_{KB} a_nB) = (IB :_{KB} a_1B + \cdots + a_nB) = (IB :_{KB} JB)$. This proves statement 1.

Now, suppose furthermore that J is regular, say, $a \in J$ is regular. Let $x \in (IB :_L JB)$. Then $ax \in IB$, so $x = a^{-1}(ax) \in K(IB) = (KI)B \subseteq KB$ and therefore $x \in KB$. Thus we have $(IB :_L JB) \subseteq (IB :_{KB} JB)$, so equality holds. This proves statement 2, and then statement 3 follows immediately.

The following proposition generalizes [9, Proposition 2.6] and comments (a) and (b) following it to rings with zerodivisors.

Proposition 7 *Let A be ring, and let B be an A-torsion-free A-algebra. Each of the following conditions implies the next:*

1. *B is flat as an A-module.*
2. *$I^{-1}B = (IB)^{-1}$ for all finitely generated $I \in \mathbf{F}(A)$.*
3. *$(I^{-1}B)^v = (IB)^{-1}$ for all finitely generated $I \in \mathbf{F}(A)$.*
4. *B is a t-compatible A-algebra.*
5. *If I is a t-closed ideal of B and $I \cap A$ is regular, then $I \cap A$ is a t-closed ideal of A, where $I \cap A = \{a \in A : a1_B \in I\}$.*

In fact, conditions 4 and 5 are equivalent.

Proof The implication $1 \Rightarrow 2$ follows immediately from Proposition 6, and $2 \Rightarrow 3$ is obvious. Suppose that condition 3 holds. Let J be a finitely generated regular subideal of a regular ideal I of A. Then $(J^{-1}B)^v = (JB)^{-1}$, so $J^vB = (J^{-1})^{-1}B \subseteq (J^{-1}B)^{-1} = ((J^{-1}B)^v)^{-1} = ((JB)^{-1})^{-1} = (JB)^v = (JB)^t \subseteq (IB)^t$. Taking the union over all such J, we see that $I^tB \subseteq (IB)^t$. Therefore $3 \Rightarrow 4$.

Next, suppose that condition 4 holds and I is a t-closed ideal of B such that $I \cap A$ is regular. Then $(I \cap A)^t \subseteq ((I \cap A)B)^t \cap A \subseteq I^t \cap A = I \cap A$, and therefore $I \cap A$ is a t-closed ideal of A. Therefore 4 implies 5. Conversely, suppose that condition 5 holds and I is a regular ideal of A. Then $(IB)^t$ is t-closed ideal of B such that $(IB) \cap A \supseteq I$ is regular, so $(IB)^t \cap A$ is a t-closed ideal of A and therefore $I^t \subseteq (IB)^t \cap A$. Thus $5 \Rightarrow 4$, so conditions 4 and 5 are equivalent.

6 An Example of Non-Functoriality

The following proposition shows by example that the property of tICL is not stable under extensions of Krull domains.

Theorem 1 *Let k be a field, and let*

$$A = k[X, Y, T, TY/X],$$

$$B = k[X, T, Y/X],$$

$$C = k[X, T, Y/X, V, TV/X].$$

Then we have the following:

1. *A, B, and C are Krull domains.*
2. *The extension $B \supseteq A$ is a subintersection and is therefore in class 0.*
3. *The extension $C \supseteq B$ is in class IVc.*
4. *The extension $C \supseteq A$ is in class VIII.*

It follows that the conditions tIC $\Longleftrightarrow$ tIL and tICC $\Longleftrightarrow$ tICL are not stable under extensions of Krull domains.

Proof By [6, Proposition 14.5] the integral domains A and B are Krull domains. Moreover, by the proof of that proposition, the extension B of A is a subintersection, equal to the intersection of all essential valuation rings of A excluding $A[X, Y, T, TY/X]_{(X,Y)}$. It follows that the extension B of A is t-compatible and therefore in class 0.

Now, B is isomorphic to $k[X, T, Z]$ and therefore $\mathbf{S}^t(B) = \mathrm{C}$. Therefore by [3, Proposition 6.2] the extension $C \supseteq B$ is tIC. Given statement 4, then, which we prove below, the extension $C \supseteq B$ must be in class IVc, because if $C \supseteq B$ were tCL, then $C \supseteq A$ would also be tCL by [3, Proposition 5.9(7)], contradicting statement 4.

Finally, we show that the extension $C \supseteq A$ is not tICL and is therefore in class VIII. By [6, Proposition 14.8] one has $\mathrm{Cl}(A) = \mathbb{Z}$ and this group is freely generated by $[(X, Y)] = [(T, U)]$. Let $I = (X, Y)$ and $J = (X, T)$. Note that $IJ = (X^2, XY, XT, TY) = X(X, Y, T, U)$ and its v-closure is XA, which implies $[I] = [J]^{-1}$. In particular one has $(IJ)^t = XA$ and therefore $((IJ)^t C)^t = XC$, so one has $[(IJ)^t C]^t = [C]$.

Note that C is isomorphic to the integral domain $k[X, T, V, TV/X][Z]$, as can be seen by mapping Y to ZX and mapping the other variables X, T, V to themselves. Thus there is an isomorphism $A[Z] \longrightarrow C$ sending X to X, Y to T, T to V, and Z to Y/X.

In particular, C is a Krull domain. Moreover, since A is integrally closed, the map $A \longrightarrow A[Z]$ induces an isomorphism $\mathrm{Cl}(A) \longrightarrow \mathrm{Cl}(A[Z]) \cong \mathrm{Cl}(C)$. Under this isomorphism of class groups, the ideal class $[(X, Y)A]$ in $\mathrm{Cl}(A)$ is sent to the ideal class $[(X, T)C]$ in $\mathrm{Cl}(C)$. Therefore, since the ideal class $[(X, Y)A]$ is of infinite order and generates the whole class group $\mathrm{Cl}(A) \cong \mathbb{Z}$, it follows that the ideal class $[(X, T)C]$ is of infinite order and generates the whole class group $\mathrm{Cl}(C) \cong \mathbb{Z}$.

Now we have $IJC = X(X, Y, T, U)C = X(X, T)C$ since Y and $U = TY/X$ are multiples of X and T, respectively, in C. Therefore one has

$$[(IJC)^t] = [(X(X,T)C)^t] = [(X,T)C] \neq [C] = [((IJ)^t C)^t].$$

This proves that the extension $C \supseteq A$ is not tICL.

In the above theorem, note that C is isomorphic to $A[Z]$ as a ring but not as an A-algebra. The inclusion $A \subseteq C$ does not induce a homomorphism of t-class groups, even though the inclusion $A \subseteq A[Z]$ induces an isomorphism of t-class groups.

References

1. Anderson, D.D., Houston, E.G., Zafrullah, M.: t-linked extensions, the t-class group, and Nagata's theorem. J. Pure Appl. Algebra **86**, 109–124 (1993)
2. Dobbs, D., Houston, E., Lucas, T., Zafrullah, M.: t-linked overrings and Prüfer v-multiplication domains. Commun. Algebra **17** 2835–2852 (1989)
3. Elliott, J.: Functorial properties of star operations. Commun. Algebra **38**(5), 1466–1490 (2010)
4. Elliott, J.: Nuclei and applications to star, semistar, and semiprime operations. arXiv:1505.06433 [math.RA]
5. Elliott, J.: Rings, Modules, and Closure Operations (in progress)
6. Fossum, R.M.: The Divisor Class Group of a Krull Domain, Ergebnisse der Mathematik und ihrer Grenzgebiete, Bd. 74. Springer, Berlin (1973)
7. Krull, W.: Idealtheorie. Springer, Berlin (1935)
8. Krull, W.: Beiträge zur Arithmetik kommutativer Integritätsbereiche. Math. Z. **41**(1), 665–679 (1936)
9. Zafrullah, M.: Putting t-invertibility to use. In: Non-Noetherian Commutative Ring Theory. Mathematics and its Applications, vol. 520. Kluwer Academic Publishers, Dordrecht (2000)

Systems of Sets of Lengths: Transfer Krull Monoids Versus Weakly Krull Monoids

Alfred Geroldinger, Wolfgang A. Schmid, and Qinghai Zhong

Abstract Transfer Krull monoids are monoids which allow a weak transfer homomorphism to a commutative Krull monoid, and hence the system of sets of lengths of a transfer Krull monoid coincides with that of the associated commutative Krull monoid. We unveil a couple of new features of the system of sets of lengths of transfer Krull monoids over finite abelian groups G, and we provide a complete description of the system for all groups G having Davenport constant $\mathsf{D}(G) = 5$ (these are the smallest groups for which no such descriptions were known so far). Under reasonable algebraic finiteness assumptions, sets of lengths of transfer Krull monoids and of weakly Krull monoids satisfy the Structure Theorem for Sets of Lengths. In spite of this common feature we demonstrate that systems of sets of lengths for a variety of classes of weakly Krull monoids are different from the system of sets of lengths of any transfer Krull monoid.

Keywords Transfer Krull monoids • Weakly Krull monoids • Sets of lengths • Zero-sum sequences

Mathematics Subject Classification (2010): 11B30, 11R27, 13A05, 13F05, 16H10, 16U30, 20M13

A. Geroldinger • Q. Zhong (✉)
Institute for Mathematics, University of Graz, Heinrichstraße 36, 8010 Graz, Austria
e-mail: alfred.geroldinger@uni-graz.at; qinghai.zhong@uni-graz.at

W.A. Schmid
Laboratoire Analyse, Géométrie et Applications (LAGA, UMR 7539, CNRS), COMUE Université Paris Lumières, Université Paris 8, CNRS, 93526 Saint-Denis cedex, France

Université Paris 13, Sorbonne Paris Cité, LAGA, CNRS, UMR 7539, Université Paris 8, F-93430 Villetaneuse, France
e-mail: schmid@math.univ-paris13.fr

M. Fontana et al. (eds.), *Rings, Polynomials, and Modules*,
https://doi.org/10.1007/978-3-319-65874-2_11

1 Introduction

By an atomic monoid we mean a cancellative semigroup with unit element such that every nonunit can be written as a finite product of irreducible elements. Let H be an atomic monoid. If $a \in H$ is a nonunit and $a = u_1 \cdot \ldots \cdot u_k$ is a factorization of a into k irreducible elements, then k is called a factorization length and the set $\mathsf{L}(a) \subset \mathbb{N}$ of all possible factorization lengths is called the set of lengths of a. Then $\mathcal{L}(H) = \{\mathsf{L}(a) \mid a \in H\}$ is the system of sets of lengths of H. Under a variety of noetherian conditions on H (e.g., H is the monoid of nonzero elements of a commutative noetherian domain) all sets of lengths are finite. Furthermore, if there is some element $a \in H$ with $|\mathsf{L}(a)| > 1$, then $|\mathsf{L}(a^N)| > N$ for all $N \in \mathbb{N}$. Sets of lengths (together with invariants controlling their structure, such as elasticities and sets of distances) are a well-studied means of describing the arithmetic structure of monoids ([13, 18]).

Let H be a transfer Krull monoid. Then, by definition, there is a weak transfer homomorphism $\theta : H \to \mathcal{B}(G_0)$, where $\mathcal{B}(G_0)$ denotes the monoid of zero-sum sequences over a subset G_0 of an abelian group, and hence $\mathcal{L}(H) = \mathcal{L}\big(\mathcal{B}(G_0)\big)$. A special emphasis has always been on the case where G_0 is a finite abelian group. Thus let G be a finite abelian group and we use the abbreviation $\mathcal{L}(G) = \mathcal{L}\big(\mathcal{B}(G)\big)$. It is well known that sets of lengths in $\mathcal{L}(G)$ are highly structured (Proposition 3.2), and the standing conjecture is that the system $\mathcal{L}(G)$ is characteristic for the group G. More precisely, if G' is a finite abelian group such that $\mathcal{L}(G) = \mathcal{L}(G')$, then G and G' are isomorphic (apart from two well-known trivial pairings; see Conjecture 3.4). This conjecture holds true, among others, for groups G having rank at most two, and its proof uses deep results from additive combinatorics which are not available for general groups. Thus there is a need for studying $\mathcal{L}(G)$ with a new approach. In Sect. 3, we unveil a couple of properties of the system $\mathcal{L}(G)$ which are first steps on a new way towards Conjecture 3.4.

In spite of all abstract work on systems $\mathcal{L}(G)$, they have been written down explicitly only for groups G having Davenport constant $\mathsf{D}(G) \le 4$, and this is not difficult to do (recall that a group G has Davenport constant $\mathsf{D}(G) \le 4$ if and only if either $|G| \le 4$ or G is an elementary 2-group of rank three). In Sect. 4 we determine the systems $\mathcal{L}(G)$ for all groups G having Davenport constant $\mathsf{D}(G) = 5$.

Commutative Krull monoids are the classic examples of transfer Krull monoids. In recent years a wide range of monoids and domains has been found which are transfer Krull but which are not commutative Krull monoids. Thus the question arose which monoids H have systems $\mathcal{L}(H)$ which are different from systems of sets of lengths of transfer Krull monoids. Commutative v-noetherian weakly Krull monoids and domains are the best investigated class of monoids beyond commutative Krull monoids (numerical monoids as well as one-dimensional noetherian domains are v-noetherian weakly Krull). Clearly, weakly Krull monoids can be half-factorial and half-factorial monoids are transfer Krull monoids. Similarly, it can happen both for weakly Krull monoids and for transfer Krull monoids that all sets of lengths are arithmetical progressions with difference 1. Apart from such extremal cases, we show in Sect. 5 that systems of sets of lengths of a variety of classes of weakly Krull monoids are different from the system of sets of lengths of any transfer Krull monoid.

2 Background on Sets of Lengths

We denote by $\mathbb{N}$ the set of positive integers, and for real numbers $a, b \in \mathbb{R}$, we denote by $[a, b] = \{x \in \mathbb{Z} \mid a \le x \le b\}$ the discrete interval between a and b, and by an interval we always mean a finite discrete interval of integers.

Let $A, B \subset \mathbb{Z}$ be subsets of the integers. Then $A + B = \{a + b \mid a \in A, b \in B\}$ is the sumset of A and B. We set $-A = \{-a \mid a \in A\}$ and for an integer $m \in \mathbb{Z}$, $m + A = \{m\} + A$ is the shift of A by m. For $m \in \mathbb{N}$, we denote by $mA = A + \ldots + A$ the m-fold subset of A and by $m \cdot A = \{ma \mid a \in A\}$ the dilation of A by m. If $A \subset \mathbb{N}$, we denote by $\rho(A) = \sup A / \min A \in \mathbb{Q}_{\ge 1} \cup \{\infty\}$ the elasticity of A and we set $\rho(\{0\}) = 1$. A positive integer $d \in \mathbb{N}$ is called a distance of A if there are $a, b \in A$ with $b - a = d$ and the interval $[a, b]$ contains no further elements of A. We denote by $\Delta(A)$ the set of distances of A. Clearly, $\Delta(A) = \emptyset$ if and only if $|A| \le 1$, and A is an arithmetical progression if and only if $|\Delta(A)| \le 1$.

Let G be an additive abelian group. An (ordered) family $(e_i)_{i \in I}$ of elements of G is said to be *independent* if $e_i \neq 0$ for all $i \in I$ and, for every family $(m_i)_{i \in I} \in \mathbb{Z}^{(I)}$,

$$\sum_{i \in I} m_i e_i = 0 \quad \text{implies} \quad m_i e_i = 0 \quad \text{for all} \quad i \in I .$$

A family $(e_i)_{i \in I}$ is called a *basis* for G if $e_i \neq 0$ for all $i \in I$ and $G = \bigoplus_{i \in I} \langle e_i \rangle$. A subset $G_0 \subset G$ is said to be independent if the tuple $(g)_{g \in G_0}$ is independent. For every prime $p \in \mathbb{P}$, we denote by $\mathsf{r}_p(G)$ the p-rank of G.

Sets of Lengths We say that a semigroup S is cancellative if for all elements $a, b, c \in S$, the equation $ab = ac$ implies $b = c$ and the equation $ba = ca$ implies $b = c$. Throughout this manuscript, a monoid means a cancellative semigroup with unit element, and we will use multiplicative notation.

Let H be a monoid. An element $a \in H$ is said to be invertible if there exists an element $a' \in H$ such that $aa' = a'a = 1$. The set of invertible elements of H will be denoted by $H^\times$, and we say that H is reduced if $H^\times = \{1\}$. For a set P, we denote by $\mathcal{F}(P)$ the free abelian monoid with basis P. Then every $a \in \mathcal{F}(P)$ has a unique representation in the form

$$a = \prod_{p \in P} p^{\mathsf{v}_p(a)} ,$$

where $\mathsf{v}_p \colon \mathcal{F}(P) \to \mathbb{N}_0$ denotes the p-adic exponent.

An element $a \in H$ is called *irreducible* (or an *atom*) if $a \notin H^\times$ and if, for all $u, v \in H$, $a = uv$ implies that $u \in H^\times$ or $v \in H^\times$. We denote by $\mathcal{A}(H)$ the set of atoms of H. The monoid H is said to be *atomic* if every $a \in H \setminus H^\times$ is a product of finitely many atoms of H. If $a \in H$ and $a = u_1 \cdot \ldots \cdot u_k$, where $k \in \mathbb{N}$ and $u_1, \ldots, u_k \in \mathscr{A}(H)$, then we say that k is the *length* of the factorization. For $a \in H \setminus H^\times$, we call

$$\mathsf{L}_H(a) = \mathsf{L}(a) = \{k \in \mathbb{N} \mid a \text{ has a factorization of length } k\} \subset \mathbb{N}$$

the *set of lengths* of a. For convenience, we set $\mathsf{L}(a) = \{0\}$ for all $a \in H^\times$. By definition, H is atomic if and only if $\mathsf{L}(a) \neq \emptyset$ for all $a \in H$. Furthermore, $\mathsf{L}(a) = \{1\}$ if and only if $a \in \mathscr{A}(H)$ if and only if $1 \in \mathsf{L}(a)$. If $a, b \in H$, then $\mathsf{L}(a) + \mathsf{L}(b) \subset \mathsf{L}(ab)$. We call

$$\mathscr{L}(H) = \{\mathsf{L}(a) \mid a \in H\}$$

the *system of sets of lengths* of H. We say that H is *half-factorial* if $|L| = 1$ for every $L \in \mathscr{L}(H)$. If H is atomic, then H is either half-factorial or for every $N \in \mathbb{N}$ there is an element $a_N \in H$ such that $|\mathsf{L}(a_N)| > N$ ([17, Lemma 2.1]). We say that H is a BF-monoid if it is atomic and all sets of lengths are finite. Let

$$\Delta(H) = \bigcup_{L \in \mathscr{L}(H)} \Delta(L) \subset \mathbb{N}$$

denote the *set of distances* of H, and if $\Delta(H) \neq \emptyset$, then $\min \Delta(H) = \gcd \Delta(H)$. We denote by $\Delta_1(H)$ the set of all $d \in \mathbb{N}$ with the following property:

For every $k \in \mathbb{N}$ there exists an $L \in \mathscr{L}(H)$ of the form $L = L' \cup \{y, y+d, \ldots, y+kd\} \cup L''$ where $y \in \mathbb{N}$ and $L', L'' \subset \mathbb{Z}$ with $\max L' < y$ and $y + kd < \min L''$.

By definition, $\Delta_1(H)$ is a subset of $\Delta(H)$. For every $k \in \mathbb{N}$ we define the kth *elasticity* of H. If $H = H^\times$, then we set $\rho_k(H) = k$, and if $H \neq H^\times$, then

$$\rho_k(H) = \sup\{\sup L \mid k \in L \in \mathscr{L}(H)\} \in \mathbb{N} \cup \{\infty\}.$$

The invariant

$$\rho(H) = \sup\{\rho(L) \mid L \in \mathscr{L}(H)\} = \lim_{k \to \infty} \frac{\rho_k(H)}{k} \in \mathbb{R}_{\geq 1} \cup \{\infty\}$$

is called the *elasticity* of H (see [17, Proposition 2.4]). Sets of lengths of all monoids, which are in the focus of the present paper, are highly structured (see Proposition 3.2 and Theorems 5.5 – 5.8). To summarize the relevant concepts, let $d \in \mathbb{N}$, $M \in \mathbb{N}_0$ and $\{0, d\} \subset \mathscr{D} \subset [0, d]$. A subset $L \subset \mathbb{Z}$ is called an *almost arithmetical multiprogression* (AAMP for short) with *difference* d, *period* $\mathscr{D}$, and *bound* M, if

$$L = y + (L' \cup L^* \cup L'') \subset y + \mathscr{D} + d\mathbb{Z}$$

where $y \in \mathbb{Z}$ is a shift parameter,

- L^* is finite nonempty with $\min L^* = 0$ and $L^* = (\mathcal{D} + d\mathbb{Z}) \cap [0, \max L^*]$, and
- $L' \subset [-M, -1]$ and $L'' \subset \max L^* + [1, M]$.

We say that *the Structure Theorem for Sets of Lengths* holds for a monoid H if H is atomic and there exist some $M \in \mathbb{N}_0$ and a finite nonempty set $\Delta \subset \mathbb{N}$ such that every $L \in \mathcal{L}(H)$ is an AAMP with some difference $d \in \Delta$ and bound M.

Monoids of Zero-Sum Sequences We discuss a monoid having a combinatorial flavor whose universal role in the study of sets of lengths will become evident at the beginning of the next section. Let G be an additive abelian group and $G_0 \subset G$ a subset. Then $\langle G_0 \rangle$ denotes the subgroup generated by G_0, and we set $G_0^\bullet = G_0 \setminus \{0\}$. In additive combinatorics, a *sequence* (over G_0) means a finite sequence of terms from G_0 where repetition is allowed and the order of the elements is disregarded, and (as usual) we consider sequences as elements of the free abelian monoid with basis G_0. Let

$$S = g_1 \cdot \ldots \cdot g_\ell = \prod_{g \in G_0} g^{\mathsf{v}_g(S)} \in \mathcal{F}(G_0)$$

be a sequence over G_0. We set $-S = (-g_1) \cdot \ldots \cdot (-g_\ell)$, and we call

- $\operatorname{supp}(S) = \{g \in G \mid \mathsf{v}_g(S) > 0\} \subset G$ the *support* of S,
- $|S| = \ell = \sum_{g \in G} \mathsf{v}_g(S) \in \mathbb{N}_0$ the *length* of S,
- $\sigma(S) = \sum_{i=1}^{l} g_i$ the *sum* of S,
- $\Sigma(S) = \left\{ \sum_{i \in I} g_i \mid \emptyset \ne I \subset [1, \ell] \right\}$ the *set of subsequence sums* of S,
- $\mathsf{k}(S) = \sum_{i=1}^{l} \frac{1}{\operatorname{ord}(g_i)}$ the *cross number* of S.

The sequence S is said to be

- *zero-sum free* if $0 \notin \Sigma(S)$,
- a *zero-sum sequence* if $\sigma(S) = 0$,
- a *minimal zero-sum sequence* if it is a nontrivial zero-sum sequence and every proper subsequence is zero-sum free.

The set of zero-sum sequences $\mathcal{B}(G_0) = \{S \in \mathcal{F}(G_0) \mid \sigma(S) = 0\} \subset \mathcal{F}(G_0)$ is a submonoid, and the set of minimal zero-sum sequences is the set of atoms of $\mathcal{B}(G_0)$. For any arithmetical invariant $*(H)$ defined for a monoid H, we write $*(G_0)$ instead of $*(\mathcal{B}(G_0))$. In particular, $\mathcal{A}(G_0) = \mathcal{A}(\mathcal{B}(G_0))$ is the set of atoms of $\mathcal{B}(G_0)$, $\mathcal{L}(G_0) = \mathcal{L}(B(G_0))$ is the system of sets of lengths of $\mathcal{B}(G_0)$, and so on. Furthermore, we say that G_0 is half-factorial if the monoid $\mathcal{B}(G_0)$ is half-factorial. We denote by

$$\mathsf{D}(G_0) = \sup\{|S| \mid S \in \mathcal{A}(G_0)\} \in \mathbb{N}_0 \cup \{\infty\}$$

the *Davenport constant* of G_0. If G_0 is finite, then $\mathsf{D}(G_0)$ is finite. Suppose that G is finite, say $G \cong C_{n_1} \oplus \ldots \oplus C_{n_r}$, with $r \in \mathbb{N}_0$, $1 < n_1 \mid \ldots \mid n_r$, then $r = \mathsf{r}(G)$ is the rank of G, and we have

$$1+\sum_{i=1}^{r}(n_i-1) \le \mathsf D(G) \le |G|\,. \tag{1}$$

If G is a p-group or $\mathsf r(G) \le 2$, then $1+\sum_{i=1}^{r}(n_i-1) = \mathsf D(G)$. Suppose that $|G| \ge 3$. We will use that $\Delta(G)$ is an interval with $\min \Delta(G) = 1$ ([22]), and that, for all $k \in \mathbb N$,

$$\begin{aligned} \rho_{2k}(G) &= k\mathsf D(G), \\ k\mathsf D(G)+1 \le \rho_{2k+1}(G) &\le k\mathsf D(G) + \lfloor \mathsf D(G)/2 \rfloor, \quad \text{and} \\ \rho(G) &= \mathsf D(G)/2\,, \end{aligned} \tag{2}$$

([18, Section 6.3]).

3 Sets of Lengths of Transfer Krull Monoids

Weak transfer homomorphisms play a critical role in factorization theory, in particular in all studies of sets of lengths. We refer to [18] for a detailed presentation of transfer homomorphisms in the commutative setting. Weak transfer homomorphisms (as defined below) were introduced in [5, Definition 2.1] and transfer Krull monoids were introduced in [17].

Definition 3.1 Let H be a monoid.

1. A monoid homomorphism $\theta\colon H \to B$ to an atomic monoid B is called a *weak transfer homomorphism* if it has the following two properties:

   (T1) $B = B^\times \theta(H) B^\times$ and $\theta^{-1}(B^\times) = H^\times$.

   (WT2) If $a \in H$, $n \in \mathbb N$, $v_1, \ldots, v_n \in \mathcal A(B)$ and $\theta(a) = v_1 \cdot \ldots \cdot v_n$, then there exist $u_1, \ldots, u_n \in \mathcal A(H)$ and a permutation $\tau \in \mathfrak S_n$ such that $a = u_1 \cdot \ldots \cdot u_n$ and $\theta(u_i) \in B^\times v_{\tau(i)} B^\times$ for each $i \in [1, n]$.

2. H is said to be a *transfer Krull monoid* (over G_0) it there exists a weak transfer homomorphism $\theta\colon H \to \mathcal B(G_0)$ for a subset G_0 of an abelian group G. If G_0 is finite, then we say that H is a *transfer Krull monoid of finite type*.

If R is a domain and $R^\bullet$ its monoid of cancellative elements, then we say that R is a transfer Krull domain (of finite type) if $R^\bullet$ is a transfer Krull monoid (of finite type). Let $\theta\colon H \to B$ be a weak transfer homomorphism between atomic monoids. It is easy to show that for all $a \in H$ we have $\mathsf L_H(a) = \mathsf L_B(\theta(a))$ and hence $\mathcal L(H) = \mathcal L(B)$. Since monoids of zero-sum sequences are BF-monoids, the same is true for transfer Krull monoids.

Let H^* be a commutative Krull monoid (i.e., H^* is commutative, completely integrally closed, and v-noetherian). Then there is a weak transfer homomorphism

$\beta: H^* \to \mathcal{B}(G_0)$ where G_0 is a subset of the class group of H^*. Since monoids of zero-sum sequences are commutative Krull monoids and since the composition of weak transfer homomorphisms is a weak transfer homomorphism again, a monoid is a transfer Krull monoid if and only if it allows a weak transfer homomorphism to a commutative Krull monoid. In particular, commutative Krull monoids are transfer Krull monoids. However, a transfer Krull monoid need neither be commutative nor v-noetherian nor completely integrally closed. To give a noncommutative example, consider a bounded HNP (hereditary noetherian prime) ring R. If every stably free left R-ideal is free, then its multiplicative monoid of cancellative elements is a transfer Krull monoid ([34]). A class of commutative weakly Krull domains which are transfer Krull but not Krull will be given in Theorem 5.8. Extended lists of commutative Krull monoids and of transfer Krull monoids, which are not commutative Krull, are given in [17].

The next proposition summarizes some key results on the structure of sets of lengths of transfer Krull monoids.

Proposition 3.2

1. *Every transfer Krull monoid of finite type satisfies the Structure Theorem for Sets of Lengths.*
2. *For every $M \in \mathbb{N}_0$ and every finite nonempty set $\Delta \subset \mathbb{N}$, there is a finite abelian group G such that the following holds : for every AAMP L with difference $d \in \Delta$ and bound M there is some $y_L \in \mathbb{N}$ such that*

$$y + L \in \mathcal{L}(G) \quad \textit{for all} \quad y \geq y_L\,.$$

3. *If G is an infinite abelian group, then*

$$\mathcal{L}(G) = \{L \subset \mathbb{N}_{\geq 2} \mid L \textit{ is finite and nonempty} \} \cup \{\{0\}, \{1\}\}.$$

Proof 1. Let H be a transfer Krull monoid and $\theta: H \to \mathcal{B}(G_0)$ be a weak transfer homomorphism where G_0 is a finite subset of an abelian group. Then $\mathcal{L}(H) = \mathcal{L}(G_0)$, and $\mathcal{B}(G_0)$ satisfies the Structure Theorem by [18, Theorem 4.4.11].

For 2. we refer to [33], and for 3. see [31] and [18, Section 7.4]. □

The inequalities in (1) and the subsequent remarks show that a finite abelian group G has Davenport constant $\mathsf{D}(G) \leq 4$ if and only if G is cyclic of order $|G| \leq 4$ or if it is isomorphic to $C_2 \oplus C_2$ or to C_2^3. For these groups an explicit description of their systems of sets of lengths has been given, and we gather this in the next proposition (in Sect. 4 we will determine the systems $\mathcal{L}(G)$ for all groups G with $\mathsf{D}(G) = 5$).

Proposition 3.3

1. *If G is an abelian group, then $\mathcal{L}(G) = \{y + L \mid y \in \mathbb{N}_0,\ L \in \mathcal{L}(G^\bullet)\} \supset \{\{y\} \mid y \in \mathbb{N}_0\}$, and equality holds if and only if $|G| \le 2$.*
2. $\mathcal{L}(C_3) = \mathcal{L}(C_2 \oplus C_2) = \{y + 2k + [0,k] \mid y,\ k \in \mathbb{N}_0\}$.
3. $\mathcal{L}(C_4) = \{y + k + 1 + [0,k] \mid y,\ k \in \mathbb{N}_0\} \cup \{y + 2k + 2 \cdot [0,k] \mid y,\ k \in \mathbb{N}_0\}$.
4. $\mathcal{L}(C_2^3) = \{y + (k+1) + [0,k] \mid y \in \mathbb{N}_0,\ k \in [0,2]\}$
$\cup \{y + k + [0,k] \mid y \in \mathbb{N}_0,\ k \ge 3\} \cup \{y + 2k + 2 \cdot [0,k] \mid y,\ k \in \mathbb{N}_0\}$.

Proof See [18, Proposition 7.3.1 and Theorem 7.3.2]. □

Let G and G' be abelian groups. Then their monoids of zero-sum sequences $\mathcal{B}(G)$ and $\mathcal{B}(G')$ are isomorphic if and only if the groups G and G' are isomorphic ([18, Corollary 2.5.7]). The standing conjecture states that the systems of sets of lengths $\mathcal{L}(G)$ and $\mathcal{L}(G')$ of finite groups coincide if and only if G and G' are isomorphic (apart from the trivial cases listed in Proposition 3.3). Here is the precise formulation of the conjecture (it was first stated in [17]).

Conjecture 3.4 Let G be a finite abelian group with $\mathsf{D}(G) \ge 4$. If G' is an abelian group with $\mathcal{L}(G) = \mathcal{L}(G')$, then G and G' are isomorphic.

The conjecture holds true for groups G having rank $\mathsf{r}(G) \le 2$, for groups of the form $G = C_n^r$ (if r is small with respect to n), and others ([21, 23, 36]). But it is far open in general, and the goal of this section is to develop new viewpoints of looking at this conjecture.

Let G be a finite abelian group with $\mathsf{D}(G) \ge 4$. If G' is a finite abelian group with $\mathcal{L}(G) = \mathcal{L}(G')$, then (2) shows that

$$\begin{aligned} \mathsf{D}(G) = \rho_2(G) &= \sup\{\sup L \mid 2 \in L \in \mathcal{L}(G)\} \\ &= \sup\{\sup L \mid 2 \in L \in \mathcal{L}(G')\} = \rho_2(G') = \mathsf{D}(G') . \end{aligned}$$

We see from the inequalities in (1) that there are (up to isomorphism) only finitely many finite abelian groups G' with given Davenport constant, and hence there are only finitely many finite abelian groups G' with $\mathcal{L}(G) = \mathcal{L}(G')$. Thus Conjecture 3.4 is equivalent to the statement that for each $m \ge 4$ and for each two non-isomorphic finite abelian groups G and G' having Davenport constant $\mathsf{D}(G) = \mathsf{D}(G') = m$ the systems $\mathcal{L}(G)$ and $\mathcal{L}(G')$ are distinct. Therefore we have to study the set

$$\Omega_m = \{\mathcal{L}(G) \mid G \text{ is a finite abelian group with } \mathsf{D}(G) = m\}$$

of all systems of sets of lengths stemming from groups having Davenport constant equal to m. If a group G' is a proper subgroup of G, then $\mathsf{D}(G') < \mathsf{D}(G)$ ([18, Proposition 5.1.11]) and hence $\mathcal{L}(G') \subsetneq \mathcal{L}(G)$. Thus if $\mathsf{D}(G) = \mathsf{D}(G')$ for some group G', then none of the groups is isomorphic to a proper subgroup of the other one. Conversely, if G' is a finite abelian group with $\mathcal{L}(G') \subset \mathcal{L}(G)$, then $\mathsf{D}(G') = \rho_2(G') \le \rho_2(G) = \mathsf{D}(G)$. However, it may happen that $\mathcal{L}(G') \subsetneq \mathcal{L}(G)$ but $\mathsf{D}(G') = \mathsf{D}(G)$. Indeed, Proposition 3.3 shows that $\mathcal{L}(C_4) \subsetneq \mathcal{L}(C_2^3)$, and we will observe this phenomenon again in Sect. 4.

Theorem 3.5 *For $m \in \mathbb{N}$, let*

$$\Omega_m = \{\mathcal{L}(G) \mid G \text{ is a finite abelian group with } \mathsf{D}(G) = m\}.$$

Then $\mathcal{L}(C_2^{m-1})$ is a maximal element and $\mathcal{L}(C_m)$ is a minimal element in Ω_m (with respect to set-theoretical inclusion). Furthermore, if G is an abelian group with $\mathsf{D}(G) = m$ and $\mathcal{L}(G) \subset \mathcal{L}(C_2^{m-1})$, then $G \cong C_m$ or $G \cong C_2^{m-1}$.

Proof If $m \in [1, 2]$, then $|\Omega_m| = 1$ and hence all assertions hold. Since C_3 and $C_2 \oplus C_2$ are the only groups (up to isomorphism) with Davenport constant three, and since $\mathcal{L}(C_3) = \mathcal{L}(C_2^2)$ by Proposition 3.3, the assertions follow. We suppose that $m \ge 4$ and proceed in two steps.

1. To show that $\mathcal{L}(C_2^{m-1})$ is maximal, we study, for a finite abelian group G, the set $\Delta_1(G)$. We define

$$\Delta^*(G) = \{\min \Delta(G_0) \mid G_0 \subset G \text{ with } \Delta(G_0) \ne \emptyset\},$$

and recall that (see [18, Corollary 4.3.16])

$$\Delta^*(G) \subset \Delta_1(G) \subset \{d_1 \in \Delta(G) \mid d_1 \text{ divides some } d \in \Delta^*(G)\}.$$

Thus $\max \Delta_1(G) = \max \Delta^*(G)$, and [25, Theorem 1.1] implies that $\max \Delta^*(G) = \max\{\exp(G)-2, \mathsf{r}(G)-1\}$. Assume to the contrary that there is a finite abelian group G with $\mathsf{D}(G) = m \ge 4$ that is not an elementary 2-group such that $\mathcal{L}(C_2^{m-1}) \subset \mathcal{L}(G)$. Then

$$\begin{aligned} m-2 = \max \Delta^*(C_2^{m-1}) &= \max \Delta_1(C_2^{m-1}) \le \max \Delta_1(G) \\ &= \max \Delta^*(G) = \max\{\exp(G)-2, \mathsf{r}(G)-1\}. \end{aligned}$$

If $\mathsf{r}(G) \ge m-1$, then $\mathsf{D}(G) = m$ implies that $G \cong C_2^{m-1}$, a contradiction. Thus $\exp(G) \ge m$, and since $\mathsf{D}(G) = m$ we infer that $G \cong C_m$. If $m = 4$, then Proposition 3.3.4 shows that $\mathcal{L}(C_2^3) \not\subset \mathcal{L}(C_4)$, a contradiction. Suppose that $m \ge 5$. Then $\Delta^*(C_2^{m-1}) = \Delta_1(C_2^{m-1}) = \Delta(C_2^{m-1}) = [1, m-2]$ by [18, Corollary 6.8.3]. For cyclic groups we have $\max \Delta^*(C_m) = m-2$ and $\max(\Delta^*(C_m) \setminus \{m-2\}) = \lfloor m/2 \rfloor - 1$ by [18, Theorem 6.8.12]. Therefore $\mathcal{L}(C_2^{m-1}) \subset \mathcal{L}(C_m)$ implies that

$$[1, m-2] = \Delta_1(C_2^{m-1}) \subset \Delta_1(C_m),$$

a contradiction to $m-3 \notin \Delta_1(C_m)$.

2. We recall some facts. Let G be a group with $\mathsf{D}(G) = m$. If $U \in \mathcal{A}(G)$ with $|U| = \mathsf{D}(G)$, then $\{2, \mathsf{D}(G)\} \subset \mathsf{L}\big(U(-U)\big)$. Cyclic groups and elementary 2-groups are the only groups G with the following property: if $L \in \mathsf{L}(G)$ with $\{2, \mathsf{D}(G)\} \subset L$, then $L = \{2, \mathsf{D}(G)\}$ ([18, Theorem 6.6.3]).

Now assume to the contrary that there is a finite abelian group G with $\mathsf{D}(G) = m$ such that $\mathcal{L}(G) \subset \mathcal{L}(C_m)$. Let $L \in \mathcal{L}(G)$ with $\{2, \mathsf{D}(G)\} \subset L$. Then $L \in \mathcal{L}(C_m)$

whence $L = \{2, \mathsf{D}(G)\}$ which implies that G is cyclic or an elementary 2-group. By 1., G is not an elementary 2-group whence G is cyclic which implies $G \cong C_m$ and hence $\mathcal{L}(G) = \mathcal{L}(C_m)$.

The furthermore assertion on groups G with $\mathsf{D}(G) = m$ and $\mathcal{L}(G) \subset \mathcal{L}(C_2^{m-1})$ follows as above by considering sets of lengths L with $\{2, \mathsf{D}(G)\} \subset L$. □

In Sect. 4 we will see that $\mathcal{L}(C_2^{m-1})$ need not be the largest element in Ω_m, and that indeed $\mathcal{L}(C_m) \subset \mathcal{L}(C_2^{m-1})$ for $m \in [2, 5]$, where the inclusion is strict for $m \geq 4$. On the other hand, it is shown in [24] that $\mathcal{L}(C_m) \not\subset \mathcal{L}(C_2^{m-1})$ for infinitely many $m \in \mathbb{N}$.

Theorem 3.6 *We have*

$$\bigcap \mathcal{L}(G) = \big\{ y + 2k + [0, k] \;\big|\; y,\, k \in \mathbb{N}_0 \big\},$$

where the intersection is taken over all finite abelian groups G with $|G| \geq 3$.

Proof By Proposition 3.3.2, the intersection on the left-hand side is contained in the set on the right-hand side. Let G be a finite abelian group with $|G| \geq 3$. If $L \in \mathcal{L}(G)$, then $y + L \in \mathcal{L}(G)$. Thus it is sufficient to show that $[2k, 3k] \in \mathcal{L}(G)$ for every $k \in \mathbb{N}$. If G contains two independent elements of order 2 or an element of order 4, then the claim follows by Proposition 3.3. Thus, it remains to consider the case when G contains an element g with $\operatorname{ord}(g) = p$ for some odd prime $p \in \mathbb{N}$. Let $k \in \mathbb{N}$ and $B_k = ((2g)^p g^p)^k$. We assert that $\mathsf{L}(B_k) = [2k, 3k]$.

We set $U_1 = g^p$, $U_2 = (2g)^p$, $V_1 = (2g)^{(p-1)/2} g$, and $V_2 = (2g) g^{p-2}$. Since $U_1 U_2 = V_1^2 V_2$ and

$$B_k = (U_1 U_2)^k = (U_1 U_2)^{k-\nu} (V_1^2 V_2)^\nu \quad \text{for all} \quad \nu \in [0, k],$$

it follows that $[2k, 3k] \subset \mathsf{L}(B_k)$.

In order to show that there are no other factorization lengths, we recall the concept of the g-norm of sequences. If $S = (n_1 g) \cdot \ldots \cdot (n_\ell g) \in \mathcal{B}(\langle g \rangle)$, where $\ell \in \mathbb{N}_0$ and $n_1, \ldots, n_\ell \in [1, \operatorname{ord}(g)]$, then

$$||S||_g = \frac{n_1 + \ldots + n_\ell}{\operatorname{ord}(g)} \in \mathbb{N}$$

is the g-norm of S. Clearly, if $S = S_1 \cdot \ldots \cdot S_m$ with $S_1, \ldots, S_m \in \mathcal{A}(G)$, then $||S||_g = ||S_1||_g + \ldots + ||S_m||_g$.

Note that $U_2 = (2g)^p$ is the only atom in $\mathcal{A}(\{g, 2g\})$ with g-norm 2, and all other atoms in $\mathcal{A}(\{g, 2g\})$ have g-norm 1. Let $B_k = W_1 \cdot \ldots \cdot W_\ell$ be a factorization of B_k, and let ℓ' be the number of $i \in [1, \ell]$ such that $W_i = (2g)^p$. We have $\|B_k\|_g = 3k$ and thus $3k = 2\ell' + (\ell - \ell') = \ell' + \ell$. Since $\ell' \in [0, k]$, it follows that $\ell = 3k - \ell' \in [2k, 3k]$. □

Theorem 3.7 *Let $L \subset \mathbb{N}_{\geq 2}$ be a finite nonempty subset. Then there are only finitely many pairwise non-isomorphic finite abelian groups G such that $L \notin \mathcal{L}(G)$.*

Proof We start with the following two assertions:

A1. There is an integer $n_L \in \mathbb{N}$ such that $L \in \mathscr{L}(C_n)$ for every $n \ge n_L$.
A2. For every $p \in \mathbb{P}$ there is an integer $r_{p,L} \in \mathbb{N}$ such that $L \in \mathscr{L}(C_p^r)$ for every $r \ge r_{p,L}$.

Proof of **A1**. By Proposition 3.2.3, there is some $B = \prod_{i=1}^k m_k \prod_{j=1}^{\ell}(-n_j) \in \mathscr{B}(\mathbb{Z})$ such that $\mathsf{L}(B) = L$, where $k, \ell, m_1, \ldots, m_k \in \mathbb{N}$ and $n_1, \ldots, n_\ell \in \mathbb{N}_0$. We set $n_L = n_1 + \ldots + n_\ell$ and choose some $n \in \mathbb{N}$ with $n \ge n_L$. If $S \in \mathscr{F}(\mathbb{Z})$ with $S \mid B$ and $f \colon \mathbb{Z} \to \mathbb{Z}/n\mathbb{Z}$ denotes the canonical epimorphism, then S has sum zero if and only if $f(S)$ has sum zero. This implies that $\mathsf{L}_{\mathscr{B}(\mathbb{Z}/n\mathbb{Z})}(f(B)) = \mathsf{L}_{\mathscr{B}(\mathbb{Z})}(B) = L$. □ [Proof of **A1**]

Proof of **A2**. Let $p \in \mathbb{P}$ be a prime and let G_p be an infinite dimensional $\mathbb{F}_p$-vector space. By Proposition 3.2.3, there is some $B_p \in \mathscr{B}(G_p)$ such that $\mathsf{L}(B_p) = L$. If $r_{p,L}$ is the rank of $\langle \operatorname{supp}(B_p) \rangle \subset G_p$, then

$$L = \mathsf{L}(B_p) \in \mathscr{L}(\langle \operatorname{supp}(B_p) \rangle) \subset \mathscr{L}(C_p^r) \quad \text{for} \quad r \ge r_{p,L} . \qquad \square\text{[Proof of}\mathbf{A2}]$$

Now let G be a finite abelian group such that $L \notin \mathscr{L}(G)$. Then **A1** implies that $\exp(G) < n_L$, and **A2** implies that $\mathsf{r}_p(G) < r_{p,L}$ for all primes p with $p \mid \exp(G)$. Thus the assertion follows. □

4 Sets of Lengths of Transfer Krull Monoids Over Small Groups

Since the very beginning of factorization theory, invariants controlling the structure of sets of lengths (such as elasticities and sets of distances) have been in the center of interest. Nevertheless, (apart from a couple of trivial cases) the full system of sets of lengths has been written down explicitly only for the following classes of monoids:

- Numerical monoids generated by arithmetical progressions: see [1].
- Self-idealizations of principal ideal domains: see [10, Corollary 4.16], [4, Remark 4.6].
- The ring of integer-valued polynomials over $\mathbb{Z}$: see [15].
- The systems $\mathscr{L}(G)$ for infinite abelian groups G and for abelian groups G with $\mathsf{D}(G) \le 4$: see Propositions 3.2 and 3.3.

The goal of this section is to determine $\mathscr{L}(G)$ for abelian groups G having Davenport constant $\mathsf{D}(G) = 5$. By inequality (1) and the subsequent remarks, a finite abelian group G has Davenport constant five if and only if it is isomorphic to one of the following groups:

$$C_3 \oplus C_3, \quad C_5, \quad C_2 \oplus C_4, \quad C_2^4 .$$

Their systems of sets of lengths are given in Theorems 4.1, 4.3, 4.5, and 4.8. We start with a brief analysis of these explicit descriptions (note that they will be needed again in Sect. 5; confer the proof of Theorem 5.7).

By Theorem 3.5, we know that $\mathscr{L}(C_2^4)$ is maximal in $\Omega_5 = \{\mathscr{L}(C_5), \mathscr{L}(C_2 \oplus C_4), \mathscr{L}(C_3 \oplus C_3), \mathscr{L}(C_2^4)\}$. Theorems 4.1, 4.3, 4.5, and 4.8 unveil that $\mathscr{L}(C_3 \oplus C_3)$, $\mathscr{L}(C_2 \oplus C_4)$, and $\mathscr{L}(C_2^4)$ are maximal in Ω_5, and that $\mathscr{L}(C_5)$ is contained in $\mathscr{L}(C_2^4)$, but it is neither contained in $\mathscr{L}(C_3 \oplus C_3)$ nor in $\mathscr{L}(C_2 \oplus C_4)$. Furthermore, Theorems 3.5, 4.3, and 4.8 show that $\mathscr{L}(C_m) \subset \mathscr{L}(C_2^{m-1})$ for $m \in [2, 5]$. It is well known that, for all $m \geq 4$, $\mathscr{L}(C_m) \neq \mathscr{L}(C_2^{m-1})$ ([16, Corollary 5.3.3]), and the standing conjecture is that $\mathscr{L}(C_m) \not\subset \mathscr{L}(C_2^{m-1})$ holds true for almost all $m \in \mathbb{N}_{\geq 2}$ (see [24]).

The group $C_3 \oplus C_3$ has been handled in [21, Theorem 4.2].

Theorem 4.1 $\mathscr{L}(C_3^2) = \{y + [2k, 5k] \mid y, k \in \mathbb{N}_0\} \cup \{y + [2k+1, 5k+2] \mid y \in \mathbb{N}_0,\, k \in \mathbb{N}\}\}$.

Remark An equivalent way to describe $\mathscr{L}(C_3^2)$ is $\{y + \lceil \frac{2k}{3} \rceil + [0, k] \mid y \in \mathbb{N}_0, k \in \mathbb{N}_{\geq 2}\} \cup \{\{y\}, y + 2 + [0, 1] \mid y \in \mathbb{N}_0\}$.

The fact that all sets of lengths are intervals is a consequence of the fact $\Delta(C_3^2) = \{1\}$. Of course, each set of lengths L has to fulfill $\rho(L) \leq 5/2 = \rho(C_3^2)$. We observe that the description shows that this is the only condition, provided $\min L \geq 2$. The following lemma is frequently helpful in the remainder of this section.

Lemma 4.2 *Let G be a finite abelian group, and let $A \in \mathscr{B}(G)$.*

1. *If* $\operatorname{supp}(A) \cup \{0\}$ *is a group, then* $\mathsf{L}(A)$ *is an interval.*
2. *If A_1 is an atom dividing A with $|A_1| = 2$, then* $\max \mathsf{L}(A) = 1 + \max \mathsf{L}(AA_1^{-1})$.
3. *If A is a product of atoms of length 2 and if every atom A_1 dividing A has length $|A_1| = 2$ or $|A_1| = 4$, then* $\max \mathsf{L}(A) - 1 \notin \mathsf{L}(A)$.

Proof

1. See [18, Theorem 7.6.8].
2. Let $\ell = \max \mathsf{L}(A)$ and $A = U_1 \cdot \ldots \cdot U_\ell$, where $U_1, \ldots, U_\ell \in \mathscr{A}(G)$. Let $A_1 = g_1 g_2$, where $g_1, g_2 \in G$. If there exists $i \in [1, \ell]$ such that $A_1 = U_i$, then $\max \mathsf{L}(A) = 1 + \max \mathsf{L}(AA_1^{-1})$. Otherwise there exist distinct $i, j \in [1, \ell]$ such that $g_1 \mid U_i$ and $g_2 \mid U_j$. Thus A_1 divides $U_i U_j$ and hence $1 + \max \mathsf{L}(AA_1^{-1}) \geq \ell$ which implies that $\max \mathsf{L}(A) = 1 + \max \mathsf{L}(AA_1^{-1})$ by the maximality of ℓ.
3. If $\max \mathsf{L}(A) - 1 \in \mathsf{L}(A)$, then $A = V_1 \cdot \ldots \cdot V_{\max \mathsf{L}(A)-1}$ with $|V_1| = 4$ and $|V_2| = \ldots = |V_{\max \mathsf{L}(A)-1}| = 2$. Thus V_1 can only be a product two atoms of length 2, a contradiction.

□

We now consider the groups C_5, $C_2 \oplus C_4$, and C_2^4, each one in its own subsection. In the proofs of the forthcoming theorems we will use Proposition 3.3 and Theorem 3.6 without further mention.

4.1 The System of Sets of Lengths of C_5

The goal of this subsection is to prove the following result.

Theorem 4.3 $\mathcal{L}(C_5) = \mathcal{L}_1 \cup \mathcal{L}_2 \cup \mathcal{L}_3 \cup \mathcal{L}_4 \cup \mathcal{L}_5 \cup \mathcal{L}_6$, *where*

$$\mathcal{L}_1 = \{\{y\} \mid y \in \mathbb{N}_0\},$$

$$\mathcal{L}_2 = \{y + 2 + \{0, 2\} \mid y \in \mathbb{N}_0\},$$

$$\mathcal{L}_3 = \{y + 3 + \{0, 1, 3\} \mid y \in \mathbb{N}_0\},$$

$$\mathcal{L}_4 = \{y + 2k + 3 \cdot [0, k] \mid y \in \mathbb{N}_0, k \in \mathbb{N}\},$$

$$\mathcal{L}_5 = \{y + 2\left\lceil \frac{k}{3} \right\rceil + [0, k] \mid y \in \mathbb{N}_0, k \in \mathbb{N} \setminus \{3\}\} \cup \{y + [3, 6] \mid y \in \mathbb{N}_0\}, \text{ and}$$

$$\mathcal{L}_6 = \{y + 2k + 3 + \{0, 2, 3\} + 3 \cdot [0, k] \mid y, k \in \mathbb{N}_0\}.$$

We observe that all sets of lengths with many elements are arithmetic multiprogressions with difference 1 or 3. Yet, there is none with difference 2. This is because $\Delta^*(C_5) = \{1, 3\}$. Moreover, we point out that the condition for an interval to be a set of lengths is different from that of the other groups with Davenport constant 5. This is related to the fact that $\rho_{2k+1}(C_5) = 5k + 1$, while $\rho_{2k+1}(G) = 5k + 2$ for the other groups with Davenport constant 5. Before we start the actual proof, we collect some results on sets of lengths over C_5.

Lemma 4.4 *Let G be cyclic of order five, and let $A \in \mathcal{B}(G)$.*

1. *If $g \in G^\bullet$ and $k \in \mathbb{N}_0$, then*

$$\mathsf{L}\big(g^{5(k+1)}(-g)^{5(k+1)}(2g)g^3\big) = 2k + 3 + \{0, 2, 3\} + 3 \cdot [0, k].$$

2. *If $2 \in \Delta(\mathsf{L}(A)) \subset [1, 2]$, then $\mathsf{L}(A) \in \{\{y, y+2\} \mid y \ge 2\} \cup \{\{y, y+1, y+3\} \mid y \ge 3\}$ or $\mathsf{L}(A) = 3 + \{0, 2, 3\} + \mathsf{L}(A')$ where $A' \in \mathcal{B}(G)$ and $\mathsf{L}(A')$ is an arithmetical progression of difference 3.*
3. *$\Delta(G) = [1, 3]$, and if $3 \in \Delta(\mathsf{L}(A))$, then $\Delta(\mathsf{L}(A)) = \{3\}$.*
4. *$\rho_{2k+1}(G) = 5k + 1$ for all $k \in \mathbb{N}$.*

Proof 1. and 2. follow from the proof of [21, Lemma 4.5].

3. See [18, Theorems 6.7.1 and 6.4.7] and [12, Theorem 3.3].

4. See [16, Theorem 5.3.1]. □

Proof (Theorem 4.3) Let G be cyclic of order five and let $g \in G^\bullet$. We first show that all the specified sets occur as sets of lengths, and then we show that no other sets occur.

Step 1. We prove that for every $L \in \mathcal{L}_2 \cup \mathcal{L}_3 \cup \mathcal{L}_4 \cup \mathcal{L}_5 \cup \mathcal{L}_5$, there exists an $A \in \mathcal{B}(G)$ such that $L = \mathsf{L}(A)$. We distinguish five cases.

If $L = \{y, y+2\} \in \mathscr{L}_2$ with $y \geq 2$, then we set $A = 0^{y-2}g^5(-g)^3(-2g)$ and obtain that $\mathsf{L}(A) = y - 2 + \{2, 4\} = L$.

If $L = \{y, y+1, y+3\} \in \mathscr{L}_3$ with $y \geq 3$, then we set $A = 0^{y-3}g^5(-g)^5g^2(-2g)$ and obtain that $\mathsf{L}(A) = y - 3 + \{3, 4, 6\} = \{y, y+1, y+3\} = L$.

If $L = y + 2k + 3 \cdot [0, k] \in \mathscr{L}_4$ with $k \in \mathbb{N}$ and $y \in \mathbb{N}_0$, then we set $A = g^{5k}(-g)^{5k}0^y \in \mathscr{B}(G)$ and hence $\mathsf{L}(A) = y + 2k + 3 \cdot [0, k] = L$.

If $L = y+2k+3+\{0, 2, 3\}+3\cdot[0, k] \in \mathscr{L}_6$ with $k \in \mathbb{N}_0$ and $y \in \mathbb{N}_0$, then we set $A = 0^y g^{5(k+1)}(-g)^{5(k+1)}(2g)g^3$ and hence $\mathsf{L}(A) = y+2k+3+\{0, 2, 3\}+3\cdot[0, k] = L$ by Lemma 4.4.1.

Now we suppose that $L \in \mathscr{L}_5$, and we distinguish two subcases. First, if $L = y + [3, 6]$ with $y \in \mathbb{N}_0$, then we set $A = 0^y(2g(-2g))g^5(-g)^5$ and hence $\mathsf{L}(A) = y + [3, 6] = L$. Second, we assume that $L = y + 2\lceil \frac{k}{3} \rceil + [0, k]$ with $y \in \mathbb{N}_0$ and $k \in \mathbb{N} \setminus \{3\}$.

If $k \in \mathbb{N}$ with $k \equiv 0 \pmod 3$, then $k \geq 6$ and by Lemma 4.2.1 we obtain that

$$\mathsf{L}\big(0^y(2g)^5(-2g)^5g^{5t}(-g)^{5t}\big) = y + [2t+2, 5t+5] = y + 2\lceil \frac{k}{3} \rceil + [0, k] = L\,,$$

where $k = 3t + 3$.

If $k \in \mathbb{N}$ with $k \equiv 1 \pmod 3$, then by Lemma 4.2.1 we obtain that

$$\mathsf{L}\big(0^y(2g(-g)^2)(g^2(-2g))g^{5t}(-g)^{5t}\big) = y+[2t+2, 5t+3] = y+2\lceil \frac{k}{3} \rceil+[0, k] = L\,,$$

where $k = 3t + 1$.

If $k \in \mathbb{N}$ with $k \equiv 2 \pmod 3$, then by Lemma 4.2.1 we obtain that

$$\mathsf{L}\big(0^y(g^3(2g))((-g)^3(-2g))g^{5t}(-g)^{5t}\big) = y+[2t+2, 5t+4] = y+2\lceil \frac{k}{3} \rceil+[0, k] = L\,,$$

where $k = 3t + 2$.

Step 2. We prove that for every $A \in \mathscr{B}(G^\bullet)$, $\mathsf{L}(A) \in \mathscr{L}_2 \cup \mathscr{L}_3 \cup \mathscr{L}_4 \cup \mathscr{L}_5 \cup \mathscr{L}_6$.

Let $A \in \mathscr{B}(G^\bullet)$. We may suppose that $\Delta(\mathsf{L}(A)) \neq \emptyset$. By Lemma 4.4.3 we distinguish three cases according to the form of the set of distances $\Delta(\mathsf{L}(A))$.

Case 1: $\Delta(\mathsf{L}(A)) = \{1\}$.

Then $\mathsf{L}(A)$ is an interval and hence we assume that $\mathsf{L}(A) = [y, y+k] = y + [0, k]$ where $y \geq 2$ and $k \geq 1$. If $k = 3$ and $y = 2$, then $\mathsf{L}(A) = [2, 5]$ and hence $\mathsf{L}(A) = \mathsf{L}(g^5(-g)^5) = \{2, 5\}$, a contradiction. Thus $k = 3$ implies that $y \geq 3$ and hence $\mathsf{L}(A) \in \mathscr{L}_5$. If $k \leq 2$, then we obviously have that $\mathsf{L}(A) \in \mathscr{L}_5$. Suppose that $k \geq 4$. If $y = 2t$ with $t \geq 2$, then $y + k \leq 5t$ and hence $y = 2t \geq 2\lceil \frac{k}{3} \rceil$ which implies that $\mathsf{L}(A) \in \mathscr{L}_5$. If $y = 2t + 1$ with $t \geq 1$, then $y + k \leq 5t + 1$ and hence $y = 2t + 1 \geq 1 + 2\lceil \frac{k}{3} \rceil$ which implies that $\mathsf{L}(A) \in \mathscr{L}_5$.

Case 2: $\Delta(\mathsf{L}(A)) = \{3\}$.

Then $\mathsf{L}(A) = y + 3 \cdot [0, k]$ where $y \geq 2$ and $k \geq 1$. If $y = 2t \geq 2$, then $y + 3k \leq 5t$ and hence $y = 2t \geq 2k$ which implies that $\mathsf{L}(A) \in \mathscr{L}_4$. If $y = 2t + 1 \geq 3$, then $y + 3k \leq 5t + 1$ and hence $y = 2t + 1 \geq 1 + 2k$ which implies that $\mathsf{L}(A) \in \mathscr{L}_4$.

Case 3: $2 \in \Delta(\mathsf{L}(A)) \subset [1,2]$.

By Lemma 4.4.2, we infer that either $\mathsf{L}(A) \in \mathscr{L}_2 \cup \mathscr{L}_3$ or that $\mathsf{L}(A) = 3 + \{0,2,3\} + \mathsf{L}(A')$, where $A' \in \mathscr{B}(G)$ and $\mathsf{L}(A')$ is an arithmetical progression of difference 3. In the latter case we obtain that $\mathsf{L}(A') = y + 2k + 3 \cdot [0\ k]$, with $y \in \mathbb{N}_0$ and $k \in \mathbb{N}_0$, and hence $\mathsf{L}(A) = y + 2k + 3 + \{0,2,3\} + 3 \cdot [0,k] \in \mathscr{L}_6$. □

4.2 The System of Sets of Lengths of $C_2 \oplus C_4$

We establish the following result, giving a complete description of the system of sets of lengths of $C_2 \oplus C_4$.

Theorem 4.5 $\mathscr{L}(C_2 \oplus C_4) = \mathscr{L}_1 \cup \mathscr{L}_2 \cup \mathscr{L}_3 \cup \mathscr{L}_4 \cup \mathscr{L}_5$, *where*

$$\begin{aligned}
\mathscr{L}_1 &= \{\{y\} \mid y \in \mathbb{N}_0\}, \\
\mathscr{L}_2 &= \{y + 2\left\lceil \frac{k}{3} \right\rceil + [0,k] \mid y \in \mathbb{N}_0, k \in \mathbb{N} \setminus \{3\}\} \cup \\
&\qquad \{y + [3,6] \mid y \in \mathbb{N}_0,\} \cup \{[2t+1, 5t+2] \mid t \in \mathbb{N}\} \\
&= \{y + \left\lceil \frac{2k}{3} \right\rceil + [0,k] \mid y \in \mathbb{N}_0, k \in \mathbb{N} \setminus \{1,3\}\} \cup \\
&\qquad \{y + 3 + [0,3], y + 2 + [0,1] \mid y \in \mathbb{N}_0\}, \\
\mathscr{L}_3 &= \{y + 2k + 2 \cdot [0,k] \mid y \in \mathbb{N}_0, k \in \mathbb{N}\}, \\
\mathscr{L}_4 &= \{y + k + 1 + (\{0\} \cup [2, k+2]) \mid y \in \mathbb{N}_0, k \in \mathbb{N} \text{ odd}\}, \textit{and} \\
\mathscr{L}_5 &= \{y + k + 2 + ([0,k] \cup \{k+2\}) \mid y \in \mathbb{N}_0, k \in \mathbb{N}\}.
\end{aligned}$$

We note that all sets of lengths are arithmetical progressions with difference 2 or almost arithmetical progressions with difference 1 and bound 2. This is related to the fact that $\Delta(C_2 \oplus C_4) = \Delta^*(C_2 \oplus C_4) = \{1,2\}$. We start with a lemma determining all minimal zero-sum sequences over $C_2 \oplus C_4$.

Lemma 4.6 *Let (e,g) be a basis of $G = C_2 \oplus C_4$ with $\operatorname{ord}(e) = 2$ and $\operatorname{ord}(g) = 4$. Then the minimal zero-sum sequences over $G^\bullet$ are given by the following list.*

1. The minimal zero-sum sequences of length 2 are:

$$\begin{aligned}
S_2^1 &= \{e^2, (e+2g)^2\}, \\
S_2^2 &= \{(2g)^2\}, \\
S_2^3 &= \{g(-g), (e+g)(e-g)\}
\end{aligned}$$

2. *The minimal zero-sum sequences of length* 3 *are :*

$$S_3^1 = \{e(2g)(e+2g)\},$$
$$S_3^2 = \{g^2(2g), (-g)^2(2g), (e+g)^2(2g), (e-g)^2(2g)\},$$
$$S_3^3 = \{eg(e-g), e(-g)(e+g), (e+2g)g(e+g), (e+2g)(-g)(e-g)\}.$$

3. *The minimal zero-sum sequences of length* 4 *are :*

$$S_4^1 = \{g^4, (-g)^4, (e+g)^4, (e-g)^4\},$$
$$S_4^2 = \{g^2(e+g)^2, (-g)^2(e-g)^2, g^2(e-g)^2, (-g)^2(e+g)^2\},$$
$$S_4^3 = \{eg^2(e+2g), e(e+g)^2(e+2g), e(-g)^2(e+2g), e(e-g)^2(e+2g)\},$$
$$S_4^4 = \{eg(2g)(e+g), e(-g)(2g)(e-g), (e+2g)g(2g)(e-g), (e+2g)(-g)(2g)(e+g)\}.$$

4. *The minimal zero-sum sequences of length* 5 *are :*

$$\begin{aligned} S_5 = \{&eg^3(e+g), e(-g)^3(e-g), e(e+g)^3g, e(e-g)^3(-g) \\ &(e+2g)g^3(e-g), (e+2g)(-g)^3(e+g), \\ &(e+2g)(e+g)^3(-g), (e+2g)(e-g)^3g\}, \end{aligned}$$

Moreover, for each two atoms W_1, W_2 in any one of the above sets, there exists a group isomorphism $\phi: G \to G$ such that $\phi(W_1) = W_2$.

Proof We give a sketch of the proof.

Since a minimal zero-sum sequence of length two is of the form $h(-h)$ for some nonzero element $h \in G$, the list given in 1. follows.

A minimal zero-sum sequence of length three contains either two elements of order four or no element of order four. If there are two elements of order four, we can have one element of order four with multiplicity two (see S_3^2) or two distinct elements of order four that are not the inverse of each other (see S_3^3). If there is no element of order four, the sequence consists of three distinct elements of order two (see S_3^1).

A minimal zero-sum sequence of length four contains either four elements of order four or two elements of order four. If there are two elements of order four, the sequence can contain one element with multiplicity two (see S_4^3) or any two distinct elements that are not each other's inverse with multiplicity one (see S_4^4). If there are four elements of order four, the sequence can contain one element with multiplicity four (see S_4^1) or two elements with multiplicity two (see S_4^2).

Since every minimal zero-sum sequence of length five contains an element with multiplicity three, the list given in 4. follows (for details, see [18, Theorem 6.6.5]).

The existence of the required isomorphism follows immediately from the given description of the sequences. □

The next lemma collects some basic results on $\mathcal{L}(C_2 \oplus C_4)$ that will be essential for the proof of Theorem 4.5.

Lemma 4.7 *Let $G = C_2 \oplus C_4$, and let $A \in \mathcal{B}(G)$.*

1. *$\Delta(G) = [1, 2]$, and if $\{2, 5\} \subset \mathsf{L}(A)$, then $\mathsf{L}(A) = \{2, 4, 5\}$.*
2. *$\rho_{2k+1}(G) = 5k + 2$ for all $k \in \mathbb{N}$.*
3. *If (e, g) is a basis of G with $\operatorname{ord}(e) = 2$ and $\operatorname{ord}(g) = 4$, then $\{0, g, 2g, e+g, e+2g\}$ and $\{0, g, 2g, e, e-g\}$ are half-factorial sets. Furthermore, if $\operatorname{supp}(A) \subset \{e, g, 2g, e+g, e+2g\}$ and $\mathsf{v}_e(A) = 1$, then $|\mathsf{L}(A)| = 1$.*

Proof

1. The first assertion follows from [18, Theorem 6.7.1 and Corollary 6.4.8]. Let $A \in \mathcal{B}(G)$ with $\{2, 5\} \subset \mathsf{L}(A)$. Then there is an $U \in \mathcal{A}(G)$ of length $|U| = 5$ such that $A = (-U)U$. By Lemma 4.6 there is a basis (e, g) of G with $\operatorname{ord}(e) = 2$ and $\operatorname{ord}(g) = 4$ such that $U = eg^3(e+g)$. This implies that $\mathsf{L}(A) = \{2, 4, 5\}$.
2. See [28, Corollary 5.2].
3. See [18, Theorem 6.7.9.1] for the first statement. Suppose that $\operatorname{supp}(A) \subset \{e, g, 2g, e+g, e+2g\}$ and $\mathsf{v}_e(A) = 1$. Then for every atom W dividing A with $e \mid W$, we have that $\mathsf{k}(W) = \frac{3}{2}$. Since $\operatorname{supp}(AW^{-1})$ is half-factorial, we obtain that $\mathsf{L}(AW^{-1}) = \{\mathsf{k}(A) - 3/2\}$ by [18, Proposition 6.7.3] which implies that $\mathsf{L}(A) = \{1 + \mathsf{k}(A) - 3/2\} = \{\mathsf{k}(A) - 1/2\}$. □

Proof (Theorem 4.5) Let (e, g) be a basis of $G = C_2 \oplus C_4$ with $\operatorname{ord}(e) = 2$ and $\operatorname{ord}(g) = 4$. We start by collecting some basic constructions that will be useful. Then, we show that all the sets in the result actually are sets of lengths. Finally, we show that there are no other sets of lengths.

Step 0. Some elementary constructions.

Let $U_1 = eg^3(e+g)$, $U_2 = (e+2g)(e+g)^3(-g)$, $U_3 = e(e-g)^3(-g)$, $U_4 = (-g)^2(e+g)^2$, and $U_5 = e(e+2g)g^2$. Then it is not hard to check that

$$\begin{aligned}
&\mathsf{L}(U_1(-U_1)) = \mathsf{L}(U_2(-U_2)) = \{2, 4, 5\}, \\
&\mathsf{L}(U_1U_3)) = [2, 4], && \mathsf{L}(U_1(-U_4)) = [2, 3], \\
&\mathsf{L}(U_1U_3U_4) = [3, 7], && \mathsf{L}(U_1(-U_1)U_2(-U_2)) = [4, 10], \\
&\mathsf{L}(U_5^2(-g)^4) = \{3, 4, 6\}, && \mathsf{L}(U_5(-U_5)g^4(-g)^4) = \{4, 5, 6, 8\}, \\
&\mathsf{L}(U_1(-U_1)(e+2g)^2) = [3, 6]. && \qquad (3)
\end{aligned}$$

Based on these results, we can obtain the sets of lengths of more complex zero-sum sequences. Let $k \in \mathbb{N}$.

Since $[2k+2, 4k+5] \supset \mathsf{L}(U_1(-U_1)g^{4k}(-g)^{4k}) \supset \mathsf{L}(U_1(-U_1)) + \mathsf{L}(g^{4k}(-g)^{4k}) = 2k+2+(\{0\} \cup [2, 2k+3])$ and $2k+3 \notin \mathsf{L}(U_1(-U_1)g^{4k}(-g)^{4k})$, we obtain that

$$\mathsf{L}(U_1(-U_1)g^{4k}(-g)^{4k}) = 2k+2+(\{0\} \cup [2, 2k+3]). \qquad (4)$$

Since $[2(k+1), 5(k+1)] \supset \mathsf{L}(U_1(-U_1)U_2^k(-U_2)^k) \supset \mathsf{L}(U_1(-U_1)U_2(-U_2)) + \mathsf{L}(U_2^{k-1}(-U_2)^{k-1}) = [2(k+1), 5(k+1)]$, we obtain that

$$\mathsf{L}(U_1(-U_1)U_2^k(-U_2)^k) = [2(k+1), 5(k+1)]\,. \tag{5}$$

Since $[2(k+1), 5(k+1)-1] \supset \mathsf{L}(U_1U_3U_2^k(-U_2)^k) \supset \mathsf{L}(U_1U_3)+\mathsf{L}(U_2^k(-U_2)^k) = [2(k+1), 5(k+1)-1]$, we obtain that

$$\mathsf{L}(U_1U_3U_2^k(-U_2)^k) = [2(k+1), 5(k+1)-1]\,. \tag{6}$$

Since $[2(k+1), 5(k+1)-2] \supset \mathsf{L}(U_1(-U_4)U_2^k(-U_2)^k) \supset \mathsf{L}(U_1(-U_4)) + \mathsf{L}(U_2^k(-U_2)^k)$ and $\mathsf{L}(U_1(-U_4)) + \mathsf{L}(U_2^k(-U_2)^k) = [2(k+1), 5(k+1)-2]$, we obtain that

$$\mathsf{L}(U_1(-U_4)U_2^k(-U_2)^k) = [2(k+1), 5(k+1)-2]\,. \tag{7}$$

Since

$$[2k+1, 5k+2] \supset \mathsf{L}(U_1U_3U_4U_2^{k-1}(-U_2)^{k-1}) \supset \mathsf{L}(U_1U_3U_4) + \mathsf{L}(U_2^{k-1}(-U_2)^{k-1})$$

and $\mathsf{L}(U_1U_3U_4) + \mathsf{L}(U_2^{k-1}(-U_2)^{k-1}) = [2k+1, 5k+2]$, we obtain that

$$\mathsf{L}(U_1U_3U_4U_2^{k-1}(-U_2)^{k-1}) = [2k+1, 5k+2]\,. \tag{8}$$

Since

$$[2k+1, 4k+2] \supset \mathsf{L}(U_5^2(-g)^4g^{4k-4}(-g)^{4k-4}) \supset \mathsf{L}(U_5^2(-g)^4) + \mathsf{L}(g^{4k-4}(-g)^{4k-4}),$$

$$\mathsf{L}(U_5^2(-g)^4) + \mathsf{L}(g^{4k-4}(-g)^{4k-4}) = [2k+1, 4k] \cup \{4k+2\} \text{ and}$$

$$4k+1 \notin \mathsf{L}(U_5^2(-g)^4g^{4k-4}(-g)^{4k-4})$$

by Lemma 4.2.3, we obtain that

$$\mathsf{L}(U_5^2(-g)^4g^{4k-4}(-g)^{4k-4}) = [2k+1, 4k] \cup \{4k+2\}\,. \tag{9}$$

Suppose that $k \ge 2$. Since

$$[2k, 4k] \supset \mathsf{L}(U_5(-U_5)g^{4k-4}(-g)^{4k-4}) \supset \mathsf{L}(U_5(-U_5)g^4(-g)^4) + \mathsf{L}(g^{4k-8}(-g)^{4k-8}),$$

$$\mathsf{L}(U_5(-U_5)g^4(-g)^4) + \mathsf{L}(g^{4k-8}(-g)^{4k-8}) = [2k, 4k-2] \cup \{4k\}, \text{ and}$$

$$4k-1 \notin \mathsf{L}(U_5(-U_5)g^{4k-4}(-g)^{4k-4})$$

by Lemma 4.2.3, we obtain that

$$\mathsf{L}(U_5(-U_5)g^{4k-4}(-g)^{4k-4}) = [2k, 4k-2] \cup \{4k\}\,. \tag{10}$$

Step 1. We prove that for every $L \in \mathscr{L}_2 \cup \mathscr{L}_3 \cup \mathscr{L}_4 \cup \mathscr{L}_5$ there exists an $A \in \mathscr{B}(G)$ such that $L = \mathsf{L}(A)$.

We distinguish four cases.

First we suppose that $L \in \mathscr{L}_2$, and we distinguish several subcases. If $L = y + [3, 6]$ with $y \in \mathbb{N}_0$, then we set $A = 0^y U_1(-U_1)(e+2g)^2$ and hence $\mathsf{L}(A) = y + [3, 6] = L$ by Equation (3). If $L = [2k+1, 5k+2]$ with $k \in \mathbb{N}$, then we set $A = U_1 U_3 U_4 U_2^{k-1}(-U_2)^{k-1}$ and hence $\mathsf{L}(A) = L$ by Equation (8). Now we assume that $L = y + 2\lceil \frac{k}{3} \rceil + [0, k]$ with $y \in \mathbb{N}_0$ and $k \in \mathbb{N} \setminus \{3\}$.

If $k \equiv 0 \pmod 3$, then $k \ge 6$ and by Equation (5) we infer that

$$\mathsf{L}(0^y U_1(-U_1)U_2^t(-U_2)^t) = y + [2t+2, 5t+5] = y + 2\lceil \frac{k}{3} \rceil + [0, k] = L\,, \text{ where } k = 3t+3\,.$$

If $k \equiv 1 \pmod 3$, then by Equation (7) we infer that

$$\mathsf{L}(0^y U_1(-U_4)U_2^t(-U_2)^t) = y + [2t+2, 5t+3] = y + 2\lceil \frac{k}{3} \rceil + [0, k] = L\,, \text{ where } k = 3t+1\,.$$

If $k \equiv 2 \pmod 3$, then by Equation (6) we infer that

$$\mathsf{L}(0^y U_1 U_3 U_2^t(-U_2)^t) = y + [2t+2, 5t+4] = y + 2\lceil \frac{k}{3} \rceil + [0, k] = L\,, \text{ where } k = 3t+2\,.$$

If $L = y + 2k + 2 \cdot [0, k] \in \mathscr{L}_3$ with $y \in \mathbb{N}_0$ and $k \in \mathbb{N}$, then we set $A = 0^y g^{4k}(-g)^{4k}$ and hence $\mathsf{L}(A) = L$.

If $L = y + 2t + 2 + (\{0\} \cup [2, 2t+3]) \in \mathscr{L}_4$ with $y, t \in \mathbb{N}_0$, then we set $A = 0^y U_1(-U_1)g^{4t}(-g)^{4t}$ and obtain that $\mathsf{L}(A) = y + 2t + 2 + (\{0\} \cup [2, 2t+3]) = L$ by Equation (4).

Finally we suppose that $L = y + k + ([0, k-2] \cup \{k\}) \in \mathscr{L}_5$ with $k \ge 3$ and $y \in \mathbb{N}_0$, and we distinguish two subcases. If $k = 2t$ with $t \ge 2$, then we set $A = 0^y U_5(-U_5)g^{4t-4}(-g)^{4t-4}$ and hence $\mathsf{L}(A) = y + k + ([0, k-2] \cup \{k\}) = L$ by Equation (10). If $k = 2t+1$ with $t \ge 1$, then we set $A = 0^y U_5^2(-g)^4 g^{4t-4}(-g)^{4t-4}$ and hence $\mathsf{L}(A) = y + k + ([0, k-2] \cup \{k\}) = L$ by Equation (9).

Step 2. We prove that for every $A \in \mathscr{B}(G^\bullet)$, $\mathsf{L}(A) \in \mathscr{L}_2 \cup \mathscr{L}_3 \cup \mathscr{L}_4 \cup \mathscr{L}_5$.

Let $A \in \mathscr{B}(G^\bullet)$. We may suppose that $\Delta(\mathsf{L}(A)) \ne \emptyset$. By Lemma 4.7.1 we have to distinguish two cases.

Case 1: $\Delta(\mathsf{L}(A)) = \{1\}$.

Then $\mathsf{L}(A)$ is an interval, say $\mathsf{L}(A) = [y, y+k] = y + [0, k]$ with $y \ge 2$ and $k \ge 1$. If $k = 3$ and $y = 2$, then $\mathsf{L}(A) = [2, 5]$, a contradiction to Lemma 4.7.1. Thus $k = 3$ implies that $y \ge 3$ and hence $\mathsf{L}(A) \in \mathscr{L}_2$. If $k \le 2$, then obviously $\mathsf{L}(A) \in \mathscr{L}_2$. Suppose that $k \ge 4$. If $y = 2t$ with $t \ge 2$, then $y + k \le 5t$ and hence $y = 2t \ge 2\lceil \frac{k}{3} \rceil$ which implies that $\mathsf{L}(A) \in \mathscr{L}_2$. Suppose that $y = 2t+1$ with $t \in \mathbb{N}$.

If $y + k \le 5t + 1$, then $y = 2t + 1 \ge 1 + 2\lceil \frac{k}{3} \rceil$ which implies that $\mathsf{L}(A) \in \mathscr{L}_2$. Otherwise $y + k = 5t + 2$ and hence $\mathsf{L}(A) = [2t + 1, 5t + 2] \in \mathscr{L}_2$.
Case 2: $2 \in \Delta(\mathsf{L}(A)) \subset [1, 2]$.

We freely use the classification of minimal zero-sum sequence given in Lemma 4.6. Since $2 \in \Delta(\mathsf{L}(A))$, there are $k \in \mathbb{N}$ and $U_1, \ldots, U_k, V_1, \ldots, V_{k+2} \in \mathscr{A}(G)$ with $|U_1| \ge |U_2| \ge \ldots \ge |U_k|$ such that

$$A = U_1 \cdot \ldots \cdot U_k = V_1 \cdot \ldots \cdot V_{k+2} \quad \text{and} \quad k + 1 \notin \mathsf{L}(A)\,,$$

and we may suppose that k is minimal with this property. Then $[\min \mathsf{L}(A), k] \in \mathsf{L}(A)$ and there exists $k_0 \in [2, k]$ such that $|U_i| \ge 3$ for every $i \in [1, k_0]$ and $|U_i| = 2$ for every $i \in [k_0 + 1, k]$. We continue with two simple assertions.

A1. For each two distinct $i, j \in [1, k_0]$, we have that $3 \notin \mathsf{L}(U_i U_j)$.
A2. $|\mathsf{L}(U_1 \cdot \ldots \cdot U_{k_0})| \ge 2$.

Proof of ***A1***. Assume to the contrary that there exist distinct $i, j \in [1, k_0]$ such that $3 \in \mathsf{L}(U_i U_j)$. This implies that $k + 1 \in \mathsf{L}(A)$, a contradiction. □

[Proof of **A1**]

Proof of ***A2***. Assume to the contrary that $|\mathsf{L}(U_1 \cdot \ldots \cdot U_{k_0})| = 1$. Then Lemma 4.2.2 implies that $\max \mathsf{L}(A) = \max \mathsf{L}(U_1 \cdot \ldots \cdot U_{k_0}) + k - k_0 = k$, a contradiction. □

[Proof of **A2**]

We use **A1** and **A2** without further mention and freely use Lemma 4.6 together with all its notation. We distinguish six subcases.
Case 2.1: $U_1 \in S_5$.

Without loss of generality, we may assume that $U_1 = eg^3(e + g)$. We choose $j \in [2, k_0]$ and start with some preliminary observations. If $|U_j| = 5$, then the fact that $3 \notin \mathsf{L}(U_1 U_j)$ implies that $U_j = -U_1$. If $|U_j| = 4$, then $3 \notin \mathsf{L}(U_1 U_j)$ implies that $U_j \in \{g^2(e + g)^2, g^4, (-g)^4, (e + g)^4\}$. If $|U_j| = 3$, then $3 \notin \mathsf{L}(U_1 U_j)$ implies that $U_j \in \{(e + 2g)g(e + g), g^2(2g), (e + g)^2(2g)\}$.

Now we distinguish three cases.

Suppose that $|U_2| = 5$. Then $U_2 = -U_1$ and by symmetry we obtain that $U_j \in \{g^4, (-g)^4\}$ for every $j \in [3, k_0]$. Let $i \in [k_0 + 1, k]$. If $U_i \ne e^2$, then $4 \in U_1 U_2 U_i$ and hence $k + 1 \in \mathsf{L}(A)$, a contradiction. Therefore we obtain that

$$A = U_1(-U_1)(g^4)^{k_1}((-g)^4)^{k_2}(e^2)^{k_3} \quad \text{where} \quad k_1, k_2, k_3 \in \mathbb{N}_0\,,$$

and without loss of generality we may assume that $k_1 \ge k_2$. Then it follows that $\mathsf{L}(A)$ is equal to

$$k_1 - k_2 + k_3 + \mathsf{L}(U_1(-U_1)(g^4)^{k_2}((-g)^4)^{k_2}) = k_3 + k_1 - k_2 + 2k_2 + 2 + (\{0\} \cup [2, 2k_2 + 3]),$$

which is an element of $\mathscr{L}_4$.

Suppose that $|U_2| = 4$ and there exists $j \in [2, k_0]$ such that $U_j = (-g)^4$, say $j = 2$. Let $i \in [3, k_0]$. If $U_i \in \{g^2(e + g)^2, g^2(2g)\}$, then $3 \in \mathsf{L}(U_2 U_i)$ and hence

$k+1 \in \mathsf{L}(A)$, a contradiction. If $U_i \in \{(e+g)^4, (e+g)^2(2g), (e+2g)g(e+g)\}$, then $4 \in \mathsf{L}(U_1U_2U_i)$ and hence $k+1 \in \mathsf{L}(A)$, a contradiction. Therefore $U_i \in \{g^4, (-g)^4\}$. Let $\tau \in [k_0+1, k]$. If $U_\tau \in \{(e+2g)^2, (2g)^2, (e+g)(e-g)\}$, then $4 \in \mathsf{L}(U_1U_2U_\tau)$ and hence $k+1 \in \mathsf{L}(A)$, a contradiction. Therefore $U_\tau \in \{e^2, g(-g)\}$. Therefore we obtain that

$$A = U_1(g^4)^{k_1}((-g)^4)^{k_2}(g(-g))^{k_3}(e^2)^{k_4} \quad \text{where} \quad k_1, k_3, k_4 \in \mathbb{N}_0 \text{ and } k_2 \in \mathbb{N}$$

and hence $\mathsf{L}(A)$ is equal to

$$\mathsf{L}((g^4)^{k_1+1}((-g)^4)^{k_2}(g(-g))^{k_3}(e^2)^{k_4}) = k_4 + \mathsf{L}(g^{4k_1+4+k_3}(-g)^{4k_2+k_3})$$

which is in $\mathscr{L}_3$

Suppose that $|U_2| \le 4$ and for every $j \in [2, k_0]$, we have $U_j \ne (-g)^4$. Then $U_j \in \{g^2(e+g)^2, g^4, (e+g)^4, (e+2g)g(e+g), g^2(2g), (e+g)^2(2g)\}$. Since $\mathrm{supp}(U_1 \cdot \ldots \cdot U_{k_0}) \subset \{e, g, 2g, e+g, e+2g\}$ and $\mathsf{v}_e(U_1 \cdot \ldots \cdot U_{k_0}) = 1$, Lemma 4.7.3 implies that $|\mathsf{L}(U_1 \cdot \ldots \cdot U_{k_0})| = 1$, a contradiction.

Case 2.2: $U_1 \in S_4^4$.

Without loss of generality, we may assume that $U_1 = eg(2g)(e+g)$. Let $j \in [2, k_0]$.

Suppose that $|U_j| = 4$. Since $3 \notin \mathsf{L}(U_1U_j)$, we obtain that $U_j \in \{g^2(e+g)^2, g^4, (e+g)^4\}$. Thus $U_1U_j = W_1W_2$ with $|W_1| = 5$, where W_1, W_2 are atoms and hence we are back to Case 2.1.

Suppose that $|U_j| = 3$. Since $3 \notin \mathsf{L}(U_1U_j)$, we obtain that $U_j \in \{(e+2g)g(e+g), g^2(2g), (e+g)^2(2g)\}$. If $U_j \in \{g^2(2g), (e+g)^2(2g)\}$, then $U_1U_j = W_1W_2$ with $|W_1| = 5$, where W_1, W_2 are atoms and hence we are back to Case 2.1. Thus it remains to consider the case where $U_j = (e+2g)g(e+g)$.

Therefore we have

$$U_1 \cdot \ldots \cdot U_{k_0} = U_1\big((e+2g)g(e+g)\big)^{k_1} \quad \text{where} \quad k_1 \in \mathbb{N}.$$

Since $\mathrm{supp}(U_1 \cdot \ldots \cdot U_{k_0}) \subset \{e, g, 2g, e+g, e+2g\}$ and $\mathsf{v}_e(U_1 \cdot \ldots \cdot U_{k_0}) = 1$, Lemma 4.7.3 implies that $|\mathsf{L}(U_1 \cdot \ldots \cdot U_{k_0})| = 1$, a contradiction.

Case 2.3: $U_1 \in S_4^3$ and for every $i \in [2, k_0]$, we have $U_i \notin S_4^4$.

Without loss of generality, we may assume that $U_1 = eg^2(e+2g)$. Let $j \in [2, k_0]$.

Suppose that $|U_j| = 4$. Since $3 \notin \mathsf{L}(U_1U_j)$, we obtain that $U_j \in \{-U_1, g^2(e+g)^2, g^2(e-g)^2, (e+g)^4, (e-g)^4, g^4\}$. If $U_j \in \{g^2(e+g)^2, g^2(e-g)^2, (e+g)^4, (e-g)^4\}$, then $U_1U_j = W_1W_2$ with $|W_1| = 5$, where W_1, W_2 are atoms and hence we are back to Case 2.1. Thus it remains to consider the cases where $U_j = -U_1$ or $U_j = g^4$.

Suppose that $|U_j| = 3$. Since $3 \notin \mathsf{L}(U_1U_j)$, we obtain that $U_j \in \{eg(e-g), (e+2g)g(e+g), g^2(2g), (e+g)^2(2g), (e-g)^2(2g)\}$. If $U_j \in \{eg(e-g), (e-2g)g(e+g)\}$, then $U_1U_j = W_1W_2$ with $|W_1| = 5$, where W_1, W_2 are atoms and hence we are back to Case 2.1. If $U_j \in \{(e+g)^2(2g), (e-g)^2(2g)\}$, then $U_1U_j = W_1W_2$ with $W_1 \in S_4^4$, where W_1, W_2 are atoms and hence we are back to Case 2.2. Thus it remains to consider the case where $U_j = g^2(2g)$.

If $U_i \neq -U_1$ for every $i \in [2, k_0]$, then $U_1 \cdot \ldots \cdot U_{k_0} = U_1(g^4)^{k_1}(g^2(2g))^{k_2}$ where $k_1, k_2 \in \mathbb{N}_0$. Since $\operatorname{supp}(U_1 \cdot \ldots \cdot U_{k_0}) \subset \{e, g, 2g, e+g, e+2g\}$ and $\mathsf{v}_e(U_1 \cdot \ldots \cdot U_{k_0}) = 1$, Lemma 4.7.3 implies that $|\mathsf{L}(U_1 \cdot \ldots \cdot U_{k_0})| = 1$, a contradiction. Thus there exists some $i \in [2, k_0]$, say $i = 2$, such that $U_2 = -U_1$. By symmetry we obtain that $k_0 = 2$. Let $\tau \in [3, k]$. If $U_\tau \in \{(2g)^2, (e+g)(e-g)\}$, then $4 \in \mathsf{L}(U_1 U_2 U_\tau)$ and hence $k+1 \in \mathsf{L}(A)$, a contradiction. Therefore $A = U_1(-U_1)(e^2)^{k_1}((e+2g)^2)^{k_2}(g(-g))^{k_3}$ where $k_1, k_2, k_3 \in \mathbb{N}_0$. Since $[\min \mathsf{L}(A), 2 + k_1 + k_2 + k_3] \subset \mathsf{L}(A)$, we obtain that $\mathsf{L}(A) = [\min \mathsf{L}(A), 2+y] \cup \{4+y\}$ where $y = k_1 + k_2 + k_3 \in \mathbb{N}_0$. For every atom V dividing A, we have that $|V| = 2$ or $|V| = 4$. Thus $\min \mathsf{L}(A) \geq 2 + \frac{y}{2}$ which implies that $\mathsf{L}(A) \in \mathscr{L}_5$.

Case 2.4: $U_1 \in S_4^2$ and for every $i \in [2, k_0]$, we have $U_i \notin S_4^4 \cup S_4^3$.

Without loss of generality, we may assume that $U_1 = g^2(e+g)^2$. Let $j \in [2, k_0]$.

Suppose that $|U_j| = 4$. If $U_j \in \{g^2(e-g)^2, (-g)^2(e+g)^2, (-g)^4, (e-g)^4\}$, then $3 \in \mathsf{L}(U_1 U_j)$, a contradiction. Thus $U_j \in \{U_1, -U_1, g^4, (e+g)^4\}$.

Suppose that $|U_j| = 3$. If $U_j \in \{(e+2g)(-g)(e-g), (-g)^2(2g), (e-g)^2(2g)\}$, then $3 \in \mathsf{L}(U_1 U_j)$, a contradiction. If $U_j \in \{eg(e-g), e(-g)(e+g)\}$, then $U_1 U_j = W_1 W_2$ with $|W_1| = 5$, where W_1, W_2 are atoms and hence we are back to Case 2.1. If $U_j = e(2g)(e+2g)$, then $U_1 U_j = \big(e(e+g)g(2g)\big)\big(g(e+g)(e+2g)\big)$ and $e(e+g)g(e+2g) \in S_4^4$, going back to Case 2.2. Thus it remains to consider the case where $U_j = g^2(2g)$ or $U_j = (e+g)^2(2g)$.

If $U_i \neq -U_1$ for every $i \in [2, k_0]$, then $\operatorname{supp}(U_1 \cdot \ldots \cdot U_{k_0}) \subset \{g, 2g, e+g, e+2g\}$ is half-factorial by Lemma 4.7.3, a contradiction. Thus there exists some $i \in [2, k_0]$, say $i = 2$, such that $U_2 = -U_1$. By symmetry we obtain that $\{U_1, \ldots, U_{k_0}\} = \{U_1, -U_1\}$. Let $\tau \in [k_0+1, k]$. If $U_\tau \in \{e^2, (2g)^2, (e+2g)^2\}$, then $4 \in \mathsf{L}(U_1 U_2 U_\tau)$ and $k+1 \in \mathsf{L}(U_1 U_2 U_\tau)$, a contradiction. Therefore $A = U_1^{k_1}(-U_1)^{k_2}(g(-g))^{k_3}((e+g)(e-g))^{k_4}$ where $k_1, k_2 \in \mathbb{N}$ and $k_3, k_4 \in \mathbb{N}_0$. If $k_1 + k_2 \geq 3$, by symmetry we assume that $k_1 \geq 2$, then $U_1^2(-U_1) = g^4(-g)^2(e+g)^2(e+g)(e-g)(e+g)(e-g)$ and hence $4 \in \mathsf{L}(U_1^2(-U_1))$ which implies that $k+1 \in \mathsf{L}(A)$, a contradiction. Thus $k_1 = k_2 = 1$ and hence $A = U_1(-U_1)(g(-g))^{k_3}((e+g)(e-g))^{k_4}$ where $k_3, k_4 \in \mathbb{N}_0$. Since $[\min \mathsf{L}(A), 2+k_3+k_4] \in \mathsf{L}(A)$, we obtain that $\mathsf{L}(A) = [\min \mathsf{L}(A), 2+y] \cup \{4+y\}$ where $y = k_3 + k_4 \in \mathbb{N}_0$. For every atom V dividing A, we have that $|V| = 2$ or $|V| = 4$. Thus $\min \mathsf{L}(A) \geq 2 + \frac{y}{2}$ which implies that $\mathsf{L}(A) \in \mathscr{L}_5$.

Case 2.5: $U_1 \in S_4^1$ and for every $i \in [2, k_0]$, we have $U_i \notin S_4^4 \cup S_4^3 \cup S_4^2$.

Without loss of generality, we may assume that $U_1 = g^4$. Let $j \in [2, k_0]$.

Suppose that $|U_j| = 4$. If $U_j \in \{(e+g)^4, (e-g)^4\}$, then $U_1 U_j = W_1 W_2$ with $W_1 \in S_4^2$, where W_1, W_2 are atoms and hence we are back to Case 2.4. Thus it remains to consider the case where $U_j = U_1$ or $U_j = -U_1$.

Suppose that $|U_j| = 3$. If $U_j \in \{(-g)^2(2g)\}$, then $3 \in \mathsf{L}(U_1 U_j)$, a contradiction. If $U_j \in \{e(-g)(e+g), (e+2g)(-g)(e-g)\}$, then $U_1 U_j = W_1 W_2$ with $|W_1| = 5$, where W_1, W_2 are atoms and hence we are back to Case 2.1. If $U_j \in \{(e+g)^2(2g), (e-g)^2(2g)\}$, then $U_1 U_j = W_1 W_2$ with $W_1 \in S_4^2$, where W_1, W_2 are atoms and hence we are back to Case 2.4. If $U_j = e(2g)(e+2g)$, then $U_1 U_j = W_1 W_2$ with $W_1 \in S_4^3$, where W_1, W_2 are atoms and hence we are back to Case 2.3. Thus it remains to consider the case where $U_j = g^2(2g)$, or $U_j = eg(e-g)$, or $U_j = (e+2g)g(e+g)$.

First, suppose that $U_i \neq -U_1$ for every $i \in [2, k_0]$. Then

$$U_1 \cdot \ldots \cdot U_{k_0} = U_1^{k_1} (eg(e-g))^{k_2} ((e+2g)g(e+g))^{k_3} (g^2(2g))^{k_4} ,$$

where $k_1 \in \mathbb{N}$ and $k_2, k_3, k_4 \in \mathbb{N}_0$. If $k_2 \geq 1$ and $k_3 \geq 1$, then $eg(e-g)(e+2g)g(e+g) = eg^2(e+2g)(e+g)(e-g)$, $eg^2(e+2g) \in S_4^3$ and hence we are back to Case 2.3. Thus we may assume that $k_2 = 0$ or $k_3 = 0$. Since $\{g, 2g, e+g, e+2g\}$ and $\{g, 2g, e, e-g\}$ are both half-factorial by Lemma 4.7.3, we obtain that $|\mathsf{L}(U_1 \cdot \ldots \cdot U_{k_0})| = 1$, a contradiction.

Second, suppose that there exists some $i \in [2, k_0]$, say $i = 2$, such that $U_2 = -U_1$. By symmetry we obtain that $\{U_1, \ldots, U_{k_0}\} = \{U_1, -U_1\}$. Since $4 \in \mathsf{L}(U_1 \cdot U_2 \cdot (2g)^2)$, $5 \in \mathsf{L}(U_1 U_2 e^2 (e-g)(e+g))$, and $5 \in \mathsf{L}(U_1 U_2 (e+2g)^2 (e-g)(e+g))$, we obtain that

$$\{U_{k_0+1}, \ldots, U_k\} \subset \{(e+g)(e-g), g(-g)\} \quad \text{or}$$

$$\{U_{k_0+1}, \ldots, U_k\} \subset \{e^2, (e+2g)^2, g(-g)\} .$$

This implies that

$$A = (g^4)^{k_1} ((-g)^4)^{k_2} ((e+g)(e-g))^{k_3} (g(-g))^{k_4} \quad \text{or}$$

$$A = (g^4)^{k_1} ((-g)^4)^{k_2} (e^2)^{k_3} ((e+2g)^2)^{k_4} (g(-g))^{k_5} ,$$

where $k_1, k_2 \in \mathbb{N}$ and $k_3, k_4, k_5 \in \mathbb{N}_0$.

Suppose that $A = (g^4)^{k_1} ((-g)^4)^{k_2} ((e+g)(e-g))^{k_3} (g(-g))^{k_4}$, where $k_1, k_2 \in \mathbb{N}$ and $k_3, k_4, k_5 \in \mathbb{N}_0$. If $k_1 \geq 2$ and $k_3 \geq 2$, then $g^4 g^4 (-g)^4 (e+g)(e-g)(e+g)(e-g) = \big(g(-g)\big)^4 g^2 (e+g)^2 g^2 (e-g)^2$ and hence $6 \in \mathsf{L}(g^4 g^4 (-g)^4 (e+g)(e-g)(e+g)(e-g))$. Thus $k+1 \in \mathsf{L}(A)$, a contradiction. Therefore by symmetry $k_3 = 1$ or $k_1 = k_2 = 1$. If $k_3 = 1$, then $\mathsf{L}(A) = 1 + \mathsf{L}((g^4)^{k_1} ((-g)^4)^{k_2} (g(-g))^{k_4}) \in \mathscr{L}_3$. If $k_1 = k_2 = 1$, then $\mathsf{L}(A) = [\min \mathsf{L}(A), 2+y] \cup \{4+y\}$ where $y = k_3 + k_4 \in \mathbb{N}_0$. For every atom V dividing A, we have that $|V| = 2$ or $|V| = 4$. Thus $\min \mathsf{L}(A) \geq 2 + \frac{y}{2}$ which implies that $\mathsf{L}(A) \in \mathscr{L}_5$.

Suppose that $A = (g^4)^{k_1} ((-g)^4)^{k_2} (e^2)^{k_3} ((e+2g)^2)^{k_4} (g(-g))^{k_5}$, where $k_1, k_2 \in \mathbb{N}$ and $k_3, k_4, k_5 \in \mathbb{N}_0$. If $k_1 \geq 2$, $k_3 \geq 1$, and $k_4 \geq 1$, then $g^4 g^4 (-g)^4 e^2 (e+2g)^2 = \big(g(-g)\big)^4 \big(e(e+2g)g^2\big)^2$ and hence $6 \in \mathsf{L}(g^4 g^4 (-g)^4 e^2 (e+2g)^2)$. Thus $k+1 \in \mathsf{L}(A)$, a contradiction. Therefore by symmetry $k_3 = 0$, or $k_4 = 0$, or $k_1 = k_2 = 1$. If $k_3 = 0$ or $k_4 = 0$, then $\mathsf{L}(A) = k_3 + k_4 + \mathsf{L}((g^4)^{k_1} ((-g)^4)^{k_2} (g(-g))^{k_5}) \in \mathscr{L}_3$. If $k_1 = k_2 = 1$, then $\mathsf{L}(A) = [\min \mathsf{L}(A), 2+y] \cup \{4+y\}$ where $y = k_3 + k_4 + k_5 \in \mathbb{N}_0$. For every atom V dividing A, we have that $|V| = 2$ or 4. Thus $\min \mathsf{L}(A) \geq 2 + \frac{y}{2}$ which implies that $\mathsf{L}(A) \in \mathscr{L}_5$.

Case 2.6: $|U_1| = 3$.

Let $j \in [2, k_0]$. We distinguish three subcases.

First, we suppose that $U_1 \in S_3^3$, and without restriction we may assume that $U_1 = eg(e-g)$. If $U_j = -U_1$, then $3 \in \mathsf{L}(U_1U_j)$, a contradiction. If $U_j \in \{(-g)^2(2g), (e+g)^2(2g), e(2g)(e+2g)\}$, then $U_1U_j = W_1W_2$ with $W_1 \in S_4^4$ where W_1, W_2 are atoms and hence we are back to Case 2.2. If $U_j \in \{(e+2g)g(e+g), (e+2g)(-g)(e-g)\}$, then $U_1U_j = W_1W_2$ with $W_1 \in S_4^3$ where W_1, W_2 are atoms and hence we are back to Case 2.3. If $U_j = U_1$, then $U_1U_j = W_1W_2$ with $W_1 \in S_4^2$ where W_1, W_2 are atoms and hence we are back to Case 2.4. Thus it remains to consider the case where $U_j = g^2(2g)$ or $(e-g)^2(2g)$. Then $U_1 \cdot \ldots \cdot U_{k_0} = U_1(g^2(2g))^{k_1}((e-g)^2(2g))^{k_2}$ where $k_1, k_2 \in \mathbb{N}_0$. Since $\{e, g, 2g, e-g\}$ is half-factorial by Lemma 4.7.3, we obtain that $|\mathsf{L}(U_1 \cdot \ldots \cdot U_{k_0})| = 1$, a contradiction.

Second, we suppose that $U_1 \in S_3^2$, and without restriction we may assume that $U_1 = g^2(2g)$ and $U_j \notin S_3^3$. If $U_j = -U_1$, then $3 \in \mathsf{L}(U_1U_j)$. If $U_j = U_1$, then $U_1U_j = W_1W_2$ with $W_1 \in S_4^1$ where W_1, W_2 are atoms and hence we are back to Case 2.5. If $U_j \in \{(e+g)^2(2g), (e-g)^2(2g)\}$, then $U_1U_j = W_1W_2$ with $W_1 \in S_4^2$ where W_1, W_2 are atoms and hence we are back to Case 2.4. If $U_j = e(2g)(e+2g)$, then $U_1U_j = W_1W_2$ with $W_1 \in S_4^3$ where W_1, W_2 are atoms and hence we are back to Case 2.3.

Third, we suppose that $U_1 \in S_3^1$, and without restriction we assume that $U_j \in S_3^1$. Thus $3 \in \mathsf{L}(U_1U_j)$, a contradiction. □

4.3 The System of Sets of Lengths of C_2^4

Now we give a complete description of the system of sets of lengths of C_2^4.

Theorem 4.8 $\mathcal{L}(C_2^4) = \mathcal{L}_1 \cup \mathcal{L}_2 \cup \mathcal{L}_3 \cup \mathcal{L}_4 \cup \mathcal{L}_5 \cup \mathcal{L}_6 \cup \mathcal{L}_7 \cup \mathcal{L}_8$, *where*

$$\mathcal{L}_1 = \big\{\{y\} \mid y \in \mathbb{N}_0\big\},$$

$$\mathcal{L}_2 = \big\{y + 2k + 3 \cdot [0, k] \mid y, k \in \mathbb{N}_0\big\},$$

$$\begin{aligned}\mathcal{L}_3 &= \big\{y + [2, 3], y + [2, 4], y + [3, 6], y + [3, 7], y + [4, 9] \mid y \in \mathbb{N}_0\big\} \cup \\ &\quad \big\{y + [m, m + k] \mid y \in \mathbb{N}_0, k \ge 6, m \text{ minimal with } m + k \le 5m/2\big\} \\ &= \Big\{y + \Big\lceil \frac{2k}{3} \Big\rceil + [0, k] \mid y \in \mathbb{N}_0, k \in \mathbb{N} \setminus \{1, 3\}\Big\} \cup \\ &\quad \{y + 3 + [0, 3], y + 2 + [0, 1] \mid y \in \mathbb{N}_0\},\end{aligned}$$

$$\mathcal{L}_4 = \big\{y + 2k + 2 \cdot [0, k] \;\big|\; y,\, k \in \mathbb{N}_0\big\},$$

$$\mathcal{L}_5 = \{y + k + 2 + ([0, k] \cup \{k + 2\}) \mid y \in \mathbb{N}_0, k \in \mathbb{N}\},$$

$$\mathcal{L}_6 = \Big\{y + 2\Big\lceil \frac{k}{3} \Big\rceil + 2 + (\{0\} \cup [2, k + 2]) \mid y \in \mathbb{N}_0, k \ge 5 \text{ or } k = 3\Big\},$$

$$\mathcal{L}_7 = \{y + 2k + 3 + \{0, 1, 3\} + 3 \cdot [0, k] \mid y, k \in \mathbb{N}_0\} \cup$$

$$\{y + 2k + 4 + \{0, 1, 3\} + 3 \cdot [0, k] \cup \{y + 5k + 8\} \mid y, k \in \mathbb{N}_0\}, \text{ and}$$

$$\mathscr{L}_8 = \{y + 2k + 3 + \{0, 2, 3\} + 3 \cdot [0, k] \mid y, k \in \mathbb{N}_0\} \cup$$
$$\{y + 2k + 4 + \{0, 2, 3\} + 3 \cdot [0, k] \cup \{y + 5k + 9\} \mid y, k \in \mathbb{N}_0\}.$$

We note that the system of sets of lengths of C_2^4 is richer than that of the other groups we considered. A reason for this is that the set $\Delta^*(C_2^4)$ is largest, namely $\{1, 2, 3\}$ (this fact was also crucial in the proof of Theorem 3.5). We recall some useful facts in the lemma below.

Lemma 4.9 *Let $G = C_2^4$, and let $A \in \mathscr{B}(G)$.*

1. *$\Delta(G) = [1, 3]$, and if $3 \in \Delta(\mathsf{L}(A))$, then $\Delta(\mathsf{L}(A)) = \{3\}$ and there is a basis $(e_1, \ldots, e_4)$ of G such that $\text{supp}(A) \setminus \{0\} = \{e_1, \ldots, e_4, e_1 + \ldots + e_4\}$.*
2. *$\rho_{2k+1}(G) = 5k + 2$ for all $k \in \mathbb{N}$.*

Proof

1. The first statement follows from [18, Theorem 6.8.3], and the second statement from [20, Lemma 3.10].
2. See [18, Theorem 6.3.4].

□

In the following result we characterize which intervals are sets of lengths for C_2^4. It turns out that, with a single exception, the sole restriction is the one implied by elasticity.

Proposition 4.10 *Let $G = C_2^4$ and let $2 \le l_1 \le l_2$ be integers. Then $[l_1, l_2] \in \mathscr{L}(G)$ if and only if $l_2/l_1 \le 5/2$ and $(l_1, l_2) \ne (2, 5)$.*

Proof Suppose that $[l_1, l_2] \in \mathscr{L}(G)$ with integers $2 \le l_1 \le l_2$. Then (2) implies that $l_2/l_1 \le \rho(G) = 5/2$. Moreover, $[2, 5] = [2, \mathsf{D}(G)] \notin \mathscr{L}(G)$ by [18, Theorem 6.6.3].

Conversely, we need to show that for integers $2 \le l_1 \le l_2$ with $(l_1, l_2) \ne (2, 5)$ and $l_2/l_1 \le 5/2$, we have $[l_1, l_2] \in \mathscr{L}(G)$. We start with an observation that reduces the problem to constructing these sets of intervals for extremal choices of the endpoints.

Let $k \in \mathbb{N}$. If $m \in \mathbb{N}$ such that $[m, m + k] \in \mathscr{L}(G)$, then $y + [m, m + k] \in \mathscr{L}(G)$ for all $y \in \mathbb{N}_0$. Thus let $m_k = \max\{2, \lceil \frac{2k}{3} \rceil\}$ if $k \in \mathbb{N} \setminus \{3\}$ and $m_3 = 3$. Therefore we only need to prove that $[m_k, m_k + k] \in \mathscr{L}(G)$.

For $k \in [1, 5]$ we are going to realize sets $[m_k, m_k + k]$ as sets of lengths. Then we handle the case $k \ge 6$.

If $k \in \{1, 3\}$, then the sets $[2, 3], [3, 6] \in \mathscr{L}(C_2^3) \subset \mathscr{L}(G)$. To handle the case $k = 2$, we have to show that $[2, 4] \in \mathscr{L}(G)$. Let $(e_1, \ldots, e_4)$ be a basis of G and $e_0 = e_1 + \ldots + e_4$. If

$$U_1 = e_0 \cdot \ldots \cdot e_4 \quad \text{and} \quad U_2 = e_1 e_2 (e_1 + e_3)(e_2 + e_4)(e_3 + e_4),$$

then $\max \mathsf{L}(U_1U_2) < 5$, and

$$U_1U_2 = \Big(e_0e_1e_2(e_3+e_4)\Big)\Big((e_1+e_3)e_1e_3\Big)\Big((e_2+e_4)e_2e_4\Big)$$
$$= \Big(e_0(e_1+e_3)(e_2+e_4)\Big)\Big(e_1^2\Big)\Big(e_2^2\Big)\Big((e_3+e_4)e_3e_4\Big),$$

shows that $\mathsf{L}(U_1U_2) = [2,4]$. It remains to verify the following assertions:

A1. $[3,7] \in \mathscr{L}(G)$ (this settles the case $k = 4$).
A2. $[4,9] \in \mathscr{L}(G)$ (this settles the case $k = 5$).
A3. Let $k \geq 6$. Then $[\lceil \frac{2k}{3} \rceil, \lceil \frac{2k}{3} \rceil + k] \in \mathscr{L}(G)$.

*Proof of **A1*** Clearly,

$$U_1 = e_0 \cdot \ldots \cdot e_4, \; U_2 = e_1e_2(e_1+e_3)(e_2+e_4)(e_3+e_4), \quad \text{and}$$
$$U_3 = (e_1+e_3)(e_2+e_4)e_3e_4(e_1+e_2)$$

are minimal zero-sum sequences of lengths 5. Since

$$U_1U_2U_3 = \Big(e_0(e_1+e_2)(e_3+e_4)\Big)\Big(e_1^2\Big)\Big(e_2^2\Big)\Big(e_3^2\Big)\Big(e_4^2\Big)\Big((e_1+e_3)^2\Big)\Big((e_2+e_4)^2\Big)$$
$$= \Big(e_0(e_1+e_2)(e_3+e_4)\Big)\Big((e_1+e_3)e_1e_3\Big)^2\Big((e_2+e_4)^2\Big)\Big(e_2^2\Big)\Big(e_4^2\Big)$$
$$= \Big(e_0(e_1+e_2)(e_3+e_4)\Big)\Big((e_1+e_3)e_1e_3\Big)^2\Big((e_2+e_4)e_2e_4\Big)^2$$
$$= U_2\Big(e_0(e_1+e_2)(e_1+e_3)e_1e_4\Big)\Big((e_2+e_4)e_2e_4\Big)\Big(e_3^2\Big),$$

it follows that $\mathsf{L}(U_1U_2U_3) = [3,7]$.

*Proof of **A2*** We use the same notation as in **A1**, set $U_4 = (e_1+e_2)(e_1+e_3)(e_2+e_4)(e_3+e_4)$, and assert that $\mathsf{L}(U_1^2U_2U_4) = [4,9]$. Clearly, $4 \in \mathsf{L}(U_1^2U_2U_4)$ and $\max \mathsf{L}(U_1^2U_2U_4) < 10$. Since

$$U_1^2U_2U_4 = \Big(e_0e_1e_2(e_3+e_4)\Big)\Big((e_1+e_3)e_1e_3\Big)\Big((e_2+e_4)e_2e_4\Big)U_1U_4$$
$$= \Big(e_0(e_1+e_3)(e_2+e_4)\Big)\Big(e_1^2\Big)\Big(e_2^2\Big)\Big((e_3+e_4)e_3e_4\Big)U_1U_4$$
$$= \prod_{\nu=0}^{4}(e_\nu^2)U_2U_4$$
$$= \Big((e_1+e_3)^2\Big)\Big((e_2+e_4)^2\Big)\Big((e_3+e_4)e_3e_4\Big)^2\Big(e_0^2\Big)\Big(e_1^2\Big)\Big(e_2^2\Big)\Big((e_1+e_2)e_1e_2\Big)$$
$$= \Big((e_1+e_3)^2\Big)\Big((e_2+e_4)^2\Big)\Big((e_3+e_4)^2\Big)\Big(e_3^2\Big)\Big(e_4^2\Big)\Big(e_0^2\Big)\Big(e_1^2\Big)\Big(e_2^2\Big)\Big((e_1+e_2)e_1e_2\Big),$$

the assertion follows.

*Proof of **A3*** We proceed by induction on k. For $k = 6$, we have to verify that $[4, 10] \in \mathcal{L}(G)$. We use the same notation as in **A1**, and assert that $\mathsf{L}(U_1^2 U_2^2) = [4, 10]$. Clearly, $\{4, 10\} \subset \mathsf{L}(U_1^2 U_2^2) \subset [4, 10]$. Since

$$\begin{aligned}
U_1^2 U_2^2 &= \Big(e_0 e_1 e_2 (e_3 + e_4)\Big)\Big((e_1 + e_3) e_1 e_3\Big)\Big((e_2 + e_4) e_2 e_4\Big) U_1 U_2 \\
&= \Big(e_0 e_1 e_2 (e_3 + e_4)\Big)^2 \Big((e_1 + e_3) e_1 e_3\Big)^2 \Big((e_2 + e_4) e_2 e_4\Big)^2 \\
&= \prod_{\nu=0}^{4} (e_\nu^2) U_2^2 \\
&= \Big(e_0 (e_1 + e_3)(e_2 + e_4)\Big)^2 \Big(e_1^2\Big)^2 \Big(e_2^2\Big)^2 \Big((e_3 + e_4) e_3 e_4\Big)^2 \\
&= \Big((e_1 + e_3)^2\Big)\Big((e_2 + e_4)^2\Big)\Big((e_3 + e_4) e_3 e_4\Big)^2 \Big(e_0^2\Big)\Big(e_1^2\Big)^2 \Big(e_2^2\Big)^2
\end{aligned}$$

it follows that $[5, 9] \subset \mathsf{L}(U_1^2 U_2^2)$, and hence $\mathsf{L}(U_1^2 U_2^2) = [4, 10]$.

If $k = 7$, then $[5, 12] \supset \mathsf{L}(U_1^3 U_2 U_3) \supset \mathsf{L}(U_1 U_2 U_3) + \mathsf{L}(U_1^2) = [3, 7] + \{2, 5\} = [5, 12]$ which implies that $[5, 12] \in \mathcal{L}(G)$. If $k = 8$, then $[6, 14] \supset \mathsf{L}(U_1^4 U_2 U_4) \supset \mathsf{L}(U_1^2 U_2 U_4) + \mathsf{L}(U_1^2) = [4, 9] + \{2, 5\} = [6, 14]$ which implies that $[6, 14] \in \mathcal{L}(G)$. Suppose that $k \geq 9$, and that the assertion holds for all $k' \in [6, k-1]$. Then the set $[\lceil \frac{2(k-3)}{3} \rceil, \lceil \frac{2(k-3)}{3} \rceil + k - 3] \in \mathcal{L}(G)$. This implies that $[\lceil \frac{2k}{3} \rceil, \lceil \frac{2k}{3} \rceil + k] = [\lceil \frac{2(k-3)}{3} \rceil, \lceil \frac{2(k-3)}{3} \rceil + k - 3] + \{2, 5\} \in \mathcal{L}(G)$. □

We now proceed to prove Theorem 4.8.

Proof (Theorem 4.8) Let (e_1, e_2, e_3, e_4) be a basis of $G = C_2^4$. We set $e_0 = e_1 + e_2 + e_3 + e_4$, $U = e_0 e_1 e_2 e_3 e_4$, and $V = e_1 e_2 e_3 (e_1 + e_2 + e_3)$.

Step 0. Some elementary constructions.

Let $t_1 \geq 2$, $t_2 \geq 2$, $t = t_1 + t_2$, and

$$L_{t_1, t_2} = \begin{cases} \{t\} \cup [t + 2, 5\lfloor t_1/2 \rfloor + 4(t/2 - \lfloor t_1/2 \rfloor)] & \text{if } t \text{ is even}, \\ \{t\} \cup [t + 2, 5\lfloor t_1/2 \rfloor + 4((t-1)/2 - \lfloor t_1/2 \rfloor) + 1] & \text{if } t \text{ is odd}. \end{cases}$$

Since $\mathsf{L}(U^2 V^2) = \{4\} \cup [6, 9]$, we have that $\mathsf{L}(U^{t_1} V^{t_2}) \supset \mathsf{L}(U^2 V^2) + \mathsf{L}(U^{t_1 - 2} V^{t_2 - 2}) = L_{t_1, t_2}$. Note that for every atom W dividing $U^{t_1} V^{t_2}$, we have

$$W = \begin{cases} U & \text{if} |W| = 5, \\ V & \text{if} |W| = 4, \\ e_0 e_4 (e_1 + e_2 + e_3) & \text{if} |W| = 3. \end{cases}$$

Assume to the contrary that $t + 1 \in \mathsf{L}(U^{t_1} V^{t_2})$. Then there exist $t_3, t_4, t_5 \in \mathbb{N}_0$ and atoms $W_1, \ldots, W_{t_3 + t_4 + 1}$ such that $U^{t_3} V^{t_4} = W_1 \ldots W_{t_3 + t_4 + 1}$ with $t_3 + t_4 \geq 2$,

$t_5 \le \min\{t_3, t_4\}$, $|W_i| = 3$ for $i \in [1, t_5]$, and $|W_i| = 2$ for $i \in [t_5+1, t_3+t_4+1]$. It follows that $5t_3 + 4t_4 = 3t_5 + 2(t_3+t_4+1-t_5) \le 3t_3 + 2t_4 + 2$ and hence $t_3 + t_4 \le 1$, a contradiction. Therefore $t+1 \notin \mathsf{L}(U^{t_1}V^{t_2})$ and

$$\mathsf{L}(U^{t_1}V^{t_2}) = L_{t_1,t_2}. \tag{11}$$

Note that for every atom W dividing U^rV with $r \ge 2$ and $e_1+e_2+e_3 \mid W$, we have $W = V$ or $W = e_0e_4(e_1+e_2+e_3)$. It follows that for all $r \ge 2$

$$\begin{aligned} &\mathsf{L}(U^rV) \\ =&\big(1+\mathsf{L}(U^r)\big) \cup \big(1+\mathsf{L}(e_1^2e_2^2e_3^2U^{r-1})\big) \qquad (12)\\ =&\begin{cases} r+1+\{0,2,3\}+3\cdot[0,r/2-1], & \text{if } r \text{ is even},\\ r+1+\{0,2,3\}+3\cdot[0,(r-1)/2-1] \cup \{r+1+(3r-3)/2+2\}, & \text{if } r \text{ is odd}. \end{cases} \end{aligned}$$

Note that for every atom W dividing $U^rVe_4^2e_0^2$ with $r \ge 2$ and $e_1+e_2+e_3 \mid W$, we have $W = V$ or $W = e_0e_4(e_1+e_2+e_3)$. It follows that for all $r \ge 2$

$$\begin{aligned} &\mathsf{L}(U^rVe_4^2e_0^2) \\ =&\big(1+\mathsf{L}(U^re_4^2e_0^2)\big) \cup \big(1+\mathsf{L}(U^{r+1})\big) \qquad (13)\\ =&\begin{cases} r+2+\{0,1,3\}+3\cdot[0,(r+1)/2-1], & \text{if } r \text{ is odd},\\ r+2+\{0,1,3\}+3\cdot[0,r/2-1] \cup \{r+2+3r/2+1\}, & \text{if } r \text{ is even}. \end{cases} \end{aligned}$$

.

Step 1. We prove that for every $L \in \mathscr{L}_2 \cup \mathscr{L}_3 \cup \mathscr{L}_4 \cup \mathscr{L}_5 \cup \mathscr{L}_6 \cup \mathscr{L}_7 \cup \mathscr{L}_8$, there exists an $A \in \mathscr{B}(G)$ such that $L = \mathsf{L}(A)$. We distinguish seven cases.

If $L = y+2k+3\cdot[0,k] \in \mathscr{L}_2$ with $y, k \in \mathbb{N}_0$, then $L = \mathsf{L}(0^yU^{2k}) \in \mathscr{L}(G)$.

If $L \in \mathscr{L}_3$, then the claim follows from Proposition 4.10.

If $L = y+2k+2\cdot[0,k] \in \mathscr{L}_4$ with $y, k \in \mathbb{N}_0$, then Proposition 3.3.4 implies that $L \in \mathscr{L}(C_2^3) \subset \mathscr{L}(G)$.

Suppose that $L = y+k+2+([0,k] \cup \{k+2\}) \in \mathscr{L}_5$ with $k \in \mathbb{N}$ and $y \in \mathbb{N}_0$. Note that $\mathsf{L}(V^2(e_1+e_4)^2(e_2+e_4)^2(e_3+e_4)^2(e_1+e_2+e_3+e_4)^2) = [4,6] \cup \{8\}$. If k is even, then we set $A = 0^yV^2(e_1+e_4)^k(e_2+e_4)^k(e_3+e_4)^k(e_1+e_2+e_3+e_4)^k$ and obtain that $\mathsf{L}(A) = L$ by Lemma 4.2.3. If k is odd, then we set $A = 0^yV^2(e_1+e_4)^{k+1}(e_2+e_4)^{k+1}(e_3+e_4)^{k-1}(e_1+e_2+e_3+e_4)^{k-1}$ and obtain that $\mathsf{L}(A) = L$ by Lemma 4.2.3.

Suppose that $L = y+2\lceil\frac{k}{3}\rceil+2+(\{0\} \cup [2,k+2]) \in \mathscr{L}_6$ with $(k \ge 5$ or $k = 3)$ and $y \in \mathbb{N}_0$. If $k \equiv 0 \mod 3$, then we set $A = 0^yU^{2k/3}V^2$ and hence $\mathsf{L}(A) = L$ by Equation (11). If $k \equiv 2 \mod 3$, then we set $A = 0^yU^{(2k-4)/3}V^4$ and hence $\mathsf{L}(A) = L$ by (11). If $k \equiv 1 \mod 3$, then we set $A = 0^yU^{(2k-8)/3}V^6$ and obtain that $\mathsf{L}(A) = L$ by Equation (11).

Suppose that $L \in \mathscr{L}_7$. If $L = y+2k+3+\{0,1,3\}+3\cdot[0,k]$ with $y \in \mathbb{N}_0$ and $k \in \mathbb{N}_0$, then we set $A = 0^yU^{2k+1}Ve_4^2(e_1+e_2+e_3+e_4)^2$ and obtain that $\mathsf{L}(A) = L$

by Equation (13). If $L = y + 2k + 4 + \{0, 1, 3\} + 3 \cdot [0, k] \cup \{y + 5k + 8\}$ with $y \in \mathbb{N}_0$ and $k \in \mathbb{N}_0$, then we set $A = 0^y U^{2k+2} V e_4^2 (e_1 + e_2 + e_3 + e_4)^2$ and obtain that $\mathsf{L}(A) = L$ by Equation (13).

Suppose that $L \in \mathscr{L}_8$. If $L = y + 2k + 3 + \{0, 2, 3\} + 3 \cdot [0, k]$ with $y \in \mathbb{N}_0$ and $k \in \mathbb{N}_0$, then we set $A = 0^y U^{2k+2} V$ and hence $\mathsf{L}(A) = L$ by Equation (12). If $L = y+2k+4+\{0, 2, 3\}+3\cdot[0, k]\cup\{y+5k+9\}$ with $y \in \mathbb{N}_0$ and $k \in \mathbb{N}_0$, then we set $A = 0^y U^{2k+3} V e_4^2 (e_1 + e_2 + e_3 + e_4)^2$ and obtain that $\mathsf{L}(A) = L$ by Equation (12).

Step 2. We prove that for every $A \in \mathscr{B}(G^\bullet)$, $\mathsf{L}(A) \in \mathscr{L}_2 \cup \mathscr{L}_3 \cup \mathscr{L}_4 \cup \mathscr{L}_5 \cup \mathscr{L}_6 \cup \mathscr{L}_7 \cup \mathscr{L}_8$.

Let $A \in \mathscr{B}(G^\bullet)$. We may suppose that $\Delta(\mathsf{L}(A)) \neq \emptyset$. By Lemma 4.9.1 we have to distinguish four cases.

Case 1: $\Delta(\mathsf{L}(A)) = \{3\}$.

By Lemma 4.9, there is a basis of G, say (e_1, e_2, e_3, e_4), such that $\text{supp}(A) = \{e_1, \ldots, e_4, e_0\}$. Let $n \in \mathbb{N}_0$ be maximal such that $U^{2n} \mid A$. Then there exist a proper subset $I \subset [0, 4]$, a tuple $(m_i)_{i \in I} \in \mathbb{N}_0^{(I)}$, and $\epsilon \in \{0, 1\}$ such that

$$A = U^\epsilon U^{2n} \prod_{i \in I} (e_i^2)^{m_i} .$$

Using [20, Lemma 3.6.1], we infer that

$$\mathsf{L}(A) = \epsilon + \sum_{i \in I} m_i + \mathsf{L}(U^{2n}) = \epsilon + \sum_{i \in I} m_i + (2n + 3 \cdot [0, n]) \in \mathscr{L}_2 .$$

Case 2: $\Delta(\mathsf{L}(A)) = \{1\}$.

Then $\mathsf{L}(A)$ is an interval, and it is a direct consequence of Proposition 4.10 that $\mathsf{L}(A) \in \mathscr{L}_3$.

Case 3: $\Delta(\mathsf{L}(A)) = \{2\}$.

The following reformulation turns out to be convenient. Clearly, we have to show that for every $L \in \mathscr{L}(G)$ with $\Delta(L) = \{2\}$ there exist $y' \in \mathbb{N}_0$ and $k' \in \mathbb{N}$ such that $L = y' + 2 \cdot [k', 2k']$, which is equivalent to $\rho(L) = \max L / \min L \le 2$. Assume to the contrary that there is an $L \in \mathscr{L}(G)$ with $\Delta(L) = \{2\}$ such that $\max L \ge 2 \min L + 1$. We choose one such $L \in \mathscr{L}(G)$ with $\min L$ being minimal, and we choose a $B \in \mathscr{B}(G)$ with $\mathsf{L}(B) = L$. Since $\min L$ is minimal, we obtain that $0 \nmid B$. Consequently, $|B| \ge 2 \max L \ge 4 \min L + 2$. Since $\mathsf{D}(G) = 5$, it follows that a factorization of minimal length of B contains at least two (possibly equal) minimal zero-sum sequences U_1, U_2 with $|U_1| = |U_2| = 5$, say $U_1 = e_0 \cdot \ldots \cdot e_4$.

If $U_1 = U_2$, then $5 \in \mathsf{L}(U_1 U_2)$ and thus $\min L + 3 \in L$, contradicting the fact that $\Delta(L) = \{2\}$. Thus $U_1 \neq U_2$. We assert that $3 \in \mathsf{L}(U_1 U_2)$, and thus obtain again a contradiction to the fact that $\Delta(L) = \{2\}$.

Let $g \in G$ with $g \mid U_2$ but $g \nmid U_1$. Then g is the sum of two elements from U_1, say $g = e_1 + e_2$. Therefore $g(e_1 e_2)^{-1} U_1$ is a minimal zero-sum sequence, whereas the sequence $(e_1 e_2) g^{-1} U_2$ cannot be a minimal zero-sum sequence because it has length 6. Since $g^{-1} U_2$ is zero-sum free, every minimal zero-sum sequence dividing $(e_1 e_2) g^{-1} U_2$ must contain e_1 or e_2. This shows that $\mathsf{L}((e_1 e_2) g^{-1} U_2) = \{2\}$ and thus $3 \in \mathsf{L}(U_1 U_2)$.

Case 4: $\Delta(\mathsf{L}(A)) = \{1, 2\}$.

Let $k \in \mathsf{L}(A)$ be minimal such that A has a factorization of the form $A = U_1 \cdot \ldots \cdot U_k = V_1 \cdot \ldots \cdot V_{k+2}$, where $k + 1 \notin \mathsf{L}(A)$ and $U_1, \ldots, U_k, V_1, \ldots, V_{k+2} \in \mathcal{A}(G)$ with $|U_1| \geq |U_2| \geq \ldots \geq |U_k|$. Without restriction we may suppose that the tuple

$$(|\{i \in [1,k] \mid |U_i|=5\}|, |\{i \in [1,k] \mid |U_i|=4\}|, |\{i \in [1,k] \mid |U_i|=3\}|) \in \mathbb{N}_0^3 \quad (14)$$

is maximal (with respect to the lexicographic order) among all factorizations of A of length k. By definition of k, we have $[\min \mathsf{L}(A), k] \in \mathsf{L}(A)$. Let $k_0 \in [2, k]$ such that $|U_i| \geq 3$ for every $i \in [1, k_0]$ and $|U_i| = 2$ for every $i \in [k_0 + 1, k]$. We start with the following assertion.

A.

1. For each two distinct $i, j \in [1, k_0]$, we have $3 \notin \mathsf{L}(U_iU_j)$.
2. For each two distinct $i, j \in [1, k_0]$ with $|U_i| = |U_j| = 5$, we have $U_i = U_j$.
3. For each two distinct $i, j \in [1, k_0]$ with $|U_i| = 5$ and $|U_j| = 4$, we have $|\gcd(U_i, U_j)| = 3$.
4. Let $i, j \in [1, k_0]$ be distinct with $|U_i| = |U_j| = 4$, say $U_i = f_1f_2f_3(f_1 + f_2 + f_3)$ where (f_1, f_2, f_3, f_4) a basis of G. Then $U_j = U_i$, or $U_j = (f_1 + f_4)(f_2 + f_4)(f_3 + f_4)(f_1 + f_2 + f_3 + f_4)$, or $U_j = f_4(f_1 + f_2 + f_4)(f_2 + f_3 + f_4)(f_1 + f_3 + f_4)$. Furthermore, if $U_i \neq U_j$, then for all $t \in [1, k_0] \setminus \{i, j\}$, we have $|U_t| \neq 4$.
5. Let $i, j \in [1, k_0]$ be distinct with $|U_i| = 5$ and $|U_j| = 3$. Then there exist $g_1, g_2, g_3 \in G$ such that $g_1g_2g_3 \mid U_i$ and $U_j = (g_1 + g_2)(g_2 + g_3)(g_3 + g_1)$. Furthermore, for all $t \in [1, k_0] \setminus \{i, j\}$, we have $|U_t| = 3$.
6. Let $i, j \in [1, k_0]$ be distinct with $|U_i| = 4$ and $|U_j| = 3$. Then $|\gcd(U_i, U_j)| = 0$, and there exist $g, g_1, g_2 \in G$ such that $g \mid U_j$, $g_1g_2 \mid U_i$ and $g = g_1 + g_2$. Furthermore, for all $t \in [1, k_0] \setminus \{i, j\}$, we have $|U_t| = 3$.
7. For each two distinct $i, j \in [1, k_0]$ with $|U_i| = |U_j| = 3$, we have $|\gcd(U_i, U_j)| = 0$.

Proof of **A.**

1. If there exist distinct $i, j \in [1, k_0]$ such that $3 \in \mathsf{L}(U_iU_j)$, then $k + 1 \in \mathsf{L}(A)$, a contradiction.
2. Since $|U_i| = 5$ and $U_j \neq U_i$, there exist $g, g_1, g_2 \in G$ with $g \mid U_j$ and $g_1g_2 \mid U_i$ such that $g = g_1 + g_2$. Thus $U_i(g_1g_2)^{-1}g$ is an atom and $U_jg^{-1}g_1g_1$ is a product of two atoms which implies that $3 \in \mathsf{L}(U_iU_j)$, a contradiction.
3. Since $|U_i| = 5$ and $U_j \neq U_i$, there exist $g, g_1, g_2 \in G$ with $g \mid U_j$ and $g_1g_2 \mid U_i$ such that $g = g_1 + g_2$. Thus gg_1g_2 is an atom and $U_iU_j(gg_1g_2)^{-1}$ is a sequence of length 6. By 1., $2 \notin \mathsf{L}(U_iU_j(gg_1g_2)^{-1})$ which implies that $\mathsf{L}(U_iU_j(gg_1g_2)^{-1}) = \{3\}$ and hence $|\gcd(U_i, U_j)| = 3$.
4. We set $G_1 = \langle f_1, f_2, f_3 \rangle$ and distinguish three cases.

   Case (i): $U_j \in \mathcal{B}(G_1)$. Since $3 \notin \mathsf{L}(U_iU_j)$, we obtain that $U_j = U_i$.

   Case (ii): $U_j = (g_1 + f_4)(g_2 + f_4)g_3g_4$ with $g_1g_2g_3g_4 \in \mathcal{B}(G_1)$.

   If $g_3, g_4 \in \{f_1, f_2, f_3, f_1 + f_2 + f_3\}$, then $3 \in \mathsf{L}(U_iU_j)$, a contradiction. Thus, without loss of generality, we may assume that $g_3 = f_1 + f_2 \notin \{f_1, f_2, f_3, f_1 + f_2 +$

$f_3\}$. Thus $g_3f_3(f_1+f_2+f_3)$ is an atom and $(g_1+f_4)(g_2+f_4)f_1f_2g_4$ is a zero-sum sequence of length 5. Since $3 \notin \mathsf{L}(U_iU_j)$, we have that $(g_1+f_4)(g_2+f_4)f_1f_2g_4$ is an atom of length 5, a contradiction to the maximality condition in Equation (14).

Case (iii): $U_j = (g_1+f_4)(g_2+f_4)(g_3+f_4)(g_4+f_4)$ with $g_1g_2g_3g_4 \in \mathcal{B}(G_1)$.

First, suppose that $g_1g_2g_3g_4$ is an atom. If $g_1g_2g_3g_4 \neq U_i$, then there exist an element $h \in \{f_1,f_2,f_3,f_1+f_2+f_3\}$ and distinct $t_1,t_2 \in [1,4]$, say $t_1 = 1, t_2 = 2$, such that $h = g_1+g_2 = (g_1+f_4)+(g_2+f_4)$. Thus $U_ih^{-1}(g_1+f_4)(g_2+f_4)$ is a zero-sum sequence of length 5 and $h(g_3+f_4)(g_4+f_4)$ is an atom. It follows that $U_ih^{-1}(g_1+f_4)(g_2f_4)$ is atom of length 5 since $3 \notin \mathsf{L}(U_iU_j)$, a contradiction to the maximality condition in Equation (14). Therefore $g_1g_2g_3g_4 = U_i$ which implies that $U_j = (f_1+f_4)(f_2+f_4)(f_3+f_4)(f_1+f_2+f_3+f_4)$.

Second, suppose that $g_1g_2g_3g_4$ is not an atom. Without loss of generality, we may assume that $g_1 = 0$ and $g_2g_3g_4$ is an atom. If $\{g_2,g_3,g_4\} \cap \{f_1,f_2,f_3,f_1+f_2+f_3\} \neq \emptyset$, say $g_2 \in \{f_1,f_2,f_3,f_1+f_2+f_3\}$, then $g_2(g_3+f_4)(g_4+f_4)$ is an atom and $U_ig_2^{-1}f_4(g_2+f_4)$ is a zero-sum sequence of length 5. It follows that $U_ig_2^{-1}f_4(g_2+f_4)$ is atom of length 5 because $3 \notin \mathsf{L}(U_iU_j)$, a contradiction to the maximality condition in Equation (14). Therefore $\{g_2,g_3,g_4\} \cap \{f_1,f_2,f_3,f_1+f_2+f_3\} = \emptyset$ which implies that $g_2g_3g_4 = (f_1+f_2)(f_2+f_3)(f_1+f_3)$ and hence $U_j = f_4(f_1+f_2+f_4)(f_2+f_3+f_4)(f_1+f_3+f_4)$.

Now suppose that $U_i \neq U_j$, and assume to the contrary there exists a $t \in [1,k_0] \setminus \{i,j\}$ such that $|U_t| = 4$. If $U_t \notin \{U_i,U_j\}$, then $U_iU_jU_t = \big(f_1f_2f_3(f_1+f_2+f_3)\big)\big((f_1+f_4)(f_2+f_4)(f_3+f_4)(f_1+f_2+f_3+f_4)\big)\big(f_4(f_1+f_2+f_4)(f_2+f_3+f_4)(f_1+f_3+f_4)\big) = \big(f_1(f_2+f_4)(f_1+f_2+f_4)\big)\big(f_2(f_3+f_4)(f_2+f_3+f_4)\big)\big(f_3(f_1+f_4)(f_1+f_3+f_4)\big)\big(f_4(f_1+f_2+f_3)(f_1+f_2+f_3+f_4)\big)$. Thus $4 \in \mathsf{L}(U_iU_jU_t)$ and hence $k+1 \in \mathsf{L}(A)$, a contradiction. If $U_t \in \{U_i,U_j\}$, then we still have that $4 \in \mathsf{L}(U_iU_jU_t)$ and hence $k+1 \in \mathsf{L}(A)$, a contradiction.

5. Since $3 \notin \mathsf{L}(U_iU_j)$, we obtain that $|\gcd(U_i,U_j)| = 0$. Every $h \in \operatorname{supp}(U_j)$ is the sum of two distinct elements from $\operatorname{supp}(U_i)$. Thus there exist $g_1,g_2,g_3 \in G$ with $g_1g_2g_3 \mid U_i$ such that $U_j = (g_1+g_2)(g_2+g_3)(g_3+g_1)$. Now we choose an element $t \in [1,k_0] \setminus \{i,j\}$, and have to show that $|U_t| = 3$. If $|U_t| = 5$, then $U_t = U_i$ by 2. and hence $4 \in \mathsf{L}(U_iU_tU_j)$ which implies that $k+1 \in \mathsf{L}(A)$, a contradiction. Suppose that $|U_t| = 4$ and let $U_i = g_1g_2g_3g_4g_5$, where $g_4,g_5 \in G$. Then $|\gcd(U_i,U_t)| = 3$ by 3. and by symmetry we only need to consider $\operatorname{supp}(U_t) \setminus \operatorname{supp}(U_i) \subset \{g_1+g_2, g_1+g_4, g_4+g_5\}$. All the three cases imply that $4 \in \mathsf{L}(U_iU_tU_j)$. It follows that $k+1 \in \mathsf{L}(A)$, a contradiction.
6. If $|\gcd(U_i,U_j)| = 2$, then $3 \in \mathsf{L}(U_iU_j)$, a contradiction. If $|\gcd(U_i,U_j)| = 1$, then $U_1U_2 = W_1W_2$ with $W_1,W_2 \in \mathcal{A}(G)$ and $|W_2| = 5$, a contradiction to the maximality condition in Equation (14). Thus we obtain that $|\gcd(U_i,U_j)| = 0$. Let (f_1,f_2,f_3,f_4) be a basis and $U_i = f_1f_2f_3(f_1+f_2+f_3)$. Since $|U_j| = 3$, there exists a $g \in \operatorname{supp}(U_j)$ such that $g \in \langle f_1,f_2,f_3\rangle$. Since $|\gcd(U_i,U_j)| = 0$, there exist $g_1,g_2 \in G$ such that $g_1g_2 \mid U_i$ and $g = g_1+g_2$.

Now we choose an element $t \in [1,k_0] \setminus \{i,j\}$ and have to show that $|U_t| = 3$. Note that 5. implies that $|U_t| \neq 5$, and we assume to the contrary that $|U_t| = 4$. Without restriction we may assume that $g = f_1+f_2$, and by 4., we distinguish three cases. If $U_t = U_i$, then $f_1^2, f_2^2, gU_i(f_1f_2)^{-1}, U_t(f_1f_2)^{-1}U_jg^{-1}$ are atoms and hence $4 \in \mathsf{L}(U_i U_t U_j)$ which implies that $k+1 \in \mathsf{L}(A)$, a contradiction. If $U_t = (f_1+f_4)(f_2+f_4)(f_3+f_4)(f_1+f_2+f_3+f_4)$, then $g(f_1+f_2+f_3)(f_1+f_2+f_3+f_4)(f_1+f_4)f_2$ is an atom of length 5 dividing $U_i U_j U_t$ and $U_i U_j U_t (g(f_1+f_2+f_3)(f_1+f_2+f_3+f_4)(f_1+f_4)f_2)^{-1}$ is a product of two atoms, a contradiction to the maximality condition in Equation (14). If $U_t = f_4(f_1+f_2+f_4)(f_2+f_3+f_4)(f_1+f_3+f_4)$, then $g f_2 f_3 f_4 (f_1+f_3+f_4)$ is an atom of length 5 dividing $U_i U_j U_t$ and $U_i U_j U_t (g f_2 f_3 f_4 (f_1+f_3+f_4))^{-1}$ is a product of two atoms, a contradiction to the maximality condition in Equation (14).

7. If $|\gcd(U_i, U_j)| \ge 2$, then $U_i = U_j$ and hence $3 \in \mathsf{L}(U_i U_j)$ which implies that $k+1 \in \mathsf{L}(A)$, a contradiction. If $|\gcd(U_i, U_j)| = 1$, then $U_i U_j = W_1 W_2$ with $W_1, W_2 \in \mathscr{A}(G)$, $|W_1| = 2$, and $|W_2| = 4$, a contradiction to the maximality condition in Equation (14). Therefore $|\gcd(U_i, U_j)| = 0$. This completes the proof of **A**. □

Note that **A**.5 implies that $\{|U_i| \mid i \in [1, k_0]\} \ne \{3, 4, 5\}$. Thus it remains to discuss the following six subcases.

Case 4.1. $\{|U_i| \mid i \in [1, k_0]\} = \{3, 5\}$.

By **A**.5 and **A**.7, we obtain that $|U_1| = 5$, $|U_2| = \ldots = |U_{k_0}| = 3$, and that $U_1 \cdot \ldots \cdot U_{k_0}$ is square-free. This implies that $\max \mathsf{L}(U_1 \cdot \ldots \cdot U_{k_0}) = k_0$, and hence $\max \mathsf{L}(A) = \max \mathsf{L}(U_0 \cdot \ldots \cdot U_{k_0}) + k - k_0 = k$, a contradiction.

Case 4.2. $\{|U_i| \mid i \in [1, k_0]\} = \{3, 4\}$.

By **A**.6 and **A**.7, we obtain that $|U_1| = 4$, $|U_2| = \ldots = |U_{k_0}| = 3$, and that $U_1 \cdot \ldots \cdot U_{k_0}$ is square-free. This implies that $\max \mathsf{L}(U_1 \cdot \ldots \cdot U_{k_0}) = k_0$, and hence $\max \mathsf{L}(A) = \max \mathsf{L}(U_0 \cdot \ldots \cdot U_{k_0}) + k - k_0 = k$, a contradiction.

Case 4.3. $\{|U_i| \mid i \in [1, k_0]\} = \{3\}$.

By **A**.7, we obtain that $U_1 \cdot \ldots \cdot U_{k_0}$ is square-free. This implies that $\max \mathsf{L}(U_1 \cdot \ldots \cdot U_{k_0}) = k_0$, and hence $\max \mathsf{L}(A) = \max \mathsf{L}(U_0 \cdot \ldots \cdot U_{k_0}) + k - k_0 = k$, a contradiction.

Case 4.4. $\{|U_i| \mid i \in [1, k_0]\} = \{5\}$.

By **A**.2, it follows that $A = U_1^{k_0} U_{k_0+1} \cdot \ldots \cdot U_k$. If $\mathrm{supp}(U_{k_0+1} \cdot \ldots \cdot U_k) \subset \mathrm{supp}(U_1)$, then $\Delta(\mathsf{L}(A)) = \{3\}$, a contradiction. Thus there exists $j \in [k_0+1, k]$ such that $U_j = g^2$ for some $g \notin \mathrm{supp}(U_1)$. Then there exist $g_1, g_2 \in G$ such that $g_1 g_2 \mid U_1$ and $g = g_1 + g_2$. It follows that $U_1^2 U_j = g_1^2 g_2^2 (U_1 (g_1 g_2)^{-1} g)^2$, where $g_1^2, g_2^2, U_1 (g_1 g_2)^{-1} g$ are atoms. Therefore $4 \in \mathsf{L}(U_1^2 U_j)$ and hence $k+1 \in \mathsf{L}(A)$, a contradiction.

Case 4.5. $\{|U_i| \mid i \in [1, k_0]\} = \{4\}$.

Assume to the contrary, that $k_0 \ge 3$. Then **A**.4 implies that $U_1 \cdot \ldots \cdot U_{k_0} = U_1^{k_0}$, and we set $G_1 = \langle \mathrm{supp}(U_1) \rangle$. If there exists $g \in \mathrm{supp}(U_{k_0+1} \cdot \ldots \cdot U_k)$ such that $g \in G_1 \setminus \mathrm{supp}(U_1)$, then $4 \in \mathsf{L}(U_1^2 g^2)$ and hence $k+1 \in \mathsf{L}(A)$, a contradiction. If there exist distinct $g_1, g_2 \in \mathrm{supp}(U_{k_0+1} \cdot \ldots \cdot U_k)$ such that $g_1 \notin G_1$ and $g_2 \notin G_1$, then $g_1 + g_2 \in G_1$. Since $g_1 + g_2 \in \mathrm{supp}(U_1)$ implies that $5 \in \mathsf{L}(U_1^2 g_1^2 g_2^2)$ and $k+1 \in \mathsf{L}(A)$, we obtain that $g_1 + g_2 \in G_1 \setminus \mathrm{supp}(U_1)$. Then $U_1^2 g_1^2 g_2^2 = W_1^2 W_2 W_3$ where $W_1, W_2, W_3 \in \mathscr{A}(G)$ with $|W_1| = 4$, $W_1 \ne U_1$, and $|W_2| = |W_3| = 2$. Say $U_{k_0+1} = g_1^2$ and $U_{k_0+2} = g_2^2$. Then $W_1^2 U_3 \cdot \ldots \cdot U_{k_0} W_1 W_2 U_{k_0+3} \cdot \ldots \cdot U_k$ is a factorization of A of length k satisfying the maximality condition of Equation (14) and hence applying **A**.4 to this factorization, we obtain a contradiction. Therefore $\mathrm{supp}(U_{k_0+1} \cdot \ldots \cdot$

$U_k) \subset \operatorname{supp}(U_1) \cup \{g\}$ where g is independent from $\operatorname{supp}(U_1)$ and hence $\operatorname{supp}(A) \subset \operatorname{supp}(U_1) \cup \{g\}$ which implies that $\Delta(\mathsf{L}(A)) = \{2\}$, a contradiction.

Therefore it follows that $k_0 = 2$. Then $U_1 = U_2$ (since otherwise we would have $\max \mathsf{L}(A) = k$ by U_1U_2 is square-free), and we obtain that $\mathsf{L}(A) = [\min \mathsf{L}(A), k] \cup \{k+2\}$. Assume to the contrary that there exists a $W \in \mathcal{A}(G)$ such that $W \mid A$ and $|W| = 5$. Then there exist $g, g_1, g_2 \in G$ such that $g \mid U_1$, $g_1g_2 \mid W$, and $g = g_1 + g_2$, and hence $|\{g_1, g_2\} \cap \operatorname{supp}(U_1)| \le 1$. If $\{g_1, g_2\} \cap \operatorname{supp}(U_1) = \emptyset$, then there exist distinct $t_1, t_2 \in [k_0+1, k]$ such that $U_{t_1} = g_1^2$ and $U_{t_2} = g_2^2$. Thus $5 \in \mathsf{L}(U_1U_2U_{t_1}U_{t_2})$ and hence $k+1 \in \mathsf{L}(A)$, a contradiction. Suppose that $|\{g_1, g_2\} \cap \operatorname{supp}(U_1)| = 1$, say $g_1 \notin \operatorname{supp}(U_1)$ and $g_2 \in \operatorname{supp}(U_1)$. Then there exists $t \in [k_0+1, k]$ such that $U_t = g_1^2$. Therefore $4 \in \mathsf{L}(U_1U_2U_t)$ and hence $k+1 \in \mathsf{L}(A)$, a contradiction.

Thus every atom W with $W \mid A$ has length $|W| < 5$. It follows that $\min \mathsf{L}(A) \ge \lceil \frac{2 \max \mathsf{L}(A)}{4} \rceil = \lceil \frac{\max \mathsf{L}(A)}{2} \rceil$ and hence $\mathsf{L}(A) \in \mathcal{L}_5$.

Case 4.6. $\{|U_i| \mid i \in [1, k_0]\} = \{4, 5\}$.

By **A**.2, **A**.3, and **A**.4, we obtain that $|\{U_1, \ldots, U_{k_0}\}| = 2$. Without restriction we may assume that $U_1 \cdot \ldots \cdot U_{k_0} = U^{k_1}V^{k_2}$ where $k_1, k_2 \in \mathbb{N}$ with $k_0 = k_1 + k_2$ and $V = e_1e_2e_3(e_1+e_2+e_3)$ (recall that $(e_1, \ldots, e_4)$ is a basis of G, $e_0 = e_1+e_2+e_3+e_4$, and $U = e_1e_2e_3e_4e_0$). We claim that

- $\operatorname{supp}(U_{k_0+1} \cdot \ldots \cdot U_k) \subset \operatorname{supp}(UV)$.
- If $k_1 \ge 2$, then $\operatorname{supp}(U_{k_0+1} \cdot \ldots \cdot U_k) \subset \operatorname{supp}(U)$, and
- if $k_2 \ge 2$, then $\{e_4, e_0\} \not\subset \operatorname{supp}(U_{k_0+1} \cdot \ldots \cdot U_k)$.

Indeed, assume to the contrary that $g \in \operatorname{supp}(U_{k_0+1} \cdot \ldots \cdot U_k) \setminus \operatorname{supp}(UV)$. By symmetry, we only need to consider $g = e_1 + e_2$ and $g = e_1 + e_4$ and both cases imply that $4 \in \mathsf{L}(UVg^2)$, a contradiction to $k+1 \notin \mathsf{L}(A)$. If $k_1 \ge 2$ and $g = e_1 + e_2 + e_3 \in \operatorname{supp}(U_{k_0+1} \cdot \ldots \cdot U_k)$, then $4 \in \mathsf{L}(U^2g^2)$ and $k+1 \in \mathsf{L}(A)$, a contradiction. Thus if $k_1 \ge 2$, then $\operatorname{supp}(U_{k_0+1} \cdot \ldots \cdot U_k) \subset \operatorname{supp}(U)$. If $k_2 \ge 2$ and $\{e_4, e_0\} \subset \operatorname{supp}(U_{k_0+1} \cdot \ldots \cdot U_k)$, then $5 \in \mathsf{L}(V^2e_4^2e_0^2)$ and hence $k+1 \in \mathsf{L}(A)$, a contradiction.

Thus all three claims are proved, and we distinguish three subcases.

Case 4.6.1. $k_1 = 1$.

If $\{e_4, e_0\} \not\subset \operatorname{supp}(U_{k_0+1} \cdot \ldots \cdot U_k)$, then $\mathsf{L}(A) = \mathsf{L}(UV^{k_2}) + k - k_0 = \mathsf{L}(V^{k_0}) + k - k_0$ and hence $\Delta(\mathsf{L}(A)) = \{2\}$, a contradiction. If $\{e_4, e_0\} \subset \operatorname{supp}(U_{k_0+1} \cdot \ldots \cdot U_k)$, then $k_2 = 1$ and we may assume that $U_{k_0+1} = e_4^2$ and that $U_{k_0+2} = e_0^2$. Then $\mathsf{L}(A) = \mathsf{L}(UVU_{k_0+1}U_{k_0+2}) + k - k_0 - 2 = \{k-1, k, k+2\}$ with $k \ge 4$, and hence $\mathsf{L}(A) \in \mathcal{L}_5$.

Case 4.6.2. $k_1 \ge 2$ and $k_2 \ge 2$.

Thus $\operatorname{supp}(U_{k_0+1} \cdot \ldots \cdot U_k)$ is independent and it follows that $\operatorname{supp}(U_{k_0+1} \cdot \ldots \cdot U_k) \subset \{e_1, e_2, e_3, e_4\}$ or $\operatorname{supp}(U_{k_0+1} \cdot \ldots \cdot U_k) \subset \{e_1, e_2, e_3, e_0\}$. Then we have $\mathsf{L}(A) = \mathsf{L}(U^{k_1}V^{k_2}) + k - k_0$. By Equation (11), $\mathsf{L}(U^{k_1}V^{k_2})$ is equal to

$$\begin{cases} \{k_0\} \cup [k_0+2, 5\lfloor k_1/2 \rfloor + 4(k_0/2 - \lfloor k_1/2 \rfloor)] & \text{if } k_0 = k_1 + k_2 \text{ is even}, \\ \{k_0\} \cup [k_0+2, 5\lfloor k_1/2 \rfloor + 4((k_0-1)/2 - \lfloor k_1/2 \rfloor) + 1] & \text{if } k_0 = k_1 + k_2 \text{ is odd}. \end{cases}$$

Let $\ell = \max \mathsf{L}(U^{k_1} V^{k_2}) - k_0 - 2$ and hence

$$\ell = \begin{cases} k_0 + \lfloor \frac{k_1}{2} \rfloor - 2 & \text{if } k_0 \ge 4 \text{ is even}, \\ k_0 + \lfloor \frac{k_1}{2} \rfloor - 3 & \text{if } k_0 \ge 5 \text{ is odd}. \end{cases}$$

Since $k_1 \ge 2$ and $k_2 \ge 2$, we obtain that $\ell \ge 3$ and $\ell \ne 4$. We also have that

$$\ell \le \begin{cases} k_0 + \lfloor \frac{k_0 - 2}{2} \rfloor - 2 = \frac{3k_0}{2} - 3 & \text{if } k_0 \text{ is even}, \\ k_0 + \lfloor \frac{k_0 - 2}{2} \rfloor - 3 = \frac{3k_0 - 9}{2} & \text{if } k_0 \text{ is odd}. \end{cases}$$

Therefore

$$k_0 \ge \begin{cases} \frac{2\ell}{3} + 2 & \text{if } k_0 \text{ is even}, \\ \frac{2\ell}{3} + 3 & \text{if } k_0 \text{ is odd}, \end{cases}$$

and hence

$$k_0 \ge \begin{cases} 2\lceil \frac{\ell}{3} \rceil + 2 & \text{if } k_0 \text{ is even}, \\ 2\lceil \frac{\ell}{3} \rceil + 2 & \text{if } k_0 \text{ is odd}. \end{cases}$$

It follows that $\mathsf{L}(U^{k_1} V^{k_2}) \in \mathscr{L}_6$ which implies that $\mathsf{L}(A) \in \mathscr{L}_6$.
Case 4.6.3. $k_1 \ge 2$ and $k_2 = 1$.

Then $\operatorname{supp}(U_{k_0+1} \cdot \ldots \cdot U_k) \subset \{e_1, e_2, e_3, e_4, e_0\}$. If $\{e_4, e_0\} \not\subset \operatorname{supp}(U_{k_0+1} \cdot \ldots \cdot U_k)$, then $\mathsf{L}(A) = \mathsf{L}(U^{k_1} V) + k - k_0$ is equal to

$$\begin{cases} k + \{0, 2, 3\} + 3 \cdot [0, k_1/2 - 1], & \text{if } k_1 \text{ is even}, \\ k + \{0, 2, 3\} + 3 \cdot [0, (k_1 - 1)/2 - 1] \cup \{k + (3k_1 - 3)/2 + 2\}, & \text{if } k_1 \text{ is odd} \end{cases}$$

by Equation (12). Therefore $\mathsf{L}(A) \in \mathscr{L}_8$.

If $\{e_4, e_0\} \subset \operatorname{supp}(U_{k_0+1} \cdot \ldots \cdot U_k)$, then we may assume that $U_{k_0+1} = e_4^2$ and that $U_{k_0+2} = e_0^2$. Thus

$$\begin{aligned} \mathsf{L}(A) &= \mathsf{L}(U^{k_1} V U_{k_0+1} U_{k_0+2}) + k - k_0 - 2 \\ &= \begin{cases} k - 1 + \{0, 1, 3\} + 3 \cdot [0, (k_1 + 1)/2 - 1], & \text{if } k_1 \text{ is odd}, \\ k - 1 + \{0, 1, 3\} + 3 \cdot [0, k_1/2 - 1] \cup \{k + 3k_1/2 + 1\}, & \text{if } k_1 \text{ is even}, \end{cases} \end{aligned}$$

by Equation (13) and hence $\mathsf{L}(A) \in \mathscr{L}_7$. □

5 Sets of Lengths of Weakly Krull Monoids

It is well known that—under reasonable algebraic finiteness conditions—the Structure Theorem for Sets of Lengths holds for weakly Krull monoids (as it is true for transfer Krull monoids of finite type, see Proposition 3.2). In spite of this common feature we will demonstrate that systems of sets of lengths for a variety of classes of weakly Krull monoids are different from the system of sets of lengths of any transfer Krull monoid (apart from well-described exceptional cases; see Theorems 5.5 to 5.8). Since half-factorial monoids are transfer Krull monoids, and since there are half-factorial weakly Krull monoids, half-factoriality is such a natural exceptional case.

So far there are only a couple of results in this direction. In [15], Frisch showed that $\mathrm{Int}(\mathbb{Z})$, the ring of integer-valued polynomials over $\mathbb{Z}$, is not a transfer Krull domain (nevertheless, the system of sets of lengths of $\mathrm{Int}(\mathbb{Z})^\bullet$ coincides with $\mathcal{L}(G)$ for an infinite abelian group G). To mention a result by Smertnig, let $\mathcal{O}$ be the ring of integers of an algebraic number field K, A a central simple algebra over K, and R a classical maximal $\mathcal{O}$-order of A. Then R is a noncommutative Dedekind domain and in particular an HNP ring (see [32, Sections 5.2 and 5.3]). Furthermore, R is a transfer Krull domain if and only if every stably free left R-ideal is free ([35, Theorems 1.1 and 1.2]).

We gather basic concepts and properties of weakly Krull monoids and domains (Propositions 5.1 and 5.2). In the remainder of this section, all monoids and domains are supposed to be commutative.

Let H be a monoid (hence commutative, cancellative, and with unit element). We denote by $\mathsf{q}(H)$ the quotient group of H, by $H_{\mathrm{red}} = H/H^\times$ the associated reduced monoid of H, by $\mathfrak{X}(H)$ the set of minimal nonempty prime s-ideals of H, and by $\mathfrak{m} = H \setminus H^\times$ the maximal s-ideal. Let $\mathcal{I}_v^*(H)$ denote the monoid of v-invertible v-ideals of H (with v-multiplication). Then $\mathcal{F}_v(H)^\times = \mathsf{q}(\mathcal{I}_v^*(H))$ is the quotient group of fractional v-invertible v-ideals, and $\mathcal{C}_v(H) = \mathcal{F}_v(H)^\times/\{xH \mid x \in \mathsf{q}(H)\}$ is the v-class group of H (detailed presentations of ideal theory in commutative monoids can be found in [18, 30]). We denote by $\widehat{H} \subset \mathsf{q}(H)$ the complete integral closure of H, and by $(H{:}\widehat{H}) = \{x \in \mathsf{q}(H) \mid x\widehat{H} \subset H\} \subset H$ the conductor of H. A submonoid $S \subset H$ is said to be saturated if $S = \mathsf{q}(S) \cap H$. For the definition and discussion of the concepts of being faithfully saturated or being locally tame we refer to [18, Sections 1.6 and 3.6].

To start with the local case, we recall that H is said to be

- *primary* if $\mathfrak{m} \neq \emptyset$ and for all $a, b \in \mathfrak{m}$ there is an $n \in \mathbb{N}$ such that $b^n \subset aH$.
- *strongly primary* if $\mathfrak{m} \neq \emptyset$ and for every $a \in \mathfrak{m}$ there is an $n \in \mathbb{N}$ such that $\mathfrak{m}^n \subset aH$. We denote by $\mathcal{M}(a)$ the smallest n having this property.
- a *discrete valuation monoid* if it is primary and contains a prime element (equivalently, $H_{\mathrm{red}} \cong (\mathbb{N}_0, +)$).

Furthermore, H is said to be

- *weakly Krull* ([30, Corollary 22.5]) if

$$H = \bigcap_{\mathfrak{p} \in \mathfrak{X}(H)} H_{\mathfrak{p}} \quad \text{and} \quad \{\mathfrak{p} \in \mathfrak{X}(H) \mid a \in \mathfrak{p}\} \quad \text{is finite for all } a \in H\,.$$

- *weakly factorial* if one of the following equivalent conditions is satisfied ([30, Exercise 22.5]) :
  - Every nonunit is a finite product of primary elements.
  - H is a weakly Krull monoid with trivial t-class group.

Clearly, every localization $H_{\mathfrak{p}}$ of H at a minimal prime ideal $\mathfrak{p} \in \mathfrak{X}(H)$ is primary, and a weakly Krull monoid H is v-noetherian if and only if $H_{\mathfrak{p}}$ is v-noetherian for each $\mathfrak{p} \in \mathfrak{X}(H)$. Every v-noetherian primary monoid H is strongly primary and v-local, and if $(H{:}\widehat{H}) \neq \emptyset$, then H is locally tame ([26, Lemma 3.1 and Corollary 3.6]). Every strongly primary monoid is a primary BF-monoid ([18, Section 2.7]). Therefore the coproduct of a family of strongly primary monoids is a BF-monoid, and every coproduct of a family of primary monoids is weakly factorial. A v-noetherian weakly Krull monoid H is weakly factorial if and only if $\mathcal{C}_v(H) = 0$ if and only if $H_{\mathrm{red}} \cong \mathcal{I}_v^*(H)$.

By a numerical monoid H we mean an additive submonoid of $(\mathbb{N}_0, +)$ such that $\mathbb{N}_0 \setminus H$ is finite. Clearly, every numerical monoid is v-noetherian primary, and hence it is strongly primary. Note that a numerical monoid is half-factorial if and only if it is equal to $(\mathbb{N}_0, +)$.

Let R be a domain. Then $R^\bullet = R \setminus \{0\}$ is a monoid, and all arithmetic and ideal theoretic concepts introduced for monoids will be used for domains in the obvious way. The domain R is weakly Krull (resp. weakly factorial) if and only if its multiplicative monoid $R^\bullet$ is weakly Krull (resp. weakly factorial). Weakly Krull domains were introduced by Anderson, Anderson, Mott, and Zafrullah [2, 3]. We recall some most basic facts and refer to an extended list of weakly Krull domains and monoids in [29, Examples 5.7]. The monoid $R^\bullet$ is primary if and only if R is one-dimensional and local. If R is one-dimensional local Mori, then $R^\bullet$ is strongly primary and locally tame ([19]). Furthermore, every one-dimensional semilocal Mori domain with nontrivial conductor is weakly factorial and the same holds true for generalized Cohen-Kaplansky domains. It can be seen from the definition that one-dimensional noetherian domains are v-noetherian weakly Krull domains.

Proposition 5.1 summarizes the main algebraic properties of v-noetherian weakly Krull monoids and Proposition 5.2 recalls that their arithmetic can be studied via weak transfer homomorphisms to weakly Krull monoids of very special form.

Proposition 5.1 *Let H be a v-noetherian weakly Krull monoid.*

1. *The monoid $\mathcal{I}_v^*(H)$ is isomorphic to the coproduct of $(H_{\mathfrak{p}})_{\mathrm{red}}$ over all $\mathfrak{p} \in \mathfrak{X}(H)$. In particular, $\mathcal{I}_v^*(H)$ is weakly factorial and v-noetherian.*
2. *Suppose that $\mathfrak{f} = (H{:}\widehat{H}) \neq \emptyset$. We set $\mathcal{P}^* = \{\mathfrak{p} \in \mathfrak{X}(H) \mid \mathfrak{p} \supset \mathfrak{f}\}$, and $\mathcal{P} = \mathfrak{X}(H) \setminus \mathcal{P}^*$.*

a. *Then $\widehat{H}$ is Krull, $\mathscr{P}^*$ is finite, and $H_\mathfrak{p}$ is a discrete valuation monoid for each $\mathfrak{p} \in \mathscr{P}$. In particular, $\mathscr{I}_v^*(H)$ is isomorphic to $\mathscr{F}(\mathscr{P}) \times \prod_{\mathfrak{p}\in\mathscr{P}^*}(H_\mathfrak{p})_{\mathrm{red}}$.*
b. *If $\mathscr{H} = \{aH \mid a \in H\}$ is the monoid of principal ideals of H, then $\mathscr{H} \subset \mathscr{I}_v^*(H)$ is saturated. Moreover, if H is the multiplicative monoid of a domain, then all monoids $H_\mathfrak{p}$ are locally tame and $\mathscr{H} \subset \mathscr{I}_v^*(H)$ is faithfully saturated.*

Proof

1. See [29, Proposition 5.3].
2. For (a) we refer to [18, Theorem 2.6.5] and for (b) we refer to [18, Theorems 3.6.4 and 3.7.1].

□

Proposition 5.2 *Let $D = \mathscr{F}(\mathscr{P}) \times \prod_{i=1}^n D_i$ be a monoid, where $\mathscr{P} \subset D$ is a set of primes, $n \in \mathbb{N}_0$, and $D_1, \ldots, D_n$ are reduced primary monoids. Let $H \subset D$ be a saturated submonoid, $G = \mathsf{q}(D)/\mathsf{q}(H)$, and $G_\mathscr{P} = \{p\,\mathsf{q}(H) \mid p \in \mathscr{P}\} \subset G$ the set of classes containing primes.*

1. *There is a saturated submonoid $B \subset F = \mathscr{F}(G_\mathscr{P}) \times \prod_{i=1}^n D_i$ and a weak transfer homomorphism $\theta: H \to B$. Moreover, if G is a torsion group, then there is a monomorphism $\mathsf{q}(F)/\mathsf{q}(B) \to G$.*
2. *If G is a torsion group, then H is weakly Krull.*

Proof

1. See [18, Propositions 3.4.7 and 3.4.8].
2. See [29, Lemma 5.2].

□

Note that, under the assumption of 5.1.2, the embedding $\mathscr{H} \hookrightarrow \mathscr{I}_v^*(H)$ fulfills the assumptions imposed on the embedding $H \hookrightarrow D$ in Proposition 5.2. Thus Proposition 5.2 applies to v-noetherian weakly Krull monoids. For simplicity and in order to avoid repetitions, we formulate the next results (including Theorem 5.7) in the abstract setting of Proposition 5.2. However, v-noetherian weakly Krull domains and their monoids of v-invertible v-ideals are in the center of our interest.

If (in the setting of Proposition 5.2) $G_\mathscr{P}$ is finite, then $F = \mathscr{F}(G_P) \times \prod_{i=1}^n D_i$ is a finite product of primary monoids and $B \subset F$ is a saturated submonoid. We formulate the main structural result for sets of lengths in v-noetherian weakly Krull monoids in this abstract setting.

Proposition 5.3 *Let $D_1, \ldots, D_n$ be locally tame strongly primary monoids and $H \subset D = D_1 \times \ldots \times D_n$ a faithfully saturated submonoid such that $\mathsf{q}(D)/\mathsf{q}(H)$ is finite.*

1. *The monoid H satisfies the Structure Theorem for Sets of Lengths.*
2. *There is a finite abelian group G such that for every $L \in \mathscr{L}(H)$ there is a $y \in \mathbb{N}$ such that $y + L \in \mathscr{L}(G)$.*

Proof 1. follows from [18, Theorem 4.5.4], and 2. follows from 1. and from Proposition 3.2.2.

□

The next lemma on zero-sum sequences will be crucial in order to distinguish between sets of lengths in weakly Krull monoids and sets of lengths in transfer Krull monoids.

Lemma 5.4 *Let G be an abelian group and $G_0 \subset G$ a non-half-factorial subset.*

1. *There exists a half-factorial subset $G_1 \subset G_0$ with $\mathcal{B}(G_1) \neq \{1\}$.*
2. *There are $M \in \mathbb{N}$ and zero-sum sequences $B_k \in \mathcal{B}(G_0)$ for every $k \in \mathbb{N}$ such that $2 \le |\mathsf{L}(B_k)| \le M$ but $\min \mathsf{L}(B_k) \to \infty$ as $k \to \infty$.*

Proof

1. Since G_0 is not half-factorial, there is a $B \in \mathcal{B}(G_0)$ such that $|\mathsf{L}(B)| > 1$. Thus $\operatorname{supp}(B)$ is finite and not half-factorial, say $\operatorname{supp}(B) = \{g_1, \ldots, g_\ell\}$ with $\ell \ge 2$. Without restriction we may suppose that every proper subset of $\{g_1, \ldots, g_\ell\}$ is half-factorial. Assume to the contrary that for every subset $G_1 \subsetneq \{g_1, \ldots, g_\ell\}$ we have $\mathcal{B}(G_1) = \{1\}$. Since $\{g_1, \ldots, g_\ell\}$ is minimal non-half-factorial, there is an atom $A_1 \in \mathcal{A}(\{g_1, \ldots, g_\ell\})$ such that $\mathsf{v}_{g_i}(A_1) > 0$ for every $i \in [1, \ell]$. Since $\{g_1, \ldots, g_\ell\}$ is not half-factorial, there is an atom $A_2 \in \mathcal{A}(\{g_1, \ldots, g_\ell\})$ distinct from A_1, say
$$A_1 = g_1^{k_1} \cdot \ldots \cdot g_\ell^{k_\ell} \quad \text{and} \quad A_2 = g_1^{t_1} \cdot \ldots \cdot g_\ell^{t_\ell}$$
where $k_i \in \mathbb{N}$ and $t_i \in \mathbb{N}_0$ for every $i \in [1, \ell]$. Let $\tau \in [1, \ell]$ such that $\frac{t_\tau}{k_\tau} = \max\{\frac{t_j}{k_j} \mid j \in [1, \ell]\}$. Then $k_j t_\tau - t_j k_\tau \ge 0$ for every $j \in [1, \ell]$ whence
$$W = A_2^{t_\tau} A_1^{-k_\tau} \in \mathcal{B}(\{g_1, \ldots, g_\ell\} \setminus \{g_\tau\}),$$
which implies that $W = 1$. Therefore $\frac{t_\tau}{k_\tau} = \frac{t_j}{k_j}$ for every $j \in [1, \ell]$ and hence $A_1 \mid A_2$ or $A_2 \mid A_1$, a contradiction.
2. Let $B \in \mathcal{B}(G_0)$ with $|\mathsf{L}(B)| > 1$. By 1., there exists a half-factorial subset $G_1 \subsetneq G_0$ such that $\mathcal{B}(G_1) \neq \{1\}$. Let $A \in \mathcal{A}(G_1)$ and $B_k = A^k B$ for every $k \in \mathbb{N}$. Obviously there exists $k_0 \in \mathbb{N}$ such that $\mathsf{L}(B_k) = \mathsf{L}(A^{k-k_0}) + \mathsf{L}(B_{k_0}) = k - k_0 + \mathsf{L}(B_{k_0})$ for every $k \ge k_0$. Thus $|\mathsf{L}(B_k)| \le \max \mathsf{L}(B_{k_0}) - \min \mathsf{L}(B_{k_0})$ and $\min \mathsf{L}(B_k) = k - k_0 + \min \mathsf{L}(B_{k_0})$. □

Now we consider strongly primary monoids and work out a feature of their systems of sets of lengths which does not occur in the system of sets of lengths of any transfer Krull monoid. To do so we study the set $\{\rho(L) \mid L \in \mathcal{L}(H)\}$ of elasticities of all sets of lengths. This set was studied first by Chapman et al. in a series of papers (see [6–8, 11]). Among others they showed that in an atomic monoid H, which has a prime element and an element $a \in H$ with $\rho(\mathsf{L}(a)) = \rho(H)$, every rational number q with $1 \le q \le \rho(H)$ can be realized as the elasticity of some $L \in \mathcal{L}(H)$ ([6, Corollary 2.2]). Primary monoids, which are not discrete valuation monoids, have no prime elements and their set of elasticities is different, as we will see in the next theorem. Statement 1. of Theorem 5.5 was proved for numerical monoids in [11, Theorem 2.2].

Theorem 5.5 *Let H be a strongly primary monoid that is not half-factorial.*

1. *There is a $\beta \in \mathbb{Q}_{>1}$ such that $\rho(L) \ge \beta$ for all $L \in \mathscr{L}(H)$ with $\rho(L) \ne 1$.*
2. *$\mathscr{L}(H) \ne \mathscr{L}(G_0)$ for any subset G_0 of any abelian group. In particular, H is not a transfer Krull monoid.*
3. *If one of the following two conditions holds, then H is locally tame.*
   - $\sup\{\min \mathsf{L}(c) \mid c \in H\} < \infty$.
   - *There exists some $u \in H \setminus H^\times$ such that $\rho_{\mathscr{M}(u)}(H) < \infty$.*

   If H is locally tame, then $\Delta(H)$ is finite, and there is an $M \in \mathbb{N}_0$ such that every $L \in \mathscr{L}(H)$ is an AAMP with period $\{0, \min \Delta(H)\}$ and bound M.

Remark If H is the multiplicative monoid of a one-dimensional local Mori domain R with nonzero conductor $(R{:}\widehat{R}) \ne \{0\}$, then one of the conditions in 3. is satisfied (see [18, Proposition 2.10.7 and Theorem 3.1.5]). However, there are strongly primary monoids for which none of the conditions holds and which are not locally tame ([26, Proposition 3.7]).

Proof

1. Let $b \in H$ such that $|\mathsf{L}(b)| \ge 2$ and let $u \in \mathscr{A}(H)$. Since H is a strongly primary monoid, we have $(H \setminus H^\times)^{\mathscr{M}(b)} \in bH$ and $(H \setminus H^\times)^{\mathscr{M}(u)} \in uH$. Thus $b \mid u^{\mathscr{M}(b)}$ and hence $|\mathsf{L}(u^{\mathscr{M}(b)})| \ge 2$. We define

$$\beta_1 = \frac{\mathscr{M}(b) + \mathscr{M}(u) + 1}{\mathscr{M}(b) + \mathscr{M}(u)}, \qquad \beta_2 = \frac{\max \mathsf{L}(u^{\mathscr{M}(b)}) + \mathscr{M}(b) + \mathscr{M}(u)}{\min \mathsf{L}(u^{\mathscr{M}(b)}) + \mathscr{M}(b) + \mathscr{M}(u)},$$

and observe that $\beta = \min\{\beta_1, \beta_2\} > 1$. Let $a \in H$ with $\rho(\mathsf{L}(a)) \ne 1$. We show that $\rho(\mathsf{L}(a)) \ge \beta$.

Let $k \in \mathbb{N}_0$ be maximal such that $u^k \mid a$, say $a = u^k u'$ with $u' \in H$. Thus $u \nmid u'$ and thus $\max \mathsf{L}(u') < \mathscr{M}(u)$. If $k < \mathscr{M}(b)$, then $\min \mathsf{L}(a) \le \min \mathsf{L}(u^k) + \min \mathsf{L}(u') \le \mathscr{M}(b) + \mathscr{M}(u)$, and hence

$$\rho(\mathsf{L}(a)) = \frac{\max \mathsf{L}(a)}{\min \mathsf{L}(a)} \ge \frac{\min \mathsf{L}(a) + 1}{\min \mathsf{L}(a)} \ge \frac{\mathscr{M}(b) + \mathscr{M}(u) + 1}{\mathscr{M}(b) + \mathscr{M}(u)} = \beta_1 \ge \beta .$$

If $k \ge \mathscr{M}(b)$, then there exist $t \in \mathbb{N}$ and $t_0 \in [0, \mathscr{M}(b) - 1]$ such that $k = t\mathscr{M}(b) + t_0$, and hence

$$\begin{aligned}\rho(\mathsf{L}(a)) = \frac{\max \mathsf{L}(a)}{\min \mathsf{L}(a)} &\ge \frac{\max \mathsf{L}(u^k) + \max \mathsf{L}(u')}{\min \mathsf{L}(u^k) + \min \mathsf{L}(u')} \\ &\ge \frac{t \max \mathsf{L}(u^{\mathscr{M}(b)}) + \max \mathsf{L}(u^{t_0}) + \max \mathsf{L}(u')}{t \min \mathsf{L}(u^{\mathscr{M}(b)}) + \min \mathsf{L}(u^{t_0}) + \min \mathsf{L}(u')} \\ &\ge \frac{t \max \mathsf{L}(u^{\mathscr{M}(b)}) + t_0 + \max \mathsf{L}(u')}{t \min \mathsf{L}(u^{\mathscr{M}(b)}) + t_0 + \max \mathsf{L}(u')}\end{aligned}$$

$$\geq \frac{t \max \mathsf{L}(u^{\mathcal{M}(b)}) + \mathcal{M}(b) + \mathcal{M}(u)}{t \min \mathsf{L}(u^{\mathcal{M}(b)}) + \mathcal{M}(b) + \mathcal{M}(u)} \geq \beta_2 \geq \beta \,.$$

2. Assume to the contrary that there are an abelian group G and a subset $G_0 \subset G$ such that $\mathcal{L}(H) = \mathcal{L}(G_0)$. Since H is not half-factorial, G_0 is not half-factorial. By 1., there exists $\beta \in \mathbb{Q}$ with $\beta > 1$ such that $\rho(L) \geq \beta$ for every $L \in \mathcal{L}(H)$. Lemma 5.4.2 implies that there are zero-sum sequences $B_k \in \mathcal{B}(G_0)$ such that $\rho(\mathsf{L}(B_k)) \to 1$ as $k \to \infty$, a contradiction.
3. This follows from [18, 3.1.1, 3.1.2, and 4.3.6]. □

Sets of lengths of numerical monoids have found wide attention in the literature (see, among others, [1, 9, 14]). As can be seen from Theorem 5.5.3, the structure of their sets of lengths is simpler than the structure of sets of lengths of transfer Krull monoids over finite abelian groups. Thus it is no surprise that there are infinitely many non-isomorphic numerical monoids whose systems of sets of lengths coincide, and that an analog of Conjecture 3.4 for numerical monoids does not hold true ([1]). It is open whether for every $d \in \mathbb{N}$ and every $M \in \mathbb{N}_0$ there is a strongly primary monoid D such that every AAMP with period $\{0, d\}$ and bound M can (up to a shift) be realized as a set of lengths in D (this would be the analog to the realization theorem given in Proposition 3.2.2). However, for every finite set $L \subset \mathbb{N}_{\geq 2}$ there is a v-noetherian primary monoid D and an element $a \in D$ such that $L = \mathsf{L}(a)$ ([26, Theorem 4.2]).

By Theorem 3.6 and Proposition 3.2.3, we know that $\{k, k+1\} \in \mathcal{L}(G)$ for every $k \geq 2$ and every abelian group G with $|G| \geq 3$. Furthermore, Theorem 3.7 is in sharp contrast to Theorem 5.6.1.

Theorem 5.6 *Let $D = D_1 \times \ldots \times D_n$ be the direct product of strongly primary monoids $D_1, \ldots, D_n$, which are not half-factorial.*

1. *For every finite nonempty set $L \subset \mathbb{N}$, there is a $y_L \in \mathbb{N}_0$ such that $y + L \notin \mathcal{L}(D)$ for any $y \geq y_L$.*
2. *We have $\mathcal{L}(D) \neq \mathcal{L}(G_0)$ for any subset G_0 of any abelian group, and hence D is not a transfer Krull monoid. If $D_1, \ldots, D_n$ are locally tame, then D satisfies the Structure Theorem for Sets of Lengths.*

Proof For every $i \in [1, n]$ we choose an element $a_i \in D_i$ such that $|\mathsf{L}(a_i)| > 1$.

1. Let $L \subset \mathbb{N}$ be a finite nonempty set and let $y_L = |L|(\mathcal{M}(a_1) + \ldots + \mathcal{M}(a_n))$. Assume to the contrary that there are $y \geq y_L$ and an element $b = b_1 \cdot \ldots \cdot b_n \in D$ such that $\mathsf{L}(b) = y + L$. Then there is an $i \in [1, n]$ such that $\min \mathsf{L}(b_i) \geq |L|\mathcal{M}(a_i)$. Then $b_i \in (D_i \setminus D_i^{\times})^{\min \mathsf{L}(b_i)} \subset (D_i \setminus D_i^{\times})^{|L|\mathcal{M}(a_i)} \subset a_i^{|L|} D_i$. Thus there is a $c_i \in D_i$ such that $a_i^{|L|} c_i = b_i$. This implies that $|L|\mathsf{L}(a_i) + \mathsf{L}(c_i) \subset \mathsf{L}(b_i)$. Since $|\mathsf{L}(a_i)| \geq 2$, we infer that $|\mathsf{L}(b_i)| \geq |L| + 1$ and hence $|L| = |y + L| = |\mathsf{L}(b)| \geq |\mathsf{L}(b_i)| \geq |L| + 1$, a contradiction.
2. By 1. and Lemma 5.4.2, the first conclusion follows.

   If $D_1, \ldots, D_n$ are locally tame, then D satisfies the Structure Theorem by Proposition 5.3.1. □

Theorem 5.7 *Let $D = \mathcal{F}(\mathcal{P}) \times D_1$ be the direct product of a free abelian monoid with nonempty basis $\mathcal{P}$ and of a locally tame strongly primary monoid D_1, and let G be an abelian group. Then D satisfies the Structure Theorem for Sets of Lengths, and the following statements are equivalent:*

(a) $\mathcal{L}(D) = \mathcal{L}(G)$.
(b) One of the following cases holds:

(b1) $|G| \le 2$ *and* $\rho(D) = 1$.
(b2) G *is isomorphic either to* C_3 *or to* $C_2 \oplus C_2$, $[2,3] \in \mathcal{L}(D)$, $\rho(D) = 3/2$, *and* $\Delta(D) = \{1\}$.
(b3) G *is isomorphic to* $C_3 \oplus C_3$, $[2,5] \in \mathcal{L}(D)$, $\rho(D) = 5/2$, *and* $\Delta(D) = \{1\}$.

Remark Let H be a v-noetherian weakly Krull monoid. If the conductor $(H{:}\widehat{H}) \in v$-$\max(H)$, then by Proposition 5.1, $\mathcal{I}_v^*(H)$ is isomorphic to a monoid D as given in Theorem 5.7.

Proof Since $\mathcal{P}$ is nonempty, $\mathcal{L}(D) = \{y + L \mid y \in \mathbb{N}_0, L \in \mathcal{L}(D_1)\}$ whence $\Delta(D) = \Delta(D_1)$ and $\rho(D) = \rho(D_1)$. In particular, D is half-factorial if and only if D_1 is half-factorial. Since D_1 satisfies the Structure Theorem of Sets of Lengths by Theorem 5.5.3, the same is true for D.

If D is half-factorial and $\mathcal{L}(D) = \mathcal{L}(G)$, then $\rho(D) = \rho(D_1) = 1$ and G is half-factorial whence $|G| \le 2$ by Proposition 3.3. Conversely, if $|G| \le 2$ and $\rho(D) = 1$, then G and D are half-factorial and $\mathcal{L}(G) = \mathcal{L}(D)$.

Thus from now on we suppose that D_1 is not half-factorial and that (b1) does not hold. Then $\Delta(D) \ne \emptyset$ and we set $\min \Delta(D) = d$.

(a) $\Rightarrow$ (b) Theorem 5.5.3 and Proposition 3.2.3 imply that G is finite. Since G is not half-factorial, it follows that $|G| \ge 3$. Theorem 5.5.3 shows that $\Delta_1(D) = \{d\}$, and since $1 \in \Delta_1(G) = \Delta_1(D)$, we infer that $d = 1$. Corollary 4.3.16 in [18] and [25, Theorem 1.1] imply that

$$\max\{\exp(G) - 2, \mathsf{r}(G) - 1\} = \max \Delta_1(G) = \max \Delta_1(D) = 1\,.$$

Therefore G is isomorphic to one of the following groups: $C_2 \oplus C_2$, C_3, $C_3 \oplus C_3$. We distinguish two cases.

Case 1: G is isomorphic to $C_2 \oplus C_2$ or to C_3.

By Proposition 3.3, we have

$$\mathcal{L}(D) = \mathcal{L}(C_2 \oplus C_2) = \mathcal{L}(C_3) = \{y + 2k + [0,k] \mid y, k \in \mathbb{N}_0\}\,.$$

In particular, we have $3/2 = \rho(G) = \rho(D)$ and $\{1\} = \Delta(G) = \Delta(D)$.

Case 2: G is isomorphic to $C_3 \oplus C_3$.

By Theorem 4.1, just using different notation, we have

$$\mathcal{L}(D) = \mathcal{L}(C_3^2) = \{[2k, \ell] \mid k \in \mathbb{N}_0, \ell \in [2k, 5k]\} \cup \{[2k+1, \ell] \mid k \in \mathbb{N}, \ell \in [2k+1, 5k+2]\} \cup \{\{1\}\}\,.$$

In particular, we have $5/2 = \rho(G) = \rho(D)$ and $\{1\} = \Delta(G) = \Delta(D)$.

(b) $\Rightarrow$ (a) First suppose that Case (b2) holds. We show that

$$\mathcal{L}(D) = \{y + 2k + [0, k] \mid y,\, k \in \mathbb{N}_0\}\,.$$

Then $\mathcal{L}(D) = \mathcal{L}(G)$ by Proposition 3.3. Since $\rho(D) = 3/2$ and $\Delta(D) = \{1\}$, it follows that $\mathcal{L}(D)$ is contained in the above family of sets. Thus we have to verify that for every $y, k \in \mathbb{N}_0$, the set $y + [2k, 3k] \in \mathcal{L}(D)$. Since $\mathcal{P}$ is nonempty, D contains a prime element and hence it suffices to show that $[2k, 3k] \in \mathcal{L}(D)$ for all $k \in \mathbb{N}$. Let $a \in D$ with $\mathsf{L}(a) = \{2, 3\}$, and let $k \in \mathbb{N}$. Then $\min \mathsf{L}(a^k) \le 2k$ and $\max \mathsf{L}(a^k) \ge 3k$. Since $\rho(\mathsf{L}(a^k)) \le \rho(D) = 3/2$, it follows that $\min \mathsf{L}(a^k) = 2k$ and $\max \mathsf{L}(a^k) = 3k$. Since $\Delta(D) = \{1\}$, we finally obtain that $\mathsf{L}(a^k) = [2k, 3k]$.

Now suppose that Case (b3) holds. We show that $\mathcal{L}(D)$ is equal to

$$\{[2k, \ell] \mid k \in \mathbb{N}_0, \ell \in [2k, 5k]\} \cup \{[2k+1, \ell] \mid k \in \mathbb{N}, \ell \in [2k+1, 5k+2]\} \cup \{\{1\}\}\,.$$

Then $\mathcal{L}(D) = \mathcal{L}(G)$ by Theorem 4.1. Since $\rho(D) = 5/2$ and $\Delta(D) = \{1\}$, it follows that $\mathcal{L}(D)$ is contained in the above family of sets. Now the proof runs along the same lines as the proof in Case (b2). □

We show that the Cases (b2) and (b3) in Theorem 5.7 can actually occur. Recall that numerical monoids are locally tame and strongly primary. Let D_1 be a numerical monoid distinct from $(\mathbb{N}_0, +)$, say $\mathcal{A}(D_1) = \{n_1, \ldots, n_t\}$ where $t \in \mathbb{N}_{\ge 2}$ and $1 < n_1 < \ldots < n_t$. Then, by [11, Theorem 2.1] and [9, Proposition 2.9],

$$\rho(D_1) = \frac{n_t}{n_1} \quad \text{and} \quad \min \Delta(D_1) = \gcd(n_2 - n_1, \ldots, n_t - n_{t-1})\,.$$

Suppose that $\rho(D_1) = m/2$ with $m \in \{3, 5\}$ and $\Delta(D_1) = \{1\}$. Then there is an $a \in D_1$ with $\mathsf{L}(a) = [2, m] \in \mathcal{L}(D_1)$. Clearly, there are non-isomorphic numerical monoids with elasticity $m/2$ and set of distances equal to $\{1\}$.

Theorem 5.8 *Let R be a v-noetherian weakly Krull domain with conductor $\{0\} \subsetneq \mathfrak{f} = (R{:}\widehat{R}) \subsetneq R$, and let $\pi\colon \mathfrak{X}(\widehat{R}) \to \mathfrak{X}(R)$ be the natural map defined by $\pi(\mathfrak{P}) = \mathfrak{P} \cap R$ for all $\mathfrak{P} \in \mathfrak{X}(\widehat{R})$.*

1. a. *$\mathcal{I}_v^*(H)$ is locally tame with finite set of distances, and it satisfies the Structure Theorem for Sets of Lengths.*
   b. *If π is not bijective, then $\mathcal{L}(\mathcal{I}_v^*(H)) \ne \mathcal{L}(G_0)$ for any finite subset G_0 of any abelian group and for any subset G_0 of an infinite cyclic group. In particular, $\mathcal{I}_v^*(H)$ is not a transfer Krull monoid of finite type.*
   c. *If R is seminormal, then the following statements are equivalent:*
      i. *π is bijective.*
      ii. *$\mathcal{I}_v^*(H)$ is a transfer Krull monoid of finite type.*
      iii. *$\mathcal{I}_v^*(H)$ is half-factorial.*

2. *Suppose that the class group $\mathcal{C}_v(R)$ is finite.*
   a. *The monoid $R^\bullet$ of nonzero elements of R is locally tame with finite set of distances, and it satisfies the Structure Theorem for Sets of Lengths.*
   b. *If π is not bijective, then $\mathcal{L}(R^\bullet) \neq \mathcal{L}(G_0)$ for any finite subset G_0 of any abelian group and for any subset G_0 of an infinite cyclic group. In particular, R is not a transfer Krull domain of finite type.*
   c. *If π is bijective, R is seminormal, every class of $\mathcal{C}_v(R)$ contains a $\mathfrak{p} \in \mathfrak{X}(R)$ with $\mathfrak{p} \not\supset \mathfrak{f}$, and the natural epimorphism $\delta: \mathcal{C}_v(R) \to \mathcal{C}_v(\widehat{R})$ is an isomorphism, then there is a weak transfer homomorphism $\theta: R^\bullet \to \mathcal{B}(\mathcal{C}_v(R))$. In particular, R is a transfer Krull domain of finite type.*

Proof Since $\mathfrak{f} \neq R$, it follows that $R \neq \widehat{R}$ and that R is not a Krull domain. We use the structural description of $\mathcal{I}_v^*(H)$ as given in Proposition 5.1.

1.(a) and 2.(a) Both monoids, $R^\bullet$ and $\mathcal{I}_v^*(H)$, are locally tame with finite set of distances by [18, Theorem 3.7.1]. Furthermore, they both satisfy the Structure Theorem for Sets of Lengths by Proposition 5.3 (use Propositions 5.1 and 5.2).

1.(b) and 2.(b) Suppose that π is not bijective. Then $\rho\big(\mathcal{I}_v^*(H)\big) = \rho(R^\bullet) = \infty$ by [18, Theorems 3.1.5 and 3.7.1]. Let G_0 be a finite subset of an abelian group G. Then $\mathcal{B}(G_0)$ is finitely generated, the Davenport constant $\mathsf{D}(G_0)$ is finite whence the set of distances $\Delta(G_0)$ and the elasticity $\rho(G_0)$ are both finite (see [18, Theorems 3.4.2 and 3.4.11]). Thus $\mathcal{L}\big(\mathcal{I}_v^*(H)\big) \neq \mathcal{L}(G_0)$ and $\mathcal{L}(R^\bullet) \neq \mathcal{L}(G_0)$. If G_0 is a subset of an infinite cyclic group, then the set of distances is finite if and only if the elasticity is finite by [27, Theorem 4.2], and hence the assertion follows again.

1.(c) Suppose that R is seminormal. By 1.(b) and since half-factorial monoids are transfer Krull monoids of finite type, it remains to show that π is bijective if and only if $\mathcal{I}_v^*(H)$ is half-factorial. Since R is seminormal, all localizations $R_\mathfrak{p}$ with $\mathfrak{p} \in \mathfrak{X}(H)$ are seminormal. Thus $\mathcal{I}_v^*(H)$ is isomorphic to a monoid of the form $\mathcal{F}(\mathcal{P}) \times D_1 \times \ldots \times D_n$, where $n \in \mathbb{N}$ and $D_1, \ldots, D_n$ are seminormal finitely primary monoids, and this monoid is half-factorial if and only if each monoid $D_1, \ldots, D_n$ is half-factorial. By [29, Lemma 3.6], D_i is half-factorial if and only if it has rank one for each $i \in [1, n]$, and this is equivalent to π being bijective (see [18, Theorem 3.7.1]).

2.(c) This follows from [29, Theorem 5.8]. □

Note that every order R in an algebraic number field is a v-noetherian weakly Krull domain with finite class group $\mathcal{C}_v(R)$ such that every class contains a $\mathfrak{p} \in \mathfrak{X}(R)$ with $\mathfrak{p} \not\supset \mathfrak{f}$. If R is a v-noetherian weakly Krull domain as above, then Theorems 5.5, 5.6, and 5.7 provide further instances of when R is not a transfer Krull domain, but a characterization of the general case remains open. We formulate the following problem (see also [17, Problem 4.7]).

Problem 5.9 Let H be a v-noetherian weakly Krull monoid with nonempty conductor $(H:\widehat{H})$ and finite class group $\mathcal{C}_v(H)$. Characterize when H and when the monoid $\mathcal{I}_v^*(H)$ are transfer Krull monoids, resp., transfer Krull monoids of finite type.

Acknowledgements This work was supported by the Austrian Science Fund FWF, Project Number P28864-N35, and the ANR project Caesar, project number ANR-12-BS01-0011

References

1. Amos, J., Chapman, S.T., Hine, N., Paixão, J.: Sets of lengths do not characterize numerical monoids. Integers **7**, Paper A50, 8p. (2007)
2. Anderson, D.D., Anderson, D.F., Zafrullah, M.: Atomic domains in which almost all atoms are prime. Commun. Algebra **20**, 1447–1462 (1992)
3. Anderson, D.D., Mott, J., Zafrullah, M.: Finite character representations for integral domains. Boll. Unione Mat. Ital. **6**, 613–630 (1992)
4. Bachman, D., Baeth, N., McQueen, A.: Factorizations of upper triangular Toeplitz matrices. Boll. Unione Mat. Ital. **8**, 131–150 (2010)
5. Baeth, N.R., Smertnig, D.: Factorization theory: from commutative to noncommutative settings. J. Algebra **441**, 475–551 (2015)
6. Baginski, P., Chapman, S.T., Crutchfield, C., Kennedy, K.G., Wright, M.: Elastic properties and prime elements. Results Math. **49**, 187–200 (2006)
7. Baginski, P., Chapman, S.T., Holden, M., Moore, T.: Asymptotic elasticity in atomic monoids. Semigroup Forum **72**, 134–142 (2006)
8. Barron, T., O'Neill, C., Pelayo, R.: On the set of elasticities in numerical monoids. Semigroup Forum **94**, 37–50 (2017)
9. Bowles, C., Chapman, S.T., Kaplan, N., Reiser, D.: On delta sets of numerical monoids. J. Algebra Appl. **5**, 695–718 (2006)
10. Chang, G.W., Smertnig, D.: Factorization in the self-idealization of a PID. Boll. Unione Mat. Ital. **IX,6**(2), 363–377 (2013)
11. Chapman, S.T., Holden, M., Moore, T.: Full elasticity in atomic monoids and integral domains. Rocky Mt. J. Math. **36**, 1437–1455 (2006)
12. Chapman, S.T., Gotti, F., Pelayo, R.: On delta sets and their realizable subsets in Krull monoids with cyclic class groups. Colloq. Math. **137**, 137–146 (2014)
13. Chapman, S.T., Fontana, M., Geroldinger, A., Olberding, B. (eds.): Multiplicative Ideal Theory and Factorization Theory. Proceedings in Mathematics and Statistics, vol. 170. Springer, Berlin (2016)
14. Colton, S., Kaplan, N.: The realization problem for delta sets of numerical monoids. J. Commut. Algebra. **9**, 313–339 (2017)
15. Frisch, S.: A construction of integer-valued polynomials with prescribed sets of lengths of factorizations. Monatsh. Math. **171**, 341–350 (2013)
16. Geroldinger, A.: Additive group theory and non-unique factorizations. In: Geroldinger, A., Ruzsa, I. (eds.) Combinatorial Number Theory and Additive Group Theory. Advanced Courses in Mathematics CRM, pp. 1–86. Birkhäuser, Barcelona (2009)
17. Geroldinger, A.: Sets of lengths. Am. Math. Monthly **123**, 960–988 (2016)
18. Geroldinger, A., Halter-Koch, F.: Non-Unique Factorizations. Algebraic, Combinatorial and Analytic Theory. Pure and Applied Mathematics, vol. 278. Chapman & Hall/CRC, Boca Raton (2006)
19. Geroldinger, A., Roitman, M.: The arithmetic of one-dimensional local domains. manuscript.
20. Geroldinger, A., Schmid, W.A.: The system of sets of lengths in Krull monoids under set addition. Rev. Mat. Iberoam. **32**, 571–588 (2016)
21. Geroldinger, A., Schmid, W.A.: A characterization of class groups via sets of lengths. http://arxiv.org/abs/1503.04679
22. Geroldinger, A., Yuan, P.: The set of distances in Krull monoids. Bull. Lond. Math. Soc. **44**, 1203–1208 (2012)

23. Geroldinger, A., Zhong, Q.: A characterization of class groups via sets of lengths II. J. Théor. Nombres Bordx. **29**, 327–346 (2017)
24. Geroldinger, A., Zhong, Q.: Long sets of lengths with maximal elasticity. Canadian J. Math. https://arxiv.org/abs/1706.06907
25. Geroldinger, A., Zhong, Q.: The set of minimal distances in Krull monoids. Acta Arith. **173**, 97–120 (2016)
26. Geroldinger, A., Hassler, W., Lettl, G.: On the arithmetic of strongly primary monoids. Semigroup Forum **75**, 567–587 (2007)
27. Geroldinger, A., Grynkiewicz, D.J., Schaeffer, G.J., Schmid, W.A.: On the arithmetic of Krull monoids with infinite cyclic class group. J. Pure Appl. Algebra **214**, 2219–2250 (2010)
28. Geroldinger, A., Grynkiewicz, D.J., Yuan, P.: On products of k atoms II. Mosc. J. Comb. Number Theory **5**, 73–129 (2015)
29. Geroldinger, A., Kainrath, F., Reinhart, A.: Arithmetic of seminormal weakly Krull monoids and domains. J. Algebra **444**, 201–245 (2015)
30. Halter-Koch, F.: Ideal Systems. An Introduction to Multiplicative Ideal Theory. Marcel Dekker, New York (1998)
31. Kainrath, F.: Factorization in Krull monoids with infinite class group. Colloq. Math. **80**, 23–30 (1999)
32. McConnell, J.C., Robson, J.C.: Noncommutative Noetherian Rings. Graduate Studies in Mathematics, vol. 30. American Mathematical Society, Providence (2001)
33. Schmid, W.A.: A realization theorem for sets of lengths. J. Number Theory **129**, 990–999 (2009)
34. Smertnig, D.: Factorizations in bounded hereditary noetherian prime rings. http://arxiv.org/abs/1605.09274
35. Smertnig, D.: Sets of lengths in maximal orders in central simple algebras. J. Algebra **390**, 1–43 (2013)
36. Zhong, Q.: Sets of minimal distances and characterizations of class groups of Krull monoids. Ramanujan J. (2017). doi:10.1007/s11139-016-9873-2

Corner's Realization Theorems from the Viewpoint of Algebraic Entropy

Brendan Goldsmith and Luigi Salce

In memoriam Rüdiger Göbel

Abstract The realization theorems for reduced torsion-free rings as endomorphism rings of reduced torsion-free Abelian groups, proved by Corner in his celebrated papers, are applied to the rings of integral polynomials $\mathbb{Z}[X]$ and the power series ring $\mathbb{Z}[[X]]$, and are compared with another realization theorem proved in Corner's paper on Hopficity in torsion-free groups, and with some variation of his results. The $\mathbb{Z}[X]$-module structure of the groups obtained from these different constructions is investigated looking at the cyclic trajectories of their endomorphisms, and at the corresponding values of the intrinsic algebraic entropy $\widetilde{\mathrm{ent}}$.

Keywords Abelian groups • Endomorphism rings • Finite topology • Intrinsic algebraic entropy

Mathematics Subject Classification: 20K30 (endomorphisms) and 37A35 (entropy)

1 Introduction

In the early 1960s, Corner proved in [2, 3] and [4] a number of realization theorems for reduced torsion-free rings as endomorphism rings of reduced torsion-free Abelian groups, which are among some of the most remarkable results obtained

B. Goldsmith (✉)
Dublin Institute of Technology, The Clock Tower 032, Lower Grangegorman, Dublin 7, Ireland
e-mail: brendan.goldsmith@dit.ie

L. Salce
Dipartimento di Matematica, Via Trieste 63, 35121 Padova, Italy
e-mail: salce@math.unipd.it

M. Fontana et al. (eds.), *Rings, Polynomials, and Modules*,
https://doi.org/10.1007/978-3-319-65874-2_12

in the theory of torsion-free Abelian groups. We start recalling these results which will then be re-examined from the viewpoint of the intrinsic algebraic entropy used for torsion-free Abelian groups. It is worthwhile remarking that these theorems provide explicit constructions of the groups with prescribed endomorphism ring. Throughout we shall deal with additively written Abelian groups usually omitting the adjective "Abelian."

Theorem 1 (Theorem A in [2]) *Every countable reduced torsion-free ring A is isomorphic to the endomorphism ring of a countable reduced torsion-free group G.*

The second result is specific for the ring of integral polynomials $\mathbb{Z}[X]$. To explain its meaning, recall that a group G is Hopfian if every epic endomorphism of G is an automorphism. A feature of Theorem 1, not explicitly mentioned in the original work [2], but highlighted in Lemma 1 of [3], is that, when a countable reduced torsion-free ring having no left zero divisors is realized as the endomorphism ring of a countable reduced torsion-free group G, the group G is necessarily Hopfian. This depends on the fact that every non-zero endomorphism of G is then injective.

In particular if R is a countable reduced torsion-free integral domain (such as $\mathbb{Z}[X]$), then any group realizing R as an endomorphism ring via the Corner construction is Hopfian. A similar phenomenon persists in more recent realizations using Black Box techniques and also in the standard realization constructions assuming the set-theoretic hypothesis $(V = L)$ – see Theorem 2.11 in [18] for some discussion of this.

Theorem 2 (Example 1 in [3]) *There exists a torsion-free non-Hopfian group of countable rank H such that the polynomial ring $\mathbb{Z}[X]$ is isomorphic to* $\mathrm{End}(H)$.

We will extend Theorem 2 to a more general result in Theorem 5.

The next result is a characterization of the endomorphism rings of countable reduced torsion-free groups (we note that, differently from Corner, we consider the endomorphisms acting on the left).

Theorem 3 (Theorem 1.1 in [4]) *A topological ring A is isomorphic to the endomorphism ring* $\mathrm{End}(G)$ *of a countable, reduced, torsion-free group G, endowed with the finite topology* fin_G, *if and only if it is Hausdorff and complete, with a basis of neighborhoods of zero consisting of a descending chain of left ideals $N_1 \geq N_2 \geq \cdots$ such that, for each integer $k \geq 1$, A/N_k is a countable, reduced, and torsion-free group.*

We will give also a detailed proof of Theorem 4 below, whose key idea goes back to Corner. The first author discovered an outline construction in a hand-written manuscript which appears to be an early draft of the final work [4]; no comparable result appeared in the printed edition. We have modified this outline to take full advantage of the simplified proof of Theorem 3 which appeared in [4]. It is, of course, true that more modern realization theorems can give similar, and indeed considerably more general, results (see, for instance, [5] and [16]), but the simpler approach of Corner has considerable advantages when applying the realization result to problems concerning entropy.

Theorem 4 *Let A be a non-discrete topological ring satisfying the equivalent conditions of Theorem 3. Then there exists a reduced torsion-free group H of cardinality* $2^{\aleph_0}$ *such that* $\mathrm{End}(H) \cong A$ *and the finite topology* fin_H *on* $\mathrm{End}(H)$ *is the discrete one.*

We remark that such a discrete realization requires the group H to be uncountable: it is an easy exercise to show that if $|G| < |\mathrm{End}(G)|$ then $\mathrm{End}(G)$ cannot be discrete in the finite topology (see Exercise 3, p. 223 in [13]).

The main goal of this paper is to apply the above theorems when the ring A coincides either with the integral polynomial ring $\mathbb{Z}[X]$ or with the power series ring $\mathbb{Z}[[X]]$. We analyze the structure of the groups constructed by means of the preceding theorems as $\mathbb{Z}[X]$-modules, and we focus on their differing behavior with respect to the intrinsic algebraic entropy $\widetilde{\mathrm{ent}}$ investigated in [10] and [17] (see next section for its definition). Recall that the intrinsic algebraic entropy, when computed for endomorphisms of torsion groups, coincides with the algebraic entropy ent, first defined in [1] and investigated in depth in [9].

There exist two more entropies normally used for torsion-free groups, namely the rank-entropy ent_{rk} introduced in [22] and investigated in [15], and the entropy h introduced by Peters for automorphisms in [20] and recently extended to endomorphisms by Dikranjan and Giordano Bruno in [8]. We will not consider them, referring to [15] for results on ent_{rk} in the same vein as those obtained in this paper, and leaving the behavior with respect to the entropy h to future investigation.

In Sect. 2 we provide the preliminary notions needed to deal with trajectories and the intrinsic algebraic entropy; we also introduce the new notion of intrinsic Pinsker subgroup of a group endowed with an endomorphism, in analogy with the notion of Pinsker subgroup defined in [7] for the entropy h.

In Sect. 3 first we derive the properties satisfied by any group whose endomorphism ring is isomorphic to $\mathbb{Z}[X]$, showing that every endomorphism of these groups different from multiplication by an integer has infinite intrinsic algebraic entropy. Then, a close analysis of the group G obtained from Corner's construction of Theorem 1, and of the group H obtained from Theorem 2, will make evident the differences between the two groups, with respect to their $\mathbb{Z}[X]$-module structure. In fact, G turns out to be a torsion-free $\mathbb{Z}[X]$-module of countable rank, while H is a mixed $\mathbb{Z}[X]$-module of finite rank, whose structure is investigated in detail. The entropy $\widetilde{\mathrm{ent}}$ will be computed for the endomorphism ω corresponding to the multiplication by the indeterminate X for both of the groups G and H, and their intrinsic Pinsker subgroups will be determined.

In Sect. 4, we compare the $\mathbb{Z}[X]$-module structure of a group G obtained from Theorem 3, whose endomorphism ring, endowed with the finite topology, is isomorphic (algebraically and topologically) to $\mathbb{Z}[[X]]$ endowed with the X-adic topology, and that of a group H obtained from Theorem 4, whose endomorphism ring, endowed with the finite topology, is isomorphic (algebraically and topologically) to $\mathbb{Z}[[X]]$ endowed with the discrete topology. The most remarkable difference is that in the first case $\widetilde{\mathrm{ent}}(\phi) = 0$ for every endomorphism ϕ of G, while H has endomorphisms ψ such that $\widetilde{\mathrm{ent}}(\psi) = \infty$, resembling what was proved for the rank-entropy ent_{rk} in [15].

Finally, in accord with what was proved in [15] and [21], we will show in Sect. 5 that, given two groups G and H with an isomorphism between $\mathrm{End}(G)$ and $\mathrm{End}(H)$ which is not only algebraic but also topological, then an endomorphism of G has zero intrinsic algebraic entropy if and only if the same happens for the corresponding endomorphism of H.

2 Preliminaries on Trajectories and Entropies

For basic concepts and facts relating to Abelian Group Theory we refer to [13]. We start by recalling the definition of the intrinsic entropy $\widetilde{\mathrm{ent}}$ and some key facts relating to this entropy; for more details we refer to [10] and [17].

The basic tool for any kind of algebraic entropy is the notion of trajectory. Given an endomorphism $\phi: G \to G$ of a group G and a subgroup H of G, the ϕ-trajectory $T(\phi, H)$ of H is the ϕ-invariant subgroup generated by H, that is, $T(\phi, H) = \sum_{n\geq 0} \phi^n H$. For an integer $n \geq 1$, the nth ϕ-trajectory $T_n(\phi, H)$ of H is the truncated sum $T_n(\phi, H) = H + \phi(H) + \cdots + \phi^{n-1}(H)$. If $H = g\mathbb{Z}$ is a cyclic group, we will denote $T(\phi, H)$ simply by $T(\phi, g)$ and we call it a cyclic trajectory.

In order to define $\widetilde{\mathrm{ent}}$, we need the notion of ϕ-inert subgroup: a subgroup H of G is *ϕ-inert* if the factor group $(H + \phi(H))/H$ is finite. Since $(H + \phi(H))/H$ is finite if and only if $T_n(\phi, H)/H$ is finite for all $n \geq 1$ (as proved in Lemma 2.1 of [10]), we can consider the limit

$$\widetilde{\mathrm{ent}}(\phi, H) = \lim_{n\to\infty} \frac{\log|T_n(\phi, H)/H|}{n}$$

which exists and is finite, by Fekete's lemma [12]. Taking the supremum of these non-negative real numbers with H ranging over the set $\mathcal{I}_\phi(G)$ of ϕ-inert subgroups of G, we have the notion of the *intrinsic algebraic entropy of ϕ*:

$$\widetilde{\mathrm{ent}}(\phi) = \sup_{H\in\mathcal{I}_\phi(G)} \widetilde{\mathrm{ent}}(\phi, H).$$

Recall that the set $\mathcal{I}_\phi(G)$ of the ϕ-inert subgroups of G contains all the finite subgroups, the subgroups of finite index, and the ϕ-invariant subgroups of G. For more information on $\widetilde{\mathrm{ent}}$ we refer to [10] and [17].

Following a classical approach in the investigation of endomorphisms of Abelian groups, we consider the category whose objects are the pairs (G, ϕ), where $\phi: G \to G$ is an endomorphism; a morphism $\alpha: (G, \phi) \to (H, \psi)$ is a homomorphism of Abelian groups $\alpha: G \to H$ satisfying the condition $\alpha \circ \phi = \psi \circ \alpha$. The pair (G, ϕ) is usually denoted by G_ϕ. This category is naturally equivalent to the category of the $\mathbb{Z}[X]$-modules; in this equivalence a polynomial $f(X) \in \mathbb{Z}[X]$ acts on G_ϕ as the endomorphism $f(\phi)$. In particular, the ϕ-invariant subgroups of G are the $\mathbb{Z}[X]$-

submodules of G_ϕ, and the ϕ-trajectory of a subgroup H of G is the $\mathbb{Z}[X]$-submodule of G_ϕ generated by H. We refer to [23] for a more detailed description of this matter.

Then the intrinsic algebraic entropy can be viewed as a map

$$\widetilde{\text{ent}}\colon \text{Mod}(\mathbb{Z}[X]) \to \mathbb{R}^* = \mathbb{R}_{\geq 0} \cup \{\infty\}$$

sending the $\mathbb{Z}[X]$-module G_ϕ to $\widetilde{\text{ent}}(\phi)$.

A fundamental fact concerning the entropy $\widetilde{\text{ent}}$ is that, in order to compute $\widetilde{\text{ent}}(G_\phi)$, it is enough to:

(1) realize G_ϕ as union of a continuous ascending chain of submodules with cyclic factors, that is, $G = \cup_{\sigma<\lambda} G_\sigma$, where λ is an ordinal, the G_σ are ϕ-invariant subgroups of G with $G_\sigma \leq G_{\sigma+1}$ for all $\sigma + 1 < \lambda$, $G_\tau = \cup_{\sigma<\tau} G_\sigma$ for all limit ordinals $\tau < \lambda$ and $G_{\sigma+1}/G_\sigma$ is a cyclic $\mathbb{Z}[X]$-module;
(2) calculate $\widetilde{\text{ent}}$ on the cyclic $\mathbb{Z}[X]$-modules $G_{\sigma+1}/G_\sigma$.

This fact is a consequence of the two properties satisfied by $\widetilde{\text{ent}}$ as an invariant on $\text{Mod}(\mathbb{Z}[X])$ of being

- upper continuous, i.e., given a $\mathbb{Z}[X]$-module M, $\widetilde{\text{ent}}(M) = sup_F\, \widetilde{\text{ent}}(F)$, ranging F over the set of the finitely generated submodules of M, and
- additive, i.e., for an exact sequence of $\mathbb{Z}[X]$-modules $0 \to A \to B \to C \to 0$, $\widetilde{\text{ent}}(B) = \widetilde{\text{ent}}(A) + \widetilde{\text{ent}}(C)$ (this property is called Addition Theorem, see [10]).

Since all groups under our consideration are torsion-free, for a cyclic ϕ-trajectory $T(\phi, g)$ we have two possible cases. In the first case, $T(\phi, g)$ is a torsion-free $\mathbb{Z}[X]$-module (isomorphic to $\mathbb{Z}[X]$, hence a free group of countable rank), therefore, in view of Example 3.7 in [10], $\widetilde{\text{ent}}(T(\phi, g)_\phi) = \infty$. In the latter case, $T(\phi, g)$ is a torsion $\mathbb{Z}[X]$-module isomorphic to $\mathbb{Z}[X]/I$ with I is a non-zero ideal of $\mathbb{Z}[X]$ and $T(\phi, g)$ is a torsion-free group of finite rank. The structure of this second type of trajectories and their intrinsic algebraic entropy will be described in Proposition 1.

We give now two lemmas which show the connection between cyclic trajectories and the finite topology of the endomorphism rings. Recall that for every finite subset F of the group G, $K_F = \{\alpha \in \text{End}(G) \mid \alpha(F) = 0\}$ denotes the left ideal of $\text{End}(G)$ consisting of the endomorphisms annihilating F. These ideals form a basis of neighborhoods of zero for the finite topology on $\text{End}(G)$, denoted by fin_G; the topological ring $(\text{End}(G), \text{fin}_G)$ is complete (see Theorem 107.1 in [13]) When $F = \{g\}$ is a singleton, we will write simply K_g.

It is well known (see p. 222 in [13]) that, for every $g \in G$, $\text{End}(G)/K_g$ is isomorphic through the evaluation map at g to the orbit O_g. The restriction of this map to the subring $\mathbb{Z}[\phi]$ of $\text{End}(G)$ generated by ϕ gives rise to the next.

Lemma 1 *Let $\phi\colon G \to G$ be an endomorphism of the group G. Then $T(\phi, g) \cong \mathbb{Z}[\phi]/(\mathbb{Z}[\phi] \cap K_g)$ for all $g \in G$.*

Passing to finitely generated trajectories, we have the following:

Lemma 2 *Let $\phi\colon G \to G$ be an endomorphism of the group G, $H = \sum_{i=1}^n g_i\mathbb{Z}$ a finitely generated subgroup of G and $F = \{g_i \mid 1 \leq i \leq n\}$. Then there is a* *monomorphism of* $\mathbb{Z}[X]$*-modules*

$$\epsilon : \mathbb{Z}[\phi]/(\mathbb{Z}[\phi] \cap K_F) \to \bigoplus_{1 \le i \le n} T(\phi, g_i)$$

Proof The injective homomorphism ϵ is the diagonal map of the isomorphisms $\epsilon_i : \mathbb{Z}[\phi]/(\mathbb{Z}[\phi] \cap K_{g_i}) \to T(\phi, g_i)$ given by Lemma 1, noting that $K_F = \cap_{1 \le i \le n} K_{g_i}$. The fact that ϵ is a $\mathbb{Z}[X]$-morphism with respect to the structure induced by ϕ is readily checked.

We can now describe the structure of the cyclic trajectories which are torsion $\mathbb{Z}[X]$-modules.

Proposition 1 *Let* $\phi : G \to G$ *be an endomorphism of a torsion-free group and let* $x \in G$ *be such that* $\mathrm{rk}_{\mathbb{Z}}(T(\phi, x)) = n$*, with* n *a positive integer. Then:*

(1) $T_n(\phi, x) = \bigoplus_{0 \le i \le n-1} \phi^i x\mathbb{Z}$*;*
(2) there exists a minimum integer $s \ge 1$ *such that* $s^k \phi^{n+k-1} x \in T_n(\phi, x)$ *for all* $k \ge 1$*, so* $f(\phi)(x) = 0$ *for a primitive polynomial* $f(X) \in \mathbb{Z}[X]$ *of degree* n *with leading coefficient* s*;*
(3) $\mathbb{Z}[X]/(f(X)) \cong T(\phi, x)$*;*
(4) $T_{k+1}(\phi, x)/T_k(\phi, x) \cong \mathbb{Z}/s\mathbb{Z}$ *for all* $k \ge n$
(5) $\widetilde{\mathrm{ent}}(T(\phi, x)_\phi) = \log s$*.*

Proof

(1) Let m be the minimum positive integer such that $t\phi^m x \in T_m(\phi, x)$ for some $0 \ne t \in \mathbb{Z}$; obviously $T_m(\phi, x) = \bigoplus_{1 \le i \le m-1} \phi^i x\mathbb{Z}$. A direct computation shows that $t^k \phi^{m+k-1} x \in T_m(\phi, x)$ for all $k \ge 1$, hence $T(\phi, x)/T_m(\phi, x)$ is a torsion group, consequently $\mathrm{rk}_{\mathbb{Z}}(T(\phi, x)) = \mathrm{rk}_{\mathbb{Z}}(T_m(\phi, x))$ and $m = n$.
(2) The proof is as in point (1), taking $s = t$ minimal positive. The primitivity of ϕ follows by the minimality of s.
(3) Let $V_x = \{g(X) \mid g(\phi)(x) = 0\}$, so $T(\phi, x) \cong \mathbb{Z}[X]/V_x$ by Lemma 1. We must show that $V_x = (f(X))$. Obviously $V_x \ge (f(X))$, and, if a non-zero polynomial $g(X) \in \mathbb{Z}[X]$ of degree k belongs to V_x, then $k \ge n$. In order to show that $g(X) \in (f(X))$, we can assume $g(X)$ primitive. The division algorithm gives $s^{k-n+1} g(X) = f(X)h(X) + r(X)$ for polynomials $h(X), r(X) \in \mathbb{Z}[X]$ and $r(X)$ of degree $\le n - 1$. So $g(X) \in V_x$ implies that $r(\phi)(x) = 0$, therefore $r(X) = 0$ and consequently $s^{k-n+1} g(X) = f(X)h(X)$. As $f(X)$ is primitive, s^{k-n+1} equals the content of $h(X)$; dividing the above equality by this content we deduce that $g(X) \in (f(X))$.
(4) The claim is obvious for $k = n$, by (2). If $k > n$, let $T_{k+1}(\phi, x)/T_k(\phi, x) \cong \mathbb{Z}/t\mathbb{Z}$. Then there exists a polynomial $g(X) \in \mathbb{Z}[X]$ of degree $k + 1$ and leading coefficient t such that $g(\phi)(x) = 0$. By point (3), $g(X)$ is a multiple of $f(X)$, hence $t \ge s$. On the other hand, since $T_{k+1}(\phi, x)/T_k(\phi, x)$ is a quotient of $T_k(\phi, x)/T_{k-1}(\phi, x)$ for all k, by [10, Lemma 1.1], $t \le s$, so we are done.

(5) From [10, Lemma 3.2] we get that $\widetilde{\mathrm{ent}}(T(\phi,x)_\phi) = \log|T_{k+1}(\phi,x)/T_k(\phi,x)|$ for all k large enough. Therefore the conclusion follows from point (4).

Point (5) of Proposition 1 can also be derived from the simplified Algebraic Yuzvinski Formula for the intrinsic algebraic entropy proved in [10], and in a simpler form in [14]. Actually, the proof in points (3) and (4) is similar to a central argument in the proof of [14, Proposition 5.6].

Given an endomorphism $\phi\colon G \to G$ of a group G and its algebraic entropy $h(\phi)$, the *Pinsker subgroup* of the $\mathbb{Z}[X]$-module G_ϕ is defined in [7] as the greatest ϕ-invariant subgroup H of G such that $h(\phi \restriction H) = 0$; this subgroup is denoted by $P(G_\phi)$. It is proved that such a subgroup does exist (Proposition 3.1 in [7]) and some different characterizations of it are furnished.

Replacing the algebraic entropy h by the intrinsic entropy $\widetilde{\mathrm{ent}}$, we define the *intrinsic Pinsker subgroup* of G_ϕ as

$$\widetilde{P}(G_\phi) = \{x \in G \mid \widetilde{\mathrm{ent}}(\phi \restriction T(\phi,x)) = 0\}.$$

Clearly $\widetilde{P}(G_\phi)$ is the greatest ϕ-invariant subgroup H of G such that $\widetilde{\mathrm{ent}}(\phi \restriction H) = 0$, and $\widetilde{P}(G_\phi) = G$ precisely when $\widetilde{\mathrm{ent}}(\phi) = 0$. Looking at Proposition 1 and using basic properties of $\widetilde{\mathrm{ent}}$ mentioned above, it is easy to check that:

$$\begin{aligned}\widetilde{P}(G_\phi) &= \{x \in G \mid f(\phi)(x) = 0 \text{ for some monic } f(X) \in \mathbb{Z}[X]\}\\ &= \{x \in G \mid T(\phi,x) = T_n(\phi,x) \text{ for some } n \geq 1\}.\end{aligned} \tag{1}$$

Given a torsion group G, in [9] the subgroup

$$t_\phi(G) = \{x \in G \mid T(\phi,x) = T_n(\phi,x) \text{ for some } n \geq 1\}$$

was defined. It follows from the above characterization of $\widetilde{P}(G_\phi)$ and from [7] that $t_\phi(G) = t(G) \cap P(G_\phi) = t(G) \cap \widetilde{P}(G_\phi)$. From the inequality $\widetilde{\mathrm{ent}} \leq h$ (see [10, Proposition 3.6]), it follows that $P(G_\phi) \leq \widetilde{P}(G_\phi)$. This inclusion may be strict, as the next two examples show.

Example 1 Let $G = \mathbb{Z}$ and let $\phi\colon G \to G$ be the endomorphism acting as the multiplication by an integer $a > 1$. Then $T(\phi,x) = T_1(\phi,x)$ and so $\widetilde{P}(G_\phi) = G$. On the other hand, ϕ is injective, hence from the characterization in [7, Theorem 6.10] it easily follows that $P(G_\phi) = 0$. Note that from Example 3.1 in [8] we get $h(\phi) = \log a$, and from the equality $\widetilde{P}(G_\phi) = G$ we get $\widetilde{\mathrm{ent}}(\phi) = 0$.

Example 2 Let $G = \mathbb{Z}e_0 \oplus \mathbb{Z}e_1$ be a free group of rank 2, and let $\phi\colon G \to G$ be the endomorphism defined by setting:

$$\phi(e_0) = e_1, \quad \phi(e_1) = e_0 + e_1.$$

The characteristic polynomial of ϕ is $p_\phi(X) = X^2 - X - 1$, so the eigenvalues of ϕ are $(1+\sqrt{5})/2$ and $(1-\sqrt{5})/2$. It follows that $\phi^2(x) = \phi(x) + x$ for all $x \in G$, consequently $T(\phi, x) = T_2(\phi, x)$ and so $\widetilde{P}(G_\phi) = G$. On the other hand, ϕ is injective, hence again from the characterization in [7, Theorem 6.10] it easily follows that $P(G_\phi) = 0$. Note that from the Algebraic Yuzvinski Formula we get that $h(\phi) = \log((1+\sqrt{5})/2)$, and from $\widetilde{P}(G_\phi) = G$ we get that $\widetilde{\mathrm{ent}}(\phi) = 0$.

3 Groups Whose Endomorphism Ring is Isomorphic to $\mathbb{Z}[X]$

We start investigating the groups G, however, constructed, such that $\mathrm{End}(G)$ is isomorphic to the integral polynomial ring $\mathbb{Z}[X]$.

Lemma 3 *Let G be a group such that* $\mathrm{End}(G) \cong \mathbb{Z}[X]$. *Then G is torsion-free, indecomposable, and reduced.*

Proof The fact that G is indecomposable is a consequence of the fact that $\mathbb{Z}[X]$ is an integral domain. If G is not torsion-free, then it has a non-zero direct summand isomorphic to a cocyclic p-group for some prime p. As G is indecomposable, this implies that G is cocyclic; but the endomorphism ring of a cocyclic p-group is either cyclic or isomorphic to the ring of the p-adic integers J_p, absurd. If G is not reduced, then $G \cong \mathbb{Q}$, which is again absurd, since $\mathrm{End}(\mathbb{Q}) = \mathbb{Q}$.

We identify $\mathrm{End}(G)$ with $\mathbb{Z}[X]$. Note that, under this identification, $\mathbb{Z}[X] = \mathbb{Z}[\omega]$, where ω is the distinguished endomorphism acting as the multiplication by X on G, and every endomorphism of G is of the form $\phi = a_0 + a_1\omega + a_2\omega^2 + \cdots + a_n\omega^n$ ($a_i \in \mathbb{Z}, n \in \mathbb{N}$).

Proposition 2 *Let G be a group such that* $\mathrm{End}(G) = \mathbb{Z}[X]$. *Then:*

(1) the finite topology of $\mathrm{End}(G)$ *is discrete;*
(2) there exist elements $g \in G$ such that $T(\omega, g)_\omega \cong \mathbb{Z}[X]$;
(3) $\widetilde{\mathrm{ent}}(\phi) = \infty$ for every endomorphism ϕ of G different from the multiplication by an integer.

Proof

(1) A linear topology on $\mathbb{Z}[X]$ has a basis of ideals; as $\mathbb{Z}[X]$ is a Noetherian ring, its ideals are finitely generated, so there are countably many different ideals. Consequently the finite topology fin_G is metrizable. Since the only metrizable linear topology making $\mathbb{Z}[X]$ a complete ring is the discrete one (see [6]), fin_G is necessarily discrete.
(2) By point (1), there exists a finite subset F of G such that $K_F = 0$. By Lemma 2, there exists an embedding $\epsilon\colon \mathbb{Z}[X] \to \bigoplus_{g\in F} T(\omega, g)$. This shows that there exists a $g \in F$ such that the subgroup $T(\omega, g)$ has infinite rank. As shown in the discussion preceding Lemma 1, this implies that $T(\omega, g)_\omega \cong \mathbb{Z}[X]$.

(3) From the same discussion we know that then $\widetilde{\mathrm{ent}}(T(\omega, g)_\omega) = \infty$, therefore $\widetilde{\mathrm{ent}}(\omega) = \infty$. A straightforward computation shows that, if $f(X) \in \mathbb{Z}[X] \setminus \mathbb{Z}$ and setting $\phi = f(\omega)$, from the equality $T(\omega, g) = \bigoplus_{n\geq 0} \omega^n g\mathbb{Z}$ it follows that $T(\phi, g) = \bigoplus_{n\geq 0} \phi^n g\mathbb{Z}$. Therefore $\widetilde{\mathrm{ent}}(T(\phi, g)_\phi) = \infty$ and consequently $\widetilde{\mathrm{ent}}(\phi) = \infty$.

Proposition 2 shows that, when $\mathrm{End}(G) = \mathbb{Z}[X]$, the entropy $\widetilde{\mathrm{ent}}$ takes infinite values on an ω-trajectory $T(\omega, g)$ of infinite rank for a suitable $g \in G$. This does not imply that the same happens for all other cyclic ω-trajectories since some of them could have finite rank, as we will see in Proposition 4; the structure of these trajectories was described in Proposition 1.

We pass now to investigate the structure as $\mathbb{Z}[X]$-modules of the torsion-free group G with endomorphism ring isomorphic to $\mathbb{Z}[X]$ obtained by means of the construction of Theorem 1, that we briefly sketch below.

Let $A = \mathbb{Z}[X]$ and let $\hat{A}$ be its completion in the natural topology. The countable reduced torsion-free group G constructed by Corner in Theorem 1, satisfying $\mathrm{End}(G) = A$, is defined according to the following steps:

- Lemma 1.5 in [2] ensures the existence of a subring P of $\hat{\mathbb{Z}}$ of cardinality $2^{\aleph_0}$, all whose non-zero elements are associated with an integer. This property implies that P is an integral domain and it is easy to see that $\hat{A}$ is a torsion-free P-module, so one can apply to subsets of A the usual notions of linear or algebraic independence over P. Using a maximal family of elements of A linearly independent over P, one can define a countable pure subring Π of P, so that P has transcendence degree $2^{\aleph_0}$ over Π;
- for each element $a \in A$ choose elements $\alpha_a, \beta_a \in P$ in such a way that the countable set $\{1\} \cup \{\alpha_a, \beta_a \mid a \in A\}$ is algebraically independent over Π;
- define G to be the pure subgroup of $\hat{A}$ generated by A and by $e_a A$ ($a \in A$), where $e_a = \alpha_a \cdot 1 + \beta_a \cdot a$, i.e., $G = \langle A,\ e_a A, (a \in A)\rangle_* \leq \hat{A}$.

Clearly A acts faithfully on G (i.e., $\mathrm{Ann}_A(G) = 0$), so A is isomorphic to a subring of $\mathrm{End}(G)$. The main part of the proof of Theorem1 consists in showing that every endomorphism of G coincides with the multiplication by an element of A. Now we can easily prove the following:

Proposition 3 *The group G such that* $\mathrm{End}(G) \cong \mathbb{Z}[X]$, *constructed by means of Theorem 1, is a torsion-free* $\mathbb{Z}[X]$*-module, and* $\mathrm{rk}_{\mathbb{Z}[X]} G = \aleph_0$. *Furthermore, if ϕ is any endomorphism different by the multiplication by an integer,* $\widetilde{\mathrm{ent}}(T(\phi, g)_\phi) = \infty$ *for all $0 \neq g \in G$, and* $\widetilde{P}(G_\phi) = 0$.

Proof Assume that $c_0 1 + e_{a_1} c_1 + \cdots + e_{a_n} c_n = 0$, where $c_0, \ldots, c_n \in A$. Then Lemma 2.1 in [2] ensures that $c_0 = \cdots = c_n = 0$, therefore G contains $F = A \oplus \bigoplus_{a\in A} e_a A$, which is a torsion-free $\mathbb{Z}[X]$-module of countable rank. As G is the pure subgroup of $\hat{A}$ generated by F, the conclusion follows. Finally, $\widetilde{\mathrm{ent}}(\phi, g) = \infty$ for every $0 \neq g \in G$, since $T(\phi, g) \cong \mathbb{Z}[X]$ as proved in Proposition 2 (3), and, as noted in Sect. 2, the torsion-freeness of G as a $\mathbb{Z}[\phi]$-module implies that the Pinsker subgroup $\widetilde{P}(G_\phi)$ is trivial.

We pass now to consider Theorem 2; we shall show that the existence of a non-Hopfian group H with $\mathrm{End}(H) \cong \mathbb{Z}[X]$ gives rise to a completely different entropic behavior. In fact, Corner's Example 1 in [3] of a non-Hopfian group with automorphism group of order two provides such a realization for the polynomial ring $\mathbb{Z}[X]$ of Theorem 2. Our next objective is to generalize Corner's idea in that example.

Recall that Corner's construction yielded a subgroup H of a $\mathbb{Q}$-vector space V of countable dimension, having basis elements $a_k, b_k, c_n (k \in \mathbb{Z}, n \in \mathbb{N})$ and utilized five distinct primes p, q, r, s, t: specifically

$$H = \langle p^{-\infty}a_k, q^{-\infty}b_k, r^{-\infty}c_k, s^{-\infty}(a_k + b_k), t^{-\infty}(a_{k-1} + b_k + c_k), (k \in \mathbb{Z})\rangle$$

where $c_k = 0$ for $k < 0$ and, as usual, $p^{-\infty}x_k$ is an abbreviation for the set of elements $p^{-m}x_k$ $(m = 0, 1, 2, \dots)$.

In fact, Corner showed by direct calculation that every endomorphism of this group H was a polynomial in the endomorphism ω, where ω was the restriction to H of the linear transformation $\bar{\omega}$ of V defined by

$$\bar{\omega}(a_k) = a_{k-1}, \quad \bar{\omega}(b_k) = b_{k-1}, \quad \bar{\omega}(c_n) = c_{n-1} \ (k \in \mathbb{Z}, n \geq 1), \quad \bar{\omega}(c_0) = 0.$$

The notation used here agrees with that used before, since $\mathrm{End}(G) = \mathbb{Z}[\omega]$ is identified with $\mathbb{Z}[X]$ under the identification of ω with the multiplication induced by X. Observe a fundamental difference in the subgroups of H: each of the subgroups

$$\langle p^{-\infty}a_k \mid k \in \mathbb{Z}\rangle, \ \langle q^{-\infty}b_k \mid k \in \mathbb{Z}\rangle$$
$$\langle s^{-\infty}(a_k + b_k) \mid k \in \mathbb{Z}\rangle, \ \langle t^{-\infty}(a_{k-1} + b_k + c_k) \mid k \in \mathbb{Z}\rangle$$

is a torsion-free $\mathbb{Z}[\omega]$-module, while the subgroup $\langle r^{-\infty}c_k \mid k \in \mathbb{N}\rangle$ is a torsion $\mathbb{Z}[\omega]$-module. In an obvious, but somewhat *ad hoc*, terminology, we say that the group H is a $(4, 1)$ realization of $\mathbb{Z}[X]$ since the five families of generators give rise to 4 torsion-free $\mathbb{Z}[X]$-modules and 1 torsion $\mathbb{Z}[X]$-module. We continue to use the notation of Example 1 in [3] for the remainder of this section.

Now let W be a $\mathbb{Q}$-vector space containing V such that a basis for W extends that of V by independent elements $d_k (k \in \mathbb{Z})$. Extend $\bar{\omega}$ to a mapping $\bar{\phi}: W \to W$ with $\bar{\phi} \restriction V = \bar{\omega}$, and $\bar{\phi}(d_k) = d_{k-1}$. Since generators are preserved, $\bar{\phi}$ induces an endomorphism ϕ on the subgroup K of W given by

$$K = \langle H, u^{-\infty}d_k, v^{-\infty}(a_k + d_k)(k \in \mathbb{Z})\rangle,$$

where u, v are distinct primes different from all of p, q, r, s, t. It is worthwhile noting that the distinctness of the primes involved ensures that the subgroups H, $\langle u^{-\infty}d_k \mid k \in \mathbb{Z}\rangle$, $\langle v^{-\infty}(a_k + d_k) \mid k \in \mathbb{Z}\rangle$ are all fully invariant in K.

Let ψ be an arbitrary endomorphism of K. Since H is fully invariant in K, ψ induces an endomorphism of H and so in Corner's notation there are integers $\alpha_n^0 (0 \leq n \leq N)$ (for some fixed integer N) such that

$$\psi(a_k) = \sum_{n=0}^{N} \alpha_n^0 \omega^n(a_k) = \sum_{n=0}^{N} \alpha_n^0 \phi^n(a_k).$$

We claim that $\psi = \sum_{n=0}^{N} \alpha_n^0 \phi^n$. By full invariance $\psi(d_k) = \sum_h \mu_h^k d_h$ for some rationals μ_h^k; full invariance again yields that there are rationals ν_h^k such that $\psi(a_k + d_k) = \sum_h \nu_h^k (a_h + d_h)$. Substituting we get

$$(\alpha_0^0 a_k + \alpha_1^0 a_{k-1} + \cdots + \alpha_N^0 a_{k-N}) + \sum_h \mu_h^k d_h = \sum_h \nu_h^k (a_h + d_h),$$

and by linear independence we conclude that $\mu_h^k = \nu_h^k$ and that $\nu_h^k = 0$ unless $h = k, k-1, \ldots, k-N$, in which case $\nu_k^k = \alpha_0^0, \nu_{k-1}^k = \alpha_1^0, \ldots, \nu_{k-N}^k = \alpha_N^0$.

Hence

$$\psi(d_k) = \sum_h \mu_h^k d_h = \sum_h \nu_h^k d_h = (\alpha_0^0 d_k + \alpha_1^0 d_{k-1} + \cdots + \alpha_N^0 d_{k-N})$$

$$= (\alpha_0^0 + \alpha_1^0 \phi + \ldots, + \alpha_N^0 \phi^N)(d_k).$$

So ψ agrees with $\alpha_0^0 + \alpha_1^0 \phi + \cdots + \alpha_N^0 \phi^N$ on all the generators of K and so $\psi = \alpha_0^0 + \alpha_1^0 \phi + \cdots + \alpha_N^0 \phi^N$. Hence the claim is established. Since ψ was arbitrary, $\mathrm{End}(K) \leq \mathbb{Z}[\phi]$ and since the reverse inequality is trivially true, we conclude that $\mathrm{End}(K) = \mathbb{Z}[\phi]$.

Note that the above argument does not depend on d_k being 0 for $k < 0$ and consequently the new group K can be made into either a $(5, 1)$ (letting k range over $\mathbb{Z}$) or a $(4, 2)$ realization (setting $d_k = 0$ for $k < 0$) of $\mathbb{Z}[X]$. A simple induction argument now yields:

Theorem 5 *For each pair (m, n) of positive integers with $m \geq 4$, there is an (m, n) realization of the ring $\mathbb{Z}[X]$.*

The next proposition describes the structure as $\mathbb{Z}[X]$-module of the original group H constructed in Theorem 2 recalled above, and the intrinsic entropies of the relevant sections of H. We leave it to the interested reader to extend the description as $\mathbb{Z}[X]$-modules of the general groups arising in Theorem 5.

Proposition 4 *Let H be the non-Hopfian group constructed in Theorem 2, whose endomorphism ring is isomorphic to $\mathbb{Z}[X]$. Then H contains a chain of $\mathbb{Z}[X]$-submodules $0 < H_1 < H_2 < H_3 < H$ such that:*

(1) H_1 is a maximal free $\mathbb{Z}[X]$-submodule of rank 2, hence $\mathrm{rk}_{\mathbb{Z}[X]} H = 2$ and $\widetilde{\mathrm{ent}}(H_1) = \infty$;
(2) $H_2 = H_1 \oplus C_0$, where C_0 is a cyclic torsion $\mathbb{Z}[X]$-submodule of H such that $\widetilde{\mathrm{ent}}(C_0) = 0$;
(3) H_3/H_2 is the direct sum of three copies of a cocyclic torsion $\mathbb{Z}[X]$-module and $\widetilde{\mathrm{ent}}(H_3/H_2) = 0$;
(4) H/H_3 is a divisible torsion group such that $\widetilde{\mathrm{ent}}(H/H_3) = \infty$;

(5) $\widetilde{\text{ent}}(\phi) = \infty$ *for every endomorphism* ϕ *of* H *different from the multiplication by an integer;*
(6) *The intrinsic Pinsker subgroup* $\widetilde{\mathbf{P}}(H_\omega)$ *equals* $\langle r^{-\infty}c_k \mid k \in \mathbb{N}\rangle$.

Proof

(1) Denote by $\mathbb{Z}[X^{\pm 1}]$ the subring of the field $\mathbb{Q}(X)$ generated by X and X^{-1}. For a fixed integer n, the ω-trajectories of a_n and b_n are

$$A_n = T(\omega, a_n) = \bigoplus_{k \le n} \mathbb{Z}a_k \ ; \ B_n = T(\omega, b_n) = \bigoplus_{k \le n} \mathbb{Z}b_k$$

and are cyclic torsion-free $\mathbb{Z}[X]$-modules. Then $H_1 = A_0 \oplus B_0$ is a maximal free $\mathbb{Z}[X]$-submodule of H_ω, hence $\text{rk}_{\mathbb{Z}[X]} H_\omega = 2$ and $\widetilde{\text{ent}}(H_1) = \infty$, by Proposition 2.

(2) Fixed an integer $n \ge 0$, the ω-trajectory of c_n is

$$C_n = T(\omega, c_n) = \bigoplus_{0 \le k \le n} \mathbb{Z}c_n;$$

it is a cyclic torsion $\mathbb{Z}[X]$-module isomorphic to $\mathbb{Z}[X]/(X^{n+1})$; in particular, $C_0 \cong \mathbb{Z}[X]/(X)$. By Proposition 1, the intrinsic algebraic entropy of every $\mathbb{Z}[X]$-module $\mathbb{Z}[X]/(X^{n+1})$ is 0.

(3) Define $A = \cup_n A_n$, $B = \cup_n B_n$, $C = \cup_n C_n$. Evidently $A \cong \mathbb{Z}[X^{\pm 1}] \cong B$, and $C \cong \mathbb{Z}[X^{\pm 1}]/(X)$. Set $H_3 = A \oplus B \oplus C$ and note that H_3/H_2 is a torsion $\mathbb{Z}[X]$-module, which is a direct sum of three copies of the cocyclic $\mathbb{Z}[X]$-module $\mathbb{Z}[X^{\pm 1}]/(X)$. As the cocyclic modules are unions of cyclic torsion modules with intrinsic entropy 0, their intrinsic entropy is also 0, by the upper continuity of $\widetilde{\text{ent}}$.

(4) The group $\bar{H} = H/H_3$ is a torsion group, with primary decomposition $\bar{H} = \bar{H}_p \oplus \bar{H}_q \oplus \bar{H}_r \oplus \bar{H}_s \oplus \bar{H}_t$. The equality $\widetilde{\text{ent}}(H/H_3) = \infty$ depends on the fact, already recalled in the Introduction, that $\widetilde{\text{ent}} = \text{ent}$ for torsion groups by [10, Proposition 3.6] and, as it is easy to see, $\bar{H}_p$ is isomorphic to $\bigoplus_{\mathbb{Z}} \mathbb{Z}(p^\infty)$ endowed with the Bernoulli shift (i.e., the map that shifts one place to the right the coordinates of the direct sum), so, by [9, Proposition 1.17], $\text{ent}(\bar{H}_p) = \infty$, and similarly for the other primary components.

(5) $\widetilde{\text{ent}}(\omega) = \infty$ since $\widetilde{\text{ent}}(H_1) = \infty$; the same argument used to prove Proposition 2 (3) gives $\widetilde{\text{ent}}(\phi) = \infty$.

(6) The calculation of the intrinsic Pinsker subgroup follows from the observation that each element $x \in \langle r^{\infty}c_k \mid k \in \mathbb{N}\rangle$ is annihilated by some power of ω, hence $T(\omega, x) = T_n(\omega, x)$ for some n, while the other generators of H give rise to torsion-free $\mathbb{Z}[X]$-modules.

Remark 1 A comparison of Proposition 2 (3) and Proposition 4 (5) shows that, for a fixed isomorphism $\Phi\colon \mathrm{End}(G) \to \mathrm{End}(H)$ and an endomorphism $\phi \in \mathrm{End}(G) \cong \mathbb{Z}[X]$, the following equivalences hold:

$$\widetilde{\mathrm{ent}}(\phi) = 0 \Leftrightarrow \phi = n \cdot 1_G \ (n \in \mathbb{Z}) \Leftrightarrow \Phi(\phi) = n \cdot 1_H \Leftrightarrow \mathrm{ent}(\Phi(\phi)) = 0.$$

This fact is generalized to arbitrary countable groups in Corollary 3.

4 Groups Whose Endomorphism Ring is Isomorphic to $\mathbb{Z}[[X]]$

In this section we investigate the groups G such that $\mathrm{End}(G) \cong \mathbb{Z}[[X]]$. The proof of the next lemma follows *verbatim* that of Lemma 3.

Lemma 4 *Let G be a group such that* $\mathrm{End}(G) \cong \mathbb{Z}[[X]]$. *Then G is torsion-free, indecomposable, and reduced.*

Before investigating the structure as $\mathbb{Z}[X]$-modules of the torsion-free groups G and H with endomorphism ring isomorphic to $\mathbb{Z}[[X]]$, obtained by means of Theorem 3 and Theorem 4, respectively, we provide the proof of the latter theorem, as promised.

Proof (Proof of Theorem 4) We follow the notation used by Corner in his proof of Theorem 3 (Theorem 1.1 in [4]). For each $k = 1, 2, \ldots$, let $C_k = A/N_k$ and denote by e_k the element $1 + N_k \in A/N_k$; note that $\mathrm{Ann}_A(e_k) = \{a \in A \mid ae_k = 0\}$, the annihilator in A of e_k, is precisely N_k. Set $C = \bigoplus_k e_k A$, so that C is a direct sum of cyclic A-modules; note that C is countable, reduced, and torsion-free as an Abelian group; let $\hat{C}$ denote the completion of C in the natural topology, which is Hausdorff since C is reduced. The elements of $\hat{C}$ may be expressed uniquely as convergent (with respect to the natural topology) countable sums $\sum_k c_k$, where $c_k \in \hat{C}_k$. The set $\{k \mid c_k \neq 0\}$ is referred to as the support of the element $\sum_k c_k$.

Let P be the subring of $\hat{\mathbb{Z}}$ used in the proof of Theorem 1. Let $P(C)$ be a countable subring of P such that P is linearly disjoint from C over $P(C)$ – the existence of such follows from Section 2 in [2]. Since P has transcendence degree $2^{\aleph_0}$ over $P(C)$, we may choose elements α_c $(c \in C)$, β'_i $(i = 1, 2, \ldots)$ of P algebraically independent over $P(C)$; replacing the elements β'_i by the elements $\beta_i = (i!)\beta'_i$ we have that the elements α_c $(c \in C)$, β_i $(i = 1, 2, \ldots)$ of P remain algebraically independent and the β_i are now a convergent sequence of elements of $\hat{\mathbb{Z}}$. Recall that in Corner's notation the group $G = \langle C, \alpha_c(cA)\ (c \in C)\rangle_* \leq \hat{C}$ was the countable group whose finitely topologized endomorphism ring was isomorphic to A with the topology τ derived from the neighborhoods N_k.

Now set

$$H = \langle C, \alpha_c(cA)\ (c \in C), (\textstyle\sum_k \beta_k e_k)A\rangle_* \leq \hat{C}$$

so that $H = \langle G, (\sum_k \beta_k e_k)A\rangle_* \leq \hat{C}$. As in [4], one sees easily that G, H are both faithful A-modules and so A may be embedded in $\mathrm{End}(G)$, $\mathrm{End}(H)$.

Since H is dense with respect to the natural topology on $\hat{C}$, we can extend an arbitrary endomorphism ϕ of H to a unique $\hat{\mathbb{Z}}$-endomorphism of $\hat{C}$, continuing to write ϕ for this extended mapping. So we can find a positive integer q and elements $x, y \in C, x_d, y_d$ $(d \in C)$, $z, w \in A$, with almost all of the x_d, y_d zero, such that

$$q\phi(c) = y + \sum_d \alpha_d(dy_d) + \sum_k (\beta_k e_k)z \tag{2}$$

$$q\phi(\alpha_c c) = x + \sum_d \alpha_d(dx_d) + \sum_k (\beta_k e_k)w. \tag{3}$$

We derive from equations (2) and (3)

$$x + \sum_d \alpha_d(dx_d) + \sum_k (\beta_k e_k)w = \alpha_c(y + \sum_d \alpha_d(dy_d) + \sum_k (\beta_k e_k)z).$$

Re-arranging terms we deduce that

$$x + \sum_d \alpha_d(dx_d) - \alpha_c(y + \sum_d \alpha_d(dy_d)) = \alpha_c(\sum_k (\beta_k e_k)z) - \sum_k (\beta_k e_k)w. \tag{4}$$

Now the term on the left-hand side of equation (4) has finite support (as an element of $\hat{C}$) and thus the term on the right- hand side must likewise have finite support. Hence there exists an integer, n_0 say, such that $\alpha_c\beta_k e_k z - \beta_k e_k w = 0$ for all $k \geq n_0$. However, the $\alpha_c\beta_k$, β_k are linearly independent over $P(C)$ and so, by linear disjointness, $e_k z = 0 = e_k w$ for all $k \geq n_0$. Hence $z, w \in \bigcap_{k \geq n_0} N_k = \{0\}$.

It follows directly from equations (2) and (3) that for an arbitrary element $g \in G$, there is an integer q such that $q\phi(g) \in G \cap qH = qG$, since G is even pure in $\hat{C}$. By torsion-freeness, we conclude that $\phi(g) \in G$ and so G is fully invariant in H. It follows from Theorem 1.1 in [4] that there is an element $a \in A$ such that $\phi \restriction G$ coincides with scalar multiplication by a. Since scalar multiplication by a is also an endomorphism of H which agrees on the dense (in the natural topology, which is Hausdorff) subgroup G with ϕ, the endomorphism ϕ must agree with scalar multiplication by a on the whole of H. Hence $\mathrm{End}(H) \leq A$ and since the reverse inequality is trivially true, we get the algebraic identity $\mathrm{End}(H) = A$.

Finally observe that the finite topology on $\mathrm{End}(H)$ is discrete since the annihilator in A of the element $\sum_k (\beta_k e_k)$ is equal to $\bigcap_k N_k = \{0\}$.

We note the following straightforward corollary of Theorem 4, which largely motivated our interest in the last result (see Proposition 5 below).

Corollary 1 *There exists a reduced torsion-free group H of cardinality $2^{\aleph_0}$ such that* $(\mathrm{End}(H), \mathrm{fin}_H) \cong (\mathbb{Z}[[X]], \delta)$*, where δ denotes the discrete topology.*

We identify End(G) with $\mathbb{Z}[[X]]$. As in Sect. 3, the multiplication by X on G acts as a distinguished endomorphism ω, and every endomorphism of G is of the form $\phi = a_0 + a_1\omega + a_2\omega^2 + \cdots$ ($a_i \in \mathbb{Z}$).

The next two results investigate the cases when the topological ring $(\mathrm{End}(G), \mathrm{fin}_G) \cong (\mathbb{Z}[[X]], \chi)$ (where χ denotes the X-adic topology on $\mathbb{Z}[[X]]$) and when the topological ring $(\mathrm{End}(H), \mathrm{fin}_H) \cong (\mathbb{Z}[[X]], \delta)$.

First we deal with the discrete topology; in this case the situation is partly similar to that of Proposition 2. Recall that a group is $\aleph_1$-free if its countable subgroups are free.

Proposition 5 *Let H be a group such that* $(\mathrm{End}(H), \mathrm{fin}_H) \cong (\mathbb{Z}[[X]], \delta)$. *Then the following facts hold:*

(1) the cardinality of H is at least $2^{\aleph_0}$;
(2) there exist elements $g \in H$ such that $T(\omega, g)_\omega \cong \mathbb{Z}[X]$, so $\widetilde{\mathrm{ent}}(\omega) = \infty$;
(3) there exists elements $g \in H$ such that the orbit $\mathbb{Z}[[\omega]]g$ is an $\aleph_1$-free non-free group of rank $2^{\aleph_0}$;
(4) if H is constructed by means of Theorem 4, then $\mathrm{rk}_{\mathbb{Z}[X]} H = 2^{\aleph_0}$. Furthermore, the intrinsic Pinsker subgroup $\widetilde{P}(H_\omega)$ equals $\langle C, \alpha_c(c\mathbb{Z}[[X]])\rangle_*$.

Proof

(1) Since fin_H is the discrete topology, there exists an embedding of $\mathbb{Z}[[X]]$ into H^k for some positive integer k, hence $|H| \geq 2^{\aleph_0}$.
(2) The proof is the same as that of point (2) of Proposition 2.
(3) There exists a finite set F in H such that $K_F = 0$, hence $\mathbb{Z}[[\omega]]$ embeds into $\bigoplus_{g\in F} \mathbb{Z}[[\omega]]g$, where $\mathbb{Z}[[\omega]]g$ is the orbit of g. Therefore at least one $g \in F$ satisfies the property that $\mathbb{Z}[[\omega]]g$ has rank $2^{\aleph_0}$. This orbit $\mathbb{Z}[[\omega]]g$ is a subgroup of $\mathbb{Z}[[X]]$, which, as a group, is isomorphic to the Baer–Specker group $\prod_{\mathbb{N}} \mathbb{Z}$. Therefore $\mathbb{Z}[[\omega]]g$ is $\aleph_1$-free; it cannot be free, since it is a quotient of $\mathbb{Z}[[X]]$ and such quotients do not have free summands of infinite rank, by a result by Nunke [19] (see also [13, Proposition 95.2]).
(4) Recall that $H = \langle C, \alpha_c(cA)\ (c \in C), (\sum_k \beta_k e_k)A\rangle_* \leq \hat{C}$, where $A = \mathbb{Z}[[X]]$ and $C = \bigoplus_k \mathbb{Z}[[X]]/X^k\mathbb{Z}[[X]]$. Now C and $\alpha_c(cA)$ are torsion $\mathbb{Z}[X]$-modules, but $(\sum_k \beta_k e_k)A$ is isomorphic to A, since $\sum_k \beta_k e_k$ has zero annihilator. We conclude by remarking that $A = \mathbb{Z}[[X]]$ has rank $2^{\aleph_0}$ as $\mathbb{Z}[X]$-module. Finally, the intrinsic Pinsker subgroup is $\widetilde{P}(H_\omega) = \langle C, \alpha_c(c\mathbb{Z}[[X]])\rangle_*$, since every element $x \in \langle C, \alpha_c(c\mathbb{Z}[[X]])\rangle_*$ is annihilated by some power of ω, hence $T(\omega, x) = T_n(\omega, x)$ for some n, while every $y \in (\sum_k \beta_k e_k)A$ has $T(\omega, y) \cong \mathbb{Z}[X]$.

Now we consider the X-adic topology χ on $\mathrm{End}(G) = \mathbb{Z}[[X]]$; a basis of neighborhoods of zero is the countable family of ideals (X^n) ($n \in \mathbb{N}$), satisfying the condition that $\mathbb{Z}[[X]]/(X^n)$ is a free group of rank n for every n. Still denoting by ω the multiplication by X on G, we will say that G is *ω-torsion* if, for every $g \in G$, there exists a positive $n \in \mathbb{N}$ such that $\omega^n(g) = 0$, and that G is *ω-bounded* if there exists a positive $n \in \mathbb{N}$ such that $\omega^n(G) = 0$.

Proposition 6 *Let G be a group such that* $(\text{End}(G), \text{fin}_G) \cong (\mathbb{Z}[[X]], \chi)$. *Then the following facts hold:*

(1) G is ω*-torsion and not* ω*-bounded, hence it is a torsion* $\mathbb{Z}[X]$*-module.*
(2) for every $g \in G$*, the orbit* $\mathbb{Z}[[\omega]]g$ *equals the trajectory* $T(\omega, g)$*, which is a free group of finite rank.*
(3) $\widetilde{\text{ent}}(\psi) = 0$ *for every endomorphism* ψ *of G and the intrinsic Pinsker subgroup* $\widetilde{P}(G_\psi)$ *equals G.*

Proof

(1) For an arbitrary fixed $g \in G$, K_g is open in χ, hence $K_g \geq (X^n)$ for some n; therefore $\omega^n(g) = 0$ and G is ω-torsion. However, G cannot be ω-bounded: suppose for a contradiction that $\omega^n(G) = 0$ for some n. From the topological ring isomorphism we deduce that there is a finite subset F of G such that $\text{Ann}_{\mathbb{Z}[[X]]}(F) = X^{n+1}\mathbb{Z}[[X]]$ and since $\omega^n(G) = 0$, we must have that $\omega^n \in \text{Ann}_{\mathbb{Z}[[X]]}(F)$, which forces $X^n \in X^{n+1}\mathbb{Z}[[X]]$ – impossible.
(2) By (1), there exists a positive integer n such that $\omega^n(g) = 0$; then, given $\phi = \sum_0^\infty a_i\omega^i \in \mathbb{Z}[[\omega]]$, we have $\phi(g) = (a_0 + a_1\omega + a_2\omega^2 + \cdots + a_{n-1}\omega^{n-1})(g) \in T(\omega, g)$; therefore $\mathbb{Z}[[\omega]]g = T(\omega, g)$. Since $\mathbb{Z}[[\omega]]g \cong \mathbb{Z}[[\omega]]/K_g$ is an epimorphic image of $\mathbb{Z}[[\omega]]/(\omega^n) \cong \mathbb{Z}^n$, $T(\omega, g)$ is a finitely generated torsion-free group, hence it is free of finite rank.
(3) Let $\psi: G \to G$ be an endomorphism of G; we must prove that $\widetilde{\text{ent}}(\psi) = 0$. As mentioned in Sect. 2, in view of the upper continuity and the Addition Theorem, which hold for $\widetilde{\text{ent}}$ (see [10]), it is enough to show that $\widetilde{\text{ent}}(T(\psi, g)_\psi) = 0$ for all $g \in G$. But $T(\psi, g)$ is a free group of finite rank, since $T(\psi, g)$ is a subgroup of $\mathbb{Z}[[\omega]]g$, so $\widetilde{\text{ent}}(\psi) = 0$ by Proposition 1. This also shows that $\widetilde{P}(G_\psi) = G$.

It is worthwhile remarking that there exist groups G satisfying the hypothesis of Proposition 6 of any cardinality $|G| = \lambda$ such that $\lambda = \lambda^{\aleph_0}$ (see [5]); Theorem 3 ensures that such a group G exists also of cardinality $|G| = \aleph_0$. Taking this into account and summarizing the results obtained in this section, we obtain a result that has clear similarities to that obtained in [15, Corollary 3.4].

Theorem 6 *There exist reduced torsion-free groups G, H such that G is countable, H has rank* $2^{\aleph_0}$ *as* $\mathbb{Z}[X]$*-module,* $\text{End}(G) \cong \mathbb{Z}[[X]] \cong \text{End}(H)$*, but* $\widetilde{\text{ent}}(\psi) = 0$ *for every endomorphism* ψ *of G, while* $\widetilde{\text{ent}}(\omega) = \infty$*, where* ω *is the endomorphism of H corresponding to the multiplication by X.*

5 Topological Isomorphisms Between Endomorphism Rings

The proof of Theorem 7 is partly similar to the proof of Proposition 3.3 in [21], showing that a topological isomorphism between the endomorphism rings of two groups endowed with the finite topology forces similar entropic behavior of the endomorphisms of the two groups. However, in the present situation we must use different arguments from those used in Proposition 3.3 in [21], which make use of the fact that $\widetilde{\text{ent}}$ is additive on Mod($\mathbb{Z}[X]$).

Theorem 7 *Let G and H be two groups and* $\Phi\colon (\text{End}(G), \text{fin}_G) \to (\text{End}(H), \text{fin}_H)$ *an isomorphism of topological rings. If* $\phi \in \text{End}(G)$*, then* $\widetilde{\text{ent}}(\phi) = 0$ *if and only if* $\widetilde{\text{ent}}(\Phi(\phi)) = 0$.

Proof By symmetry, it is enough to prove that $\widetilde{\text{ent}}(\phi) = 0$ implies $\widetilde{\text{ent}}(\Phi(\phi)) = 0$. Set $\psi = \Phi(\phi)$. It is enough to show that $\widetilde{\text{ent}}(\psi \restriction T_x) = 0$ for all $x \in H$, where $T_x = T(\psi, x)$. As Φ^{-1} is an homeomorphism, $\Phi^{-1}(K_x)$ is open in fin_G, therefore $\Phi^{-1}(K_x)$ contains K_F for F a suitable finitely generated subgroup of G. Now we have, by Lemma 1:

$$T(\psi, x\mathbb{Z}) \cong \mathbb{Z}[\psi]/(\mathbb{Z}[\psi] \cap K_x) \cong \mathbb{Z}[\phi]/(\mathbb{Z}[\phi] \cap \Phi^{-1}(K_x)).$$

But $\mathbb{Z}[\phi]/(\mathbb{Z}[\phi] \cap \Phi^{-1}(K_x))$ is a quotient of $\mathbb{Z}[\phi]/(\mathbb{Z}[\phi] \cap K_F)$, so the additivity of $\widetilde{\text{ent}}$ ensures that

$$\widetilde{\text{ent}}(\mathbb{Z}[\phi]/(\mathbb{Z}[\phi] \cap \Phi^{-1}(K_x))) \leq \widetilde{\text{ent}}(\mathbb{Z}[\phi]/(\mathbb{Z}[\phi] \cap K_F)).$$

Now Lemma 2 ensures that if $F = \sum_{1 \leq i \leq n} g_i\mathbb{Z}$, then $\mathbb{Z}[\phi]/(\mathbb{Z}[\phi] \cap K_F)$ is embedded into $\bigoplus_{1 \leq i \leq n} T(\phi, g_i)$. But $\widetilde{\text{ent}}(\phi) = 0$ implies that $\widetilde{\text{ent}}(T(\phi, g_i) = 0$ for all $i \leq n$, therefore we deduce that $\widetilde{\text{ent}}(\mathbb{Z}[\phi]/(\mathbb{Z}[\phi] \cap K_F)) = 0$ and consequently $\widetilde{\text{ent}}(\psi \restriction T_x) = 0$, as desired.

As an immediate application of Theorem 7 and Proposition 2 (1), we get a result that we already pointed out in Remark 1.

Corollary 2 *Let G and H be two torsion-free groups such that* $\text{End}(G) \cong \mathbb{Z}[X] \cong \text{End}(H)$*. Let* $\Phi\colon \text{End}(G) \to \text{End}(H)$ *be an isomorphism. Then, given* $\phi \in \text{End}(G)$, $\widetilde{\text{ent}}(\phi) = 0$ *if and only if* $\widetilde{\text{ent}}(\Phi(\phi)) = 0$.

We will apply Theorem 7 to arbitrary countable reduced torsion-free groups. First we demonstrate a result that was only stated by Corner in his paper [4].

Theorem 8 *If G, H are countable reduced torsion-free groups with* $\text{End}(G) \cong \text{End}(H)$*, then* $(\text{End}(G), \text{fin}_G) \cong (\text{End}(H), \text{fin}_H)$.

Proof Enumerate the elements of G and H as $\{g_1, g_2, \ldots, g_n, \ldots\}$ and $\{h_1, h_2, \ldots, h_m, \ldots\}$, respectively. For $n, m \in \mathbb{N}$, set $G_n = \{g_1, g_2, \ldots, g_n\}$ and $H_m = \{h_1, h_2, \ldots, h_m\}$, so that the left ideals $N_n = \{\phi \in \text{End}(G) \mid \phi(G_n) = 0\}$ and $M_m = \{\psi \in \text{End}(H) \mid \psi(H_m) = 0\}$ form subbases of 0 in the respective topologies $\text{fin}_G, \text{fin}_H$. Denote the (algebraic) isomorphism between $\text{End}(G)$ and $\text{End}(H)$ by Φ and its inverse by Ψ.

Observe firstly that $\text{End}(G)$ contains a subgroup P isomorphic to the direct product $P \cong \prod_{i=1}^{\infty} N_i$. To see this, note that for an arbitrary element g of G, there is an integer n with $g = g_n$ and so if $i \geq n$, $\phi_i(g_n) = 0$ since $g_n \in G_i$. Hence the sum $(\sum_{i=1}^{\infty} \phi_i)(g_n)$ reduces to $(\sum_{i=1}^{n-1} \phi_i)(g_n)$ and so the element $\sum_{i=1}^{\infty} \phi_i$ is a well-defined endomorphism of G. Regarding elements of the direct product $\prod_{i=1}^{\infty} N_i$ as formal sums $\sum_{i=1}^{\infty} \phi_i$, we obtain the desired subgroup $P \cong \prod_{i=1}^{\infty} N_i$.

Let η_m be the canonical projection $\mathrm{End}(H) \to \mathrm{End}(H)/M_m$, so that the composition $\eta_m \Phi \restriction P$ maps the group P into the group $\mathrm{End}(H)/M_m$. Since M_m is the kernel of the homomorphism $\mathrm{End}(H) \to H^m$ given by $\psi \mapsto (\psi(x))_{x \in M_m}$, the group $\mathrm{End}(H)/M_m$ is a subgroup of a finite direct sum of countable reduced torsion-free groups; in particular $\mathrm{End}(H)/M_m$ is *slender* (for the notion of slenderness we refer to [11, Chapter III], [13, §94] or [19]).

So for each integer m, there exists an integer k_m such that $\eta_m \Phi(\prod_{i=k_m}^{\infty} N_i) = 0$ – this is a standard property of slender groups, see, for example, Theorem 1.2 in [11, Chapter III]. Thus $\Phi(\prod_{i=k_m}^{\infty} N_i) \leq \ker(\eta_m) = M_m$ for each m. In particular $\Phi(N_{k_m}) \leq M_m$ and so $N_{k_m} \leq \Psi(M_m)$. Thus Ψ is an open mapping with respect to the finite topologies on $\mathrm{End}(H)$ and $\mathrm{End}(G)$.

An identical argument replacing $P \cong \prod_{i=1}^{\infty} N_i$ with $Q \cong \prod_{i=1}^{\infty} M_i$, shows that Φ is open with respect to the finite topologies on $\mathrm{End}(G)$ and $\mathrm{End}(H)$ and so Φ and Ψ are inverse topological isomorphisms, as required.

Note that the proof of Theorem 8 actually shows that any algebraic isomorphism between the endomorphism rings of two countable, reduced, and torsion-free groups is in fact a homeomorphism with respect to the finite topologies.

Now an immediate application of Theorem 7 and Theorem 8 gives the following:

Corollary 3 *Let G and H be two countable reduced torsion-free groups such that* $\mathrm{End}(G) \cong \mathrm{End}(H)$. *Let* $\Phi\colon \mathrm{End}(G) \to \mathrm{End}(H)$ *be an isomorphism. Then, given* $\phi \in \mathrm{End}(G)$, $\widetilde{\mathrm{ent}}(\phi) = 0$ *if and only if* $\widetilde{\mathrm{ent}}(\Phi(\phi)) = 0$.

Acknowledgements The authors would like to thank Simone Virili for the careful reading of a first draft of the paper and the anonymous referee for useful comments and suggestions.

Research supported by "Progetti di Eccellenza 2011/12" of Fondazione CARIPARO.

References

1. Adler, R.L., Konheim, A.G., McAndrew, M.H.: Topological entropy. Trans. Am. Math. Soc. **114**, 309–319 (1965)
2. Corner, A.L.S.: Every countable reduced torsion-free ring is an endomorphism ring. Proc. Lond. Math. Soc. **s3–13**, 687–710 (1963)
3. Corner, A.L.S.: Three examples on Hopficity on torsion-free Abelian groups. Acta Math. Acad. Sci. Hungar. **16**, 304–310 (1965)
4. Corner, A.L.S.: Endomorphism rings of torsion-free abelian groups. In: Proceedings of International Conference Theory of Groups, Canberra 1965, Gordon and Breach, London, pp. 59–69 (1967)
5. Corner, A.L.S., Göbel, R.: Prescribing endomorphism algebras—a unified treatment. Proc. London. Math. Soc. **s3–50**, 447–479 (1985)
6. De Marco, G., Orsatti, A.: Complete linear topologies on abelian groups. Symposia Math. **XIII**, 153–161 (1974)
7. Dikranjan, D., Giordano Bruno, A.: The Pinsker subgroup of an algebraic flow. J. Pure Appl. Algebra **216**, 364–376 (2012)

8. Dikranjan, D., Giordano Bruno, A.: Entropy in Abelian groups. Adv. Math. **298**, 612–653 (2016)
9. Dikranjan, D., Goldsmith, B., Salce, L., Zanardo, P.: Algebraic entropy for Abelian groups. Trans. Am. Math. Soc. **361**, 3401–3434 (2009)
10. Dikranjan, D., Giordano Bruno, A., Salce, L., Virili, S.: Intrinsic algebraic entropy. J. Pure Appl. Algebra **219**(7), 2933–2961 (2015)
11. Eklof, P., Mekler, A.: Almost Free Modules — Set-Theoretic Methods. North-Holland, Amsterdam (2002)
12. Fekete, M.: Über die Verteilung der Wurzeln bei gewisser algebraicher Gleichungen mit ganzzahlingen Közienten. Math. Zeitschr. **17**, 228–249 (1923)
13. Fuchs, L.: Infinite Abelian Groups, Vol. II. Academic Press, New York (1973)
14. Giordano Bruno, A., Virili, S.: About the algebraic Yuzvinski formula. Topol. Algebra Appl. **3**, 86–103 (2015)
15. Göbel, R., Salce, L.: Endomorphism rings with different rank-entropy supports. Quart. J. Math. **63**(2), 381–397 (2012)
16. Göbel, R., Trlifaj, J.: Approximations and Endomorphism Algebras of Modules. Volume 2. Predictions. Second revised and extended edition. de Gruyter Expositions in Mathematics, 41. de Gruyter, Berlin (2012)
17. Goldsmith, B., Salce, L.: When the intrinsic algebraic entropy is not really intrinsic. Topol. Algebra Appl. **3**, 45–56 (2015)
18. Goldsmith, B., Strüngmann, L.: Torsion-free weakly transitive Abelian groups. Commun. Algebra **33**, 1177–1191 (2005)
19. Nunke, R.J.: Slender groups. Acta Sci. Math. Szeged **23**, 67–73 (1962)
20. Peters, J.: Entropy on discrete Abelian groups. Adv. Math. **33**, 1–13 (1979)
21. Salce, L.: Some results on the algebraic entropy. In: Proceedings of the 2011 Muelheim Conference, Contemporary Mathematics, vol. 576, pp. 297–304. AMS (2012)
22. Salce, L., Zanardo, P.: A general notion of algebraic entropy and the rank-entropy. Forum Math. **21**(4), 579–599 (2009)
23. Salce, L., Vámos, P., Virili, S.: Length functions, multiplicities and algebraic entropies. Forum Math. **25**(2), 255–282 (2013)

Directed Unions of Local Quadratic Transforms of Regular Local Rings and Pullbacks

Lorenzo Guerrieri, William Heinzer, Bruce Olberding, and Matthew Toeniskoetter

Abstract Let $\{R_n, \mathfrak{m}_n\}_{n\geq 0}$ be an infinite sequence of regular local rings with R_{n+1} birationally dominating R_n and $\mathfrak{m}_n R_{n+1}$ a principal ideal of R_{n+1} for each n. We examine properties of the integrally closed local domain $S = \bigcup_{n\geq 0} R_n$.

Keywords Regular local ring • Local quadratic transform • Valuation ring • Pullback construction

Mathematics Subject Classification (2010) 13H05, 13A15, 13A18

1 Introduction

Let R be a regular local ring with maximal ideal $\mathfrak{m} = (x_1, \dots, x_n)R$, where $n = \dim R$ is the Krull dimension of R. Choose $i \in \{1, \dots, n\}$, and consider the overring $R[\frac{x_1}{x_i}, \dots, \frac{x_n}{x_i}]$ of R. Choose any prime ideal P of $R[\frac{x_1}{x_i}, \dots, \frac{x_n}{x_i}]$ that contains $\mathfrak{m}$. Then the ring $R_1 := R[\frac{x_1}{x_i}, \dots, \frac{x_n}{x_i}]_P$ is a *local quadratic transform* of R, R_1 is again a regular local ring and $\dim R_1 \leq n$. Iterating the process we obtain a sequence $R = R_0 \subseteq R_1 \subseteq R_2 \subseteq \cdots$ of regular local overrings of R such that for each i, R_{i+1} is a local quadratic transform of R_i. The sequence of positive integers $\{\dim R_i\}_{i\in\mathbb{N}}$ stabilizes, and $\dim R_i = \dim R_{i+1}$ for all sufficiently large i. If $\dim R_i = 1$, then necessarily $R_i = R_{i+1}$, while if $\dim R_i \geq 2$, then $R_i \subsetneq R_{i+1}$.

L. Guerrieri (✉)
Università di catania, Viale Andrea Doria 6, 95125 Catania, Italy
e-mail: guerrieri@dmi.unict.it

W. Heinzer • M. Toeniskoetter
Department of Mathematics, Purdue University, West Lafayette, IN 47907, USA
e-mail: heinzer@purdue.edu; mtoenisk@purdue.edu

B. Olberding
Department of Mathematical Sciences, New Mexico State University,
Las Cruces, NM 88003-8001, USA
e-mail: olberdin@nmsu.edu

M. Fontana et al. (eds.), *Rings, Polynomials, and Modules*,
https://doi.org/10.1007/978-3-319-65874-2_13

The process of iterating local quadratic transforms of the same Krull dimension is the algebraic expression of following a closed point through a sequence of blow-ups of a nonsingular point of an algebraic variety, with each blow-up occurring at a closed point in the fiber of the previous blow-up. This geometric process plays a central role in embedded resolution of singularities for curves on surfaces (see, for example, [2] and [6, Sections 3.4 and 3.5]), as well as factorization of birational morphisms between nonsingular surfaces ([1, Theorem 3] and [33, Lemma, p. 538]). These applications depend on properties of iterated sequences of local quadratic transforms of a two-dimensional regular local ring. For a two-dimensional regular local ring R, Abhyankar [1, Lemma 12] shows that the limit of this process of iterating local quadratic transforms $R = R_0 \subseteq R_1 \subseteq R_2 \subseteq \cdots$ results in a valuation ring that birationally dominates R; i.e., $\mathcal{V} = \bigcup_{i=0}^{\infty} R_i$ is a valuation ring with the same quotient field as R and the maximal ideal of $\mathcal{V}$ contains the maximal ideal of R.

Moving beyond dimension two, examples due to David Shannon [30, Examples 4.7 and 4.17] show that the union $S = \bigcup_i R_i$ of an iterated sequence of local quadratic transforms of a regular local ring of Krull dimension > 2 need not be a valuation ring. The recent articles [21, 22] address the structure of such rings S and how this structure encodes asymptotic properties of the sequence $\{R_i\}_{i=0}^{\infty}$. We call S a *quadratic Shannon extension* of R. In general, a quadratic Shannon extension need not be a valuation ring nor a Noetherian ring, although it is always an intersection of two such rings (see Theorem 2.2).

The class of quadratic Shannon extensions separates naturally into two cases, the archimedean and non-archimedean cases. A quadratic Shannon extension S is non-archimedean if there is an element x in the maximal ideal of S such that $\bigcap_{i>0} x^i S \neq 0$. The class of non-archimedean quadratic Shannon extensions is analyzed in detail in [21] and [22]. We carry this analysis further in the present article by using techniques from multiplicative ideal theory to classify a non-archimedean quadratic Shannon extension as the pullback of a valuation ring of rational rank one along a homomorphism from a regular local ring onto its residue field. We present several variations of this classification in Lemma 4.3 and Theorems 4.8 and 5.1.

The pullback description leads in Theorem 4.5 to existence results for both archimedean and non-archimedean quadratic Shannon extensions contained in a localization of the base ring at a nonmaximal prime ideal. As another application, in Theorem 5.2 we use pullbacks to characterize the quadratic Shannon extensions S of regular local rings R such that R is essentially finitely generated over a field of characteristic 0 and S has a principal maximal ideal.

That non-archimedean quadratic Shannon extensions occur as pullbacks is also useful because of the extensive literature on transfer properties between the rings in a pullback square. In Sect. 6 we use the pullback classification along with structural results for archimedean quadratic Shannon extensions from [21] to show in Theorem 6.2 that a quadratic Shannon extension is coherent if and only if it is a valuation domain.

Our methods sometimes involve local quadratic transforms of Noetherian local domains that need not be regular local rings. To formalize these notions as well as those mentioned above, let $(R, \mathfrak{m})$ be a Noetherian local domain and let $(V, \mathfrak{m}_V)$ be a valuation domain birationally dominating R. Then $\mathfrak{m}V = xV$ for some $x \in \mathfrak{m}$. The ring $R_1 = R[\mathfrak{m}/x]_{\mathfrak{m}_V \cap R[\mathfrak{m}/x]}$ is called a *local quadratic transform* (LQT) of R along V. The ring R_1 is a Noetherian local domain that dominates R with maximal ideal $\mathfrak{m}_1 = \mathfrak{m}_V \cap R_1$. Since V birationally dominates R_1, we may iterate this process to obtain an infinite sequence $\{R_n\}_{n \geq 0}$ of LQTs of $R_0 = R$ along V. If $R_n = V$ for some n, then V is a DVR and the sequence stabilizes with $R_m = V$ for all $m \geq n$. Otherwise, $\{R_n\}$ is an infinite strictly ascending sequence of Noetherian local domains.

If R is a regular local ring (RLR), it is well known that R_1 is an RLR; cf. [28, Corollary 38.2]. Moreover, $R = R_1$ if and only if $\dim R \leq 1$. Assume that R is an RLR with $\dim R \geq 2$ and V is minimal as a valuation overring of R. Then $\dim R_1 = \dim R$, and the process may be continued by defining R_2 to be the LQT of R_1 along V. Continuing the procedure yields an infinite strictly ascending sequence $\{R_n\}_{n \in \mathbb{N}}$ of RLRs all dominated by V.

In general, our notation is as in Matsumura [26]. Thus a local ring need not be Noetherian. An element x in the maximal ideal $\mathfrak{m}$ of a regular local ring R is said to be a *regular parameter* if $x \notin \mathfrak{m}^2$. It then follows that the residue class ring R/xR is again a regular local ring. We refer to an extension ring B of an integral domain A as an *overring of* A if B is a subring of the quotient field of A. If, in addition, A and B are local and the inclusion map $A \hookrightarrow B$ is a local homomorphism, we say that B *birationally dominates* A. We use UFD as an abbreviation for unique factorization domain, and DVR as an abbreviation for rank 1 discrete valuation ring. If P is a prime ideal of a ring A, we denote by $\kappa(P)$ the residue field A_P/PA_P of A_P.

2 Quadratic Shannon Extensions

Let $(R, \mathfrak{m})$ be a regular local ring with $\dim R \geq 2$ and let F denote the quotient field of R. David Shannon's work in [30] on sequences of quadratic and monoidal transforms of regular local rings motivates our terminology quadratic Shannon extension in Definition 2.1.

Definition 2.1 Let $(R, \mathfrak{m})$ be a regular local ring with $\dim R \geq 2$. With $R = R_0$, let $\{R_n, \mathfrak{m}_n\}$ be an infinite sequence of RLRs, where $\dim R_n \geq 2$ for each n. If R_{n+1} is an LQT of R_n for each n, then the ring $S = \bigcup_{n \geq 0} R_n$ is called a *quadratic Shannon extension*[1] of R.

If $\dim R = 2$, then the quadratic Shannon extensions of R are precisely the valuation rings that birationally dominate R and are minimal as a valuation overring of R [1, Lemma 12]. If $\dim R > 2$, then, examples due to Shannon [30, Examples 4.7

[1] In [21] and [22], the authors call S a Shannon extension of R. We have made a distinction here with monoidal transforms. Since $\dim R_n \geq 2$, we have $R_n \subsetneq R_{n+1}$ for each positive integer n and $\bigcup_n R_n$ is an infinite ascending union.

and 4.17] show that there are quadratic Shannon extensions that are not valuation rings. Similarly, if $\dim R > 2$, then there are valuations rings V that birationally dominate R with V minimal as a valuation overring of R, but V is not a Shannon extension of R. Indeed, if V has rank > 2, then V is not a quadratic Shannon extension of R; see [13, Proposition 7].

These observations raise the question of the ideal-theoretic structure of a quadratic Shannon extension of a regular local ring R with $\dim R > 2$, a question that was taken up in [21] and [22]. In this section we recall some of the results from [21] and [22] with special emphasis on non-archimedean quadratic Shannon extensions, a class of Shannon extensions that we classify in Sects. 3 and 6.

To each quadratic Shannon extension there is an associated collection of rank 1 discrete valuation rings. Let $S = \bigcup_{i \geq 0} R_i$ be a quadratic Shannon extension of $R = R_0$. For each i, let V_i be the DVR defined by the *order function* ord_{R_i}, where for $x \in R_i$, $\operatorname{ord}_{R_i}(x) = \sup\{n \mid x \in \mathfrak{m}_i^n\}$ and ord_{R_i} is extended to the quotient field of R_i by defining $\operatorname{ord}_{R_i}(x/y) = \operatorname{ord}_{R_i}(x) - \operatorname{ord}_{R_i}(y)$ for all $x, y \in R_i$ with $y \neq 0$. The family $\{V_i\}_{i=0}^{\infty}$ determines a unique set

$$V = \bigcup_{n \geq 0} \bigcap_{i \geq n} V_i = \{a \in F \mid \operatorname{ord}_{R_i}(a) \geq 0 \text{ for } i \gg 0\}.$$

The set V consists of the elements in F that are in all but finitely many of the V_i. In [21, Corollary 5.3], the authors prove that V is a valuation domain that birationally dominates S, and call V the *boundary valuation ring* of the Shannon extension S.

Theorem 2.2 records properties of a quadratic Shannon extension.

Theorem 2.2 *[21, Theorems 4.1, 5.4 and 8.1] Let $(S, \mathfrak{m}_S)$ be a quadratic Shannon extension of a regular local ring R. Let T be the intersection of all the DVR overrings of R that properly contain S, and let V be the boundary valuation ring of S. Then:*

(1) $\dim S = 1$ *if and only if S is a rank 1 valuation ring.*
(2) $S = V \cap T$.
(3) There exists $x \in \mathfrak{m}_S$ such that xS is $\mathfrak{m}_S$-primary, and $T = S[1/x]$ for any such x. It follows that the units of T are precisely the ratios of $\mathfrak{m}_S$-primary elements of S and $\dim T = \dim S - 1$.
(4) T is a localization of R_i for $i \gg 0$. In particular, T is a Noetherian regular UFD.
(5) T is the unique minimal proper Noetherian overring of S.

In light of item 5 of Theorem 2.2, the ring T is called the *Noetherian hull* of S.

The boundary valuation ring is given by a valuation from the nonzero elements of the quotient field of R to a totally ordered abelian group of rank at most 2 [22, Theorem 6.4 and Corollary 8.6]. In [22] the following two mappings on the quotient field of R are introduced as invariants of a quadratic Shannon extension. The first, e, takes values in $\mathbb{Z} \cup \{\infty\}$, while the second, w, takes values in $\mathbb{R} \cup \{-\infty, +\infty\}$. Both e and w are used in [22] to decompose the boundary valuation v of the quadratic Shannon extension into a function that takes its values in $\mathbb{R} \oplus \mathbb{R}$ with the lexicographic ordering. The function e is defined in terms of the transform $(aR_n)^{R_{n+i}}$ of a principal ideal aR_n in R_{n+i} for $i > n$; see [25] for the general definition of the (weak) transform of an ideal and [21] for more on the properties of the transform in our setting.

Definition 2.3 Let $S = \bigcup_{i \geq 0} R_i$ be a quadratic Shannon extension of a regular local ring R.

(1) Let $a \in S$ be nonzero. Then $a \in R_n$ for some $n \geq 0$. Define

$$\mathrm{e}(a) = \lim_{i \to \infty} \mathrm{ord}_{n+i}((aR_n)^{R_{n+i}}).$$

For a, b nonzero elements in S, let $n \in \mathbb{N}$ be such that $a, b \in R_n$ and define $\mathrm{e}(\frac{a}{b}) = \mathrm{e}(a) - \mathrm{e}(b)$. That e is well defined is given by [22, Lemma 5.2].

(2) Fix $x \in S$ such that xS is primary for the maximal ideal of S, and define

$$w : F \to \mathbb{R} \cup \{-\infty, +\infty\}$$

by defining $w(0) = +\infty$, and for each $q \in F^{\times}$,

$$w(q) = \lim_{n \to \infty} \frac{\mathrm{ord}_n(q)}{\mathrm{ord}_n(x)}.$$

The structure of Shannon extensions naturally separate into those that are archimedean and those that are non-archimedean as in the following definition.

Definition 2.4 An integral domain A is *archimedean* if $\bigcap_{n>0} a^n A = 0$ for each nonunit $a \in A$.

An integral domain A with $\dim A \leq 1$ is archimedean. A valuation ring V with $\dim V \geq 2$ is non-archimedean.

Theorem 2.5, which characterizes quadratic Shannon extensions in several ways, shows that there is a prime ideal Q of a non-archimedean quadratic Shannon extension S such that S/Q is a rational rank one valuation ring and Q is a prime ideal of the Noetherian hull T of S. In the next section this fact serves as the basis for the classification of non-archimedean quadratic Shannon extensions via pullbacks.

Theorem 2.5 *Let $S = \bigcup_{n \geq 0} R_n$ be a quadratic Shannon extension of a regular local ring R with quotient field F, and let x be an element of S that is primary for the maximal ideal $\mathfrak{m}_S$ of S (see Theorem 2.2). Assume that* $\dim S \geq 2$. *Let* $Q =$

$\bigcap_{n\geq 1} x^n S$, and let $T = S[1/x]$ be the Noetherian hull of S. Then the following are equivalent:

(1) S is non-archimedean.
(2) $T = (Q :_F Q)$.
(3) Q is a nonzero prime ideal of S.
(4) Every nonmaximal prime ideal of S is contained in Q.
(5) T is a (regular) local ring.
(6) $\sum_{n=0}^{\infty} w(\mathfrak{m}_n) = \infty$, where w is as in Definition 2.3 and $\mathfrak{m}_n$ is the maximal ideal of R_n for each $n \geq 0$.

Moreover if (1)–(6) hold for S and Q, then $T = S_Q$, $Q = QS_Q$ is a common ideal of S and T, and S/Q is a rational rank 1 valuation domain on the residue field T/Q of T. In particular, Q is the unique maximal ideal of T.

Proof The equivalence of items 1–5 can be found in [22, Theorem 8.3]. That statement 1 is equivalent to 6 follows from [22, Theorem 6.1]. To prove the moreover statement, define $Q_\infty = \{a \in S \mid w(a) = +\infty\}$, where w is as in Definition 2.3. By [22, Theorem 8.1], Q_∞ is a prime ideal of S and T, and by [22, Remark 8.2], Q_∞ is the unique prime ideal of S of dimension 1. Since also item 4 implies every nonmaximal prime ideal of S is contained in Q, it follows that $Q = Q_\infty$. By item 5, $T = S[1/x]$ is a local ring. Since xS is $\mathfrak{m}_S$-primary, we have that $T = S_Q$. Since Q is an ideal of T, we conclude that $QS_Q = Q$ and Q is the unique maximal ideal of T. By [22, Corollary 8.4], S/Q is a valuation domain, and by [22, Theorem 8.5], S/Q has rational rank 1. □

Remark 2.6 If statements (1) – (6) hold for S, then Theorem 2.5 and [11, Theorem 2.3] imply that any principal ideal of S that is primary for $\mathfrak{m}_S$ is comparable to every other ideal of S with respect to set inclusion. Conversely, if a Shannon extension S has a principal ideal that is primary for $\mathfrak{m}_S$ and is comparable to every other ideal of S, then by [11, Theorem 2.3], S satisfies statement 3 of Theorem 2.5, and hence S decomposes as in the statement of Theorem 2.5.

We can further separate the case where S is archimedean to whether or not S is completely integrally closed. We recall the definition and result.

Definition 2.7 Let A be an integral domain. An element x in the field of fractions of A is called *almost integral* over A if $A[x]$ is contained in a principal fractional ideal of A. The ring A is called *completely integrally closed* if it contains all of the almost integral elements over it.

For a Noetherian domain, an element of the field of fractions is almost integral if and only if it is integral.

Theorem 2.8 *[21, Theorems 6.1, 6.2] Let S be an archimedean quadratic Shannon extension. Then the function w as in Definition 2.3 is a rank* 1 *nondiscrete valuation. Its valuation ring W is the rank* 1 *valuation overring of V and W also dominates S. The following are equivalent:*

(1) S is completely integrally closed.

(2) The boundary valuation V has rank 1; that is, $V = W$.

In Theorem 2.9 we recall from [22] the decomposition of the boundary valuation of a non-archimedean quadratic Shannon extension in terms of the functions w and e in Definition 2.3. For a decomposition of the boundary valuation in terms of w and e in the archimedean case, see [22, Theorem 6.4].

Theorem 2.9 *[22, Theorem 8.5 and Corollary 8.6] Assume that S is a non-archimedean quadratic Shannon extension of a regular local ring R with quotient field F. Let Q be as in Theorem 2.5, and let e and w be as in Definition 2.3. Then:*

(1) e is a rank 1 discrete valuation on F whose valuation ring E contains V. If in addition $R/(Q \cap R)$ is a regular local ring, then E is the order valuation ring of T.
(2) w induces a rational rank 1 valuation w' on the residue field $E/\mathfrak{m}_E$ of E. The valuation ring W' defined by w' extends the valuation ring S/Q, and the value group of W' is the same as the value group of S/Q.
(3) V is the valuation ring defined by the composite valuation of e and w'.
(4) Let $z \in E$ such that $\mathfrak{m}_E = zE$. Then V is defined by the valuation v given by

$$v : F \setminus \{0\} \to \mathbb{Z} \oplus \mathbb{Q} : a \mapsto \left(\frac{\mathrm{e}(a)}{\mathrm{e}(z)}, \frac{w(a)\mathrm{e}(z)}{w(z)\mathrm{e}(a)} \right),$$

where the direct sum is ordered lexicographically.

3 The Relation of Shannon Extensions to Pullbacks

Let $\alpha : A \to C$ be an extension of rings, and let B be a subring of C. The subring $D = \alpha^{-1}(B)$ of A is the *pullback* of B along $\alpha : A \to C$.

$$\begin{array}{ccc} D & \longrightarrow & B \\ \downarrow & & \downarrow \\ A & \xrightarrow{\alpha} & C \end{array}$$

Alternatively, D is the fiber product $A \times_C B$ of α and the inclusion map $\iota : B \to C$; see, for example, [24, page 43].

The pullback construction has been extensively studied in multiplicative ideal theory, where it serves as a source of examples and generalizes the classical "$D+M$" construction. (For more on the latter construction, see [10].) We will be especially interested in the case in which A, B, C, D are domains, α is a surjection, and B has quotient field C. In this case, following [9], we say the diagram above is of type $\square^*$. For a diagram of type $\square^*$, the kernel of α is a maximal ideal of A that is contained in D. The quotient field C of B can be identified with the residue field of this maximal ideal. If A is local with $\dim A \geq 1$ and $\dim B \geq 1$, then $A = D_M$ is a localization of D and D is non-archimedean. These observations have a number of consequences for transfer properties between the ring D and the rings A and B; see, for example, [7–9].

While pullback diagrams of type $\square^*$ are often used to construct examples in non-Noetherian commutative ring theory, there are also instances where the pullback construction is used as a classification tool. A simple example is given by the observation that a local domain D has a principal maximal ideal if and only if D occurs in a pullback diagram of type $\square^*$, where B is a DVR [23, Exercise 1.5, p. 7]. A second example is given by the fact that for nonnegative integers $k < n$, a ring D is a valuation domain of rank n if and only if D occurs in pullback diagram of type $\square^*$, where A is a valuation ring of rank $n - k$ and B is a valuation ring of rank k; see [7, Theorem 2.4]. Theorem 2.5 provides an instance of this decomposition in the present context. In the theorem, V is the pullback of E and W':

$$\begin{array}{ccc} V = \alpha^{-1}(W') & \twoheadrightarrow & W' \\ \downarrow & & \downarrow \\ E & \xrightarrow{\ \alpha\ } & E/\mathfrak{m}_E \end{array}$$

A third example of classification via pullbacks of the form $\square^*$ is given by the classification of local rings of global dimension 2 by Greenberg [16, Corollary 3.7] and Vasoncelos [31]: A local ring D has global dimension 2 if and only if D satisfies one of the following:

(a) D is a regular local ring of Krull dimension 2,
(b) D is a valuation ring of global dimension 2, or
(c) D has countably many principal prime ideals and D occurs in a pullback diagram of type $\square^*$, where A is a valuation ring of global dimension 1 or 2 and B is a regular local ring of global dimension 2.

Motivated by these examples, we use the pullback construction in this and the next section to classify among the overrings of a regular local ring R those that are non-archimedean quadratic Shannon extensions of R. We prove in Theorem 4.8 that these are precisely the overrings of R that occur in pullback diagrams of type $\square^*$, where A is a localization of an iterated quadratic transform R_i of R at a prime ideal P and B is a rank 1 valuation overring of R_i/P having a divergent multiplicity sequence. Thus a non-archimedean quadratic Shannon extension is determined by a rank 1 valuation ring and a regular local ring.

As a step towards this classification, in Theorem 3.1 we restate part of Theorem 2.5 as an assertion about how a non-archimedean quadratic Shannon extension can be decomposed using pullbacks. Much of the rest of this section and the next is devoted to a converse of this assertion, which is given in Theorems 4.8 and 5.1.

Theorem 3.1 *Let S be a non-archimedean quadratic Shannon extension. Then there is a prime ideal P of S and a rational rank 1 valuation ring $\mathscr{V}$ of $\kappa(P)$ such that S_P is the Noetherian hull of S and S is the pullback of $\mathscr{V}$ along the residue map $\alpha : S_P \to \kappa(P)$, as in the following diagram:*

$$\begin{array}{ccc} S = \alpha^{-1}(\mathscr{V}) & \longrightarrow\!\!\!\!\to & \mathscr{V} \\ \downarrow & & \downarrow \\ S_P & \xrightarrow{\ \alpha\ }\!\!\!\!\to & \kappa(P) \end{array}$$

Proof Theorem 2.5 implies that there is a prime ideal P of S such that S_P is the Noetherian hull of S, $P = PS_P$ and S/P is a rational rank 1 valuation ring. Theorem 3.1 follows from these observations. □

Definition 3.2 Let R be a Noetherian local domain, let $\mathscr{V}$ be a rank 1 valuation ring dominating R with corresponding valuation ν, and let $\{(R_i, \mathfrak{m}_i)\}_{i=0}^{\infty}$ be the infinite sequence[2] of LQTs along $\mathscr{V}$. Then the sequence $\{\nu(\mathfrak{m}_i)\}_{i=0}^{\infty}$ is the *multiplicity sequence of* $(R, \mathscr{V})$; see [15, Section 5]. We say the multiplicity sequence is *divergent* if $\sum_{i\geq 0} \nu(\mathfrak{m}_i) = \infty$.

Remark 3.3 Let R be a regular local ring and let $\mathscr{V}$ be a rank 1 valuation ring birationally dominating R. If the multiplicity sequence of $(R, \mathscr{V})$ is divergent, then $\mathscr{V}$ is a quadratic Shannon extension of R [15, Proposition 23] and $\mathscr{V}$ has rational rank 1 [20, Proposition 7.3]. This is observed in [21, Corollary 3.9] in the case where $\mathscr{V}$ is a DVR. In Proposition 3.4, we observe that $\mathscr{V}$ is the union of the rings in the LQT sequence of R along $\mathscr{V}$ for every Noetherian local domain $(R, \mathfrak{m})$ birationally dominated by $\mathscr{V}$.

Proposition 3.4 *Let $(R, \mathfrak{m})$ be a Noetherian local domain, let $\mathscr{V}$ be a rank 1 valuation ring that birationally dominates R, and let $\{R_i\}_{i=0}^{\infty}$ be the infinite sequence of LQTs of R along $\mathscr{V}$. If the multiplicity sequence of $(R, \mathscr{V})$ is divergent, then $\mathscr{V} = \bigcup_{n\geq 0} R_n$. Thus if $\mathscr{V}$ is a DVR, then $\mathscr{V} = \bigcup_{n\geq 0} R_n$.*

Proof Let ν be a valuation for $\mathscr{V}$ and let y be a nonzero element in $\mathscr{V}$. Suppose we have an expression $y = a_n/b_n$, where $a_n, b_n \in R_n$. Since $R_n \subseteq \mathscr{V}$, it follows that $\nu(b_n) \geq 0$. If $\nu(b_n) = 0$, then since $\mathscr{V}$ dominates R_n, we have $1/b_n \in R_n$ and $y \in R_n$.

Assume otherwise, that is, $\nu(b_n) > 0$. Then $b_n \in \mathfrak{m}_n$, and since $\nu(a_n) \geq \nu(b_n)$, also $a_n \in \mathfrak{m}_n$. Let $x_n \in \mathfrak{m}_n$ be such that $x_n R_{n+1} = \mathfrak{m}_n R_{n+1}$. Then $a_n, b_n \in x_n R_{n+1}$, so the elements $a_{n+1} = a_n/x$ and $b_{n+1} = b_n/x$ are in R_{n+1}. Thus we have the expression $y = a_{n+1}/b_{n+1}$, where $\nu(b_{n+1}) = \nu(b_n) - \nu(\mathfrak{m}_n)$.

Consider an expression $y = a_0/b_0$, where $a_0, b_0 \in R_0$. Then we iterate this process to obtain a sequence of expressions $\{a_n/b_n\}$ of y, with $a_n, b_n \in R_n$, where this process halts at some $n \geq 0$ if $\nu(b_n) = 0$, implying $y \in R_n$. Assume by way of contradiction that this sequence is infinite. For $N \geq 0$, it follows that $\nu(b_0) = \nu(b_N) + \sum_{n=0}^{N-1} \nu(\mathfrak{m}_n)$. Then $\nu(b_0) \geq \sum_{n=0}^{N} \nu(\mathfrak{m}_n)$ for any $N \geq 0$, so $\nu(b_0) \geq \sum_{n=0}^{\infty} \nu(\mathfrak{m}_n) = \infty$, which contradicts $\nu(b_0) < \infty$. This shows that the sequence $\{a_n/b_n\}$ is finite and hence $y \in \bigcup_n R_n$. □

Remark 3.5 Examples of $(R, \mathscr{V})$ with divergent multiplicity sequence such that $\mathscr{V}$ is not a DVR are given in [20, Examples 7.11 and 7.12].

[2] If $R_n = R_{n+1}$ for some integer n, then $R_n = \mathscr{V}$ is a DVR and $R_n = R_m$ for all $m \geq n$.

Discussion 3.6 Let $(R, \mathfrak{m})$ be a Noetherian local domain, let $\mathscr{V}$ be a rational rank 1 valuation ring that birationally dominates R, and let $\{R_i\}_{i=0}^{\infty}$ be the infinite sequence of LQTs of R along $\mathscr{V}$. The divergence of the multiplicity sequence in Proposition 3.4 is a sufficient condition for $\mathscr{V} = \bigcup_{n \geq 0} R_n$, but not a necessary condition; see Example 3.7. It would be interesting to understand more about conditions in order that the multiplicity sequence of $(R, \mathscr{V})$ is divergent. Example 3.7 illustrates that an archimedean Shannon extension S of a 3-dimensional regular local ring R may be birationally dominated by a rational rank 1 valuation ring V, where $S \subsetneq V$. In this case by Proposition 3.4, the multiplicity sequence of (R, V) must be convergent.

Example 3.7 Let x, y, z be indeterminates over a field k. We first construct a rational rank 1 valuation ring V' on the field $k(x, y)$. We do this by describing an infinite sequence $\{(R'_n, \mathfrak{m}'_n)\}_{n \geq 0}$ of local quadratic transforms of $R'_0 = k[x, y]_{(x,y)}$. To indicate properties of the sequence, we define a rational valued function v on specific generators of the $\mathfrak{m}'_n$. The function v is to be additive on products. We set $v(x) = v(y) = 1$. This indicates that y/x is a unit in every valuation ring birationally dominating R'_1.

Step 1. Let R'_1 have maximal ideal $\mathfrak{m}'_1 = (x_1, y_1)R_1$, where $x_1 = x$, $y_1 = (y/x) - 1$. Define $v(y_1) = 1/2$.
Step 2. The local quadratic transform R'_2 of R'_1 has maximal ideal $\mathfrak{m}'_2$ generated by $x_2 = x_1/y_1$, $y_2 = y_1$. We have $v(x_2) = 1/2$, $v(y_2) = 1/2$.
Step 3. Define $y_3 = (y_2/x_2) - 1$ and assign $v(y_3) = 1/4$. Then $x_3 = x_2$, $v(x_3) = 1/2$.
Step 4. The local quadratic transform R'_4 of R'_3 has maximal ideal $\mathfrak{m}'_4$ generated by $x_4 = x_3/y_3$, $y_4 = y_3$. Then $v(x_4) = v(y_4) = 1/4$.

Continuing this 2-step process yields an infinite directed union $(R'_n, \mathfrak{m}'_n)$ of local quadratic transforms of 2-dimensional RLRs. Let $V' = \bigcup_{n \geq 0} R'_n$. Then V' is a valuation ring by [1, Lemma 12]. Let v' be a valuation associated with V' such that $v'(x) = 1$. Then $v'(y) = 1$ and v' takes the same rational values on the generators of $\mathfrak{m}'_n$ as defined by v. Since there are infinitely many translations as described in Steps $2n + 1$ for each integer $n \geq 0$, it follows that V' has rational rank 1, e.g., see [20, Remark 5.1(4)].

The multiplicity values of $\{R'_n, \mathfrak{m}'_n)\}$ are $1, \frac{1}{2}, \frac{1}{2}, \frac{1}{4}, \frac{1}{4}, \frac{1}{8}, \frac{1}{8} \ldots$, the sum of which converges to 3.

Define $V = V'(\frac{z}{x^2y^2})$, the localization of the polynomial ring $V'[\frac{z}{x^2y^2}]$ at the prime ideal $\mathfrak{m}_{V'}V'[\frac{z}{x^2y^2}]$. One sometimes refers to V as a Gaussian or trivial or Nagata extension of V' to a valuation ring on the simple transcendental field extension generated by $\frac{z}{x^2y^2}$ over $k(x, y)$. It follows that V has the same value group as V' and the residue field of V is a simple transcendental extension of the residue field of V' that is generated by the image of $\frac{z}{x^2y^2}$ in $V/\mathfrak{m}_V$.

Let v denote the associated valuation to V such that $v(x) = 1$. It follows that $v(y) = 1$ and $v(z) = v(x^2y^2) = 4$. Let $R_0 = k[x, y, z]_{(x,y,z)}$. Then R_0 is birationally dominated by V. Let $\{(R_n, \mathfrak{m}_n)\}_{n \geq 0}$ be the sequence of local quadratic transforms of R_0 along V.

We describe the first few steps:

Step 1. R_1 has maximal ideal $\mathfrak{m}_1 = (x_1, y_1, z_1)R_1$, where $x_1 = x$, $y_1 = (y/x) - 1$, and $z_1 = z/x$. Also $v(y_1) = 1/2$.
Step 2. The local quadratic transform R_2 of R_1 along V has maximal ideal $\mathfrak{m}_2$ generated by $x_2 = x_1/y_1$, $y_2 = y_1$ and $z_2 = z_1/y_1$. We have $v(x_2) = 1/2$, $v(y_2) = 1/2$ and $v(z_2) = 4 - 3/2 > 3/2$.
Step 3. The local quadratic transform R_3 of R_2 along V has $y_3 = (y_2/x_2) - 1$, where $v(y_3) = 1/4$, and $x_3 = x_2$, $v(x_3) = 1/2$ and $v(z_3) > 1/2$.
Step 4. The local quadratic transform R_4 of R_3 along V has maximal ideal $\mathfrak{m}_4$ generated by $x_4 = x_3/y_3$, $y_4 = y_3$ and $z_4 = z_3/y_3$.

The multiplicity values of the sequence $\{(R_n, \mathfrak{m}_n)\}_{n\geq 0}$ along V are the same as that for $\{R'_n, \mathfrak{m}'_n)\}$, namely $1, \frac{1}{2}, \frac{1}{2}, \frac{1}{4}, \frac{1}{4}, \frac{1}{8}, \frac{1}{8} \ldots$. Let $S = \bigcup_{i\geq 0} R_n$. Since S is birationally dominated by the rank 1 valuation ring V, it follows that S is an archimedean Shannon extension. Since we never divide in the z-direction, we have $S \subseteq R_{zR}$, and S is not a valuation ring.

4 Quadratic Shannon Extensions Along a Prime Ideal

Let R be a Noetherian local domain and let $\{R_n\}_{n\geq 0}$ be an infinite sequence of LQTs of $R = R_0$. Using the terminology of Granja and Sanchez-Giralda [14, Definition 3 and Remark 4], for a prime ideal P of R, we say the quadratic sequence $\{R_n\}$ is along R_P if $\bigcup_{n\geq 0} R_n \subseteq R_P$.

Let P be a nonzero, nonmaximal prime ideal of a Noetherian local domain $(R, \mathfrak{m})$. Proposition 4.1 establishes a one-to-one correspondence between sequences $\{R_n\}$ of LQTs of $R = R_0$ along R_P and sequences $\{\overline{R_n}\}$ of LQTs of $\overline{R_0} = R/P$.

Proposition 4.1 *Let R be a Noetherian local domain and let P be a nonzero nonmaximal prime ideal of R. Then there is a one-to-one correspondence between:*

(1) Infinite sequences $\{R_n\}_{n\geq 0}$ of LQTs of $R_0 = R$ along R_P.
(2) Infinite sequences $\{\overline{R_n}\}_{n\geq 0}$ of LQTs of $\overline{R_0} = R/P$.

Given such a sequence $\{R_n\}_{n\geq 0}$, the corresponding sequence is $\{R_n/(PR_P \cap R_n)\}$. Denote $S = \bigcup_{n\geq 0} R_n$ and $\overline{S} = \bigcup_{n\geq 0} \overline{R_n}$, and let $\widetilde{S}$ be the pullback of $\overline{S}$ with respect to the quotient map $R_P \to \kappa(P)$ as in the following diagram:

$$\begin{array}{ccc} \widetilde{S} = \alpha^{-1}(\overline{S}) & \twoheadrightarrow & \overline{S} \\ \downarrow & & \downarrow \\ R_P & \xrightarrow{\alpha} & \kappa(P) \end{array}$$

Then $\widetilde{S} = S + PR_P$ and $\widetilde{S}$ is non-archimedean.

Proof The correspondence follows from [17, Corollary II.7.15, p. 165]. The fact that $\widetilde{S} = S + PR_P$ is a consequence of the fact that $\widetilde{S}$ is a pullback of $\overline{S}$ and R_P. That

$\widetilde{S}$ is non-archimedean is a consequence of the observation that for each $x \in \mathfrak{m}_{\widetilde{S}} \setminus PR_P$, the fact that $PR_P \subseteq \widetilde{S} \subseteq R_P$ implies $PR_P \subseteq x^k\widetilde{S}$ for all $k > 0$.

Lemma 4.2 *Assume notation as in Proposition 4.1. If $\overline{S}$ is a rank* 1 *valuation ring and the multiplicity sequence of $(\overline{R}, \overline{S})$ is divergent, then $S = \tilde{S}$.*

Proof Let ν be a valuation for $\overline{S}$ and assume that ν takes values in $\mathbb{R}$. Let $f \in \tilde{S}$. We claim that $f \in S$. Since $\tilde{S} = S + PR_P$, we may assume $f \in PR_P$. Write $f = \frac{g_0}{h_0}$, where $g_0 \in P$ and $h_0 \in R \setminus P$.

Suppose we have an expression of the form $f = \frac{g_n}{h_n}$, where $g_n \in PR_P \cap R_n$ and $h_n \in R_n \setminus PR_P$. Write $\mathfrak{m}_n R_{n+1} = xR_{n+1}$ for some $x \in \mathfrak{m}_n$. Since $PR_P \cap R_n \subseteq \mathfrak{m}_n$, it follows that $g_n = xg_{n+1}$ for $g_{n+1} = \frac{g_n}{x} \in R_{n+1}$. Denote the image of $h \in R_n$ in $\overline{R_n}$ by $\overline{h}$. Since $h_n \in R_n \setminus PR_P$, we have that $\overline{h_n} \neq 0$ and $\nu(\overline{h_n})$ is a finite nonnegative real number. If $\nu(\overline{h_n}) > 0$, then $h_n \in \mathfrak{m}_n$, so $h_n = xh_{n+1}$ for $h_{n+1} = \frac{h_n}{x} \in R_{n+1}$. Thus we have written $f = \frac{g_{n+1}}{h_{n+1}}$, where $g_{n+1} \in PR_P \cap R_{n+1}$ and $h_{n+1} \in R_{n+1} \setminus PR_P$, such that $\nu(\overline{h_{n+1}}) = \nu(\overline{h_n}) - \nu(\overline{\mathfrak{m}_n})$.

Since $\sum_{n\geq 0} \nu(\overline{\mathfrak{m}_n}) = \infty$ and $\nu(\overline{h_0})$ is finite, this process must halt with $f = \frac{g_n}{h_n}$ as before such that $\nu(\overline{h_n}) = 0$. Since $\nu(\overline{h_n}) = 0$, $\overline{h_n}$ is a unit in $\overline{R_n}$, so h_n is a unit in R_n, and thus $f \in R_n$.

Lemma 4.3 *Let P be a nonzero nonmaximal prime ideal of a regular local ring R. Let $\{R_n\}_{n\geq 0}$ be a sequence of LQTs of $R_0 = R$ along R_P and let $\{\overline{R_n}\}$ be the induced sequence of LQTs of $\overline{R_0} = R/P$ as in Proposition 4.1. Denote $S = \bigcup_{n\geq 0} R_n$ and $\overline{S} = \bigcup_{n\geq 0} \overline{R_n}$. Then the following are equivalent:*

(1) S is the pullback of $\overline{S}$ along the surjective map $R_P \to \kappa(P)$.
(2) The Noetherian hull of S is R_P.
(3) $\overline{S}$ is a rank 1 *valuation ring and the multiplicity sequence of $(\overline{R}, \overline{S})$ is divergent.*

If these conditions hold, then $\overline{S}$ has rational rank 1.

Proof (1) $\implies$ (2): As a pullback, the quadratic Shannon extension S is non-archimedean (see the proof of Proposition 4.1). Let $x \in S$ be such that xS is $\mathfrak{m}_S$-primary (see Theorem 2.2). By Theorem 2.5, the ideal $Q = \bigcap_{n\geq 0} x^n S$ is a nonzero prime ideal of S, every nonmaximal prime ideal of S is contained in Q and $T = S_Q$. Assumption (1) implies that PR_P is a nonzero ideal of both S and R_P. Hence R_P is almost integral over S. We have $S \subseteq S_Q = T \subseteq R_P$, and S_Q is an RLR and therefore completely integrally closed. It follows that $S_Q = R_P$ is the Noetherian hull of S.

(2) $\implies$ (3): Since the Noetherian hull R_P of S is local, Theorem 2.5 implies that S is non-archimedean and $PR_P \subseteq S$. By Theorem 3.1, $\overline{S} = S/PR_P$ is a rational rank 1 valuation ring. The valuation ν associated with $\overline{S}$ is equal to the valuation w' of Theorem 2.9. By item 2 of Theorem 2.9 and item 6 of Theorem 2.5, we have

$$\sum_{n=0}^{\infty} \nu(\overline{\mathfrak{m}_n}) = \sum_{n=0}^{\infty} w(\mathfrak{m}_n) = \infty.$$

(3) $\implies$ (1): This is proved in Lemma 4.2.

Remark 4.4 Let P be a nonzero nonmaximal prime ideal of R and let S be a non-archimedean quadratic Shannon extension of R with Noetherian hull R_P. The proof of Lemma 4.3 shows that the multiplicity sequence of $(R/P, S/PR_P)$ is given by $\{w(\mathfrak{m}_i)\}$, where w is as in Definition 2.3.

With notation as in Lemma 4.3, examples where $\overline{S}$ is a rank 1 valuation ring that is not discrete are given in [20, Examples 7.11 and 7.12].

Theorem 4.5 (Existence of Shannon Extensions) *Let P be a nonzero nonmaximal prime ideal of a regular local ring R.*

(1) *There exists a non-archimedean quadratic Shannon extension of R with R_P as its Noetherian hull.*
(2) *If there exists an archimedean quadratic Shannon extension of R contained in R_P, then* $\dim R/P \geq 2$.

Proof To prove item 1, we use a result of Chevalley that every Noetherian local domain is birationally dominated by a DVR [5]. Let V be a DVR birationally dominating R/P. We apply Lemma 4.3 with this R and P. Let $\{\overline{R_n}\}$ be the sequence of LQTs of $\overline{R_0} = R/P$ along V. Let S be the union of the corresponding sequence of LQTs of R given by Proposition 4.1. Proposition 3.4 implies that $\overline{S} = V$ and Lemma 4.3 implies that $S = \widetilde{S}$ is a non-archimedean Shannon extension with R_P as its Noetherian hull.

For item 2, if $\dim R/P = 1$, then $\dim R_P = \dim R - 1$ since an RLR is catenary. If S is an archimedean Shannon extension of R, then $\dim S \leq \dim R - 1$ by [21, Lemma 3.4 and Corollary 3.6]. Therefore R_P does not contain the Noetherian hull of an archimedean Shannon extension of R if $\dim R/P = 1$. □

Discussion 4.6 Let P be a nonzero nonmaximal prime of a regular local ring R such that $\dim R/P \geq 2$. We ask:

Question Does there exists an archimedean quadratic Shannon extension of R contained in R_P?

The question reduces to the case where $\dim R/P = 2$, for if Q is a prime ideal of R with $\dim R/Q \geq 2$, then there exists a prime ideal P of R such that $Q \subseteq P$ and $\dim R/P = 2$. Then $R_P \subseteq R_Q$. Hence a quadratic Shannon extension of R contained in R_P is contained in R_Q.

Assume that P is a nonzero prime ideal of R such that $\dim R/P = 2$. It is not difficult to see that the 2-dimensional Noetherian local domain $\overline{R_0} = R/P$ is birationally dominated by a rank 1 valuation domain V of rational rank 2. Consider the infinite sequence of LQTs $\{\overline{R_n}\}_{n \geq 0}$ of $\overline{R_0} = R/P$ along V and let $\overline{S} = \bigcup_{n \geq 0} \overline{R_n}$. Then $\overline{S}$ is birationally dominated by V. Each of the $\overline{R_n}$ is a 2-dimensional Noetherian local domain and $\dim \overline{S}$ is either 1 or 2.

Let $\{R_n\}$ be the sequence of LQTs of R given by Proposition 4.1 that corresponds to $\{\overline{R_n}\}$, and let $S = \bigcup_n R_n$. Then $\dim R_n > 2$ for all n. Hence S is a quadratic Shannon extension of R and $S \subseteq R_P$. Let $\mathfrak{p} = PR_P \cap S$. Then $S/\mathfrak{p} = \overline{S}$.

If $\dim \overline{S} = 1$, then there are no prime ideals of S strictly between $\mathfrak{p}$ and $\mathfrak{m}_S$. Since V has rational rank 2, the multiplicity sequence of $(\overline{R}, \overline{S})$ is convergent. Lemma 4.3 implies that the Noetherian hull of S is not R_P. Hence if $\dim \overline{S} = 1$, there exists an archimedean quadratic Shannon extension S of R contained in R_P.

Theorem 4.5 implies the following:

Corollary 4.7 (Lipman [25, Lemma 1.21.1]) *Let P be a nonmaximal prime ideal of a regular local ring R. Then there exists a quadratic Shannon extension of R contained in R_P.*

In Theorem 4.8 we use Lemma 4.3 to characterize the overrings of a regular local ring R that are Shannon extensions of R with Noetherian hull R_P, where P is a nonzero nonmaximal prime ideal of R. Note that by Theorem 2.5 such a Shannon extension is necessarily non-archimedean.

Theorem 4.8 (Shannon Extensions with Specified Local Noetherian Hull) *Let P be a nonzero nonmaximal prime ideal of a regular local ring R. The quadratic Shannon extensions of R with Noetherian hull R_P are precisely the rings S such that S is a pullback along the residue map $\alpha : R_P \to \kappa(P)$ of a rational rank 1 valuation ring birationally dominating R/P whose multiplicity sequence is divergent.*

$$\begin{array}{ccc} S = \alpha^{-1}(\mathscr{V}) & \twoheadrightarrow & \mathscr{V} \\ \downarrow & & \downarrow \\ R_P & \xrightarrow{\ \alpha\ } & \kappa(P) \end{array}$$

Proof If S is a quadratic Shannon extension with Noetherian hull R_P, then by Lemma 4.3, S is a pullback along the map $R_P \to \kappa(P)$ of a rational rank 1 valuation ring birationally dominating R/P whose multiplicity sequence is divergent.

Conversely, let S be such a pullback. Let $\{\overline{R_n}\}_{n\geq 0}$ denote the sequence of LQTs of $\overline{R_0} = R/P$ along $\mathscr{V}$ and let $\{R_n\}_{n\geq 0}$ denote the induced sequence of LQTs of $R_0 = R$ as in Proposition 4.1. Then Lemma 4.3 implies that $S = \bigcup_{n\geq 0} R_n$, so S is a quadratic Shannon extension.

Corollary 4.9 *Let P be a prime ideal of the regular local ring R with $\dim R/P = 1$.*

(1) *The quadratic Shannon extensions of R with Noetherian hull R_P are precisely the pullbacks along the residue map $R_P \to \kappa(P)$ of the finitely many DVR overrings $\mathscr{V}$ of R/P.*
(2) *If R/P is a DVR, then $R + PR_P$ is the unique quadratic Shannon extension of R with Noetherian hull R_P.*

Proof The Krull–Akizuki Theorem [26, Theorem 11.7] implies that R/P has finitely many valuation overrings, each of which is a DVR. By Theorem 4.8 there is a one-to-one correspondence between these DVRs and the Shannon extensions of R with Noetherian hull R_P. This proves item 1. If R/P is a DVR, then by item 1, the pullback $R+PR_P$ of R/P along the map $R_P \to \kappa(P)$ is the unique quadratic Shannon extension of R with Noetherian hull R_P. This verifies item 2. □

5 Classification of Non-Archimedean Shannon Extensions

In Theorem 5.1 we classify the non-archimedean quadratic Shannon extensions S that occur as overrings of a given regular local ring R. The classification is extrinsic to S in the sense that a prime ideal of an iterated quadratic transform of R is needed for the description of the overring S as a pullback. In Theorem 5.2 we give an intrinsic rather than extrinsic characterization of certain of the non-archimedean quadratic Shannon extensions with principal maximal ideal that occur in an algebraic function field of characteristic 0. In this case, we are able to characterize such rings in terms of pullbacks without the explicit requirement of a regular local "underring" of S. This allows us to give an additional source of examples of non-archimedean quadratic Shannon extensions in Example 5.4.

Theorem 5.1 (Classification of non-archimedean Shannon extensions) *Let R be a regular local ring with* $\dim R \geq 2$, *and let S be an overring of R. Then S is a non-archimedean quadratic Shannon extension of R if and only if there is a ring $\mathscr{V}$, a nonnegative integer i, and a prime ideal P of R_i such that*

(a) $\mathscr{V}$ is a rational rank 1 *valuation ring of $\kappa(P)$ that contains the image of R_i/P in $\kappa(P)$ and has divergent multiplicity sequence over this image, and*
(b) S is a pullback of $\mathscr{V}$ along the residue map $\alpha : (R_i)_P \to \kappa(P)$.

$$\begin{array}{ccc} S = \alpha^{-1}(\mathscr{V}) & \twoheadrightarrow & \mathscr{V} \\ \downarrow & & \downarrow \\ (R_i)_P & \xrightarrow{\alpha} & \kappa(P) \end{array}$$

Proof Suppose S is a non-archimedean quadratic Shannon extension, and let $\{R_i\}$ be the sequence of iterated QDTs such that $S = \bigcup_i R_i$. By Theorem 2.5, the Noetherian hull T of S is a local ring, and by Theorem 2.2 there is $i > 0$ and a prime ideal P of R_i such that $T = (R_i)_P$. Since S is a non-archimedean quadratic Shannon extension of R_i, Theorem 4.8 implies there is a valuation ring $\mathscr{V}$ such that (a) and (b) hold for i, P, S, and $\mathscr{V}$.

Conversely, suppose there is a ring $\mathscr{V}$, a nonnegative integer i and a prime ideal P of R_i that satisfy (a) and (b). By Theorem 4.8, S is a quadratic Shannon extension of R_i with Noetherian hull $(R_i)_P$. Thus S is a quadratic Shannon extension of R that is non-archimedean by Theorem 2.5. □

In contrast to Theorem 5.1, the pullback description in Theorem 5.2 is without reference to a specific regular local underring of S. Instead, the proof constructs one using resolution of singularities. Because our use of this technique is elementary, we frame our proof in terms of projective models rather than projective schemes. For more background on projective models, see [3, Sections 1.6–1.8] and [34, Chapter VI, §17]. Let F be a field and let k be a subfield of F. Let $t_0 = 1$ and assume that $t_1, \ldots, t_n$ are nonzero elements of F such that $F = k(t_1, \ldots, t_n)$. For each $i \in \{0, 1, \ldots, n\}$, define $D_i = k[t_0/t_i, \ldots, t_n/t_i]$. The *projective model* of F/k with respect to $t_0, \ldots, t_n$ is the collection of local rings given by

$$X = \{(D_i)_P : i \in \{0, 1, \ldots, n\},\ P \in \text{Spec}\,(D_i)\}.$$

If k has characteristic 0, then by resolution of singularities (see, for example, [6, Theorem 6.38, p. 100]) there is a projective model Y of F/k such that every regular local ring in X is in Y, every local ring in Y is a regular local ring, and every local ring in X is dominated by a (necessarily regular) local ring in Y.

By a *valuation ring of* F/k we mean a valuation ring V with quotient field F such that k is a subring of V.

Theorem 5.2 *Let S be a local domain containing as a subring a field k of characteristic* 0. *Assume that* $\dim S \geq 2$ *and that the quotient field F of S is a finitely generated extension of k. Then the following are equivalent:*

(1) S has a principal maximal ideal and S is a quadratic Shannon extension of a regular local ring R that is essentially finitely generated over k.
(2) There is a regular local overring A of S and a DVR $\mathscr{V}$ of $(A/\mathfrak{m}_A)/k$ such that

(a) $\text{tr.deg}_k\, A/\mathfrak{m}_A + \dim A = \text{tr.deg}_k\, F$, and
(b) S is the pullback of $\mathscr{V}$ along the residue map $\alpha : A \to A/\mathfrak{m}_A$.

$$\begin{array}{ccc} S = \alpha^{-1}(\mathscr{V}) & \twoheadrightarrow & \mathscr{V} \\ \downarrow & & \downarrow \\ A & \xrightarrow{\ \alpha\ } & A/\mathfrak{m}_A \end{array}$$

Proof (1) $\Longrightarrow$ (2): Let $x \in S$ be such that $\mathfrak{m}_S = xS$. By Theorem 2.2, $S[1/x]$ is the Noetherian hull of S and $S[1/x]$ is a regular ring. Since $\dim S > 1$, the ideal $P = \bigcap_{k>0} x^k S$ is a nonzero prime ideal of S [23, Exercise 1.5, p. 7]. Hence S is non-archimedean. By Theorem 2.5, S_P is the Noetherian hull of S and hence $S_P = S[1/x]$. Let $A = S_P$ and $\mathscr{V} = S/P$. By Theorem 3.1, S is a pullback of the DVR $\mathscr{V}$ with respect to the map $A \to A/\mathfrak{m}_A$. By assumption, S is a quadratic Shannon extension of a regular local ring R that is essentially finitely generated over k. For sufficiently large i, we have $A = S_P = (R_i)_{P \cap R_i}$ by [21, Proposition 3.3]. Since R_i is essentially finitely generated over R, and R is essentially finitely generated over k, we have that A is essentially finitely generated over k. By the Dimension Formula [26, Theorem 15.6, p. 118],

$$\text{tr.deg}_k\, A/\mathfrak{m}_A + \dim A = \text{tr.deg}_k\, F.$$

This completes the proof that statement 1 implies statement 2.

(2) $\Longrightarrow$ (1): Let $P = \mathfrak{m}_A$. By item 2b, P is a prime ideal of S, $A = S_P$, $P = PS_P$ and $\mathscr{V} = S/P$. Let $x \in \mathfrak{m}_S$ be such that the image of x in the DVR S/P generates the maximal ideal. Since $P = PS_P$, we have $P \subseteq xS$. Consequently, $\mathfrak{m}_S = xS$, and so S has a principal maximal ideal.

To prove that S is a quadratic Shannon extension of a regular local ring that is essentially finitely generated over k, it suffices by Theorem 4.8 to prove:

(i) There is a subring R of S that is a regular local ring essentially finitely generated over k.
(ii) A is a localization of R at the prime ideal $P \cap R$.
(iii) $\mathscr{V}$ is a valuation overring of $(R+P)/P$ with divergent multiplicity sequence.

Since F is a finitely generated field extension of k and A (as a localization of S) has quotient field F, there is a finitely generated k-subalgebra D of A such that the quotient field of D is F. By item 2a, A/P has finite transcendence degree over k. Let $a_1, \ldots, a_n$ be elements of A whose images in A/P form a transcendence basis for A/P over k. Replacing D with $D[a_1, \ldots, a_n]$, and defining $p = P \cap D$, we may assume that A/P is algebraic over $\kappa(p) = D_p/pD_p$. In fact, since the normalization of an affine k-domain is again an affine k-domain, we may assume also that D is an integrally closed finitely generated k-subalgebra of A with quotient field F. Since D is a finitely generated k-algebra, D is universally catenary. By the Dimension Formula [26, Theorem 15.6, p. 118], we have

$$\dim D_p + tr.deg_k\, \kappa(p) = tr.deg_k\, F.$$

Therefore, item 2a implies

$$\dim D_p + tr.deg_k\, \kappa(p) = \dim A + tr.deg_k\, A/P.$$

Since A/P is algebraic over $\kappa(p)$, we conclude that $\dim D_p = \dim A$.

The normal ring A birationally dominates the excellent normal ring D_p, so A is essentially finitely generated over D_p [19, Theorem 1]. Therefore A is essentially finitely generated over k.

Since A is essentially finitely generated over k, the local ring A is in a projective model X of F/k. As discussed before the theorem, resolution of singularities implies that there exists a projective model Y of F/k such that every regular local ring in X is in Y, every local ring in Y is a regular local ring, and every local ring in X is dominated by a local ring in Y.

Since A is a regular local ring in X, A is a local ring in the projective model Y. Let $x_0, \ldots, x_n \in F$ be nonzero elements such that with $D_i := k[x_0/x_i, \ldots, x_n/x_i]$ for each $i \in \{0, 1, \ldots, n\}$, we have

$$Y = \bigcup_{i=0}^{n} \{(D_i)_Q : Q \in \text{Spec}\,(D_i)\}.$$

Since S has quotient field F, we may assume that $x_0, \ldots, x_n \in S$. Since A is in Y, there is $i \in \{0, 1, \ldots, n\}$ such that $A = (D_i)_{P \cap D_i}$.

By item 2b, $\mathscr{V} = S/P$ is a valuation ring with quotient field A/P. For $a \in A$, let $\overline{a}$ denote the image of a in the field A/P. Since S/P is a valuation ring of A/P, there exists $j \in \{0, 1, \ldots, n\}$ such that

$$(\,\{\overline{x_k/x_i}\}_{k=0}^{n}\,)(S/P) \;=\; (\overline{x_j/x_i})(S/P). \tag{1}$$

Notice that $x_i/x_i = 1 \notin P$. Hence at least one of the $x_k/x_i \notin P$, and Equation (1) implies $x_j/x_i \notin P$. Since $A = S_P$ and $P = PS_P$, every fractional ideal of S contained in A is comparable to P with respect to set inclusion. Therefore $P \subsetneq (x_j/x_i)S$. This and Equation (1) imply that

$$(x_0/x_i, \dots, x_n/x_i)S = (x_j/x_i)S. \tag{2}$$

Multiplying both sides of Equation (2) by x_i/x_j we obtain

$$D_j = k[x_0/x_j, \dots, x_n/x_j] \subseteq S.$$

Let $R = (D_j)_{\mathfrak{m}_S \cap D_j}$. Since Y is a nonsingular model, R is a regular local ring with $R \subseteq S \subseteq A$.

We observe next that $A = R_{P \cap R}$. Since $R \subseteq A$, we have that A dominates the local ring $A' := R_{P \cap R}$. The local ring A' is a member of the projective model Y, and every valuation ring dominating the local ring A in Y dominates also the local ring A' in Y. Since Y is a projective model of F/k, the Valuative Criterion for Properness [17, Theorem II.4.7, p. 101] implies no two distinct local rings in Y are dominated by the same valuation ring. Therefore, $A = A'$, so that $A = R_{P \cap R}$.

Finally, observe that since $\mathscr{V} = S/P$ is a DVR overring of $(R + P)/P$, the multiplicity sequence of S/P over $(R + P)/P$ is divergent. By Theorem 4.8, S is a quadratic Shannon extension of R with Noetherian hull $A = R_{P \cap R}$. By Theorem 2.5, S is non-archimedean, so the proof is complete. □

As an application of Theorem 5.2, we describe for a finitely generated field extension F/k of characteristic 0 the valuation rings with principal maximal ideal that arise as quadratic Shannon extensions of regular local rings that are essentially finitely generated over k, i.e., the valuation rings on the Zariski–Riemann surface of F/k that arise from desingularization followed by infinitely many successive quadratic transforms of projective models. Recall that a valuation ring V of F/k is a *divisorial* valuation ring if

$$\text{tr.deg}_k\, V/\mathfrak{m}_V = \text{tr.deg}_k\, F - 1.$$

Such a valuation ring is necessarily a DVR (apply, e.g., [1, Theorem 1]).

Corollary 5.3 *Let F/k be a finitely generated field extension where k has characteristic 0, and let S be a valuation ring of F/k with principal maximal.*

(1) Suppose rank $S = 1$. Then there is a sequence $\{R_i\}$ (possibly finite) of LQTs of a regular local ring R essentially finitely type over k such that $S = \bigcup_i R_i$. This sequence is finite if and only if S is a divisorial valuation ring.

(2) Suppose rank $S > 1$. Then S is a quadratic Shannon extension of a regular local ring essentially finitely generated over k if and only if S has rank 2 and S is contained in a divisorial valuation ring of F/k.

Proof For item 1, assume rank $S = 1$. By resolution of singularities, there is a nonsingular projective model X of F/k with function field F. Let R be the regular local ring in X that is dominated by S. Let $\{R_i\}$ be the sequence of LQTs of R along S. If $\{R_i\}$ is finite, then $\dim R_i = 1$ for some i, so that R_i is a DVR. Since S is a DVR between R_i and its quotient field, we have $R_i = S$. Otherwise, if $\{R_i\}$ is infinite, then Proposition 3.4 implies $S = \bigcup_i R_i$ since S is a DVR. That the sequence is finite if and only if S is a divisorial valuation ring follows from [1, Proposition 4].

For item 2, suppose rank $S > 1$. Assume first that S is a Shannon extension of a regular local ring essentially finitely generated over k. By [21, Theorem 8.1], $\dim S = 2$. By Theorem 5.2, S is a contained in a regular local ring $A \subseteq F$ such that $A/\mathfrak{m}_A$ is the quotient field of a proper homomorphic image of S and

$$\text{tr.deg}_k\, A/\mathfrak{m}_A + \dim A = \text{trdeg}_k\, F. \tag{3}$$

We claim A is a divisorial valuation ring of F/k. Since $A/\mathfrak{m}_A$ is the quotient field of a proper homomorphic image of S, it follows that

$$\text{tr.deg}_k\, A/\mathfrak{m}_A < \text{trdeg}_k F. \tag{4}$$

From Equations (3) and (4) we conclude that $\dim A \geq 1$. As an overring of the valuation ring S, A is also a valuation ring. Since A is a regular local ring that is not a field, it follows that A is a DVR. Thus $\dim A = 1$ and Equation (3) implies that

$$\text{tr.deg}_k\, A/\mathfrak{m}_A = \text{trdeg}_k\, F - 1,$$

which proves that A is a divisorial valuation ring.

Conversely, suppose rank $S = 2$ and S is a quadratic Shannon extension of a regular local ring that is essentially finitely generated over k. Theorem 5.2 and rank $S = 2$ imply S is contained in a regular local ring A with $\dim A = 1$ and

$$\text{tr.deg}_k\, A/\mathfrak{m}_A + 1 = \text{trdeg}_k\, F.$$

Thus A is a divisorial valuation ring.

Finally, suppose rank $S = 2$ and S is contained in a divisorial valuation ring A of F/k. Since S is a valuation ring of rank 2 with principal maximal ideal it follows that $\mathfrak{m}_A \subseteq S$ and $S/\mathfrak{m}_A$ is DVR. Since A is a divisorial valuation ring, we have

$$\text{tr.deg}_k\, A/\mathfrak{m}_A + \dim A = \text{trdeg}_k\, F.$$

As a DVR, A is a regular local ring, so Theorem 5.2 implies S is a quadratic Shannon extension of a regular local ring that is essentially finitely generated over k. □

Example 5.4 Let k be a field of characteristic 0, let $x_1, \ldots, x_n, y_1, \ldots, y_m$ be algebraically independent over k, and let

$$A = k(x_1, \ldots, x_n)[y_1, \ldots, y_m]_{(y_1, \ldots, y_m)}.$$

Let $\alpha : A \to k(x_1, \dots, x_n)$ be the canonical residue map. For every DVR V of $k(x_1, \dots, x_n)/k$, the ring $S = \alpha^{-1}(V)$ is by Theorem 5.2 a quadratic Shannon extension of a regular local ring that is essentially finitely generated over k. As in the proof that statement 2 implies statement 1 of Theorem 5.2, the Noetherian hull of S is A.

Conversely, suppose S is a k-subalgebra of F with principal maximal ideal such that S is a quadratic Shannon extension of a regular local ring that is essentially finitely generated over k and S has Noetherian hull A. As in the proof that statement 1 implies statement 2 of Theorem 5.2, there is a DVR V of $k(x_1, \dots, x_n)/k$ such that $S = \alpha^{-1}(V)$.

It follows that there is a one-to-one correspondence between the DVRs of $k(x_1, \dots, x_n)/k$ and the quadratic Shannon extensions S of regular local rings that are essentially finitely generated over k, have Noetherian hull A, and have a principal maximal ideal.

Theorem 5.2 concerns quadratic Shannon extensions of regular local rings that are essentially finitely generated over k. Example 5.5 is a quadratic Shannon extension of a regular local ring R in a function field for which R is not essentially finitely generated over k.

Example 5.5 Let $F = k(x, y, z)$, where k is a field and x, y, z are algebraically independent over k. Let $\tau \in xk[[x]]$ be a formal power series in x such that x and τ are algebraically independent over k. Set $y = \tau$ and define $V = k[[x]] \cap k(x, y)$. Then V is a DVR on the field $k(x, y)$ with maximal ideal xV and residue field $V/xV = k$. Let $V(z) = V[z]_{xV[z]}$. Then $V(z)$ is a DVR on the field F with residue field $k(z)$, and $V(z)$ is not essentially finitely generated over k. Let $R = V[z]_{(x,z)V[z]}$. Notice that R is a 2-dim RLR. The pullback diagram of type $\square^*$

$$\begin{array}{ccc} S = \alpha^{-1}(k[z]_{zk[z]}) & \twoheadrightarrow & k[z]_{zk[z]} \\ \downarrow & & \downarrow \\ V(z) & \xrightarrow{\ \alpha\ } & k(z) \end{array}$$

defines a rank 2 valuation domain S on F that is by Theorem 5.1 a quadratic Shannon extension of R. For each positive integer n, define $R_n = R[\frac{x}{z^n}]_{(z, \frac{x}{z^n})R[\frac{x}{z^n}]}$. Then $S = \bigcup_{n \geq 1} R_n$.

6 Quadratic Shannon Extensions and GCD Domains

As an application of the pullback description of non-archimedean quadratic Shannon extensions given in Sect. 6, we show in Theorem 6.2 that a quadratic Shannon extension S is coherent, a GCD domain or a finite conductor domain if and only if S is a valuation domain. We extend this fact to all quadratic Shannon extensions S, regardless of whether S is archimedean, by applying structural results for archimedean quadratic Shannon extensions from [21].

Definition 6.1 Following McAdam in [27], an integral domain D is a *finite conductor domain* if for elements a, b in the field of fractions of D, the D-module $aD \cap bD$ is finitely generated. A ring is said to be *coherent* if every finitely generated ideal is finitely presented. Chase [4, Theorem 2.2] proves that an integral domain D is coherent if and only if the intersection of two finitely generated ideals of D is finitely generated. Thus a coherent domain is a finite conductor domain. An integral domain D is a *GCD domain* if for all $a, b \in D$, $aD \cap bD$ is a principal ideal of D [10, page 76 and Theorem 16.2, p. 174]. It is clear from the definitions that a GCD domain is a finite conductor domain.

Examples of GCD domains and finite conductor domains that are not coherent are given by Glaz in [12, Example 4.4 and Example 5.2] and by Olberding and Saydam in [29, Prop. 3.7]. Every Noetherian integral domain is coherent, and a Noetherian domain D is a GCD domain if and only if it is a UFD. Noetherian domains that are not UFDs are examples of coherent domains that are not GCD domains.

Theorem 6.2 *Let S be a quadratic Shannon extension of a regular local ring. The following are equivalent:*

(1) S is coherent.
(2) S is a GCD domain.
(3) S is a finite conductor domain.
(4) S is a valuation domain.

Proof [3] It is true in general that if S is a valuation domain, then S satisfies each of the first three items. As noted above, if S is coherent or a GCD domain, then S is a finite conductor domain. To complete the proof of Theorem 6.2, it suffices to show that if S is not a valuation domain, then S is not a finite conductor domain. Specifically, we assume S is not a valuation domain and we consider three cases. In each case, we find a pair of principal fractional ideals of S whose intersection is not finitely generated.

Case 1: S is non-archimedean. By Theorem 2.5, there is a unique dimension 1 prime ideal Q of S, $QS_Q = Q$, and S_Q is the Noetherian hull of S. If $\dim S_Q = 1$, then, as a regular local ring, S_Q is a DVR; this, along with the fact that $Q = QS_Q$, implies S is a DVR, contrary to the assumption that S is not a valuation domain. Therefore $\dim S_Q \geq 2$, and there exist elements $f, g \in Q$ that have no common factors in the UFD S_Q. Consider $I = fS \cap gS$, let $x \in \mathfrak{m}_S$ such that $\sqrt{xS} = \mathfrak{m}_S$

[3]Muhammad Zafrullah has shown us a different proof of Theorem 6.2, an outline of which is as follows. Observe that a Shannon extension S is a Schreier domain, i.e., S is an integrally closed domain such that if $r|xy$, then $r = st$ where $s|x$ and $t|y$. A finite conductor Schreier domain is a GCD domain by [32, Theorem 3.6]. By Theorem 2.2, $\mathfrak{m}_S$ is the radical of a principal ideal, and so S is t-local [18, Proposition 1.1(5)]. In particular, $(((x, y)S)^{-1})^{-1} \subseteq \mathfrak{m}_S$ for all $x, y \in S$. If S is a GCD domain that is not a valuation domain, then there exist $x, y \in \mathfrak{m}_S$ such that $xS \cap yS = xyS$. However, this implies $(((x, y)S)^{-1})^{-1} = S$, a contradiction. Thus if S is a GCD domain, S is a valuation domain. These observations imply the theorem.

(see Theorem 2.2), and let $a \in I$. Since $a \in fS_Q \cap gS_Q = fgS_Q$, we can write $a = fgy$ for some $y \in S_Q$. Now $gy \in QS_Q = Q$ and $fy \in QS_Q = Q$, so $\frac{gy}{x} \in S$ and $\frac{fy}{x} \in S$. Thus $\frac{a}{x} = f\frac{gy}{x} = g\frac{fy}{x} \in I$. This shows that $xI = I$ and so $\mathfrak{m}_S I = I$. Since $I \neq (0)$, Nakayama's Lemma implies that I is not finitely generated.

Case 2: S is archimedean, but not completely integrally closed. By Theorem 2.2, $\dim S \geq 2$. We claim that $\mathfrak{m}_S$ is not finitely generated as an ideal of S. Since $\dim S > 1$, if $\mathfrak{m}_S$ is a principal ideal, then $\bigcap_i \mathfrak{m}_S^i$ is a nonzero prime ideal of S, a contradiction to the assumption that S is archimedean. Thus $\mathfrak{m}_S$ is not principal. By [21, Proposition 3.5], this implies $\mathfrak{m}_S^2 = \mathfrak{m}_S$. From Nakayama's Lemma it follows that $\mathfrak{m}_S$ is not finitely generated. Since S is not completely integrally closed, there is an almost integral element θ over S that is not in S. By [21, Corollary 6.6], $\mathfrak{m}_S = \theta^{-1}S \cap S$.

Case 3: S is archimedean and completely integrally closed. By Theorem 2.2, $\dim S \geq 2$. By Theorems 2.2 and 2.8, $S = T \cap W$, where W is the rank 1 nondiscrete valuation ring with associated valuation $w(-)$ as in Definition 2.3 and T is a UFD that is a localization of S. Since $\sum_{n\geq 0} w(\mathfrak{m}_n) < \infty$ by Theorem 2.5, and since $\mathfrak{m}_n S$ is principal and generated by a unit of T for $n \gg 0$, the w-values of units of T generate a nondiscrete subgroup of $\mathbb{R}$.

Since S is archimedean, Theorem 2.5 implies T is a non-local UFD. Therefore there exist elements $f, g \in S$ that have no common factors in T. As in Case 1, we consider $I = fS \cap gS$. Since $S = T \cap W$, it follows that

$$\begin{aligned} I &= (fT \cap gT) \cap (fW \cap gW) \\ &= fT \cap gT \cap \{a \in W \mid w(a) \geq \max\{w(f), w(g)\}\}. \end{aligned}$$

Assume without loss of generality that $w(f) \geq w(g)$.

For $a \in I$, write $a = (\frac{a}{f})f$ in S and consider $w(a)$. Since $\frac{a}{f}$ is divisible by g in T, it is a non-unit in T, and thus it is a non-unit in S. Since W dominates S, it follows that $w(\frac{a}{f}) > 0$ and thus $w(a) > w(f)$.

We claim that $\mathfrak{m}_S I = I$. Since the w-values of the units of T generate a nondiscrete subgroup of $\mathbb{R}$, for any $\epsilon > 0$, there exists a unit x in T with $0 < w(x) < \epsilon$. Then for $a \in I$ and for some x with $0 < w(x) < w(a) - w(f)$, we have $\frac{a}{x} \in I$ and thus $a \in \mathfrak{m}_S I$. Since $\mathfrak{m}_S I = I$ and $I \neq (0)$, Nakayama's Lemma implies that I is not finitely generated.

In every case, we have constructed a pair of principal fractional ideals of S whose intersection is not finitely generated. We conclude that if S is not a valuation domain, then S is not a finite conductor domain. □

Acknowledgements We thank Muhammad Zafrullah for helpful comments and for showing us Remark 2.6 and the different proof of Theorem 6.2 described in footnote 3.

References

1. Abhyankar, S.: On the valuations centered in a local domain. Am. J. Math. **78**, 321–348 (1956)
2. Abhyankar, S.: Desingularization of plane curves. AMS Proc. Symp. Pure Math. (I) **40**, 1–45 (1983)
3. Abhyankar, S.: Resolution of Singularities of Embedded Algebraic Surfaces. Springer, Berlin (1998)
4. Chase, S.: Direct products of modules. Trans. Am. Math. Soc. **97**, 457–473 (1960)
5. Chevalley, C.: La notion d'anneau de décomposition. Nagoya Math. J. **7**, 21–33 (1954)
6. Cutkosky, D.: Resolution of Singularities. Graduate Studies in Mathematics, vol. 63. American Mathematical Society, Providence, RI (2004)
7. Fontana, M.: Topologically defined classes of commutative rings. Ann. Mat. Pura Appl. (4) **123**, 331–355 (1980)
8. Fontana, M., Huckaba, J., Papick, I.: Prüfer Domains. Marcel Dekker, New York (1997)
9. Gabelli, S., Houston, E.: Ideal theory in pullbacks. In: Chapman, S., Glaz, S. (eds.) Non-Noetherian Commutative Ring Theory. Kluwer, Dordrecht (2000)
10. Gilmer, R.: Multiplicative Ideal Theory. Marcel Dekker, New York (1972)
11. Gilmer, R., Mott, J., Zafrullah, M.: *t*-invertibility and comparability. In: Cahen, P.-J., Costa, D., Fontana M., Kabbaj, S.-E. (eds.) Commutative Ring Theory, pp. 141–150. Marcel Dekker, New York (1994)
12. Glaz, S.: Finite conductor rings. Proc. Am. Math. Soc. **129**(10), 2833–2843 (2001)
13. Granja, A.: Valuations determined by quadratic transforms of a regular ring. J. Algebra **280**(2), 699–718 (2004)
14. Granja, A., Sánchez-Giralda, T.: Valuations, equimultiplicity and normal flatness. J. Pure Appl. Algebra **213**(9), 1890–1900 (2009)
15. Granja, A., Martinez, M.C., Rodriguez, C.: Valuations dominating regular local rings and proximity relations. J. Pure Appl. Algebra **209**(2), 371–382 (2007)
16. Greenberg, B.: Global dimension of cartesian squares. J. Algebra **32**, 31–43 (1974)
17. Hartshorne, R.: Algebraic Geometry. Springer, New York (1977)
18. Hedstrom, J., Houston, E.: Some remarks on star operations. J. Pure Appl. Algebra **18**, 37–44 (1980)
19. Heinzer, W., Huneke, C., Sally, J.D.: A criterion for spots. J. Math. Kyoto Univ. **26**(4), 667–671 (1986)
20. Heinzer, W., Kim, M.-K., Toeniskoetter, M.: Directed unions of local quadratic transforms of a regular local ring. arXiv:1512.03848
21. Heinzer, W., Loper, K.A., Olberding, B., Schoutens, H., Toeniskoetter, M.: Ideal theory of infinite directed unions of local quadratic transforms. J. Algebra **474**, 213–239 (2017)
22. Heinzer, W., Olberding, B., Toeniskoetter, M.: Asymptotic properties of infinite directed unions of local quadratic transforms. J. Algebra **479**, 216–243 (2017)
23. Kaplansky, I.: Commutative Rings. Allyn and Bacon, Boston (1970)
24. Kunz, E.: Introduction to Commutative Algebra and Algebraic Geometry. Birkhäuser, Boston (1985)
25. Lipman, J.: On complete ideals in regular local rings. In: Algebraic Geometry and Commutative Algebra, vol. I, pp. 203–231. Kinokuniya, Tokyo (1988)
26. Matsumura, H.: Commutative Ring Theory. Cambridge University Press, Cambridge (1986)
27. McAdam, S.: Two conductor theorems. J. Algebra **23**, 239–240 (1972)
28. Nagata, M.: Local Rings. Wiley, New York (1962)

29. Olberding, B., Saydam, S.: Ultraproducts of commutative rings. In: Commutative Ring Theory and Applications (Fez, 2001), pp. 369–386. Lecture Notes in Pure and Appl. Math., vol. 231, Dekker, New York (2003)
30. Shannon, D.: Monoidal transforms of regular local rings. Am. J. Math. **95**, 294–320 (1973)
31. Vasconcelos, W.: The local rings of global dimension two. Proc. Am. Math. Soc. **35**, 381–386 (1972)
32. Zafrullah, M.: On a property of pre-Schreier domains. Commun. Algebra **15**, 1895–1920 (1987)
33. Zariski, O.: Reductions of the singularities of algebraic three dimensional varieties. Ann. Math. (2) **40**, 639–689 (1939)
34. Zariski, O., Samuel, P.: Commutative Algebra. Vol. II. Reprint of the 1960 edition. Graduate Texts in Mathematics, vol. 29. Springer, New York/Heidelberg (1975)

Divisorial Prime Ideals in Prüfer Domains

Thomas G. Lucas

Abstract For an integral domain R with quotient field $K \neq R$, the inverse of a nonzero fractional ideal I of R is the set $(R : I) = \{t \in K \in tI \subseteq R\}$. The divisorial closure of I with respect to R is the fractional ideal $(R : (R : I))$. In addition I is divisorial as an ideal of R if $I = (R : (R : I))$. Of concern here are divisorial prime ideals in Prüfer domains. In some cases one can have a pair of comparable Prüfer domains $T \subsetneq R$ with a common nonzero prime ideal P such that P is divisorial as an ideal of T but is not divisorial as an ideal of R. For example, if $P = P^2$ is a nonzero nonmaximal prime of a valuation domain V, then P is divisorial as an ideal of V but $P = PV_P$ is not divisorial as an ideal of V_P. We review several relevant results on divisorial primes and present some new sufficient conditions on when P is divisorial as an ideal of R, and if not when a $T \subsetneq R$ exists such that $P = P \cap T$ is divisorial as an ideal of T.

Keywords Prüfer domain • Divisorial ideal

Subject Classifications [MSC 2010] Primary 13F05, 13A15

1 Introduction

All the rings considered below are assumed to be integral domains, in most cases Prüfer domains. For a domain R with quotient field $K = qf(R)$ and nonzero fractional ideal J of R, $(R : J) = \{t \in K \mid tJ \subseteq R\}$, and $J_v = (R : (R : J))$ is the *divisorial closure of J with respect to R*. The fractional ideal J is a *divisorial ideal of* R if $J = (R : (R : J))$. As we will see, if I is a common ideal of comparable Prüfer domains $R \subsetneq S$, it may be that $I = (R : (R : I)) \subsetneq (S : (S : I))$ so one must be careful when using the "I_v" notation.

T.G. Lucas (✉)
Department of Mathematics and Statistics, University of North Carolina Charlotte, Charlotte, NC 28223, USA
e-mail: tglucas@uncc.edu

M. Fontana et al. (eds.), *Rings, Polynomials, and Modules*,
https://doi.org/10.1007/978-3-319-65874-2_14

We are primarily interested in the divisoriality behavior of nonzero prime ideals in Prüfer domains. In Sect. 4, we look at various sufficient conditions for a prime to be divisorial and provide a "new" characterization of when a nonzero nonmaximal prime of a Prüfer domain is divisorial (for the latter, see Corollary 4.3). We begin with a review of earlier related work and conclude with examples to illustrate how erratic the divisoriality status of a "starting" prime can be in chains of Prüfer domains. In particular, we give a pair of examples of strictly ascending chains of Prüfer domains both starting with the same nondivisorial nonmaximal prime P of a Prüfer domain R. Both chains $\{S_n\}$ and $\{D_n\}$ are indexed over the nonnegative integers beginning with $S_0 = R = D_0$. For n odd, PS_n and PD_n are divisorial primes of S_n and D_n, respectively. On the other hand for m even, $(S_m : PS_m) = S_m$ and $(D_m : PD_m) = D_m$, so neither PS_m nor PD_m is divisorial. The difference with the chains is the behavior of PS and PD for $S = \bigcup S_k (\subsetneq R_P)$ and $D = \bigcup D_k (\subsetneq R_P)$. In S, $(S : PS) = S$ with PS not maximal. On the other hand, PD is both divisorial and nonmaximal.

We make use of the usual notation of $\mathrm{Max}(R)$ for the set of maximal ideals of R and $\mathrm{Spec}(R)$ for the set of prime ideals. Also, we let $\mathrm{Spec}^*(R)$ denote the set of nonzero prime ideals of R.

2 Background

In 1966, Robert Gilmer introduced property (#) for integral domains [7]. His motivation was to find a property that could allow one to determine exactly when an almost Dedekind domain was a Dedekind domain. An integral domain D is said to be (#) if for each pair of distinct nonempty sets of maximal ideals $\mathscr{S} \neq \mathscr{T}$, the domains $\bigcap\{D_M \mid M \in \mathscr{S}\}$ and $\bigcap\{D_N \mid N \in \mathscr{T}\}$ are also distinct. Later, he and Bill Heinzer consider the property that each overring of D is (also) (#) [9]. Such a domain is now referred to as a (##) domain.

Here are characterizations of these two properties for Prüfer domains.

Theorem 2.1 (*[9, Theorem 1 and 3]*) *Let R be a Prüfer domain.*

1. *R is a (#) domain if and only if each maximal ideal contains an invertible ideal that is contained in no other maximal ideal.*
2. *R is a (##) domain if and only if each nonzero prime ideal P contains an invertible ideal I such that each maximal ideal that contains I also contains P.*

For a domain R, we let $\mathrm{Max}(R)$ denote the set of maximal ideals of R and for each ideal I, we let $\mathrm{Max}(I, R)$ denote the set of maximal ideals of R that contain I. In the case $\mathrm{Max}(I, R)$ is a proper subset of $\mathrm{Max}(R)$, we have an overring $\Theta_R(I) = \bigcap\{R_N \mid N \in \mathrm{Max}(R)\backslash\mathrm{Max}(I, R)\}$. In the case $\mathrm{Max}(I, R) = \mathrm{Max}(R)$, we let $\Theta_R(I) = K$, the quotient field of R.

With regard to the second statement in Theorem 2.1, Gilmer and Heinzer first proved that for a given nonzero prime P of a Prüfer domain R, R_P does not contain $\Theta_R(P)$ if and only if P contains an invertible ideal I such that $\mathrm{Max}(I,R) = \mathrm{Max}(P,R)$ [9, Corollary 2]. They did not give this property a name. Current terminology is that a nonzero prime P of a domain R (not necessarily Prüfer) is a *sharp* prime of R if R_P does not contain $\Theta_R(P)$.

The next three results are due to Jim Huckaba and Ira Papick. The first two concern when $(R : I)$ is a ring. The first applies to any nonzero ideal in an integral domain, the second is specific to nonzero prime ideals in a Prüfer domain. The third result provides a few sufficient conditions for a nonzero nonmaximal prime (of a Prüfer domain) to be divisorial.

Theorem 2.2 [13, Proposition 2.2] *The following are equivalent for a nonzero ideal I of a domain R:*

1. *$(R : I)$ is a ring.*
2. $(R : I) = (I_v : I_v)$.
3. $(R : I) = (I(R : I) : I(R : I))$.

Theorem 2.3 [13, Theorems 3.2 & 3.8, Proposition 3.9 & Corollary 3.4] *Let P be a nonzero nonmaximal prime of a Prüfer domain R and let M be a maximal ideal of R.*

1. $(R : P) = (P : P) = \Theta_R(P) \cap R_P$.
2. *The following are equivalent:*
   a. *M is divisorial.*
   b. *M is invertible.*
   c. $(R : M) \supsetneq R$.

Theorem 2.4 [13, Propositions 3.10 & 3.11] *Let P be a nonzero nonmaximal prime of a Prüfer domain R.*

1. *If R is a valuation domain, then P is divisorial.*
2. *If the set* $\mathrm{Max}(R)\backslash\mathrm{Max}(P,R)$ *is finite, then P is divisorial.*
3. *If R is a (##) domain, then P is divisorial.*

In the mid 1980s, Huckaba and Papick wrote a pair of papers with Marco Fontana on divisorial prime ideals in Prüfer domains. In [2], we find the following sufficient condition for a nonzero nonmaximal prime to be divisorial. In statement (2), $T(P) = \bigcup(R : P^n)$, the Nagata ideal transform of P (with respect to R).

Theorem 2.5 [2, Lemma 2.2, Theorems 2.1 & 3.1] *Let P be a nonzero nonmaximal prime of a Prüfer domain R.*

1. *If $(R : P) \subsetneq \Theta_R(P)$, then P is divisorial.*
2. *If $(R : P) \subsetneq T(P)$, then P is divisorial.*
3. *$(R : P) \subsetneq \Theta_R(P)$ if and only if there is an invertible ideal $I \subsetneq P$ such that* $\mathrm{Max}(I,R) = \mathrm{Max}(P,R)$ *(in other words, P is sharp).*

In the second paper, they were able to establish a necessary and sufficient condition for a nonzero prime of a Prüfer domain to be divisorial. This characterization involves the Nagata ideal transform, and it is valid for all nonzero primes including maximal ideals.

Theorem 2.6 [3, Proposition 9] *Let P be a nonzero prime ideal of a Prüfer domain R. Then P is a divisorial ideal of R if and only if either $(R : P) \subsetneq T(P)$ or $P = (R : T(P))$.*

For the special case that R is two-dimensional we have the following characterization.

Theorem 2.7 [3, Proposition 14] *Let P be a nonzero prime of a two-dimensional Prüfer domain R. Then P is a divisorial ideal of R if and only if $(R : P) \supsetneq R$.*

We need to skip ahead many years to find other results about divisorial prime ideals in Prüfer domains.

The notion of an antesharp prime was introduced in 2010 by Fontana, Evan Houston, and the author of this article. A nonzero prime ideal P of a domain R is said to be *antesharp* if each prime of $(P : P)$ that contains P contracts to P when intersected with R. In the case R is a Prüfer domain, this simply means that P is a maximal ideal of $(P : P)$. A maximal ideal is always antesharp. With regard to nonzero nonmaximal primes we have the following.

Theorem 2.8 [5, Proposition 2.3, Corollary 2.4] *Let P be a nonzero nonmaximal prime of a domain R.*

1. *P is antesharp if and only if each prime that properly contains P contains an invertible ideal that properly contains P.*
2. *If P is antesharp, then it is divisorial.*
3. *If R is a Prüfer domain and P is sharp, then it is antesharp (and thus divisorial).*

For Prüfer domains, “sharp" and “antesharp" are stable in the following sense: if P is sharp (antesharp) as an ideal of R, then PS is sharp (antesharp) in each overring $R \subseteq S \subseteq R_P$. As we will see, “divisorial" can be very unstable.

3 A Few Questions

Let M be a maximal ideal of a Prüfer domain R and let $T \subsetneq R \subsetneq S \subseteq R_M \subsetneq qf(T)$. From the Huckaba/Papick result mentioned above, we know that M is a divisorial ideal of R if and only if it is an invertible ideal of R. So if M is a divisorial ideal of R, then MS is a divisorial maximal ideal of S. Hence, if M is not a divisorial ideal of R, then there is an $R \subsetneq S \subseteq R_M$ such that MS is a divisorial ideal of S if and only if MR_M is principal. Also in the case that M is not a divisorial ideal of R, it is possible to determine when there is no T such that $M \cap T$ is a divisorial prime ideal of T. Specifically, no such T exists if and only if R/M is algebraic over a finite field [14, Theorem 2.3].

Next consider the case of a nonzero nonmaximal prime P of R. For each maximal ideal N that contains P, PR_N is a nonmaximal prime of the valuation domain R_N and thus is a divisorial ideal of R_N [13, Propositions 3.10]. So there is always an $R \subsetneq S \subsetneq R_P$ such that PS is a divisorial ideal of S. Here are several results from [14] related to when there is a Prüfer domain $T \subsetneq R \subsetneq qf(T)$ such that $P \cap T$ is a divisorial ideal of T.

Theorem 3.1 [14, Theorem 3.1] *Let P be a nonzero nonmaximal prime of a Prüfer domain R such that P is not a divisorial ideal of R. Let T be the pullback of D over P where $D \subsetneq R/P$ is a Prüfer domain with quotient field R_P/PR_P.*

1. *T is a Prüfer domain and P is a nonmaximal prime of T.*
2. *P is a divisorial ideal of T if and only if D and $(P : P)/P$ have no common nonzero ideals.*

Let $Un(R)$ be the set of Prüfer domains $T \subseteq R \subsetneq qf(T)$ and let $Un(R, P) = \{T \in Un(R) \mid P \cap T = P\}$. Similarly let $Ov(R, P) = \{S \mid R \subseteq S \subseteq qf(R)$ such that $PS = P\}$. By [14, Theorem 3.2], if P is a divisorial ideal of R, then it is divisorial with respect to each $T \in Un(R, P)$. On the other hand, if P is not a divisorial ideal of R, then for each $S \in Ov(R, P)$, P is not a divisorial ideal of S. As noted above, if P is a nonzero nonmaximal prime of R, then PR_N is a divisorial ideal of R_N for each maximal ideal N that contains P. So P not divisorial as an ideal of R is preserved for those $S \in Ov(R, P)$, but not for all PS' where $R \subsetneq S' \subseteq R_P$. In contrast, underrings exhibit the following behavior.

Theorem 3.2 [14, Theorem 3.7] *The following are equivalent for a nonzero nonmaximal prime ideal P of a Prüfer ring R.*

1. *There is no Prüfer domain $T' \subseteq R \subsetneq qf(T')$ where $P = P \cap T'$ is a divisorial ideal of T'.*
2. *There is no Prüfer domain $T \subseteq R \subsetneq qf(T)$ where $P \cap T$ is a divisorial ideal of T.*

From the proof provided for [14, Theorem 3.7], we have the following.

Theorem 3.3 *Let P be a nonzero nonmaximal prime ideal of a Prüfer domain R and let $T \in Un(R)$. Then $P \cap T$ is a divisorial ideal of T if and only if P is a divisorial ideal of $T' = T + P$.*

The next result from [14] provides a sufficient condition for the existence of $T \subsetneq R$ such that $P = P \cap T$ is a divisorial ideal of T.

Theorem 3.4 [14, Theorem 3.12] *Let P be a nonzero nonmaximal prime of a Prüfer domain R such that $R = (P : P)$. Also let $T \in Un(R, P)$ be a proper Prüfer underring of R and let $D = T/P (\subsetneq R/P)$. If there is a maximal (or prime) ideal N' of domain D such that each nonzero prime $Q' \subseteq N'$ blows up in R/P, then P is a divisorial ideal of T.*

In the next section we consider two variations of this theorem.

The next result is for two very specific cases (with negative answers).

Theorem 3.5 [14, Theorem 3.4] *Let P be a nonzero prime of a Prüfer domain R.*

1. *If $(P : P)/P$ is isomorphic to $\mathbb{Z}$, then P is neither a maximal ideal nor a divisorial ideal of R, $(P : P) = R = P_v$ and for each $T \in Un(R)$, $P \cap T$ is such that $(P \cap T : P \cap T) = T = (P \cap T)_v$ and $P \cap T$ is neither maximal nor divisorial as an ideal of T.*
2. *If $(P : P)/P$ is isomorphic to $F[X]$ where F is an algebraic extension of the finite field $\mathbb{Z}_p$ for some prime p, then P is neither a maximal ideal nor a divisorial ideal of R, $(P : P) = R = P_v$ and for each $T \in Un(R)$, $P \cap T$ is such that $(P \cap T : P \cap T) = T = (P \cap T)_v$ and $P \cap T$ is neither maximal nor divisorial as an ideal of T.*

Corollary 3.6 [14, Corollary 3.5] *Let P be a nonzero prime of a Prüfer domain R.*

1. *If $(P : P)/P$ is the integral closure of $\mathbb{Z}$ in some algebraic extension of $\mathbb{Q}$, then $(P \cap T : P \cap T) = (T : P \cap T) = T$ for each Prüfer underring T of R and so there is no such T where $P \cap T$ is a divisorial ideal of T.*
2. *If $(P : P)/P$ is isomorphic to the integral closure of $\mathbb{Z}_p[X]$ in some algebraic extension of $\mathbb{Z}_p(X)$, then $(P \cap T : P \cap T) = (T : P \cap T) = T$ for each Prüfer underring T of R and so there is no such T where $P \cap T$ is a divisorial ideal of T.*

Essentially this takes us up to the present time. In the next section, we provide new results about divisorial prime ideals in Prüfer domains.

4 Divisorial Primes

For a nonzero ideal I of a domain R, we let $\mathrm{Inv}(I, R)$ denote the set of invertible integral ideals, including R, that contain I.

It is well known that for a nonzero ideal I of a domain R with quotient field K, $I_v = \bigcap\{tR \mid I \subseteq tR, t \in K\}$. In general, one cannot restrict to principal integral ideals that contain I, it is not uncommon that the only such ideal is the domain R even if I is divisorial. For example, the maximal ideal M of $R = \mathbb{Q}+X\mathbb{R}[[X]]$ is divisorial (with inverse $\mathbb{R}[X]$) but R is the only principal integral ideal (of R) that contains M. Also if R is a Dedekind domain that is not a PID, then there are (invertible) maximal ideals that are not principal, and certainly none of these is the intersection of principal integral ideals. However, it is the case that I_v is the intersection of the invertible integral ideals in $\mathrm{Inv}(I, R)$ when I is a nonzero (integral) ideal of a Prüfer domain R.

Lemma 4.1 (cf. [4, Lemma 4.9]) *Let R be a Prüfer domain. If tR is a principal fractional ideal of R, then $R \cap tR$ is an invertible integral ideal of R.*

Proof Simply use that a finite intersection of finitely generated fractional ideals is finitely generated when R is a Prüfer domain [8, Proposition 25.4]. Also note that

Lemmas 4.8 and 4.9 (together with the proof of Proposition 1.1) in [4] provide a way to obtain an explicit pair of elements that generates the ideal $R \cap tR$.

Please note that the fact that $R \cap tR$ is an invertible integral ideal of R (when R is Prüfer) was also used in the proof of [11, Theorem 3.1].

The following result provides a characterization for the divisorial closure of a nonzero ideal in a Prüfer domain. It is an easy consequence of the previous lemma.

Theorem 4.2 *Let R be a Prüfer domain and let J be a nonzero ideal of R.*

1. $J_v = \bigcap\{I \in \mathrm{Inv}(J,R)\}$.
2. *J is divisorial if and only if* $J = \bigcap\{I \in \mathrm{Inv}(J,R)\}$.
3. $J_v = R$ *if and only if* $\mathrm{Inv}(J,R) = \{R\}$.

Proof The proof of (1) follows easily from Lemma 4.1 and the fact that $J_v = \bigcap\{tR \mid J \subseteq tR, t \in K\}$. The other two statements follow easily from (1).

Thus for a nonzero prime ideal we have the following (this result contains the "new" characterization of when $P = P_v$ mentioned above).

Corollary 4.3 *Let R be a Prüfer domain and let P be a nonzero prime ideal of R.*

1. $P_v = \bigcap\{I \in \mathrm{Inv}(P,R)\}$.
2. *P is divisorial if and only if* $P = \bigcap\{I \in \mathrm{Inv}(P,R)\}$.
3. $P_v = R$ *if and only if* $\mathrm{Inv}(P,R) = \{R\}$.

Corollary 4.4 *Let R be a Prüfer domain.*

1. *If P is a nonzero prime ideal of R, then $(R : P) \supsetneq R$ if and only if some proper invertible ideal contains P.*
2. *$(R : P) = R$ for each nonzero prime ideal P of R if and only if no proper invertible ideal contains a nonzero prime.*

For comparable distinct primes $P \subsetneq Q$ of a domain R, let $(P, Q]$ denote the set of primes P' such that $P \subsetneq P' \subseteq Q$. If R is a Prüfer domain, then $(P, Q]$ is a chain. In the case there is a prime $N \supsetneq P$ such that each prime in $(P, N]$ is sharp, then P is a divisorial ideal of R [14, Lemma 3.10]. The next few results present a few variations with regard to this sufficient condition.

Theorem 4.5 *Let $P \subsetneq Q$ be a pair of nonzero prime ideals of a Prüfer domain R.*

1. *Each prime in $(P, Q]$ blows up in $(P : P)$ if and only if each prime in $(P, Q]$ contains an invertible ideal that contains P.*
2. *If each prime in $(P, Q]$ blows up in $(P : P)$, then P is divisorial and equal to the intersection $\{I \in \mathrm{Inv}(P,R) \mid I \subseteq Q\}$.*

Proof Let $P' \in (P, Q]$.

First suppose P' blows up in $(P : P)$. Then from the proof of Corollary 4.4, there is an invertible ideal $P \subsetneq I \subseteq P'$.

Conversely, if P' contains an invertible ideal J that contains P, then $(R : J) \subsetneq (R : P) = (P : P)$ and thus J blows up in $(P : P)$. It follows that P' blows up in $(P : P)$.

Now assume that each prime in $(P, Q]$ blows up in $(P : P)$ and consider the ideal $B = \bigcap\{I \in \mathrm{Inv}(P, R) \mid I \subseteq Q\}$. Clearly the ideal B is contained in each prime in $(P, Q]$. Also B contains P and P_v. All we need to do to prove $B = P$ is show $PR_Q = BR_Q$.

By way of contradiction, suppose $P \subsetneq B$. Then there is a prime $Q' \subseteq Q$ that is minimal over B. We have $P \subsetneq B \subseteq Q'$. Also there is an invertible ideal $P \subsetneq I \subseteq Q'$. Since Q' is minimal over B, there is a positive integer n such that $I^n R_{Q'} \subsetneq BR_{Q'}$. But I^n also contains P and thus $B \subseteq I^n$, a contradiction. Hence $B = P$.

Let $\mathrm{Inv}^*(P, R) = \{N \in \mathrm{Spec}(P, R) \mid N$ minimal over some $I \in \mathrm{Inv}(P, R)\}$.

Lemma 4.6 *Let P be a nonzero nonmaximal prime in a Prüfer domain R.*

1. *If Q is a sharp prime of R that properly contains P, then Q contains a sharp prime in the set* $\mathrm{Inv}^*(P, R)$.
2. *If Q is a sharp prime of R that properly contains P and Q is branched, then* $Q \in \mathrm{Inv}^*(P, R)$ *and* $P' = \bigcap\{I \in \mathrm{Inv}(P, R) \mid \sqrt{I} = Q\}$ *is a divisorial prime such that* $P \subseteq P' \subsetneq Q$ *and there are no primes properly between* P' *and* Q.

Proof Suppose Q is a sharp prime that contains P. Then there is an invertible ideal $J \subseteq Q$ such that $\mathrm{Max}(J, R) = \mathrm{Max}(Q, R)$ [9, Corollary 2]. Hence checking locally reveals that $J \supsetneq P$ and so $J \in \mathrm{Inv}(P, R)$. Let N be such that N is minimal over J and contained in Q. As $\mathrm{Max}(Q, R) = \mathrm{Max}(J, R) \supseteq \mathrm{Max}(N, R) \supseteq \mathrm{Max}(Q, R)$, another local check shows that $N = \sqrt{J}$ and therefore N is a sharp and in the set $\mathrm{Inv}^*(P, R)$.

Note that if Q is both sharp and branched, we may assume Q is minimal over J and thus $Q \in \mathrm{Inv}^*(P, R)$. If I is an invertible ideal such that $\sqrt{I} = Q$ and Q' is a prime that is properly contained in Q, then $IR_M \supsetneq Q'R_M$ for each maximal ideal $M \in \mathrm{Max}(Q, R) = \mathrm{Max}(I, R)$. It follows that I contains Q' and therefore $Q' \subseteq P' = \bigcap\{B \in \mathrm{Inv}(P, R) \mid \sqrt{B} = Q\}$. We have $P'R_M \subsetneq IR_M$ (since $P'R_M \subseteq I^2R_M \subsetneq IR_M$). Thus P' is a divisorial prime that is properly contained in Q and contains each prime ideal that is properly contained in Q.

In the next result we consider the special case that $\mathrm{Inv}^*(P, R)$ is a finite nonempty set.

Theorem 4.7 *Let P be a nonzero nonmaximal prime in a Prüfer domain R. If the set* $\mathrm{Inv}^*(P, R)$ *is finite and nonempty, then the following hold:*

1. *Each prime in* $\mathrm{Inv}^*(P, R)$ *is sharp and the radical of an invertible ideal that contains P.*
2. *For each* $Q \in \mathrm{Inv}^*(P, R)$, *each prime that contains Q is in* $\mathrm{Inv}^*(P, R)$ *so all are sharp.*
3. *Each sharp prime that contains P is in the set* $\mathrm{Inv}^*(P, R)$.
4. *The following are equivalent:*

   a. *P is divisorial.*
   b. *At least one* $Q \in \mathrm{Inv}^*(P, R)$ *is such that there are no primes properly between P and Q.*
   c. *There is a prime* $N \in \mathrm{Inv}^*(P, R)$ *such that* $P = \bigcap\{I \in \mathrm{Inv}(P, R) \mid \sqrt{I} = N\}$.

Proof To start assume $\mathrm{Inv}^*(P,R) = \{N_1, N_2, \ldots, N_m\}$.

For each N_j, let $I'_j \in \mathrm{Inv}(P,R)$ be such that N_j is minimal over I'_j. By prime avoidance, we may select an element $b_j \in N_j$ that is in no N_i that does not contain N_j. The ideal $I_j = b_jR + I'_j$ is an invertible ideal that contains P. As N_j is the only prime in $\mathrm{Inv}^*(P,R)$ that is minimal over I_j, $N_j = \sqrt{I_j}$ and therefore N_j is a sharp prime of R.

Continue with N_j and suppose Q is a prime that properly contains N_j. We are done if Q is one of the N_is. By way of contradiction assume it is not. Then by prime avoidance there is an element $d \in Q$ that is not in $\bigcup N_i$. The ideal $J = dR + I_j$ is invertible and contains P. We have a contradiction since no prime in $\mathrm{Inv}^*(P,R)$ contains J. Hence $Q = N_i$ for some i.

By Lemma 4.6, each sharp prime of R that properly contains P contains a prime in the set $\mathrm{Inv}^*(P,R)$, hence it is a prime in the set $\mathrm{Inv}^*(P,R)$.

Also by Lemma 4.6, for each N_j, $P'_j = \bigcap\{I \in \mathrm{Inv}(P,R) \mid \sqrt{I} = N_j\}$ is a divisorial prime that contains P and all other primes that are properly contained in N_j. The equivalence of the three statements in (4) now follow from Corollary 4.3.

Corollary 4.8 *Let Q be a sharp prime of a Prüfer domain R. If Q is branched and not height one, then $J = \bigcap\{I \in \mathrm{Inv}(R) \mid \sqrt{I} = Q\} \subsetneq Q$ is a divisorial prime ideal of R that contains each prime that is properly contained in Q.*

Proof Assume Q is branched and let $\mathscr{I}$ be the set of invertible ideals with radical equal to Q. This set is nonempty. For $I \in \mathscr{I}$, we have $IR_M \supsetneq PR_M$ for each prime P that is properly contained in Q and maximal ideal $M \in \mathrm{Max}(Q,R) = \mathrm{Max}(I,R)$. For $N \in \mathrm{Max}(R)\backslash\mathrm{Max}(Q,R)$, we have $IR_N = R_N$. Hence I contains each prime that is properly contained in Q.

It follows that $J = \bigcap\{I \in \mathscr{I}\}$ contains each prime ideal that is properly contained in Q. For $M \in \mathrm{Max}(Q,R)$, we have $JR_M \subseteq I^2R_M \subsetneq IR_M$ for each $I \in \mathscr{I}$. So it must be that JR_M is a prime that is properly contained in Q. As J is an intersection of invertible ideals, it is divisorial.

The next theorem is related to [14, Theorem 3.12]. One difference is instead of assuming $(P:P) = R$, we simply have that P is not a divisorial ideal of R (although this is used only to emphasize that P becomes divisorial when viewed as an ideal of T). Another is that here we are not starting with $T \subsetneq R$, instead T is built as the pullback of $D = V \cap R/P \subsetneq R/P$ for some valuation domain V which satisfies very particular restrictions with respect to R/P.

Theorem 4.9 *Let P be a nonzero nonmaximal prime of a Prüfer domain R that is not a divisorial ideal of R. Also let L be the quotient field of R/P and let V be a (proper) valuation domain of L such that (i) $D = V \cap R/P$ is a Prüfer domain with quotient field L, and $V(R/P) = L$. Then the following occur for T, the pullback of V over P.*

1. *T is a Prüfer domain.*
2. *For each nonzero prime Q' of V, $Q = Q' \cap R/P$ is a sharp prime of D that blows up in R/P.*

3. *For Q as in (2), the pullback of Q over P is a sharp prime of T.*
4. *P is a divisorial ideal of T.*

Proof By [12, Proposition 1.3], T is a Prüfer domain. From basic properties of pullbacks, each prime of T that properly contains P is the pullback of a nonzero prime ideal of D. Let N' be the maximal ideal of V and let $N = N' \cap R/P$.

Let $Q = Q' \cap R/P$ where Q' is a nonzero prime of V. Since $V(R/P) = L$, $V_{Q'}$ is incomparable with R/P. As D is a Prüfer domain, $D_Q = V_{Q'}$. It follows that Q blows up in R/P. Hence there is a finitely generated ideal $I \subseteq Q$ such that $IR/P = R/P$. In particular, there is a finitely generated ideal $B \subseteq N$ such that $BR/P = R/P$.

Let J be a finitely generated proper ideal of D that blows up in R/P. Since D is a Prüfer domain, $(D : J) = (V : J) \cap (R/P : J) = (V : J) \cap R/P$ properly contains D and thus JV is a proper ideal of V. It follows that $J \subseteq N$ and therefore N is a maximal ideal of D.

Next consider a finitely generated proper ideal H of D that blows up in V. As with J, $D \subsetneq (D : H) = (V : H) \cap (R/P : H)$ implies that HR/P is a proper ideal of R/P. It follows that each maximal ideal of R/P contracts to a maximal ideal of D. We also have that $\mathrm{Max}(D) = \{N\} \cup \{M \mid M = M' \cap D, M' \in \mathrm{Max}(R/P)\}$. Therefore N and each Q above are sharp primes of D. In addition, $\mathrm{Spec}(D)\backslash\{(0)\}$ is the disjoint union of $\{Q \mid Q = Q' \cap R/P$ some nonzero $Q' \in \mathrm{Spec}(V)\}$ and $\{P'' \mid P'' = P' \cap V$ some nonzero $P' \in \mathrm{Spec}(R/P)\}$. For each $Q_\alpha \in \{Q \mid Q = Q' \cap R/P$ some nonzero $Q' \in \mathrm{Spec}(V)\}$, we abuse notation and let $Q_\alpha T$ denote the pullback of Q_α over P. In particular, NT denotes the pullback of N over P.

For the domain T, each maximal ideal of R contracts to a maximal ideal of T. Moreover for each such maximal ideal M' of R, $T_M = R_{M'}$ where $M = M' \cap T$. The only other maximal ideal of T is the pullback of N over P. Thus $\Theta_T(NT) = R \supsetneq T$ and therefore NT is a sharp prime of T. We also have $Q_\alpha T$ a sharp prime for each $Q_\alpha \in \{Q \mid Q = Q' \cap R/P$ some nonzero $Q' \in \mathrm{Spec}(V)\}$.

By [14, Lemma 3.10], P is a divisorial prime of T.

Above, we saw that if $(P : P)/P$ is isomorphic to $\mathbb{Z}$, then $R = (P : P)$ and not only is P not a divisorial ideal of R, but there is no Prüfer underring $T \subsetneq R \subsetneq qf(T)$ such that $P \cap T$ is a divisorial ideal of T. In the next result, we show that we can make the opposite conclusion in the case $\mathbb{Z} \subsetneq R/P \subsetneq \mathbb{Q}$.

Theorem 4.10 *Let P be a nonzero nonmaximal prime of a Prüfer domain R such that $\mathbb{Z} \subsetneq R/P \subsetneq \mathbb{Q}$. Then the set $B = \{p \in \mathbb{Z} \mid p$ a prime that is a unit in $R/P\}$ is nonempty.*

1. *For each $p \in B$, the pullback of $\mathbb{Z}_{(p)} \cap R/P$ over P results in a Prüfer domain T_p such that P is a divisorial prime ideal.*
2. *P is a divisorial ideal of the Prüfer domain T that is the pullback of $\mathbb{Z}$ over P.*

Proof By [12, Theorem 1.3], T and each T_p are Prüfer domains. Since $\mathbb{Z}$, $\mathbb{Z}_{(p)} \cap R/P$ and R/P are comparable Dedekind domains with $\mathbb{Z} \subseteq \mathbb{Z}_{(p)} \cap R/P \subsetneq R/P$, neither $\mathbb{Z}$ nor $\mathbb{Z}_{(p)} \cap R/P$ have a nonzero ideal in common with R/P. Thus by [14, Theorem 3.1], P is a divisorial ideal of T and each T_p.

Together with Theorem 3.5 and Corollary 3.6 above, the previous result suggests a pair of problems to solve.

Problem 1 Let R be a one-dimensional Prüfer domain with quotient field K. Characterize those R such that there is no proper one-dimensional Prüfer underring with quotient field K.

An equivalent problem is the following.

Problem 2 Let R be a one-dimensional Prüfer domain with quotient field K. Characterize those R such that there is no proper Prüfer underring S with quotient field K which has a height one prime that blows up in R.

For both, we simply need that there is no one-dimensional valuation domain V with quotient field K such that $VR = K$ and $V \cap R$ is a Prüfer domain with quotient field K. To establish the equivalence of Problems 1 and 2 with this statement about the nonexistence of V, we first consider the "positive" versions of these three problems, when there do exist these Prüfer underrings.

Theorem 4.11 *Let R be a one-dimensional Prüfer domain with quotient field K. Then the following are equivalent:*

1. *There is a proper one-dimensional Prüfer underring $R' \subsetneq R$ with quotient field K.*
2. *There is a proper Prüfer underring S with quotient field K which has a height one prime that blows up in R.*
3. *There is a one-dimensional valuation domain V with quotient field K such that $VR = K$ and $V \cap R$ is a Prüfer domain with quotient field K.*

Proof First note that if $R' \subsetneq R$ is a one-dimensional Prüfer domain with quotient field K, then at least one nonzero, hence height one, prime of R' must blow up in R. Thus (1) implies (2).

Next we suppose there is a one-dimensional valuation domain V with quotient field K such that $VR = K$ and $T = V \cap R$ is a Prüfer domain with quotient field K. We will show that in this case T is one-dimensional and for the maximal ideal M of V, $M \cap T = M \cap R$ is a height one prime of T that blows up in R.

For each nonzero ideal J of T, $(T : I) = (V : I) \cap (R : I)$. If J is such that $JV = V$ and $JR = R$, there is a finitely generated ideal $I \subseteq J$ such that $IV = V$ and $IR = R$. As T is a Prüfer domain, I is invertible and thus must be equal to T. Hence for each proper ideal B of T, as least one of $BV \subseteq V$ and $BR \subseteq R$ is a proper containment. As T is a Prüfer domain, if Q is a nonzero prime ideal of T such that $QV \neq V$, then $QV = M$ the maximal ideal of V and in this case $T_Q = V$. On the other hand, if P is a nonzero prime of T such that $PR \neq R$, then PR is a prime ideal of R and $T_P = R_{PR}$. Thus each nonzero prime of T survives in exactly one of V and R. [By [6, Theorem 6.3.5 & Corollary 6.3.2], $\{V, R\}$ is a Jaffard family and so $\text{Spec}^*(T)$ is the disjoint union of $\{M \cap T\}$ and $\{P \cap T \mid P \in \text{Max}(R) = \text{Spec}^*(R)\}$.] As both V and R are one-dimensional, it must be that T is one-dimensional with a unique height one maximal ideal $N = M \cap T = M \cap R$ which extends to M in V and blows up in R. All other maximal ideals extend to maximal ideals of R and blow up in V. Hence (3) implies (1).

Finally, suppose there is a Prüfer domain $S \subsetneq R$ with quotient field K and a height one prime P such that $PR = R$. Then $W = S_P$ is a one-dimensional valuation domain such that $WR = K$ and $W \cap R$ is a Prüfer domain with quotient field K (as it sits between S and R). In addition $W \cap R \subsetneq R$ and $PS_P \cap R$ is a height one prime of $W \cap R$ that blows up in R. Thus (2) implies (3).

With regard to the exact wording of the two problems we have the following corollary.

Corollary 4.12 *Let R be a one-dimensional Prüfer domain with quotient field K. Then the following are equivalent:*

1. *There is no proper one-dimensional Prüfer underring $R' \subsetneq R$ with quotient field K.*
2. *There is no proper Prüfer underring S with quotient field K which has a height one prime that blows up in R.*
3. *There is no one-dimensional valuation domain V with quotient field K such that $VR = K$ and $V \cap R$ is a Prüfer domain with quotient field K.*

The final result of this section looks at things a little differently, instead of concentrating on a specific prime P, we look at a pair of comparable Prüfer domains $R \subsetneq S(\subsetneq qf(R))$ and consider certain Prüfer domains T where $R \subseteq T \subsetneq S$.

Theorem 4.13 *Suppose $R \subsetneq S \subsetneq qf(R)$ are Prüfer domains and Q is a prime ideal of R that blows up in S. Then $T = S \cap R_Q$ is a Prüfer domain such that QT is a sharp maximal ideal of T. In addition,*

1. *if $P \subseteq Q$ is such that $PS = S$, then PT is a sharp prime of T, and*
2. $\mathrm{Max}(T) = \{QT\} \cup \{N \cap T \mid N \in \mathrm{Max}(S)$ *with* $N \cap R \not\subseteq Q\}$.

Proof Since $R \subseteq T$, T is a Prüfer domain. Also since $QS = S$, there are elements $a_1, a_2, \ldots, a_n \in Q$ and $s_1, s_2, \ldots, s_n \in S$ such that $\sum s_i a_i = 1$. Thus the ideal $I = \sum a_i T$ is contained in QT and blows up in S.

Suppose J is a finitely generated proper ideal of T that contains I. Since $T = R_Q \cap S$ and $IS = S$, $(T : J) = (R_Q : JR_Q) \cap (S : JS) = (R_Q : JRQ_) \cap S$. As J is invertible, it must be that $(R_Q : JR_Q)$ properly contains R_Q and therefore $J \subseteq QT$. It follows that QT is a maximal ideal of T.

Next, let B be a finitely generated ideal in T that is not contained in QT. Then $(T : BT) = R_Q \cap (S : BS) \supsetneq T$ implies BS is a proper ideal of S. It follows that no $M \in \mathrm{Max}(T)\backslash\{QT\}$ blows up in S. Hence each $M \in \mathrm{Max}(T)\backslash\{QT\}$ is the contraction of a maximal ideal $M' \in \mathrm{Max}(S)$ such that $M' \cap R$ is not contained in Q.

That QT and each prime $P \subseteq QT$ such that $PS = S$ are sharp primes of T.

In the previous theorem, QT is the unique maximal ideal of T if and only if S is a valuation domain with maximal ideal N such that $N \cap R \subsetneq Q$.

We close this section with a pair of related questions (conjectures).

Question 1 Let R be a Prüfer domain with $dim(R) \geq 2$ and quotient field K. Is there a valuation domain V with quotient field K such that $V \subsetneq VR = K$ and $V \cap R$ is a Prüfer domain with quotient field K? Conjecture 1: there is such a valuation domain.

As we saw above, if P is a prime ideal of a Prüfer domain R such that R/P is one-dimensional, then in certain situations there is no Prüfer underring $T \subseteq R \subsetneq qf(T)$ such that $P \cap T$ is a divisorial ideal of T. The obvious question is whether this can occur in the case $dim(R/P) \geq 2$.

Question 2 Let P be a nonzero nonmaximal prime of a Prüfer domain R such that P is not a divisorial ideal of R and $dim(R/P) \geq 2$. Is there a Prüfer domain $T \subsetneq R \subsetneq qf(T)$ such that $T \cap P$ is a divisorial ideal of T?
Conjecture 2: there is such a Prüfer domain T.

A consequence of Theorem 4.9 is that if the answer to Question 1 is "Yes," then so is the answer to Question 2.

5 Examples

The domain $\text{Int}(\mathbb{Z}) = \{f(X) \in \mathbb{Q}[X] \mid f(\mathbb{Z}) \subseteq \mathbb{Z}\}$ is referred to as the *ring of integer valued polynomials*. It is well known to be a two-dimensional Prüfer domain. In addition it is completely integrally closed [1, Proposition VI.1.2]. There are no invertible maximal ideals and so for each nonzero prime ideal P, $(\text{Int}(\mathbb{Z}) : P) = (P : P) = \text{Int}(\mathbb{Z})$.

The primes of $\text{Int}(\mathbb{Z})$ are of two types [1, Propositions V.2.7 and V.2.8]. For an irreducible polynomial $f(X) \in \mathbb{Q}[X]$, $P_f = f(X)\mathbb{Q}[X] \cap \text{Int}(\mathbb{Z})$ is a height one nonmaximal prime. These are the only height one nonmaximal primes. Each maximal ideal has the form $M_{p,\alpha} = \{f(X) \in IntZ \mid f(\alpha) \in p\widehat{\mathbb{Z}_p}\}$ for a prime p and $\alpha \in \widehat{\mathbb{Z}_p}$ (the p-adic completion of $\mathbb{Z}$). If α is transcendental over $\mathbb{Z}$, then $M_{p,\alpha}$ has height one. Otherwise, $M_{p,\alpha}$ is height two and contains P_f where $f(\alpha) = 0$. For all primes p and $\alpha \in \widehat{\mathbb{Z}_p}$, $\text{Int}(\mathbb{Z})/M_{p,\alpha} = \mathbb{Z}_p$. Note that since $\text{Int}(\mathbb{Z})$ is a Prüfer ring, each height two maximal ideal contains a unique P_f.

For the prime $P_x = X\mathbb{Q}[X] \cap \text{Int}(\mathbb{Z})$, $\text{Int}(\mathbb{Z})/P_x \cong \mathbb{Z}$. Thus by Theorem 3.5 there is no Prüfer underring $T \subsetneq \text{Int}(\mathbb{Z})$ with quotient field $\mathbb{Q}(X)$ such that $P_x \cap T$ is a divisorial ideal of T.

Our first two constructions involving $\text{Int}(\mathbb{Z})$ are to make Prüfer underrings of the form $R = V \cap \text{Int}(\mathbb{Z})$ where the maximal ideal of V contracts to a sharp maximal ideal of R and the other maximal ideals of R are contracted from maximal ideals of $\text{Int}(\mathbb{Z})$.

For both, we make use of the Prüfer domain $D = \text{Int}((1/2)\mathbb{Z}, \mathbb{Z})$, the set of polynomials $f(X) \in \mathbb{Q}[X]$ such that $f((1/2)\mathbb{Z}) \subseteq \mathbb{Z}$. That D is a Prüfer domain follows from knowing that the map $\varphi : \text{Int}(\mathbb{Z}) \to D$ given by $\varphi(f(X)) = f(2X))$ is an isomorphism (see [1, Remark 1.11]). Clearly X is not in D so $D \subsetneq \text{Int}(\mathbb{Z})$. For each maximal ideal M of $\text{Int}(\mathbb{Z})$, $M \cap D$ is also a maximal ideal of D. In addition, $M = (M \cap D)\text{Int}(\mathbb{Z})$.

As with $\text{Int}(\mathbb{Z})$, each maximal ideal of D has the form $N_{p,\beta} = \{f(X) \in D \mid f(\beta) \in p\widehat{\mathbb{Z}_p}\}$ for some prime p and $\beta \in (1/2)\widehat{\mathbb{Z}_p}$ (the p-adic closure of $(1/2)\mathbb{Z}$ in the quotient field of $\widehat{\mathbb{Z}_p}$). Note that for p odd, $1/2$ is a unit of $\widehat{\mathbb{Z}_p}$ and thus in this case $N_{p,\beta} = M_{p,\beta} \cap D$ with $\beta \in \widehat{\mathbb{Z}_p}$. Also note that if $\beta \in \widehat{\mathbb{Z}_2}$, then $N_{2,\beta} = M_{2,\beta} \cap D$. Thus the maximal ideals of D that are not contracted from maximal ideals of $\mathrm{Int}(\mathbb{Z})$ are those of the form $N_{2,\beta}$ where $\beta \in (1/2)\widehat{\mathbb{Z}_2} \backslash \widehat{\mathbb{Z}_2}$. As with the maximal ideals of $\mathrm{Int}(\mathbb{Z})$, the height of $N_{2,\beta}$ is one if β is transcendental over $\mathbb{Z}$, and two if β is algebraic over $\mathbb{Z}$. A simple way to obtain $\beta \notin \widehat{\mathbb{Z}_2}$ is to simply choose an $\alpha \in \widehat{\mathbb{Z}_2}$ and then set $\beta = \alpha + 1/2$. For such a β, it is transcendental if and only if α is transcendental.

Example 5.1 Let $V = \mathrm{Int}((1/2)\mathbb{Z}, \mathbb{Z})_{N_{2,\gamma}}$ where $\gamma \in (1/2)\widehat{\mathbb{Z}_2} \backslash \widehat{\mathbb{Z}_2}$ is transcendental over $\mathbb{Z}$. Then $S = V \cap \mathrm{Int}(\mathbb{Z})$ is a two-dimensional Prüfer domain with $\mathrm{Max}(S) = \{N_{2,\gamma}V \cap S\} \bigcup \{M_{p,\alpha} \cap S \mid M_{p,\alpha} \in \mathrm{Max}(\mathrm{Int}(\mathbb{Z}))\}$. For $N = N_{2,\gamma}$, $S_N = V$ is a discrete rank one valuation domain such that $V\mathrm{Int}(\mathbb{Z}) = \mathbb{Q}(X)$. Also N is a height one sharp maximal ideal of S. For all other nonzero primes Q of S we have $(S : Q) = S$.

Example 5.2 For the second construction, consider the polynomial $g(X) = 2X-1 \in D$. The maximal ideal $N_{2,1/2}$ properly contains the height one prime $P_g \cap D$. Hence $W = \mathrm{Int}((1/2)\mathbb{Z}, \mathbb{Z})_{N_{1,1/2}}$ is a two-dimensional valuation domain with principal maximal ideal $2W$ and height one prime $P_g W = g\mathbb{Q}[X]_{(g)}$. The domain $T = W \cap \mathrm{Int}(\mathbb{Z})$ is a Prüfer domain with maximal ideal $N_{2,1/2}T = 2T + (2X-1)T$ that contains P_g. Moreover, by checking locally, one can show that $N_{2,1/2}T$ is an invertible ideal of T. Thus $N_{2,1/2}$ is sharp and from this we get that P_g is a divisorial ideal of T.

An alternate way to construct T is to use the fact that $\mathrm{Int}(\mathbb{Z})/P_g = \mathbb{Z}[1/2]$, a minimal ring extension of $\mathbb{Z}$. Now simply define T as the pullback of $\mathbb{Z}(= Z_{(2)} \cap Z[1/2])$ over P_g. With this view, we can see that P_g is a divisorial ideal of T by Theorem 4.10.

For a multiplicative subset S of the integers, $\mathrm{Int}(\mathbb{Z}_S) = \mathrm{Int}(\mathbb{Z})_S$ is completely integrally closed. If p is a prime not contained in the saturation of S, $M_{p,\alpha}\mathrm{Int}(\mathbb{Z})_S$ is a maximal ideal of $\mathrm{Int}(\mathbb{Z})_S$ for each $\alpha \in \widehat{\mathbb{Z}_p}$. The ring $\mathrm{Int}(\mathbb{Z})_S$ is contained in $\mathbb{Q}[X]$, thus $P_f\mathrm{Int}(\mathbb{Z})_S$ is a height one prime of $\mathrm{Int}(\mathbb{Z})_S$ for each irreducible polynomial $f(X)$. Unlike what happens in $\mathrm{Int}(\mathbb{Z})$, such a prime may be a maximal ideal of $\mathrm{Int}(\mathbb{Z})_S$. In fact, for a given $f(X)$, $P_f\mathrm{Int}(\mathbb{Z})_S$ is a maximal ideal of $\mathrm{Int}(\mathbb{Z})_S$ if and only if each prime q such that $f(X)$ has a zero in $\widehat{\mathbb{Z}_q}$ is in the saturation of S. For example, if $f(X) = X^2 + 1$, then $P_f\mathrm{Int}(\mathbb{Z})_S$ is a maximal ideal of $\mathrm{Int}(\mathbb{Z})_S$ if and only if the saturation of S contains each prime of the form $4k + 1$. No matter whether $P_f\mathrm{Int}(\mathbb{Z})_S$ is a maximal ideal of $\mathrm{Int}(\mathbb{Z})_S$ or not, $(P_f\mathrm{Int}(\mathbb{Z})_S : P_f\mathrm{Int}(\mathbb{Z})_S) = \mathrm{Int}(\mathbb{Z})_S$.

Our next two constructions using $\mathrm{Int}(\mathbb{Z})$ involve making ascending chains $R_0 = \mathrm{Int}(\mathbb{Z}) \subsetneq R_1 \subsetneq R_2 \subsetneq \cdots \subsetneq \bigcup R_n \subsetneq \mathrm{Int}(\mathbb{Z})_{P_x}$ such that P_xR_n is a divisorial prime ideal of R_n when n is odd and is not divisorial (so has trivial inverse) when n is even. For these two constructions we start with the set $B = \{p_n \mid n \in \mathbb{N}\}$ consisting of the primes of the form $4k + 1$. For each $n \geq 1$, we let $B_n = \{p_1, p_2, \ldots, p_n\}$ and $A_n = B \backslash B_n$. Next let $\overline{B}_n$ be the multiplicative set generated by B_n and $\overline{B}$ be the multiplicative set generated by B. We also define sets $C_n = \{h_1, h_2, \ldots, h_n\}$ where

$h_i = \dfrac{x(x-1)(x_2)\cdots(x-i+1)}{i!}$ for each positive integer i and let C_n' be the ideal generated by C_n. Finally, to simplify notation, we let $Q_n = M_{p_n,0}$ for each n.

Example 5.3 Assume the notation in the previous paragraph. Next set $S_0 = \text{Int}(\mathbb{Z})$ and then for $n \geq 1$, let $S_{2n} = \text{Int}(\mathbb{Z})_{\overline{B}_n}$ and $S_{2n-1} = S_{2n} \cap \text{Int}(\mathbb{Z})_{Q_n}$. The resulting chain is $S_0 = \text{Int}(\mathbb{Z}) \subsetneq S_1 \subsetneq S_2 \subsetneq \cdots \subsetneq S = \bigcup S_n = \text{Int}(\mathbb{Z})_{\overline{B}}$. This is a specific example of the chain created in [14, Example 4.2]. What occurs is that P_xS_m is a divisorial prime ideal of S_m when m is odd, and P_xS_m is not a divisorial as an ideal of S_m when m is even. It is also the case that P_xS is not divisorial as an ideal of S. For details see [14, Example 4.2].

Example 5.4 For the next chain, we make a slight variation in the sets B_n (and B) and the domains S_n. First, we let $B' = B \cup \{2\}$ and $B_n' = B_n \cup \{2\}$, then let $S_n' = \text{Int}(\mathbb{Z})_{B_n'}$ (so no longer distinguishing odd indices from even ones). Also for $n \geq 1$ we let $W_n = \bigcap\{\text{Int}(\mathbb{Z})_{M_{2,\alpha}} \mid C_n \subsetneq M_{2,\alpha},\ \alpha \in \widehat{\mathbb{Z}_2}\}$. To create an ascending chain $D_0 = \text{Int}(\mathbb{Z}) \subsetneq D_1 \subsetneq D_2 \subsetneq \cdots \subsetneq D = \bigcup D_m$ where P_xD_m is divisorial when m is odd, not divisorial when m is even and P_xD is divisorial and not maximal, we let $D_{2n} = S_n' \cap W_n$ and $D_{2n-1} = D_{2n} \cap \text{Int}(\mathbb{Z})_{Q_n}$.

Proof To verify the divisorial behavior of P_x in these extensions, first note that $\bigcup C_n$ generates P_x. Thus the only maximal ideal of the form $M_{2,\beta}$ that contains $\bigcup C_n$ is $M_{2,0}$. For all other $\alpha \in \widehat{\mathbb{Z}_2}$, some h_n is not contained in $M_{2,\alpha}$. By [5, Lemma 1.5.1], the only maximal ideals of Int($\mathbb{Z}$) that survive in W_n are those that contain 2 and the set C_n. It follows that $\bigcup W_n = \text{Int}(\mathbb{Z})_{M_{2,0}}$. Also note that 2 is a unit in S_n' for each $n \geq 1$. Similarly, each odd prime is a unit in W_n for each n.

For each n, $(S_n' : P_xS_n') = S_n'$ since $S_n' = \text{Int}(\mathbb{Z})_{\overline{B}_n} = \text{Int}(\mathbb{Z}_{\overline{B}_n})$ is completely integrally closed and P_xS_n' is not maximal.

Each maximal ideal of W_n contains 2 and the set C_n. In addition, $M_{2,0}W_n$ is the only maximal ideal of W_n that contains P_xW_n. Hence to see that $(W_n, P_xW_n) = W_n$ for each n we simply show that $W_n = \bigcap\{\text{Int}(\mathbb{Z})_{M_{2,\alpha}} \mid \alpha \neq 0\}$. For this, it suffices to show that each finitely generated ideal contained in $M_{2,0}$ is contained in infinitely many other $M_{2,\beta}$. Suppose otherwise and let I be a finitely generated ideal that is contained in $M_{2,0}$ and no other $M_{2,\beta}$ that survives in W_n. Then $M_{2,0}$ is the radical of the invertible ideal $2\text{Int}(\mathbb{Z}) + C_n' + I$ which would mean $M_{2,0}$ is sharp (and P_x divisorial in Int($\mathbb{Z}$)), a contradiction. So for P_xW_n we have $(W_n : P_xW_n) = W_n$.

Since $D_{2n} = S_n' \cap W_n$, $(D_{2n} : P_xD_{2n}) = (S_n' : P_xS_n') \cap (W_n : P_xW_n) = S_n' \cap W_n = D_{2n}$. Thus P_xD_{2n} is not divisorial.

For odd indices, the prime p_n is a unit in both S_n' and W_n, but it is not a unit in $\text{Int}(\mathbb{Z})_{Q_n}$. As the intersection to get D_{2n-1} results in a Prüfer domain, Q_nD_{2n-1} is a sharp maximal ideal of D_{2n-1} that contains P_xD_{2n-1}. It follows that P_xD_{2n-1} is a divisorial ideal of D_{2n-1}. A similar analysis yields that P_xD is a divisorial ideal of D since $M_{2,0}D$ is the only maximal ideal of D that contains 2.

With regard to descending chains, [14, Example 4.3] starts with $T_0 = \text{Int}(\mathbb{Z})_{\overline{B}}$ and uses the sets A_n defined above as $A_n = B \backslash B_n$ (so a descending chain of cofinite subsets of the set of primes of the form $4m + 1$ $(m \geq 1)$). Specifically, $T_{2m-1} =$

$T_{2m-2} \cap \mathrm{Int}(\mathbb{Z})_{Q_m}$ and $T_{2m} = \mathrm{Int}(\mathbb{Z})_{\overline{A}_m}$ for all positive m. The result is that P_xT_k is divisorial when k is odd and not divisorial when k is even. In this case $T = \bigcap T_k = \mathrm{Int}(\mathbb{Z})$ so that $P_xT = P_x$ is not divisorial. More generally we have the following for certain types of descending chains.

Theorem 5.5 *Let $T_0 \supsetneq T_1 \supsetneq T_2 \supsetneq \cdots \supsetneq \bigcap T_n = T$ be a chain of Prüfer domains where $qf(T) \supsetneq T_0$. If P is a nonzero nonmaximal prime of T_0 such that $(T_n : P \cap T_n) = T_n$ for infinitely many n, then $(T : P \cap T) = T$ even if $P \cap T_m$ is a divisorial ideal for infinitely many m.*

Proof Clearly we have $(T_n \cap P : T_n \cap P) = T_n$ for all n such that $(T_n : P \cap T_n) = T_n$. As T and each T_k have the same quotient field, $(T_i : T_k \cap P) \supseteq (T_j : P \cap T_j) \supseteq (T : P \cap T)$ for all $i \leq j$. Hence if there are infinitely many n such that $(T_n : P \cap T_n) = T_n$, then $T = \bigcap (T_i : P \cap T_i) \supseteq= (T : P \cap T) \supseteq T$. Therefore $P \cap T$ is not a divisorial ideal of T.

In the specific examples above and the earlier theorems about the existence of a Prüfer domain $T \subsetneq R$ where P becomes divisorial when viewed as an ideal of T, there has been a sharp prime $Q \supsetneq P$ of T where there are no primes properly between Q and P. For our final example, we construct a pair of Prüfer domains $T \subsetneq R$ with a common nonzero prime P where P is divisorial as an ideal of T, but is not divisorial as an ideal of R and, in addition, for each sharp prime $Q \supsetneq P$ in T there is a prime properly between P and Q. We again use P_x and $\mathrm{Int}(\mathbb{Z})$ as a base, but make a much larger Prüfer domain R.

Example 5.6 Start with a set $\mathscr{X} = \{X_p\}$ of indeterminates over $\mathbb{Q}(X)$ indexed over the prime numbers and let Y be an indeterminate over $\mathbb{Q}(X, \mathscr{X})$. Then $R = \mathrm{Int}(\mathbb{Z})(\mathscr{X}, Y)$ is a Bezout domain such that each ideal is extended from an ideal of $\mathrm{Int}(\mathbb{Z})$. Also, for each nonzero ideal I of $\mathrm{Int}(\mathbb{Z})$, $(\mathrm{Int}(\mathbb{Z}) : I)R = (R : IR)$. In particular, $(R : P_xR) = R$.

Modding out by P_xR yields the following isomorphism: $R/P_xR \cong \mathbb{Z}(\mathscr{X}, Y)$. For each prime number p, $p\mathbb{Z}(\mathscr{X}, Y)$ is a principal maximal ideal.

For each prime p, let $W_p = \mathbb{Z}_{(p)}(\mathscr{X})$ and let $\mathscr{X}_p = \mathscr{X} \backslash \{X_p\}$. Each W_p is a rank one discrete valuation domain with maximal ideal pW_p. The corresponding residue field is $\mathbb{Z}_p(\mathscr{X})$. Consider the polynomial ring $\mathbb{Z}_p(\mathscr{X}_p)[X_p]$ and let V_p' be the valuation overring $\mathbb{Z}_p(\mathscr{X}_p)[X_p]_{(X_p)}$. The pullback of V_p' over pW_p is a discrete rank two valuation domain V_p with principal maximal ideal X_pV_p.

Next let $\{V_p(Y)\}$ be the family of rank two valuation domains of $\mathbb{Q}(\mathscr{X}, Y)$ obtained by using the trivial extension to the polynomial ring $\mathbb{Q}(\mathscr{X})[Y]$. Hence each $V_p(Y)$ is a valuation domain with the same quotient field as R/P_xR. For a given prime p, the maximal ideal of $V_p(Y)$ is $X_pV_p(Y)$ and the height one prime is $pW_p(Y)$. By [8, Theorems 32.7 and 32.11], the intersection $D = \bigcap V_p(Y)$ is a Bezout domain. In addition, since X_p is a unit in $V_q(Y)$ for each prime $q \neq p$, the intersection is irredundant. Since $\mathbb{Q}(\mathscr{X}, Y)$ is the quotient field of D, each nonzero nonunit is the quotient of a pair of nonzero polynomials. From the definition of the V_ps, it is clear that such a quotient is a unit in all but at most finitely many $V_p(Y)$s. Hence the intersection has finite character. In addition, for a nonzero ideal I of D, if $IV_p(Y)$ is a proper ideal of $V_p(Y)$, then $IV_p(Y)$ contains a positive power of p. Hence $I_P = IV_p(Y) \cap D$ contains this same power of p and therefore $I_pV_q(Y) = V_q(Y)$ for each prime $q \neq p$. By [5, Theorem 6.3.5], the family $\{V_p(Y)\}$ is a "Jaffard family" for D. Hence by [5, Theorem 6.31. and Corollary 6.3.2], $\mathrm{Max}(D) = \{X_pV_p(Y) \cap D \mid p \text{ prime}\}$ and $\{pW_p(Y) \cap D \mid p \text{ prime}\}$ is the complete set of nonzero nonmaximal primes of D. Clearly, each nonzero prime ideal is contained in a unique maximal ideal. Also it is clear that the intersection has finite character. Hence D is two-dimensional and h-local. It follows that each nonzero prime ideal of D is sharp. By checking locally, we also have that $X_pD(= X_pV_p(Y) \cap D)$ is a maximal ideal of D.

Finally, we let T be the pullback of D over P_xR.

For each prime p, X_pT is a maximal ideal of T. It contains the chain of primes $M_{p,0}R \cap T$ and P_xR. Since $\bigcap X_pD = (0)$, $\bigcap X_pT = P_xR$. Therefore P_xR is a divisorial ideal of T. Since $(R : P_xR) = R$, P_xR is not an antesharp prime of T and thus it is not sharp either. The primes $M_{p,0}R \cap T$ are not sharp as $M_{p,0}R$ is not a sharp prime of R. For $\alpha \neq 0$ in $\widehat{\mathbb{Z}_p}$, $M_{p,\alpha}R \cap T$ is a maximal ideal of T. The only other maximal ideals are the X_pT. Hence X_pT is the only maximal ideal of T that contains $M_{p,0}R \cap R$. It follows that $M_{p,0}R \cap T$ is an antesharp (nonmaximal) prime of T, so it is divisorial as an ideal of T.

In several of the results and proofs above involving pullback constructions, a sharp prime in the residue ring pulls back to a sharp prime in its preimage/pullback. This is not always the case. Consider the rings R and T from Example 5.6. The image of $M_{p,0}R \cap T$ in D is sharp, but since the maximal ideal $M_{p,0}R$ is not sharp, $M_{p,0}R \cap T$ is not a sharp prime of T. However, it is antesharp as an ideal of T since X_pT is the only prime of T that properly contains $M_{p,0} \cap T$. Note that $M_{p,0}R$ is antesharp in R "for free" since it is a maximal ideal of R.

References

1. Cahen, P.-J., Chabert, J.-L.: Integer-Valued Polynomials. Mathematical Surveys and Monographs. American Mathematical Society, Providence (1997)
2. Fontana, M., Huckaba, J., Papick, I.: Divisorial prime ideals in Prüfer domains. Can. Math. Bull. **27**, 324–328 (1984)
3. Fontana, M., Huckaba, J., Papick, I.: Some properties of divisorial prime ideals in Prüfer domains. J. Pure Appl. Algebra **39**, 95–103 (1986)
4. Fontana, M., Huckaba, J., Papick, I., Roitman, M.: Prüfer domains and endomorphism rings of their ideals. J. Algebra **157**, 489–516 (1993)
5. Fontana, M., Houston, E., Lucas, T.G.: Toward a classification of prime ideals in Prüfer domains. Forum Math. **22**, 741–766 (2010)
6. Fontana, M., Houston, E., Lucas, T.G.: Factoring Ideals in Integral Domains. Lecture Notes of the Unione Matematica Italiana, vol. 14. Springer, Heidelberg; UMI, Bologna (2013)
7. Gilmer, R.: Overrings of Prüfer domains. J. Algebra **4**, 331–340 (1966)
8. Gilmer, R.: Multiplicative Ideal Theory. Queen's Papers in Pure and Applied Mathematics, vol. 90, Queen's University Press, Kingston (1992)
9. Gilmer, R., Heinzer, W.: Overrings of Prüfer domains. II. J. Algebra **7**, 281–302 (1967)

10. Heinzer, W.: Integral domains in which each non-zero ideal is divisorial. Mathematatika **15**, 164–170 (1968)
11. Heubo-Kwenga, O., Olberding, B., Reinhart, A.: Group-theoretic and topological invariants of completely integrally closed Prüfer domains. J. Pure Appl. Algebra **220**, 3927–3947 (2016)
12. Houston, E., Taylor, J.R.: Arithmetic properties in pullbacks. J. Algebra **310**, 235–260 (2007)
13. Huckaba, J., Papick, I.: When the dual of an ideal is a ring. Manuscr. Math. **37**, 67–85 (1982)
14. Lucas, T.G.: Divisorial Behavior of prime ideals in Prüfer domains. Rend. Circ. Mat. Palermo **64**, 277–290 (2015)

A gg-Cancellative Semistar Operation on an Integral Domain Need Not Be gh-Cancellative

Ryûki Matsuda

Abstract Let D be an integral domain with quotient field K, let h (resp., g, f) be the non-zero D-submodules of K (resp., the non-zero fractional ideals of D, the finitely generated non-zero fractional ideals of D), and let $\{x, y\}$ be a subset of the set {f, g, h} of symbols. For a semistar operation $\star$ on D, if $(EE_1)^\star = (EE_2)^\star$ implies $E_1^\star = E_2^\star$ for every $E \in$ x and every $E_1, E_2 \in$ y, then $\star$ is called xy-cancellative. We prove that a gg-cancellative semistar operation on an integral domain need not be gh-cancellative.

Keywords Star operation • Semistar operation

Mathematics Subject Classification (2010) 13A15

1 Introduction

The notions of a.b. and e.a.b. star operations and semistar operations are important in the study of topics such as Kronecker function rings and generalized integral closure. This has led to the study of various abstractions of the cancellative property. Thus, an fh- (resp., ff-) cancellative semistar operation is the same thing as an a.b. (resp., e.a.b.) semistar operation. We refer to Fontana and Loper [2] and Halter-Koch [6] for star and semistar operations and their Kronecker function rings.

Let D be a domain with quotient field K. If a mapping $E \longmapsto E^\star$ from $\bar{\mathrm{F}}(D)$ to $\bar{\mathrm{F}}(D)$ satisfies the following conditions, $\star$ is called a semistar operation on D: For every $0 \neq x \in K$ and every $E, H \in \bar{\mathrm{F}}(D)$, we have $(xE)^\star = xE^\star$, $E \subseteq E^\star$, $(E^\star)^\star = E^\star$, and $E \subseteq H$ implies $E^\star \subseteq H^\star$. If a mapping $E \longmapsto E^\star$ from $\mathrm{F}(D)$ to $\mathrm{F}(D)$ satisfies the following conditions, $\star$ is called a star operation on D: For every $0 \neq x \in K$ and every $E, H \in \mathrm{F}(D)$, we have $D^\star = D$, $(xE)^\star = xE^\star, E \subseteq E^\star, (E^\star)^\star = E^\star$, and $E \subseteq H$ implies $E^\star \subseteq H^\star$. If we set $E^e = K$ for every $E \in \bar{\mathrm{F}}(D)$, the mapping

R. Matsuda (✉)
Ibaraki University, Mito, Ibaraki 310-0903, Japan
e-mail: rmazda@adagio.ocn.ne.jp

M. Fontana et al. (eds.), *Rings, Polynomials, and Modules*,
https://doi.org/10.1007/978-3-319-65874-2_15

$E \longmapsto E^e$ is a semistar operation on D, called the e-semistar operation. Also, a star (resp., semistar) operation $\star$ on D is called of finite type, if $E^\star = \bigcup\{F^\star \mid F \in \mathrm{f}(D)$ with $F \subseteq E\}$ for every $E \in \mathrm{F}(D)$ (resp., $E \in \bar{\mathrm{F}}(D)$).

Let $\{V_\lambda \mid \lambda \in \Lambda\}$ be a non-empty set of valuation overrings of D. Then the mapping $E \longmapsto \bigcap\{EV_\lambda \mid \lambda \in \Lambda\}$ from $\bar{\mathrm{F}}(D)$ to $\bar{\mathrm{F}}(D)$ is a semistar operation on D, called the semistar operation defined by the set $\{V_\lambda \mid \lambda \in \Lambda\}$. This semistar operation is fh-cancellative (cf., [5, Theorem 32.5]).

Various implications hold among the cancellation properties of semistar operations:

Remark 1 1 [1]. We have the following diagram of implications:

$$\begin{array}{ccccccc} \mathrm{hh} = \mathrm{hg} = \mathrm{hf} \longrightarrow & \mathrm{gh} & \longrightarrow & \mathrm{gg} & \longrightarrow & \mathrm{gf} \\ & \downarrow & & \downarrow & & \downarrow \\ & \mathrm{fh} & \longrightarrow & \mathrm{fg} & \longrightarrow & \mathrm{ff} \end{array}$$

A semistar operation is hh-cancellative if and only if it coincides with the e-semistar operation.

A gh-cancellative semistar operation of finite type need not be hf-cancellative.

A gf-cancellative semistar operation of finite type need not be gg-cancellative.

An fh-cancellative semistar operation of finite type need not be gf-cancellative.

2 [4] and [7]. A gf-cancellative semistar operation need not be fg-cancellative.

Remark 2 (cf., [3, Lemma 3]) A finite type ff-cancellative semistar operation is fh-cancellative. So we have a simplified diagram of implications in the case of finite type semistar operations:

$$\mathrm{hh} = \mathrm{hg} = \mathrm{hf} \longrightarrow \mathrm{gh} \longrightarrow \mathrm{gg} \longrightarrow \mathrm{gf} \longrightarrow \mathrm{fh} = \mathrm{fg} = \mathrm{ff}$$

It follows that every example for Remark 1, 2 is not of finite type.

2 A gg not gh Semistar Operation

Lemma 1 (cf., [5, Proposition 18.4]) *Let v_0 be a valuation on a field k_0 with value group Γ, let $\{X_\lambda \mid \lambda \in \Lambda\}$ be a set of indeterminates over k_0, and let $\{\gamma_\lambda \mid \gamma_\lambda \in \Lambda\}$ be a subset of Γ', a totally ordered abelian additive group containing Γ as a subgroup. Then the mapping v: $k_0[X_\lambda \mid \lambda \in \Lambda] \longrightarrow \Gamma'$ defined by*

$v(\sum \alpha_{i_1\cdots i_n} X_{\lambda_1}^{i_1} \cdots X_{\lambda_n}^{i_n}) = min_{i_1\cdots i_n}\{v_0(\alpha_{i_1\cdots i_n}) + i_1\gamma_{\lambda_1} + \cdots + i_n\gamma_{\lambda_n}\}$

gives a valuation on $k_0(X_\lambda \mid \lambda \in \Lambda)$ which is an extension of v_0, where $\alpha_{i_1\cdots i_n} \in k_0$.

The valuation v is called the canonical extension of v_0, and is denoted by

$< v|_{k_0} = v_0;\ v(X_\lambda) = \gamma_\lambda$ for $\lambda \in \Lambda >$, where $v|_{k_0}$ is the restriction of v to k_0.

Hereafter throughout the paper, let k be a field, let $u_1, u_2, \cdots$ be an infinite set of indeterminates over k, and let

$\mathfrak{M} := \{f(u_1, u_2, \cdots) \in k[u_1, u_2, \cdots] \mid f(0, 0, \cdots) = 0\}$.

Let $D := k[u_1, u_2, \cdots]_{\mathfrak{M}}$, let M be the maximal ideal of D, and let K be the quotient field of D.

Let $\boldsymbol{Q}$ be the additive group of rational numbers, let $\boldsymbol{Q}_0 := \{\gamma \mid 0 \leq \gamma \in \boldsymbol{Q}\}$, and let $\mathbf{Z}_0 := \{i \mid 0 \leq i \in \mathbf{Z}\}$.

Let v_0 be the trivial valuation on k, that is, $v_0(\alpha) = \infty$ (resp., $v_0(\alpha) = 0$) if $\alpha = 0$ (resp., $\alpha \neq 0$), and let $\gamma_i \in \boldsymbol{Q}_0$ for every $i = 1, 2, \cdots$. Then the valuation

$< v|_k = v_0;\ v(u_i) = \gamma_i$ for $i = 1, 2, \cdots >$

is simply denoted by

$< v(u_i) = \gamma_i$ for $i = 1, 2, \cdots >$

or

$< v(u_1) = \gamma_1, v(u_2) = \gamma_2, \cdots >$,

and is called a q-valuation on K. The valuation ring belonging to a q-valuation on K is a valuation overring of D.

The following Lemma 2 is obvious.

Lemma 2 *1.Let $v := \ < v(u_i) = \gamma_i$ for $i = 1, 2, \cdots >$ be a q-valuation on K. Let $0 \neq x \in k[u_1, u_2, \cdots]$, and let*

$x = \sum \alpha_{i_1 i_2 \cdots} u_1^{i_1} u_2^{i_2} \cdots$ be the canonical expression of x, where $\alpha_{i_1 i_2 \cdots} \in k$

Then we have

$$v(x) = min_{\ i_1 i_2 \cdots} \{v(\alpha_{i_1 i_2 \cdots}) + \sum_{j=1}^{\infty} i_j \gamma_j\}$$

$= min_{\ i_1 i_2 \cdots} \{i_1\gamma_1 + i_2\gamma_2 + i_3\gamma_3 + \cdots \mid i_j \geq 0$ *for every* $j\}$.

2. Let $X := \{\gamma_{ab} \mid a \in \Lambda$ and $b \in \Sigma\}$ be a subset of $\boldsymbol{Q}_0 \cup \{\infty\}$, where almost all γ_{ab} are ∞. Then $\min_a\{\min_b \gamma_{ab}\} = \min_b\{\min_a \gamma_{ab}\} = \min(X)$.

Lemma 2 is used in the following,

Lemma 3 *Let v be a q-valuation on K, let $\gamma_i := v(u_i)$ for every i, let n be a positive integer, let $v_0 := v|_{k(u_1,\cdots,u_n)}$, i.e. the restriction of v to $k(u_1, \cdots, u_n)$, let $\gamma'_j \in \boldsymbol{Q}_0$ for every $j \geq n+1$, and let w be the canonical extension of v_0 to K with*

$w := \ < w|_{k(u_1,\cdots,u_n)} = v_0;\ w(u_{n+1}) = \gamma'_{n+1}, w(u_{n+2}) = \gamma'_{n+2}, \cdots >$

Then w is a q-valuation with

$w = \ < w(u_1) = \gamma_1, \cdots, w(u_n) = \gamma_n, w(u_{n+1}) = \gamma'_{n+1},$

$w(u_{n+2}) = \gamma'_{n+2}, \cdots >$.

Proof Clearly,

$w(u_1) = \gamma_1, \cdots, w(u_n) = \gamma_n, w(u_{n+1}) = \gamma'_{n+1}, \cdots$,

and the valuation $w|_k$ is of trivial. Let

$0 \neq x = \sum a_{i_{n+1} i_{n+2} \cdots} u_{n+1}^{i_{n+1}} u_{n+2}^{i_{n+2}} \cdots \in k[u_1, u_2, \cdots]$

and let

$$a_{i_{n+1} i_{n+2} \cdots} = \sum_{i_1 \cdots i_n} \alpha_{i_1 \cdots i_n}^{(i_{n+1} i_{n+2} \cdots)} u_1^{i_1} \cdots u_n^{i_n}$$

be canonical expressions, where $\alpha_{i_1 \cdots i_n}^{(i_{n+1} i_{n+2} \cdots)} \in k$ for every $i_1, \cdots, i_n, i_{n+1}, \cdots$

Then
$x = \sum_{i_1 \cdots i_n i_{n+1} \cdots} \alpha_{i_1 \cdots i_n}^{(i_{n+1} i_{n+2} \cdots)} u_1^{i_1} \cdots u_n^{i_n} u_{n+1}^{i_{n+1}} \cdots$
is the canonical expression of x. Since
$w =< w|_{k(u_1, \cdots, u_n)} = v_0; w(u_{n+1}) = \gamma'_{n+1}, w(u_{n+2}) = \gamma'_{n+2}, \cdots >$
we have,
$w(x) = \min_{i_{n+1} i_{n+2} \cdots} \{v_0(a_{i_{n+1} i_{n+2} \cdots}) + w(u_{n+1}^{i_{n+1}} u_{n+2}^{i_{n+2}} \cdots)\}$
Since $v = \ < v(u_i) = \gamma_i$ for $i = 1, 2, \cdots >$, we have,
$v_0(a_{i_{n+1} i_{n+2}} \cdots) = v(a_{i_{n+1} i_{n+2} \cdots}) = \min_{i_1 \cdots i_n} v(\alpha_{i_1 \cdots i_n}^{(i_{n+1} i_{n+2} \cdots)} u_1^{i_1} \cdots u_n^{i_n})$ for every $i_{n+1}, i_{n+2}, \cdots$.
It follows that
$w(x) = \min_{i_{n+1} i_{n+2} \cdots} \{\min_{i_1 \cdots i_n} v(\alpha_{i_1 \cdots i_n}^{(i_{n+1} i_{n+2} \cdots)} u_1^{i_1} \cdots u_n^{i_n}) \ + w(u_{n+1}^{i_{n+1}} u_{n+2}^{i_{n+2}} \cdots)\}$
$= \min_{i_{n+1} i_{n+2} \cdots} \{\min_{i_1 \cdots i_n} \{v(\alpha_{i_1 \cdots i_n}^{(i_{n+1} i_{n+2} \cdots)} u_1^{i_1} \cdots u_n^{i_n}) + w(u_{n+1}^{i_{n+1}} u_{n+2}^{i_{n+2}} \cdots)\}\}$
$= \min_{i_1 \cdots i_n i_{n+1} \cdots} \{v(\alpha_{i_1 \cdots i_n}^{(i_{n+1} i_{n+2} \cdots)} u_1^{i_1} \cdots u_n^{i_n}) + w(u_{n+1}^{i_{n+1}} u_{n+2}^{i_{n+2}} \cdots)\}$
$= \min_{i_1 \cdots i_n i_{n+1} \cdots} \{w(\alpha_{i_1 \cdots i_n}^{(i_{n+1} i_{n+2} \cdots)} u_1^{i_1} \cdots u_n^{i_n} u_{n+1}^{i_{n+1}} u_{n+2}^{i_{n+2}} \cdots)\}$.
The proof is complete.

Lemma 4 *Let $\gamma_i \in \boldsymbol{Q}_0$ for $1 \leq i \leq n$, and let S be a non-empty subset of $\sum_{i=1}^n \mathbf{Z}_0 \gamma_i$. Then the set S has a minimal element.*

The proof is obvious. Because, any non-empty set of positive integers has a minimal element.

Lemma 5 *Let I be a non-zero ideal of D with $I \subseteq M$. Let v be a q-valuation on K with valuation ring V. Set $v(u_i) := \gamma_i \in \boldsymbol{Q}_0$ for every i. Set $I_0 := k[u_1, u_2, \cdots] \cap I$, and let n be a positive integer such that $k[u_1, \cdots, u_n] \cap I \neq 0$, let*
$k[u_1, \cdots, u_n] \cap I_0 = \sum_{i=1}^m g_i k[u_1, \cdots, u_n]$
with $0 \neq g_i$ for every i. Let v_0 be the restriction of v to $k(u_1, \cdots, u_n)$, and let w be the canonical extension of v_0 to K with
$w :=< w|_{k(u_1, \cdots, u_n)} = v_0; w(u_{n+1}) = \gamma'_{n+1}, w(u_{n+2}) = \gamma'_{n+2}, \cdots >$,
where max $\{v(g_1), \cdots, v(g_m), \gamma_l\} < \gamma'_l \in \boldsymbol{Q}_0$ for every $l \geq n+1$. Then,

1. *Every $w(g_i)$ is of the form $i_1 \gamma_1 + \cdots + i_n \gamma_n$, where $i_j \in \mathbf{Z}_0$ for every j.*
2. *Let $I_0 \ni g \neq 0$. Then, $w(g)$ is either $w(g) \geq min_i\ w(g_i)$, or of the form $i_1 \gamma_1 + \cdots + i_n \gamma_n$, where $i_j \in \mathbf{Z}_0$ for every j.*
3. *The set $w(I)$ has a minimal element.*

Proof w is a q-valuation by Lemma 3. Also, every element in I is a non-unit of D.

1 follows from the fact that $g_i \in k[u_1, \cdots, u_n]$ for every $1 \leq i \leq n$.
2 Let $g = \sum \alpha_{i_1 i_2 \cdots} u_1^{i_1} u_2^{i_2} \cdots$ be the canonical expression. We have $g \notin k$, and $w(g) = \min_{i_1 i_2 \cdots} w(\alpha_{i_1 i_2 \cdots} u_1^{i_1} u_2^{i_2} \cdots)$.

There are the following two cases:

(2.1) $g \in k[u_1, \cdots, u_n]$,
(2.2) $g \notin k[u_1, \cdots, u_n]$.

The case (2.1): Then $g = \sum_{i=1}^m g_i h_i$ for every $h_i \in k[u_1, \cdots, u_n]$. Hence $w(g) \geq \min_i w(g_i)$.

The case (2.2): There are the following two cases:

(2.2.1) $w(g) = w(u_1^{i_1} \cdots u_n^{i_n})$ for some $i_1, \cdots, i_n$,
(2.2.2) $w(g) = w(u_1^{i_1} \cdots u_n^{i_n} \cdots u_l^{i_l} \cdots)$ for some $i_1, i_2, \cdots$, where $n < l$ and $1 \le i_l$.

The case (2.2.1): Then $w(g) = i_1\gamma_1 + \cdots + i_n\gamma_n$.

The case (2.2.2): Then

$w(g) = i_1\gamma_1 + \cdots + i_n\gamma_n + \cdots + i_l\gamma_l' + \cdots \ge \gamma_l' > \max_i v(g_i) \ge \min_i v(g_i) = \min_i w(g_i)$.

3 Set

$S := \{\gamma \in \sum_{i=1}^n \mathbf{Z}_0\gamma_i \mid \gamma = w(g) \text{ for some } g \in I\}$,

and set

$S' := \{\gamma \in \boldsymbol{Q}_0 \mid \gamma = w(g) \text{ for some } g \in I\}$.

Then $S \subseteq S'$, and S has a minimal element by Lemma 4. By 2, every element of S' is greater than or equal to some element of S. It follows that S' has a minimal element.

Let $\{v_\lambda \mid \lambda \in \Lambda\}$ be the set of all q-valuations on K, and let V_λ be the valuation overring of D belonging to v_λ for every λ. Then we denote the semistar operation on D defined by the set $\{V_\lambda \mid \lambda \in \Lambda\}$ by b'.

Proposition 1 *The semistar operation b' is a gg-cancellative semistar operation on D.*

Proof Let $If \subseteq (IJ)^{b'}$, where I, J are non-zero ideals of D, and $0 \neq f \in D$. We must show that $f \in J^{b'}$. Suppose that $f \notin J^{b'}$. We will derive a contradiction.

We may assume that $f \in k[u_1, u_2, \cdots], I \subseteq M$, and $J \subseteq M$. Since $f \notin J^{b'}$, there is a q-valuation overring V of D with $f \notin JV$. Let v be a q-valuation on K belonging to V. Set $v(u_i) := \gamma_i \in \boldsymbol{Q}_0$ for every i. We have $v(f) < v(x)$ for every $x \in J$. We have $f \in k[u_1, \cdots, u_r]$ for some r.

Set $I_0 := k[u_1, u_2, \cdots] \cap I$. There is a positive integer $n > r$ such that $k[u_1, \cdots, u_n] \cap I \neq 0$. Since $k[u_1, \cdots, u_n]$ is a Noetherian ring, we have

$k[u_1, \cdots, u_n] \cap I_0 = \sum_{i=1}^m g_i k[u_1, \cdots, u_n]$

with $g_i \neq 0$ for every i. We may assume that $0 \le v(g_1) \le \cdots \le v(g_m)$. Let v_0 be the restriction of v to $k(u_1, \cdots, u_n)$, and let w be the canonical extension of v_0 to K with

$w := \; < w|_{k(u_1, \cdots, u_n)} = v_0; w(u_{n+1}) = \gamma'_{n+1}, w(u_{n+2}) = \gamma'_{n+2}, \cdots >$,

where $\max\{v(g_m), \gamma_l\} < \gamma_l' \in \boldsymbol{Q}_0$ for every $l \ge n+1$. We have $w(u_i) = v(u_i)$ for every $i \le n$, $w(f) = v(f)$, and w is a q-valuation on K by Lemma 3.

We have that $w(f) < w(x)$ for every $x \in J$.

For, we may assume that $0 \neq x \in k[u_1, u_2, \cdots]$. If $x \in k[u_1, \cdots, u_n]$, then $w(f) = v(f) < v(x) = w(x)$.

Thus, assume that $x \notin k[u_1, \cdots, u_n]$. Let $x = \sum_{i_1 i_2 \cdots} \alpha_{i_1 i_2 \cdots} u_1^{i_1} u_2^{i_2} \cdots$ be the canonical expression, where $\alpha_{i_1 i_2 \cdots} \in k$. Then we have

$w(f) = v(f) < v(x)$

$= \min_{i_1 i_2 \cdots} v(\alpha_{i_1 i_2 \cdots} u_1^{i_1} u_2^{i_2} \cdots)$

$= \min_{i_1 i_2 \cdots} \{v(\alpha_{i_1 i_2 \cdots}) + i_1\gamma_1 + \cdots + i_n\gamma_n + i_{n+1}\gamma_{n+1} + \cdots\}$

$\leq \min_{i_1 i_2 \cdots}\{v(\alpha_{i_1 i_2 \cdots}) + i_1\gamma_1 + \cdots + i_n\gamma_n + i_{n+1}\gamma'_{n+1} + \cdots\}$
$= \min_{i_1 i_2 \cdots} w(\alpha_{i_1 i_2 \cdots} u_1^{i_1} \cdots u_n^{i_n} u_{n+1}^{i_{n+1}} \cdots)$
$= w(x)$.

On the other hand, by Lemma 5, 3, $w(I)$ has a minimal element. Let $\min w(I) = w(g_0)$ with $g_0 \in I$. Since $If \subseteq (IJ)^{b'}$, we have $If \subseteq IJW$, where W is the valuation ring belonging to w. Hence $g_0 f = gxy$ for some $g \in I, x \in J$, and $y \in W$. It follows that $w(f) \geq w(x)$; a contradiction to the above assertion.

Proposition 2 *The b'-semistar operation on D is not gh-cancellative.*

Proof Let H be the D-submodule of K generated by the set $\{\frac{u_1}{u_2}, \frac{u_2}{u_3}, \cdots\}$. Let $\{\gamma_1, \gamma_2, \cdots\}$ be a set of positive rational numbers with $\gamma_i < \gamma_{i-1}$ for every i, let v be the q-valuation $< v(u_i) = \gamma_i$ for $i = 1, 2, \cdots >$ on K, and let V be the valuation overring belonging to v. Since $v(\frac{u_{i-1}}{u_i}) > 0$ for every i, we have $1 \notin HV$, hence $1 \notin H^{b'}$. Since $u_i = u_{i+1} \cdot \frac{u_i}{u_{i+1}} \in MH$ for every i, we have $M \subseteq MH \subseteq (MH)^{b'}$. It follows that the b'-semistar operation on D is not gh-cancellative.

Propositions 1 and 2 imply the following,

Proposition 3 *A gg-cancellative semistar operation on a domain need not be gh-cancellative.*

Remark 3

1. We have that $D^{b'} \supsetneqq D$, and that $D \not\subseteq M^{b'} \supsetneqq M$.
2. $M^{b'} = \bigcup\{F^{b'} \mid F$ is a finitely generated ideal of D with $F \subseteq M\}$.

Proof

1. Let $v := \; < v(u_i) = \gamma_i$ for $i = 1, 2, \cdots >$ be a q-valuation on K with valuation ring V. Then

$v(\frac{u_1 u_2}{u_1 + u_2}) = v(u_1 u_2) - v(u_1 + u_2) = \gamma_1 + \gamma_2 - \min\{\gamma_1, \gamma_2\} \geq 0$

Hence we have $\frac{u_1 u_2}{u_1 + u_2} \in V$. Therefore $\frac{u_1 u_2}{u_1 + u_2} \in D^{b'}$. It follows that

$\frac{u_1 u_2 u_3}{u_1 + u_2} \in u_3 D^{b'} = (u_3 D)^{b'} \subseteq M^{b'}$

The proof is complete.

2. Let

$$0 \neq z = \frac{\Sigma \alpha_{i_1 \cdots i_n} u_1^{i_1} \cdots u_n^{i_n}}{\Sigma \beta_{j_1 \cdots j_n} u_1^{j_1} \cdots u_n^{j_n}} \in M^{b'},$$

where every $\alpha_{i_1 \cdots i_n} \in k$ and $\beta_{j_1 \cdots j_n} \in k$. We will prove that $z \in (u_1, \cdots, u_n)^{b'}$.

Let v be a q-valuation on K with valuation ring V. Let

$v := \; < v(u_i) = \gamma_i$

for $i = 1, 2, \cdots >$. We may assume that $\gamma_1 \leq \cdots \leq \gamma_n$. Then $(u_1, \cdots, u_n)V = u_1 V$. It is sufficient to show that $v(z) \geq v(u_1)$.

Let

$w := \; < w(u_1) = \gamma_1, \cdots, w(u_n) = \gamma_n, w(u_{n+1}) = \gamma'_{n+1}, w(u_{n+2}) = \gamma'_{n+2}, \cdots >$,

where $\max\{\gamma_n, \gamma_j\} < \gamma'_j$ for every $j \geq n + 1$. Let W be the valuation ring of w. Since

$w(u_1) = v(u_1), \cdots, w(u_n) = v(u_n)$,

we have $w(z) = v(z)$. Since $z \in M^{b'}$, we have $z \in MW$. By the choice of γ_j', we have $MW = u_1W$, hence $w(z) \geq w(u_1)$, and hence $v(z) \geq v(u_1)$.

We do not know if a gg- (resp., fg-) cancellative semistar operation is fh-cancellative.

References

1. Chapman, S.T., Matsuda, R.: A note on the cancellation properties of semistar operations. Int. Electron. J. Algebra **7**, 78–86 (2010)
2. Fontana, M., Loper, K.A.: Kronecker function rings: a general approach. Lect. Notes Pure Appl. Math. **220**, 189–205 (2001)
3. Fontana, M., Loper, K.A.: Cancellative properties in ideal systems: a classification of e.a.b. semistar operations. J. Pure Appl. Algebras **213**, 2095–2103 (2009)
4. Fontana, M., Loper, K.A., Matsuda, R.: Cancellation properties in ideal systems: an e.a.b. not a.b. star operation. Arab. J. Sci. Eng. **35**, 45–49 (2010)
5. Gilmer, R.: Multiplicative Ideal Theory. Marcel Dekker, New York (1972)
6. Halter-Koch, F.: Ideal Systems: An Introduction to Multiplicative Ideal Theory. Marcel Dekker, New York (1998)
7. Matsuda, R.: Note on cancellation properties in ideal systems. Commun. Algebra **43**, 23–28 (2015)

Quasi-Prüfer Extensions of Rings

Gabriel Picavet and Martine Picavet-L'Hermitte

Abstract We introduce quasi-Prüfer ring extensions, in order to relativize quasi-Prüfer domains and to take also into account some contexts in recent papers. An extension is quasi-Prüfer if and only if it is an INC pair. The class of these extensions has nice stability properties. We also define almost-Prüfer extensions that are quasi-Prüfer, the converse being not true. Quasi-Prüfer extensions are closely linked to finiteness properties of fibers. Applications are given for FMC extensions, because they are quasi-Prüfer.

Keywords Flat epimorphism • FIP • FCP Extension • Minimal extension • Integral extension • Morita • Prüfer hull • Support of a module • Fiber

2010 *Mathematics Subject Classification* Primary:13B02; Secondary:13B22, 13A18, 13F05, 14A05

1 Introduction and Notation

We consider the category of commutative and unital rings. An epimorphism is an epimorphism of this category. Let $R \subseteq S$ be a (ring) extension. The set of all R-subalgebras of S is denoted by $[R, S]$. A *chain* of R-subalgebras of S is a set of elements of $[R, S]$ that are pairwise comparable with respect to inclusion. We say that the extension $R \subseteq S$ has FCP (for the "finite chain property") if each chain in $[R, S]$ is finite. Dobbs and the authors characterized FCP extensions [13]. An extension $R \subseteq S$ is called FMC if there is a finite maximal chain of extensions from R to S.

G. Picavet (✉) • M. Picavet-L'Hermitte
Laboratoire de Mathématiques, Université Blaise Pascal, 24, avenue des Landais, UMR6620 CNRS, BP 80026, 63177 Aubière Cedex, France
e-mail: Gabriel.Picavet@math.univ-bpclermont.fr

M. Fontana et al. (eds.), *Rings, Polynomials, and Modules*,
https://doi.org/10.1007/978-3-319-65874-2_16

We begin by explaining our motivations and aims. The reader who is not familiar with the notions used will find some Scholia in the sequel, as well as necessary definitions that exist in the literature. Knebusch and Zhang introduced Prüfer extensions in their book [25]. Actually, these extensions are nothing but normal pairs, that are intensively studied in the literature. We do not intend to give an extensive list of recent papers, written by Ayache, Ben Nasr, Dobbs, Jaballah, Jarboui, and some others. We are indebted to these authors because their papers are a rich source of suggestions. We observed that some of them are dealing with FCP (FMC) extensions, followed by a Prüfer extension, perhaps under a hidden form. These extensions reminded us quasi-Prüfer domains (see [18] for a comprehensive study). Therefore, we introduced in [38] *quasi-Prüfer* extensions $R \subseteq S$ as extensions that can be factored $R \subseteq R' \subseteq S$, where the first extension is integral and the second is Prüfer. Note that FMC extensions are quasi-Prüfer.

We give a systematic study of quasi-Prüfer extensions in Sects. 2 and 3. The class of quasi-Prüfer extensions has a nice behavior with respect to the classical operations of commutative algebra. An important result is that quasi-Prüfer extensions coincide with INC-pairs. Another one is that this class is stable under forming subextensions and composition. A striking result is the stability of the class of quasi-Prüfer extensions by absolutely flat base change, like localizations and Henselizations. An arbitrary ring extension $R \subseteq S$ admits a quasi-Prüfer closure, contained in S. Examples are provided by Laskerian pairs, open pairs, and the pseudo-Prüfer pairs of Dobbs-Shapiro [12].

Section 4 deals with *almost-Prüfer* extensions, a special kind of quasi-Prüfer extensions. They are of the form $R \subseteq T \subseteq S$, where the first extension is Prüfer and the second is integral. An arbitrary ring extension $R \subseteq S$ admits an almost-Prüfer closure, contained in S. The class of almost-Prüfer extensions seems to have less properties than the class of quasi-Prüfer extensions but has the advantage that almost-Prüfer closures commute with localizations at prime ideals. We examine the transfer of the quasi (almost)-Prüfer properties to subextensions. It is noteworthy that the class of FCP almost-Prüfer extensions is stable under the formation of subextensions, although this does not hold for arbitrary almost-Prüfer extensions.

In Sect. 5, we complete and generalize the results of Ayache-Dobbs in [5], with respect to the finiteness of fibers. These authors have evidently considered particular cases of quasi-Prüfer extensions. A main result is that if $R \subseteq S$ is quasi-Prüfer with finite fibers, then so is $R \subseteq T$ for $T \in [R, S]$. In particular, we recover a result of [5] about FMC extensions.

1.1 Recalls About Some Results and Definitions

The reader is warned that we will mostly use the definition of Prüfer extensions by flat epimorphic subextensions investigated in [25]. The results needed may be found in Scholium A for flat epimorphic extensions and some results of [25] are summarized in Scholium B. Their powers give quick proofs of results that are generalizations of results of the literature.

As long as FCP or FMC extensions are concerned, we use minimal (ring) extensions, a concept introduced by Ferrand-Olivier [17]. An extension $R \subset S$ is called *minimal* if $[R, S] = \{R, S\}$. It is known that a minimal extension is either module-finite or a flat epimorphism [17] and these conditions are mutually exclusive. There are three types of integral minimal (module-finite) extensions: ramified, decomposed, or inert [36, Theorem 3.3]. A minimal extension $R \subset S$ admits a crucial ideal $\mathcal{C}(R, S) =: M$ which is maximal in R and such that $R_P = S_P$ for each $P \neq M, P \in \mathrm{Spec}(R)$. Moreover, $\mathcal{C}(R, S) = (R : S)$ when $R \subset S$ is an integral minimal extension. The key connection between the above ideas is that if $R \subseteq S$ has FCP or FMC, then any maximal (necessarily finite) chain of R-subalgebras of S, $R = R_0 \subset R_1 \subset \cdots \subset R_{n-1} \subset R_n = S$, with *length* $n < \infty$, results from juxtaposing n minimal extensions $R_i \subset R_{i+1}$, $0 \leq i \leq n-1$.

We define the *length* $\ell[R, S]$ of $[R, S]$ as the supremum of the lengths of chains in $[R, S]$. In particular, if $\ell[R, S] = r$, for some integer r, there exists a maximal chain in $[R, S]$ with length r.

As usual, Spec(R), Max(R), Min(R), U(R), Tot(R) are, respectively, the set of prime ideals, maximal ideals, minimal prime ideals, units, total ring of fractions of a ring R and $\kappa(P) = R_P/PR_P$ is the residual field of R at $P \in \mathrm{Spec}(R)$.

If $R \subseteq S$ is an extension, then $(R : S)$ is its conductor and if $P \in \mathrm{Spec}(R)$, then S_P is the localization $S_{R\setminus P}$. We denote the integral closure of R in S by $\overline{R}^S$ (or $\overline{R}$).

A local ring is here what is called elsewhere a quasi-local ring. The *support* of an R-module E is $\mathrm{Supp}_R(E) := \{P \in \mathrm{Spec}(R) \mid E_P \neq 0\}$ and $\mathrm{MSupp}_R(E) := \mathrm{Supp}_R(E) \cap \mathrm{Max}(R)$. Finally, $\subset$ denotes proper inclusion and $|X|$ the cardinality of a set X.

Scholium A We give some recalls about flat epimorphisms (see [26, Chapitre IV], except (2) which is [30, Proposition 2]).

(1) $R \to S$ is a flat epimorphism $\Leftrightarrow$ for all $P \in \mathrm{Spec}(R)$, either $R_P \to S_P$ is an isomorphism or $S = PS \Leftrightarrow R_P \subseteq S_P$ is a flat epimorphism for all $P \in \mathrm{Spec}(R) \Leftrightarrow R_{(Q\cap R)} \to S_Q$ is an isomorphism for all $Q \in \mathrm{Spec}(S)$ and $\mathrm{Spec}(S) \to \mathrm{Spec}(R)$ is injective.
(2) (S) A flat epimorphism, with a zero-dimensional domain, is surjective.
(3) If $f : A \to B$ and $g : B \to C$ are ring morphisms such that $g \circ f$ is injective and f is a flat epimorphism, then g is injective.
(4) Let $R \subseteq T \subseteq S$ be a tower of extensions, such that $R \subseteq S$ is a flat epimorphism. Then $T \subseteq S$ is a flat epimorphism but $R \subseteq T$ does not need. A Prüfer extension remedies this defect.
(5) (L) A faithfully flat epimorphism is an isomorphism. Hence, $R = S$ if $R \subseteq S$ is an integral flat epimorphism.
(6) If $f : R \to S$ is a flat epimorphism and J an ideal of S, then $J = f^{-1}(J)S$.
(7) If $f : R \to S$ is an epimorphism, then f is spectrally injective (i.e., ${}^af : \mathrm{Spec}(S) \to \mathrm{Spec}(R)$ is an injection) and its residual extensions are isomorphisms.

(8) Flat epimorphisms remain flat epimorphisms under base change (in particular, after a localization with respect to a multiplicatively closed subset).
(9) Flat epimorphisms are descended by faithfully flat morphisms.

1.2 Recalls and Results on Prüfer Extensions

There are a lot of characterizations of Prüfer extensions. We keep only those that are useful in this paper. Let $R \subseteq S$ be an extension.

Scholium B

(1) [25] $R \subseteq S$ is called Prüfer if $R \subseteq T$ is a flat epimorphism for each $T \in [R, S]$.
(2) $R \subseteq S$ is called a *normal* pair if $T \subseteq S$ is integrally closed for each $T \in [R, S]$.
(3) $R \subseteq S$ is Prüfer if and only if it is a normal pair [25, Theorem 5.2(4)].
(4) R is called Prüfer if its finitely generated regular ideals are invertible, or equivalently, $R \subseteq \mathrm{Tot}(R)$ is Prüfer [21, Theorem 13((5)(9))].

Hence Prüfer extensions are a relativization of Prüfer rings. Clearly, a minimal extension is a flat epimorphism if and only if it is Prüfer. We will then use for such extensions the terminology: *Prüfer minimal* extensions. The reader may find some properties of Prüfer minimal extensions in [36, Proposition 3.2, Lemma 3.4 and Proposition 3.5], where in addition R must be supposed local. The reason why is that this word has disappeared during the printing process of [36].

We will need the two next results. Some of them do not explicitly appear in [25] but deserve to be emphasized. We refer to [25, Definition 1, p.22] for a definition of Manis extensions and remark that Proposition 1.1(1) was also noted in [12].

Proposition 1.1 *Let $R \subseteq S$ be a ring extension.*

(1) $R \subseteq S$ is Prüfer if and only if $R_P \subseteq S_P$ is Prüfer for each $P \in \mathrm{Spec}(R)$ (respectively, $P \in \mathrm{Supp}(S/R)$).
(2) $R \subseteq S$ is Prüfer if and only if $R_M \subseteq S_M$ is Manis for each $M \in \mathrm{Max}(R)$.

Proof (1) The class of Prüfer extensions is stable under localization [25, Proposition 5.1(ii), p.46-47]. To get the converse, use Scholium A(1). (2) follows from [25, Proposition 2.10, p.28, Definition 1, p.46]. □

Proposition 1.2 *Let $R \subseteq S$ be a ring extension, where R is local.*

(1) $R \subseteq S$ is Manis if and only if $S \setminus R \subseteq \mathrm{U}(S)$ and $x \in S \setminus R \Rightarrow x^{-1} \in R$. In that case, $R \subseteq S$ is integrally closed.
(2) $R \subseteq S$ is Manis if and only if $R \subseteq S$ is Prüfer.
(3) $R \subseteq S$ is Prüfer if and only if there exists $P \in \mathrm{Spec}(R)$ such that $S = R_P$, $P = SP$ and R/P is a valuation domain. Under these conditions, S/P is the quotient field of R/P.

Proof (1) is [25, Theorem 2.5, p.24]. (2) is [25, Scholium 10.4, p. 147]. Then (3) is [13, Theorem 6.8]. □

Next result shows that Prüfer FCP extensions can be described in a special manner.

Proposition 1.3 *Let $R \subset S$ be a ring extension.*

(1) If $R \subset S$ has FCP, then $R \subset S$ is integrally closed $\Leftrightarrow$ $R \subset S$ is Prüfer $\Leftrightarrow$ $R \subset S$ is a composite of Prüfer minimal extensions.

(2) If $R \subset S$ is integrally closed, then $R \subset S$ has FCP $\Leftrightarrow$ $R \subset S$ is Prüfer and $\mathrm{Supp}(S/R)$ *is finite.*

Proof (1) Assume that $R \subset S$ has FCP. If $R \subset S$ is integrally closed, then, $R \subset S$ is composed of Prüfer minimal extensions by [13, Lemma 3.10]. We know that a composite of Prüfer extensions is a Prüfer extension [25, Theorem 5.6]. Thus, by [25], $R \subset S$ is a normal pair. Conversely, if $R \subset S$ is composed of Prüfer minimal extensions, $R \subset S$ is integrally closed, since so is each Prüfer minimal extension. A Prüfer extension is obviously integrally closed, and an FCP integrally closed extension is Prüfer by [13, Theorem 6.3].

(2) The logical equivalence is [13, Theorem 6.3]. □

Definition 1.4 [25] A ring extension $R \subseteq S$ has:

(1) a greatest flat epimorphic subextension $R \subseteq \widehat{R}^S$, called the **Morita hull** of R in S.

(2) a greatest Prüfer subextension $R \subseteq \widetilde{R}^S$, called the **Prüfer hull** of R in S.

We set $\widehat{R} := \widehat{R}^S$ and $\widetilde{R} := \widetilde{R}^S$, if no confusion can occur. $R \subseteq S$ is called Prüfer-closed if $R = \widetilde{R}$.

Note that $\widetilde{R}^S$ is denoted by $\mathrm{P}(R, S)$ in [25] and $\widehat{R}^S$ is the weakly surjective hull $\mathrm{M}(R, S)$ of [25]. Our terminology is justified because Morita's work is earlier [29, Corollary 3.4]. The Morita hull can be computed by using a (transfinite) induction [29]. Let S' be the set of all $s \in S$ such that there is some ideal I of R, such that $IS = S$ and $Is \subseteq R$. Then $R \subseteq S'$ is a subextension of $R \subseteq S$. We set $S_1 := S'$ and $S_{i+1} := (S_i)' \subseteq S_i$. By [29, p. 36], if $R \subset S$ is an FCP extension, then $\widehat{R} = S_n$ for some integer n.

At this stage it is interesting to point out a result showing again that integral closedness and Prüfer extensions are closely related.

Proposition 1.5 *Olivier [32, Corollary, p. 56] An extension $R \subseteq S$ is integrally closed if and only if there is a pullback square:*

$$\begin{array}{ccc} R & \longrightarrow & S \\ \downarrow & & \downarrow \\ V & \longrightarrow & K \end{array}$$

where V is a semi-hereditary ring and K its total quotient ring.

In that case $V \subseteq K$ is a Prüfer extension, since V is a Prüfer ring, whose localizations at prime ideals are valuation domains and K is an absolutely flat ring. As there exist integrally closed extensions that are not Prüfer, we see in passing that the pullback construction may not descend Prüfer extensions. The above result has a companion for minimal extensions that are Prüfer [20, Proposition 3.2].

Proposition 1.6 *Let $R \subseteq S$ be an extension and $T \in [R, S]$, then $\widetilde{R}^T = \widetilde{R} \cap T$. Therefore, for $T, U \in [R, S]$ with $T \subseteq U$, then $\widetilde{R}^T \subseteq \widetilde{R}^U$.*

Proof Obvious, since the Prüfer hull $\widetilde{R}^T$ is the greatest Prüfer extension $R \subseteq V$ contained in T. □

We will show later that in some cases $\widetilde{T} \subseteq \widetilde{U}$ if $R \subseteq S$ has FCP.

2 Quasi-Prüfer Extensions

We introduced the following definition in [38, p. 10].

Definition 2.1 An extension of rings $R \subseteq S$ is called quasi-Prüfer if one of the following equivalent statements holds:

(1) $\overline{R} \subseteq S$ is a Prüfer extension;
(2) $R \subseteq S$ can be factored $R \subseteq T \subseteq S$, where $R \subseteq T$ is integral and $T \subseteq S$ is Prüfer. In that case $\overline{R} = T$.

To see that (2) $\Rightarrow$ (1) observe that if (2) holds, then $T \subseteq \overline{R}$ is integral and a flat injective epimorphism, so that $\overline{R} = T$ by (L) (Scholium A(5)).

We observe that quasi-Prüfer extensions are akin to quasi-finite extensions if we refer to Zariski Main Theorem. This will be explored in Sect. 5, see, for example, Theorem 5.2.

Hence integral or Prüfer extensions are quasi-Prüfer. An extension is clearly Prüfer if and only if it is quasi-Prüfer and integrally closed. Quasi-Prüfer extensions allow us to avoid FCP hypotheses.

We give some other definitions involved in ring extensions $R \subseteq S$. The *fiber* at $P \in \mathrm{Spec}(R)$ of $R \subseteq S$ is $\mathrm{Fib}_{R,S}(P) := \{Q \in \mathrm{Spec}(S) \mid Q \cap R = P\}$. The subspace $\mathrm{Fib}_{R,S}(P)$ of $\mathrm{Spec}(S)$ is homeomorphic to the spectrum of the fiber ring at P, $\mathrm{F}_{R,S}(P) := \kappa(P) \otimes_R S$. The homeomorphism is given by the spectral map of $S \to \kappa(P) \otimes_R S$ and $\kappa(P) \to \kappa(P) \otimes_R S$ is the *fiber morphism* at P.

Definition 2.2 A ring extension $R \subseteq S$ is called:

(1) *incomparable* if for each pair $Q \subseteq Q'$ of prime ideals of S, then $Q \cap R = Q' \cap R \Rightarrow Q = Q'$, or equivalently, $\kappa(P) \otimes_R T$ is a zero-dimensional ring for each $T \in [R, S]$ and $P \in \mathrm{Spec}(R)$, such that $\kappa(P) \otimes_R T \neq 0$.
(2) an *INC-pair* if $R \subseteq T$ is incomparable for each $T \in [R, S]$ $\Leftrightarrow$ $T \subseteq U$ is incomparable for all $T \subseteq U$ in $[R, S]$.
(3) *residually algebraic* if $R/(Q \cap R) \subseteq S/Q$ is algebraic for each $Q \in \mathrm{Spec}(S)$.

(4) a *residually algebraic pair* if the extension $R \subseteq T$ is residually algebraic for each $T \in [R, S]$.

An extension $R \subseteq S$ is an INC-pair if and only if $R \subseteq S$ is a residually algebraic pair. This fact is an easy consequence of [10, Theorem] (via a short proof that was explicitly given in [9]). This fact was given for the particular case where S is an integral domain in [4].

The following characterization was announced in [38]. We were unaware that this result is also proved in [6, Corollary 1], when we presented it in ArXiv. However, our proof is largely shorter because we use the powerful results of [25].

Theorem 2.3 *An extension $R \subseteq S$ is quasi-Prüfer if and only if $R \subseteq S$ is an INC-pair and, if and only if, $R \subseteq S$ is a residually algebraic pair.*

Proof Suppose that $R \subseteq S$ is quasi-Prüfer and let $T \in [R, S]$. We set $U := \overline{R}T$. Then $\overline{R} \subseteq U$ is a flat epimorphism by definition of a Prüfer extension and hence is incomparable as is $R \subseteq \overline{R}$. It follows that $R \subseteq U$ is incomparable. Since $T \subseteq U$ is integral, it has going-up. It follows that $R \subseteq T$ is incomparable. Conversely, if $R \subseteq S$ is an INC-pair, then so is $\overline{R} \subseteq S$. Since $\overline{R} \subseteq S$ is integrally closed, $\overline{R} \subseteq S$ is Prüfer [25, Theorem 5.2,(9'), p. 48]. The second equivalence is given by the above comments about [10] and [9]. □

Corollary 2.4 *An extension $R \subseteq S$ is quasi-Prüfer if and only if $\overline{R} \subseteq \overline{T}$ is Prüfer for each $T \in [R, S]$. In this case, $\overline{R}$ is the least $T \in [R, S]$ such that $T \subseteq S$ is Prüfer.*

It follows that most of the properties described in [4] for integrally closed INC-pairs of domains are valid for arbitrary ring extensions. Moreover, a result of Dobbs is easily gotten as a consequence of Corollary 2.4: an INC-pair $R \subseteq S$ is an integral extension if and only if $\overline{R} \subseteq S$ is spectrally surjective [11, Theorem 2.2]. This follows from Corollary 2.4 and Scholium A, Property (L).

Example 2.5 Quasi-Prüfer domains R with quotient fields K can be characterized by $R \subseteq K$ is quasi-Prüfer. The reader may consult [7, Theorem 1.1] or [18].

We give here another example of quasi-Prüfer extension. An extension $R \subset S$ is called a *going-down pair* if each of its subextensions has the going-down property. For such a pair, $R \subseteq T$ has incomparability for each $T \in [R, S]$, at each non-maximal prime ideal of R [2, Lemma 5.8](ii). Now let M be a maximal ideal of R, whose fiber is not void in T. Then $R \subseteq T$ is a going-down pair, and so is $R/M \subseteq T/MT$ because $MT \cap R = M$. By [2, Corollary 5.6], the dimension of T/MT is ≤ 1. Therefore, if $R \subset S$ is a going-down pair, then $R \subset S$ is quasi-Prüfer if and only if $\dim(T/MT) \neq 1$ for each $T \in [R, S]$ and $M \in \mathrm{Max}(R)$.

Also *open-ring pairs* $R \subset S$ are quasi-Prüfer by [8, Proposition 2.13].

An *i-pair* is an extension $R \subseteq S$ such that $\mathrm{Spec}(T) \to \mathrm{Spec}(R)$ is injective for each $T \in [R, S]$, or equivalently if and only if $R \subseteq S$ is quasi-Prüfer and $R \subseteq \overline{R}$ is spectrally injective [38, Proposition 5.8]. These extensions appear frequently in the integral domains context. Another examples are given by some extensions $R \subseteq S$, such that $\mathrm{Spec}(S) = \mathrm{Spec}(R)$ as sets, as we will see later.

We proved that Δ-extensions $R \subseteq S$ (such that $U, V \in [R, S] \Rightarrow U + V \in [R, S]$) are quasi-Prüfer [38, Proposition 5.15].

3 Properties of quasi-Prüfer Extensions

We now develop the machinery of quasi-Prüfer extensions.

Proposition 3.1 *An extension $R \subset S$ is (quasi-)Prüfer if and only if $R_P \subseteq S_P$ is (quasi-)Prüfer for any $P \in \mathrm{Spec}(R)$ ($P \in \mathrm{MSupp}(S/R)$).*

Proof The proof is easy if we use the INC-pair property definition of quasi-Prüfer extension (see also [4, Proposition 2.4]). □

Proposition 3.2 *Let $R \subseteq S$ be a quasi-Prüfer extension and $\varphi : S \to S'$ an integral ring morphism. Then $\varphi(R) \subseteq S'$ is quasi-Prüfer and $S' = \overline{\varphi(S)\varphi(\overline{R})}$, where $\overline{\varphi(\overline{R})}$ is the integral closure of $\varphi(\overline{R})$ in S'.*

Proof It is enough to apply [25, Theorem 5.9] to the Prüfer extension $\overline{R} \subseteq S$ and to use Definition 2.1. □

This result applies with $S' := S \otimes_R R'$, where $R \to R'$ is an integral morphism. Therefore integrality ascends the quasi-Prüfer property.

Recall that a composite of Prüfer extensions is Prüfer [25, Theorem 5.6, p. 51]. We next give a result that will be used frequently. The following Corollary 3.3 contains [6, Theorem 3].

Corollary 3.3 *Let $R \subseteq T \subseteq S$ be a tower of extensions. Then $R \subseteq S$ is quasi-Prüfer if and only if $R \subseteq T$ and $T \subseteq S$ are quasi-Prüfer. Hence, $R \subseteq T$ is quasi-Prüfer if and only if $R \subseteq \overline{R}T$ is quasi-Prüfer.*

Proof Consider a tower $(\mathcal{T})$ of extensions $R \subseteq \overline{R} \subseteq S := R' \subseteq \overline{R'} \subseteq S'$ (a composite of two quasi-Prüfer extensions). By using Proposition 3.2 we see that $\overline{R} \subseteq S = R' \subseteq \overline{R'}$ is quasi-Prüfer. Then $(\mathcal{T})$ is obtained by writing on the left an integral extension and on the right a Prüfer extension. Therefore, $(\mathcal{T})$ is quasi-Prüfer. We prove the converse.

If $R \subseteq T \subseteq S$ is a tower of extensions, then $R \subseteq T$ and $T \subseteq S$ are INC-pairs whenever $R \subseteq S$ is an INC-pair. The converse is then a consequence of Theorem 2.3.

The last statement is [6, Corollary 4]. □

Using the above corollary, we can exhibit new examples of quasi-Prüfer extensions. We recall that a ring R is called *Laskerian* if each of its ideals is a finite intersection of primary ideals and a ring extension $R \subset S$ a *Laskerian pair* if each $T \in [R, S]$ is a Laskerian ring. Then [41, Proposition 2.1] shows that if R is an integral domain with quotient field $F \neq R$ and $F \subset K$ is a field extension, then $R \subset K$ is a Laskerian pair if and only if K is algebraic over R and $\overline{R}$ (in K) is a Laskerian Prüfer domain. It follows easily that $R \subset K$ is quasi-Prüfer under these conditions.

Next result generalizes [24, Proposition 1].

Corollary 3.4 *An FMC extension $R \subset S$ is quasi-Prüfer.*

Proof Because $R \subset S$ is a composite of finitely many minimal extensions, by Corollary 3.3, it is enough to observe that a minimal extension is either Prüfer or integral. □

Corollary 3.5 *Let $R \subseteq S$ be a quasi-Prüfer extension and a tower $R \subseteq T \subseteq S$, where $R \subseteq T$ is integrally closed. Then $R \subseteq T$ is Prüfer.*

Proof Observe that $R \subseteq T$ is quasi-Prüfer and then that $R = \overline{R}^T$. □

Next result deals with the Dobbs-Shapiro *pseudo-Prüfer* extensions of integral domains [12], that they called pseudo-normal pairs. Suppose that R is local, we call here pseudo-Prüfer an extension $R \subseteq S$ such that there exists $T \in [R, S]$ with $\mathrm{Spec}(R) = \mathrm{Spec}(T)$ and $T \subseteq S$ is Prüfer [12, Corollary 2.5]. If R is arbitrary, the extension $R \subseteq S$ is called pseudo-Prüfer if $R_M \subseteq S_M$ is pseudo-Prüfer for each $M \in \mathrm{Max}(R)$. In view of the Corollary 3.3, it is enough, if one wishes to characterize quasi-Prüfer extensions, to characterize quasi-Prüfer extensions of the type $R \subseteq T$ with $\mathrm{Spec}(R) = \mathrm{Spec}(T)$.

Corollary 3.6 *Let $R \subseteq T$ be an extension with $\mathrm{Spec}(R) = \mathrm{Spec}(T)$ and (R, M) local. Then $R \subseteq T$ is quasi-Prüfer if and only if $\mathrm{Spec}(R) = \mathrm{Spec}(U)$ for all $U \in [R, T]$ and, if and only if $R/M \subseteq T/M$ is an algebraic field extension. In such a case, $R \subseteq T$ is integral, hence Prüfer-closed.*

Proof It follows from [1] that $M \in \mathrm{Max}(T)$. Part of the proof is gotten by observing that $R/M \subseteq T/M$ is an algebraic field extension $\Rightarrow$ $\mathrm{Spec}(R) = \mathrm{Spec}(U)$ for all $U \in [R, T] \Rightarrow R \subseteq T$ is quasi-Prüfer $\Rightarrow$ ($R \subseteq T$ is integral and) $R/M \subseteq T/M$ is an algebraic field extension. Now $R \subseteq \widetilde{R}$ is a spectrally surjective flat epimorphism and then, by Scholium A, $R = \widetilde{R}$. □

Let $R \subseteq S$ be an extension and I an ideal shared by R and S. It is easy to show that $R \subseteq S$ is quasi-Prüfer if and only if $R/I \subseteq S/I$ is quasi-Prüfer by using [25, Proposition 5.8] in the Prüfer case. We are able to give a more general statement.

Lemma 3.7 *Let $R \subseteq S$ be a (quasi-)Prüfer extension and J an ideal of S with $I = J \cap R$. Then $R/I \subseteq S/J$ is a (quasi-)Prüfer extension. If $R \subseteq S$ is Prüfer and N is a maximal ideal of S, then $R/(N \cap R)$ is a valuation domain with quotient field S/N.*

Proof It follows from [25, Proposition 5.8] that if $R \subseteq S$ is Prüfer, then $R/I \cong (R + J)/J \subseteq S/J$ is Prüfer. Then the quasi-Prüfer case is an easy consequence. □

With this lemma we generalize and complete [23, Proposition 1.1].

Proposition 3.8 *Let $R \subseteq S$ be an extension of rings. The following statements are equivalent:*

(1) $R \subseteq S$ is quasi-Prüfer;
(2) $R/(Q \cap R) \subseteq S/Q$ is quasi-Prüfer for each $Q \in \mathrm{Spec}(S)$;

(3) $(X - s)S[X] \cap R[X] \not\subseteq M[X]$ *for each* $s \in S$ *and* $M \in \mathrm{Max}(R)$*;*
(4) For each $T \in [R, S]$*, the fiber morphisms of* $R \subseteq T$ *are integral.*

Proof (1) $\Rightarrow$ (2) is entailed by Lemma 3.7. Assume that (2) holds and let $M \in \mathrm{Max}(R)$ that contains a minimal prime ideal P lain over by a minimal prime ideal Q of S. Then (2) $\Rightarrow$ (3) follows from [23, Proposition 1.1(1)], applied to $R/(Q \cap R) \subseteq S/Q$. If (3) holds, argue as in the paragraph before [23, Proposition 1.1] to get that $R \subseteq S$ is a $\mathcal{P}$-extension, whence an INC-pair, cf. [11]. Then $R \subseteq S$ is quasi-Prüfer by Theorem 2.3, giving (3) $\Rightarrow$ (1). Because integral extensions have incomparability, we see that (4) $\Rightarrow$ (1). Corollary 3.3 shows that the reverse implication holds, if any quasi-Prüfer extension $R \subseteq S$ has integral fiber morphisms. For $P \in \mathrm{Spec}(R)$, the extension $R_P/PR_P \subseteq S_P/PS_P$ is quasi-Prüfer by Lemma 3.7. The ring $\overline{R}_P/P\overline{R}_P$ is zero-dimensional and $\overline{R}_P/P\overline{R}_P \to S_P/PS_P$, being a flat epimorphism, is therefore surjective by Scholium A (S). It follows that the fiber morphism at P is integral. □

Remark 3.9 The logical equivalence (1) $\Leftrightarrow$ (2) is still valid if we replace quasi-Prüfer with integral in the above proposition. It is enough to show that an extension $R \subseteq S$ is integral when $R/P \subseteq S/Q$ is integral for each $Q \in \mathrm{Spec}(S)$ and $P := Q \cap R$. We can suppose that $S = R[s] \cong R[X]/I$, where X is an indeterminate, I an ideal of $R[X]$, and Q varies in $\mathrm{Min}(S)$, because for an extension $A \subseteq B$, any element of $\mathrm{Min}(A)$ is lain over by some element of $\mathrm{Min}(B)$. If Σ is the set of unitary polynomials of $R[X]$, the assumptions show that any element of $\mathrm{Spec}(R[X])$, containing I, meets Σ. As Σ is a multiplicatively closed subset, $I \cap \Sigma \neq \emptyset$, whence s is integral over R.

But a similar result does not hold if we replace quasi-Prüfer with Prüfer, except if we suppose that $R \subseteq S$ is integrally closed. To see this, apply the above proposition to get a quasi-Prüfer extension $R \subseteq S$ if each $R/P \subseteq S/Q$ is Prüfer. Actually, this situation already occurs for Prüfer rings and their factor domains, as Lucas's paper [28] shows. More precisely, [28, Proposition 2.7] and the third paragraph of [28, p. 336] shows that if R is a ring with $\mathrm{Tot}(R)$ absolutely flat, then R is a quasi-Prüfer ring if R/P is a Prüfer domain for each $P \in \mathrm{Spec}(R)$. Now example [28, Example 2.4] shows that R is not necessarily Prüfer.

We observe that if $R \subseteq S$ is quasi-Prüfer, then R/M is a quasi-Prüfer domain for each $N \in \mathrm{Max}(S)$ and $M := N \cap R$ (in case $R \subseteq S$ is integral, R/M is a field). To prove this, observe that $R/M \subseteq S/N$ can be factored $R/M \subseteq \kappa(M) \subseteq S/N$. As we will see, $R/M \subseteq \kappa(M)$ is quasi-Prüfer because $R/M \subseteq S/N$ is quasi-Prüfer.

The class of Prüfer extensions is not stable by (flat) base change. For example, let V be a valuation domain with quotient field K. Then $V[X] \subseteq K[X]$ is not Prüfer [25, Example 5.12, p. 53].

Proposition 3.10 *Let* $R \subseteq S$ *be a (quasi)-Prüfer extension and* $R \to T$ *a flat epimorphism, then* $T \subseteq S \otimes_R T$ *is (quasi)-Prüfer. If in addition* S *and* T *are both subrings of some ring and* $R \subseteq T$ *is an extension, then* $T \subseteq TS$ *is (quasi)-Prüfer.*

Proof For the first part, it is enough to consider the Prüfer case. It is well known that the following diagram is a pushout if $Q \in \mathrm{Spec}(T)$ is lying over P in R:

$$\begin{array}{ccc} R_P & \longrightarrow & S_P \\ \downarrow & & \downarrow \\ T_Q & \longrightarrow & (T \otimes_R S)_Q \end{array}$$

As $R_P \to T_Q$ is an isomorphism since $R \to T$ is a flat epimorphism by Scholium A (1), it follows that $R_P \subseteq S_P$ identifies to $T_Q \to (T \otimes_R S)_Q$. The first assertion follows because Prüfer extensions localize and globalize.

The final assertion is then a special case because, under its hypotheses, $TS \cong T \otimes_R S$ canonically. □

The reader may find in [25, Corollary 5.11, p. 53] that if $R \subseteq A \subseteq S$ and $R \subseteq B \subseteq S$ are extensions and $R \subseteq A$ and $R \subseteq B$ are both Prüfer, then $R \subseteq AB$ is Prüfer.

Proposition 3.11 *Let $R \subseteq A$ and $R \subseteq B$ be two extensions, where A and B are subrings of a ring S. If they are both quasi-Prüfer, then $R \subseteq AB$ is quasi-Prüfer.*

Proof Let U and V be the integral closures of R in A and B. Then $R \subseteq A \subseteq AV$ is quasi-Prüfer because $A \subseteq AV$ is integral and Corollary 3.3 applies. Using again Corollary 3.3 with $R \subseteq V \subseteq AV$, we find that $V \subseteq AV$ is quasi-Prüfer. Now Proposition 3.10 entails that $B \subseteq AB$ is quasi-Prüfer because $V \subseteq B$ is a flat epimorphism. Finally $R \subseteq AB$ is quasi-Prüfer, since a composite of quasi-Prüfer extensions. □

It is known that an arbitrary direct product of extensions is Prüfer if and only if each of its components is Prüfer [25, Proposition 5.20, p. 56] The following result is an easy consequence.

Proposition 3.12 *Let $\{R_i \subseteq S_i | i = 1, \dots, n\}$ be a finite family of quasi-Prüfer extensions, then $R_1 \times \cdots \times R_n \subseteq S_1 \times \cdots \times S_n$ is quasi-Prüfer. In particular, by Corollary 3.3, if $\{R \subseteq S_i | i = 1, \dots, n\}$ is a finite family of quasi-Prüfer extensions, then $R \subseteq S_1 \times \cdots \times S_n$ is quasi-Prüfer.*

In the same way we have the following result deduced from [25, Remark 5.14, p. 54].

Proposition 3.13 *Let $R \subseteq S$ be an extension of rings and an upward directed family $\{R_\alpha | \alpha \in I\}$ of elements of $[R, S]$ such that $R \subseteq R_\alpha$ is quasi-Prüfer for each $\alpha \in I$. Then $R \subseteq \cup[R_\alpha | \alpha \in I]$ is quasi-Prüfer.*

Proof It is enough to use [25, Proposition 5.13, p. 54] where A_α is the integral closure of R in R_α. □

Here are some descent results used later on.

Proposition 3.14 *Let $R \subseteq S$ be a ring extension and $R \to R'$ a spectrally surjective ring morphism (for example, either faithfully flat or injective and integral). Then $R \subseteq S$ is quasi-Prüfer if $R' \to R' \otimes_R S$ is injective (for example, if $R \to R'$ is faithfully flat) and quasi-Prüfer.*

Proof Let $T \in [R, S]$ and $P \in \mathrm{Spec}(R)$ and set $T' := T \otimes_R R'$. There is some $P' \in \mathrm{Spec}(R')$ lying over P, because $R \to R'$ is spectrally surjective. By [22, Corollaire 3.4.9], there is a faithfully flat morphism $\mathrm{F}_{R,T}(P) \to \mathrm{F}_{R',T'}(P') \cong \mathrm{F}_{R,T}(P) \otimes_{\mathbf{k}(P)} \kappa(P')$, inducing a surjective map $\mathrm{Fib}_{R',T'}(P') \to \mathrm{Fib}_{R,T}(P)$ since it satisfies lying over. By Theorem 2.3, the result follows from the faithful flatness of $\mathrm{F}_{R,T}(P) \to \mathrm{F}_{R',T\otimes_R R'}(P')$. □

Corollary 3.15 *Let $R \subseteq S$ be an extension of rings, $R \to R'$ a faithfully flat ring morphism and set $S' := R' \otimes_R S$. If $R' \subseteq S'$ is (quasi-) Prüfer (respectively, FCP), then so is $R \subseteq S$.*

Proof The Prüfer case is clear, because faithfully flat morphisms descend flat epimorphisms (Scholium A (9)). For the quasi-Prüfer case, we use Proposition 3.14. The FCP case is proved in [15, Theorem 2.2]. □

The integral closure of a ring morphism $f : R \to T$ is the integral closure of the extension $f(R) \subseteq T$. By definition, a ring morphism $R \to T$ preserves the integral closure of ring morphisms $R \to S$ if $\overline{T}^{T\otimes_R S} \cong T \otimes_R \overline{R}$ for every ring morphism $R \to S$. An absolutely flat morphism $R \to T$ ($R \to T$ and $T \otimes_R T \to T$ are both flat) preserves integral closure [32, Theorem 5.1]. Flat epimorphisms, Henselizations, and étale morphisms are absolutely flat. Another examples are morphisms $R \to T$ that are essentially of finite type and (absolutely) reduced [34, Proposition 5.19](2). Such morphisms are flat if R is reduced [27, Proposition 3.2].

We will prove an ascent result for absolutely flat ring morphisms. This will be proved by using base changes. For this we need to introduce some concepts. A ring A is called an AIC ring if each monic polynomial of $A[X]$ has a zero in A. The first author recalled in [35, p. 4662] that any ring A has a faithfully flat integral extension $A \to A^*$, where A^* is an AIC ring. Moreover, if A is an AIC ring, each localization A_P at a prime ideal P of A is a strict Henselian ring [35, Lemma II.2].

Theorem 3.16 *Let $R \subseteq S$ be a (quasi-) Prüfer extension and $R \to T$ an absolutely flat ring morphism. Then $T \to T \otimes_R S$ is a (quasi-) Prüfer extension.*

Proof We can suppose that R is an AIC ring. To see this, it is enough to use the base change $R \to R^*$. We set $T^* := T \otimes_R R^*$, $S^* := S \otimes_R R^*$. We first observe that $R^* \subseteq S^*$ is quasi-Prüfer for the following reason: the composite extension $R \subseteq S \subseteq S^*$ is quasi-Prüfer by Corollary 3.3 because the last extension is integral. Moreover, $R^* \to T^*$ is absolutely flat. In case $T^* \subseteq T^* \otimes_{R^*} S^*$ is quasi-Prüfer, so is $T \subseteq T \otimes_R S$, because $T \to T^* = T \otimes_R R^*$ is faithfully flat and $T^* \subseteq T^* \otimes_{R^*} S^*$ is deduced from $T \subseteq T \otimes_R S$ by the faithfully flat base change $T \to T \otimes_R R^*$. It is then enough to apply Proposition 3.14.

We thus assume from now on that R is an AIC ring.

Let $N \in \mathrm{Spec}(T)$ be lying over M in R. Then $R_M \to T_N$ is absolutely flat [31, Proposition f] and $R_M \subseteq S_M$ is quasi-Prüfer. Now observe that $(T \otimes_R S)_N \cong T_N \otimes_{R_M} S_M$. Therefore, we can suppose that R and T are local and $R \to T$ is local and injective. We deduce from [32, Theorem 5.2] that $R_M \to T_N$ is an isomorphism because R_M is a strict Henselian ring. Therefore the proof is complete in the quasi-Prüfer case. For the Prüfer case, we need only to observe that absolutely flat morphisms preserve integral closure and a quasi-Prüfer extension is Prüfer if it is integrally closed. □

Lemma 3.17 *Let $R \subseteq S$ be an extension of rings and $R \to T$ a base change which preserves integral closure. If $T \subseteq T \otimes_R S$ has FCP and $R \subseteq S$ is Prüfer, then $T \subseteq T \otimes_R S$ is Prüfer.*

Proof The result holds because an FCP extension is Prüfer if and only if it is integrally closed. □

We observe that $T \otimes_R \widetilde{R} \subseteq \widetilde{T}$ need not to be an isomorphism, since this property may fail even for a localization $R \to R_P$, where P is a prime ideal of R.

Theorem 3.18 *Let $R \subseteq S$ be a ring extension.*

(1) $R \subseteq S$ has a greatest quasi-Prüfer subextension $R \subseteq \overset{\Longrightarrow}{R} = \widetilde{\overline{R}}$.
(2) $R \subseteq \overline{R}\widetilde{R} =: \vec{R}$ is quasi-Prüfer and then $\vec{R} \subseteq \overset{\Longrightarrow}{R}$.
(3) $\overline{R}^{\overset{\Longrightarrow}{R}} = \overline{R}$ and $\widetilde{R}^{\overset{\Longrightarrow}{R}} = \widetilde{R}$.

Proof To see (1), use Proposition 3.11 which tells us that the set of all quasi-Prüfer subextensions is upward directed and then use Proposition 3.13 to prove the existence of $\overset{\Longrightarrow}{R}$. Then let $R \subseteq T \subseteq \overset{\Longrightarrow}{R}$ be a tower with $R \subseteq T$ integral and $T \subseteq \overset{\Longrightarrow}{R}$ Prüfer. From $T \subseteq \overline{R} \subseteq \widetilde{\overline{R}} \subseteq \overset{\Longrightarrow}{R}$, we deduce that $T = \overline{R}$ and then $\overset{\Longrightarrow}{R} = \widetilde{\overline{R}}$.

(2) Now $R \subseteq \overline{R}\widetilde{R}$ can be factored $R \subseteq \widetilde{R} \subseteq \overline{R}\widetilde{R}$ and is a tower of quasi-Prüfer extensions, because $\widetilde{R} \to \overline{R}\widetilde{R}$ is integral.

(3) Clearly, the integral closure and the Prüfer closure of R in $\overset{\Longrightarrow}{R}$ are the respective intersections of $\overline{R}$ and $\widetilde{R}$ with $\overset{\Longrightarrow}{R}$, and $\overline{R}, \widetilde{R} \subseteq \overset{\Longrightarrow}{R}$. □

This last result means that, as far as properties of integral closures and Prüfer closures of subsets of $\overset{\Longrightarrow}{R}$ are concerned, we can suppose that $R \subseteq S$ is quasi-Prüfer.

4 Almost-Prüfer Extensions

We next give a definition "dual" of the definition of a quasi-Prüfer extension.

4.1 Arbitrary Extensions

Definition 4.1 A ring extension $R \subseteq S$ is called an *almost-Prüfer* extension if it can be factored $R \subseteq T \subseteq S$, where $R \subseteq T$ is Prüfer and $T \subseteq S$ is integral.

Proposition 4.2 *An extension $R \subseteq S$ is almost-Prüfer if and only if $\widetilde{R} \subseteq S$ is integral. It follows that the subring T of the above definition is $\widetilde{R} = \widehat{R}$ when $R \subseteq S$ is almost-Prüfer.*

Proof If $R \subseteq S$ is almost-Prüfer, there is a factorization $R \subseteq T \subseteq \widetilde{R} \subseteq \widehat{R} \subseteq S$, where $T \subseteq \widehat{R}$ is both integral and a flat epimorphism by Scholium A (4). Therefore, $T = \widetilde{R} = \widehat{R}$ by Scholium A (5) (L). □

Corollary 4.3 *Let $R \subseteq S$ be a quasi-Prüfer extension, and let $T \in [R, S]$. Then, $T \cap \overline{R} \subseteq T\overline{R}$ is almost-Prüfer and $T = \widetilde{\overline{R} \cap T}^{T\overline{R}}$. Moreover, if $T \cap \overline{R} = R$, then, $T = T\overline{R} \cap \widetilde{R}$.*

Proof $T \cap \overline{R} \subseteq T$ is quasi-Prüfer by Corollary 3.3. Being integrally closed, it is Prüfer by Corollary 3.5. Moreover, $T \subseteq T\overline{R}$ is an integral extension. Then, $T \cap \overline{R} \subseteq T\overline{R}$ is almost-Prüfer and $T = \widetilde{\overline{R} \cap T}^{T\overline{R}}$. If $T \cap \overline{R} = R$, then $T \subseteq T\overline{R} \cap \widetilde{R}$ is both Prüfer and integral, so that $T = T\overline{R} \cap \widetilde{R}$. □

We note that integral extensions and Prüfer extensions are almost-Prüfer and hence minimal extensions are almost-Prüfer. There are quasi-Prüfer extensions that are not almost-Prüfer. It is enough to consider [37, Example 3.5(1)]. Let $R \subseteq T \subseteq S$ be two minimal extensions, where R is local, $R \subseteq T$ integral and $T \subseteq S$ is Prüfer. Then $R \subseteq S$ is quasi-Prüfer but not almost-Prüfer, because $S = \widehat{R}$ and $R = \widetilde{R}$. The same example shows that a composite of almost-Prüfer extensions may not be almost-Prüfer.

But the reverse implication holds.

Theorem 4.4 *Let $R \subseteq S$ be an almost-Prüfer extension. Then $R \subseteq S$ is quasi-Prüfer. Moreover, $\widetilde{R} = \widehat{R}$, $(\widetilde{R})_P = \widetilde{R_P}$ for each $P \in \mathrm{Spec}(R)$. In this case, any flat epimorphic subextension $R \subseteq T$ is Prüfer.*

Proof Let $R \subseteq \widetilde{R} \subseteq S$ be an almost-Prüfer extension, that is $\widetilde{R} \subseteq S$ is integral. The first assertion follows from Corollary 3.3 because $R \subseteq \widetilde{R}$ is Prüfer. Now the Morita hull and the Prüfer hull coincide by Proposition 4.2. In the same way, $(\widetilde{R})_P \to \widetilde{R_P}$ is a flat epimorphism and $(\widetilde{R})_P \to S_P$ is integral. □

We could define almost-Prüfer rings as the rings R such that $R \subseteq \mathrm{Tot}(R)$ is almost-Prüfer. But in that case $\widetilde{R} = \mathrm{Tot}(R)$ (by Theorem 4.4), so that R is a Prüfer ring. The converse evidently holds. Therefore, this concept does not define something new.

It was observed in [13, Remark 2.9(c)] that there is an almost-Prüfer FMC extension $R \subseteq S \subseteq T$, where $R \subseteq S$ is a Prüfer minimal extension and $S \subseteq T$ is minimal and integral, but $R \subseteq T$ is not an FCP extension.

Proposition 4.5 *Let $R \subseteq S$ be an extension verifying the hypotheses:*

(i) $R \subseteq S$ is quasi-Prüfer.
(ii) $R \subseteq S$ can be factored $R \subseteq T \subseteq S$, where $R \subseteq T$ is a flat epimorphism.

(1) Then the following commutative diagram (D) is a pushout,

$$\begin{array}{ccc} R & \longrightarrow & \overline{R} \\ \downarrow & & \downarrow \\ T & \longrightarrow & T\overline{R} \end{array}$$

$T\overline{R} \subseteq S$ is Prüfer and $R \subseteq T\overline{R}$ is quasi-Prüfer. Moreover, $\mathrm{F}_{R,\overline{R}}(P) \cong \mathrm{F}_{T,T\overline{R}}(Q)$ *for each $Q \in \mathrm{Spec}(T)$ and $P := Q \cap R$.*
(2) If in addition $R \subseteq T$ is integrally closed, (D) is a pullback, $T \cap \overline{R} = R$, $(R : \overline{R}) = (T : T\overline{R}) \cap R$ and $(T : T\overline{R}) = (R : \overline{R})T$.

Proof (1) Consider the injective composite map $\overline{R} \to \overline{R} \otimes_R T \to T\overline{R}$. As $\overline{R} \to \overline{R} \otimes_R T$ is a flat epimorphism, because deduced by a base change of $R \to T$, we get that the surjective map $\overline{R} \otimes_R T \to T\overline{R}$ is an isomorphism by Scholium A (3). By fibers transitivity, we have $\mathrm{F}_{T,\overline{R}T}(Q) \cong \kappa(Q) \otimes_{\mathbf{k}(P)} \mathrm{F}_{R,\overline{R}}(P)$ [22, Corollaire 3.4.9]. As $\kappa(P) \to \kappa(Q)$ is an isomorphism by Scholium A, we get that $\mathrm{F}_{R,\overline{R}}(P) \cong \mathrm{F}_{T,\overline{R}T}(Q)$.

(2) As in [5, Lemma 3.5], $R = T \cap \overline{R}$. The first statement on the conductors has the same proof as in [5, Lemma 3.5]. The second holds because $R \subseteq T$ is a flat epimorphism (see Scholium A (6)). □

Theorem 4.6 *Let $R \subset S$ be a quasi-Prüfer extension and the diagram (D'):*

$$\begin{array}{ccc} R & \longrightarrow & \overline{R} \\ \downarrow & & \downarrow \\ \widetilde{R} & \longrightarrow & \widetilde{R}\overline{R} \end{array}$$

(1) (D') is a pushout and a pullback, such that $\overline{R} \cap \widetilde{R} = R$ and $(R : \overline{R}) = (\widetilde{R} : \widetilde{R\overline{R}}) \cap R$ so that $(\widetilde{R} : \widetilde{R\overline{R}}) = (R : \overline{R})\widetilde{R}$.
(2) $R \subset S$ can be factored $R \subseteq \widetilde{R\overline{R}} = \overline{\widetilde{R}} = \vec{R} \subseteq \overrightarrow{R} = \widetilde{\overline{R}} = S$, where the first extension is almost-Prüfer and the second is Prüfer.
(3) $R \subset S$ is almost-Prüfer $\Leftrightarrow S = \widetilde{R\overline{R}} \Leftrightarrow \widetilde{\overline{R}} = \overline{\widetilde{R}}$.
(4) $R \subseteq \widetilde{R\overline{R}} = \overline{\widetilde{R}} = \vec{R}$ is the greatest almost-Prüfer subextension of $R \subseteq S$ and $\widetilde{R} = \widetilde{R}^{\vec{R}}$.
(5) $\mathrm{Spec}(\vec{R})$ *is homeomorphic to* $\mathrm{Spec}(\overline{R}) \times_{\mathrm{Spec}(R)} \mathrm{Spec}(\widetilde{R})$.
(6) $\mathrm{Supp}(S/R) = \mathrm{Supp}(\widetilde{R}/R) \cup \mathrm{Supp}(\overline{R}/R)$ *if $R \subseteq S$ is almost-Prüfer. (*Supp *can be replaced with* MSupp*).*

Proof To show (1), (2), in view of Theorem 3.18, it is enough to apply Proposition 4.5 with $T = \widetilde{R}$ and $S = \overset{\Longrightarrow}{R}$, because $R \subseteq \overline{R}\widetilde{R}$ is almost-Prüfer whence quasi-Prüfer, keeping in mind that a Prüfer extension is integrally closed, whereas an integral Prüfer extension is trivial. Moreover, $\overline{\widetilde{R}} = \overline{R}\widetilde{R}$ because $\overline{R}\widetilde{R} \subseteq \overline{\widetilde{R}}$ is both integral and integrally closed.

(3) is obvious.

(4) Now consider an almost-Prüfer subextension $R \subseteq T \subseteq U$, where $R \subseteq T$ is Prüfer and $T \subseteq U$ is integral. Applying (3), we see that $U = \overline{R}^U \widetilde{R}^U \subseteq \overline{R}\widetilde{R}$ in view of Proposition 1.6.

(5) Recall from [33] that a ring morphism $A \to A'$ is called a subtrusion if for each pair of prime ideals $P \subseteq Q$ of A, there is a pair of prime ideals $P' \subseteq Q'$ above $P \subseteq Q$. A subtrusion defines a submersion $\mathrm{Spec}(A') \to \mathrm{Spec}(A)$. We refer to [33, First paragraph of p. 570] for the definition of the property $P(\Delta)$ of a pushout diagram (Δ). Then [33, Lemme 2,(b), p. 570] shows that $P(D')$ holds, because $R \to \widetilde{R}$ is a flat epimorphism. Now [33, Proposition 2, p. 576] yields that $\mathrm{Spec}(\vec{R}) \to \mathrm{Spec}(\overline{R}) \times_{\mathrm{Spec}(R)} \mathrm{Spec}(\widetilde{R})$ is subtrusive. This map is also injective because $R \to \widetilde{R}$ is spectrally injective. Observing that an injective submersion is an homeomorphism, the proof is complete.

(6) Obviously, $\mathrm{Supp}(\widetilde{R}/R) \cup \mathrm{Supp}(\overline{R}/R) \subseteq \mathrm{Supp}(S/R)$. Conversely, let $M \in \mathrm{Spec}(R)$ be such that $R_M \neq S_M$, and $R_M = (\widetilde{R})_M = \overline{R}_M$. Then (3) entails that $S_M = (\overline{R}\widetilde{R})_M = (\overline{R})_M(\widetilde{R})_M = R_M$, which is absurd. □

Corollary 4.7 *Let $R \subseteq S$ be an almost-Prüfer extension. The following conditions are equivalent:*

(1) $\mathrm{Supp}(S/\overline{R}) \cap \mathrm{Supp}(\overline{R}/R) = \emptyset$.
(2) $\mathrm{Supp}(S/\widetilde{R}) \cap \mathrm{Supp}(\widetilde{R}/R) = \emptyset$.
(3) $\mathrm{Supp}(\widetilde{R}/R) \cap \mathrm{Supp}(\overline{R}/R) = \emptyset$.

Proof Since $R \subseteq S$ is almost-Prüfer, we get $(\widetilde{R})_P = \widetilde{R_P}$ for each $P \in \mathrm{Spec}(R)$. Moreover, $\mathrm{Supp}(S/R) = \mathrm{Supp}(\widetilde{R}/R) \cup \mathrm{Supp}(\overline{R}/R) = \mathrm{Supp}(S/\overline{R}) \cup \mathrm{Supp}(\overline{R}/R) = \mathrm{Supp}(S/\widetilde{R}) \cup \mathrm{Supp}(\widetilde{R}/R)$.

(1) ⇒ (2): Assume that there exists $P \in \mathrm{Supp}(S/\widetilde{R}) \cap \mathrm{Supp}(\widetilde{R}/R)$. Then, $(\widetilde{R})_P \neq S_P, R_P$, so that $R_P \subset S_P$ is neither Prüfer nor integral. But, $P \in \mathrm{Supp}(S/R) = \mathrm{Supp}(S/\overline{R}) \cup \mathrm{Supp}(\overline{R}/R)$. If $P \in \mathrm{Supp}(S/\overline{R})$, then $P \notin \mathrm{Supp}(\overline{R}/R)$, so that $(\overline{R})_P = R_P$ and $R_P \subset S_P$ is Prüfer, a contradiction. If $P \in \mathrm{Supp}(\overline{R}/R)$, then $P \notin \mathrm{Supp}(S/\overline{R})$, so that $(\overline{R})_P = S_P$ and $R_P \subset S_P$ is integral, a contradiction.

(2) ⇒ (3): Assume that there exists $P \in \mathrm{Supp}(\widetilde{R}/R) \cap \mathrm{Supp}(\overline{R}/R)$. Then, $R_P \neq (\widetilde{R})_P, (\overline{R})_P$, so that $R_P \subset S_P$ is neither Prüfer nor integral. But, $P \in \mathrm{Supp}(S/R) = \mathrm{Supp}(S/\widetilde{R}) \cup \mathrm{Supp}(\widetilde{R}/R)$. If $P \in \mathrm{Supp}(S/\widetilde{R})$, then $P \notin \mathrm{Supp}(\widetilde{R}/R)$, so that $(\widetilde{R})_P = R_P$ and $R_P \subset S_P$ is integral, a contradiction. If $P \in \mathrm{Supp}(\widetilde{R}/R)$, then $P \notin \mathrm{Supp}(S/\widetilde{R})$, so that $(\widetilde{R})_P = S_P$ and $R_P \subset S_P$ is Prüfer, a contradiction.

(3) ⇒ (1): Assume that there exists $P \in \mathrm{Supp}(S/\overline{R}) \cap \mathrm{Supp}(\overline{R}/R)$. Then, $(\overline{R})_P \neq R_P, S_P$, so that $R_P \subset S_P$ is neither Prüfer nor integral. But, $P \in \mathrm{Supp}(S/R) = \mathrm{Supp}(\overline{R}/R) \cup \mathrm{Supp}(\widetilde{R}/R)$. If $P \in \mathrm{Supp}(\widetilde{R}/R)$, then $P \notin \mathrm{Supp}(\overline{R}/R)$, so that $(\overline{R})_P =$

R_P and $R_P \subset S_P$ is Prüfer, a contradiction. If $P \in \mathrm{Supp}(\overline{R}/R)$, then $P \notin \mathrm{Supp}(\widetilde{R}/R)$, so that $(\widetilde{R})_P = R_P$ and $R_P \subset S_P$ is integral, a contradiction. □

Proposition 4.5 has the following similar statement proved by Ayache and Dobbs. It reduces to Theorem 4.6 in case $R \subseteq S$ has FCP because of Proposition 1.3.

Proposition 4.8 *Let $R \subseteq T \subseteq S$ be a quasi-Prüfer extension, where $T \subseteq S$ is an integral minimal extension and $R \subseteq T$ is integrally closed. Then the diagram (D) is a pullback, $S = T\overline{R}$ and $(T : S) = (R : \overline{R})T$.*

Proof [5, Lemma 3.5]. □

Proposition 4.9 *Let $R \subseteq U \subseteq S$ and $R \subseteq V \subseteq S$ be two towers of extensions, such that $R \subseteq U$ and $R \subseteq V$ are almost-Prüfer. Then $R \subseteq UV$ is almost-Prüfer and $\widetilde{UV} = \widetilde{U}\widetilde{V}$.*

Proof Denote by U', V', and W' the Prüfer hulls of R in U, V, and $W = UV$. We deduce from [25, Corollary 5.11, p. 53], that $R \subseteq U'V'$ is Prüfer. Moreover, $U'V' \subseteq UV$ is clearly integral and $U'V' \subseteq W'$ because the Prüfer hull is the greatest Prüfer subextension. We deduce that $R \subseteq UV$ is almost-Prüfer and that $\widetilde{UV} = \widetilde{U}\widetilde{V}$. □

Proposition 4.10 *Let $R \subseteq U \subseteq S$ and $R \subseteq V \subseteq S$ be two towers of extensions, such that $R \subseteq U$ is almost-Prüfer and $R \subseteq V$ is a flat epimorphism. Then $U \subseteq UV$ is almost-Prüfer.*

Proof Mimic the proof of Proposition 4.9, using [25, Theorem 5.10, p. 53]. □

Proposition 4.11 *Let $R \subseteq S$ be an almost-Prüfer extension and $R \to T$ a flat epimorphism. Then $T \subseteq T \otimes_R S$ is almost-Prüfer.*

Proof It is enough to use Proposition 3.10 and Definition 4.1. □

Proposition 4.12 *An extension $R \subseteq S$ is almost-Prüfer if and only if $R_P \subseteq S_P$ is almost-Prüfer and $\widetilde{R_P} = (\widetilde{R})_P$ for each $P \in \mathrm{Spec}(R)$.*

Proof For an arbitrary extension $R \subseteq S$ we have $(\widetilde{R})_P \subseteq \widetilde{R_P}$. Suppose that $R \subseteq S$ is almost-Prüfer, then so is $R_P \subseteq S_P$ and $(\widetilde{R})_P = \widetilde{R_P}$ by Theorem 4.4. Conversely, if $R \subseteq S$ is locally almost-Prüfer, whence locally quasi-Prüfer, then $R \subseteq S$ is quasi-Prüfer. If $\widetilde{R_P} = (\widetilde{R})_P$ holds for each $P \in \mathrm{Spec}(R)$, we have $S_P = (\widetilde{R}\overline{R})_P$ so that $S = \widetilde{R}\overline{R}$ and $R \subseteq S$ is almost-Prüfer by Theorem 4.6. □

Corollary 4.13 *An FCP extension $R \subseteq S$ is almost-Prüfer if and only if $R_P \subseteq S_P$ is almost-Prüfer for each $P \in \mathrm{Spec}(R)$.*

Proof It is enough to show that $R \subseteq S$ is almost-Prüfer if $R_P \subseteq S_P$ is almost-Prüfer for each $P \in \mathrm{Spec}(R)$ using Proposition 4.12. Any minimal extension $\widetilde{R} \subset R_1$ is integral by definition of $\widetilde{R}$. Assume that $(\widetilde{R})_P \subset (\widetilde{R_P})$, so that there exists $R_2' \in [\widetilde{R}, S]$ such that $(\widetilde{R})_P \subset (R_2')_P$ is a Prüfer minimal extension with crucial maximal ideal $Q(\widetilde{R})_P$, for some $Q \in \mathrm{Max}(\widetilde{R})$ with $Q \cap R \subseteq P$. In particular, $\widetilde{R} \subset R_2'$ is not integral. We may assume that there exists $R_1' \in [\widetilde{R}, R_2']$ such that $R_1' \subset R_2'$ is a Prüfer minimal extension with $P \notin \mathrm{Supp}(R'_1/\widetilde{R})$. Using [37, Lemma 1.10], there exists $R_2 \in [\widetilde{R}, R'_2]$ such that $\widetilde{R} \subset R_2$ is a Prüfer minimal extension with crucial maximal ideal Q, a contradiction. Then, $(\widetilde{R})_P \subset S_P$ is integral for each P, whence $(\widetilde{R})_P = \widetilde{(R_P)}$. □

We now intend to demonstrate that our methods allow us to prove easily some results. For instance, next statement generalizes [5, Corollary 4.5] and can be fruitful in algebraic number theory.

Proposition 4.14 *Let (R, M) be a one-dimensional local ring and $R \subseteq S$ a quasi-Prüfer extension. Suppose that there is a tower $R \subset T \subseteq S$, where $R \subset T$ is integrally closed. Then $R \subseteq S$ is almost-Prüfer, $T = \widetilde{R}$ and S is zero-dimensional.*

Proof Because $R \subset T$ is quasi-Prüfer and integrally closed, it is Prüfer. If some prime ideal of T is lying over M, $R \subset T$ is a faithfully flat epimorphism, whence an isomorphism by Scholium A, which is absurd. Now let N be a prime ideal of T and $P := N \cap R$. Then R_P is zero-dimensional and isomorphic to T_N. Therefore, T is zero-dimensional. It follows that $T\overline{R}$ is zero-dimensional. Since $R\overline{T} \subseteq S$ is Prüfer, we deduce from Scholium A, that $\overline{R}T = S$. The proof is now complete. □

We also generalize [5, Proposition 5.2] as follows.

Proposition 4.15 *Let $R \subset S$ be a quasi-Prüfer extension, such that $\overline{R}$ is local with maximal ideal $N := \sqrt{(R : \overline{R})}$. Then R is local and $[R, S] = [R, \overline{R}] \cup [\overline{R}, S]$. If in addition R is one-dimensional, then either $R \subset S$ is integral or there is some minimal prime ideal P of $\overline{R}$, such that $S = (\overline{R})_P$, $P = SP$ and $\overline{R}/P$ is a one-dimensional valuation domain with quotient field S/P.*

Proof R is obviously local. Let $T \in [R, S] \setminus [R, \overline{R}]$ and $s \in T \setminus \overline{R}$. Then $s \in \mathrm{U}(S)$ and $s^{-1} \in \overline{R}$ by Proposition 1.2 (1). But $s^{-1} \notin \mathrm{U}(\overline{R})$, so that $s^{-1} \in N$. It follows that there exists some integer n such that $s^{-n} \in (R : \overline{R})$, giving $s^{-n}\overline{R} \subseteq R$, or, equivalently, $\overline{R} \subseteq Rs^n \subseteq T$. Then, $T \in [\overline{R}, S]$ and we obtain $[R, S] = [R, \overline{R}] \cup [\overline{R}, S]$.

Assume that R is one-dimensional. If $R \subset S$ is not integral, then $\overline{R} \subset S$ is Prüfer and $\overline{R}$ is one-dimensional. To complete the proof, use Proposition 1.2 (3). □

4.2 FCP Extensions

In case we consider only FCP extensions, we obtain more results.

Proposition 4.16 *Let $R \subseteq S$ be an FCP extension. The following statements are equivalent:*

(1) $R \subseteq S$ *is almost-Prüfer.*
(2) $R_P \subseteq S_P$ *is either integral or Prüfer for each* $P \in \mathrm{Spec}(R)$.
(3) $R_P \subseteq S_P$ *is almost-Prüfer for each* $P \in \mathrm{Spec}(R)$ *and* $\mathrm{Supp}(S/\widetilde{R}) \cap \mathrm{Supp}(\widetilde{R}/R) = \emptyset$.
(4) $\mathrm{Supp}(\overline{R}/R) \cap \mathrm{Supp}(S/\overline{R}) = \emptyset$.

Proof The equivalence of Proposition 4.12 shows that (2) $\Leftrightarrow$ (1) holds because $\widehat{T} = \widetilde{T}$ and over a local ring T, an almost-Prüfer FCP extension $T \subseteq U$ is either integral or Prüfer [37, Proposition 2.4] . Moreover when $R_P \subseteq S_P$ is either integral or Prüfer, it is easy to show that $(\widetilde{R})_P = \widetilde{R_P}$

Next we show that (3) is equivalent to (2) of Proposition 4.12.

Let $P \in \mathrm{Supp}(S/\widetilde{R}) \cap \mathrm{Supp}(\widetilde{R}/R)$ be such that $R_P \subsetneq S_P$ is almost-Prüfer. Then, $(\widetilde{R})_P \neq R_P, S_P$, so that $R_P \subset (\widetilde{R})_P \subset S_P$. Since $R \subset \widetilde{R}$ is Prüfer, so is $R_P \subset (\widetilde{R})_P$, giving $(\widetilde{R})_P \subseteq \widetilde{R_P}$ and $R_P \neq \widetilde{R_P}$. It follows that $\widetilde{R_P} = S_P$ in view of the dichotomy principle [37, Proposition 3.3] since R_P is a local ring, and then $\widetilde{R_P} \neq (\widetilde{R})_P$.

Conversely, assume that $\widetilde{R_P} \neq (\widetilde{R})_P$, *i.e.* $P \in \mathrm{Supp}(S/R)$. Then, $R_P \neq \widetilde{R_P}$, so that $\widetilde{R_P} = S_P$, as we have just seen. Hence $R_P \subset S_P$ is integrally closed. It follows that $\overline{R_P} = \overline{R}_P = R_P$, so that $P \notin \mathrm{Supp}(\overline{R}/R)$ and $P \in \mathrm{Supp}(\widetilde{R}/R)$ by Theorem 4.6(5). Moreover, $\widetilde{R_P} \neq S_P$ implies that $P \in \mathrm{Supp}(S/\widetilde{R})$. To conclude, $P \in \mathrm{Supp}(S/\widetilde{R}) \cap \mathrm{Supp}(\widetilde{R}/R)$.

(1) $\Leftrightarrow$ (4) An FCP extension is quasi-Prüfer by Corollary 3.4. Suppose that $R \subseteq S$ is almost-Prüfer. By Theorem 4.6, letting $U := \widetilde{R}$, we get that $U \cap \overline{R} = R$ and $S = \overline{R}U$. We deduce from [37, Proposition 3.6] that $\mathrm{Supp}(\overline{R}/R) \cap \mathrm{Supp}(S/\overline{R}) = \emptyset$. Suppose that this last condition holds. Then by [37, Proposition 3.6] $R \subseteq S$ can be factored $R \subseteq U \subseteq S$, where $R \subseteq U$ is integrally closed, whence Prüfer by Proposition 1.3, and $U \subseteq S$ is integral. Therefore, $R \subseteq S$ is almost-Prüfer. □

Proposition 4.17 *Let $R \subset S$ be an FCP almost-Prüfer extension. Then, $\widetilde{R} = \widehat{R}$ and $\widetilde{R}$ is the least $T \in [R, S]$ such that $T \subseteq S$ is integral.*

Proof Let $T \in [R, S]$ be such that $T \subseteq S$ is integral. So is $T_M \subseteq S_M$ for each $M \in \mathrm{Max}(R)$. But $R_M \subseteq S_M$ is either integral (1) or Prüfer (2). In case (1), we get $R_M = \widetilde{R_M} \subseteq T_M$ and in case (2), we get $\widetilde{R_M} = S_M = T_M$, so that $\widetilde{R}_M \subseteq T_M$. By globalization, $\widetilde{R} \subseteq T$. □

We will need a relative version of the support. Let $f : R \to T$ be a ring morphism and E a T-module. The relative support of E over R is $\mathscr{S}_R(E) := {}^af(\mathrm{Supp}_T(E))$ and $\mathrm{M}\mathscr{S}_R(E) := \mathscr{S}_R(E) \cap \mathrm{Max}(R)$. In particular, for a ring extension $R \subset S$, we have $\mathscr{S}_R(S/R) := \mathrm{Supp}_R(S/R))$.

Proposition 4.18 *Let $R \subseteq S$ be an FCP extension. The following statements hold:*

(1) $\mathrm{Supp}(\overline{\overline{R}}/\overline{R}) \cap \mathrm{Supp}(\overline{R}/R) = \emptyset$.
(2) $\mathrm{Supp}(\widetilde{R}/R) \cap \mathrm{Supp}(\overline{R}/R) = \mathrm{Supp}(\overline{\overline{R}}/\widetilde{R}) \cap \mathrm{Supp}(\widetilde{R}/R) = \emptyset$.
(3) $\mathrm{MSupp}(S/R) = \mathrm{MSupp}(\widetilde{R}/R) \cup \mathrm{MSupp}(\overline{R}/R)$.

Proof (1) is a consequence of Proposition 4.16(4) because $R \subseteq \overline{\overline{R}}$ is almost-Prüfer.

We prove the first part of (2). If some $M \in \mathrm{Supp}(\widetilde{R}/R) \cap \mathrm{Supp}(\overline{R}/R)$, it can be supposed in $\mathrm{Max}(R)$ because supports are stable under specialization. Set $R' := R_M$, $U := (\widetilde{R})_M$, $T := (\overline{R})_M$ and $M' := MR_M$. Then, $R' \neq U, T$, with $R' \subset U$ FCP Prüfer and $R' \subset T$ FCP integral, an absurdity [37, Proposition 3.3].

To show the second part, assume that some $P \in \mathrm{Supp}(\overline{\widetilde{R}}/\widetilde{R}) \cap \mathrm{Supp}(\widetilde{R}/R)$. Then, $P \notin \mathrm{Supp}(\overline{R}/R)$ by the first part of (2), so that $\overline{R}_P = R_P$, giving $(\overline{\widetilde{R}})_P = \overline{R}_P\widetilde{R}_P = \widetilde{R}_P$, a contradiction.

(3) Obviously, $\mathrm{MSupp}(S/R) = \mathrm{M}\mathscr{S}(S/R) = \mathrm{M}\mathscr{S}(S/\overline{T}^S) \cup \mathrm{M}\mathscr{S}(\overline{T}^S/T)$ $\cup \mathrm{M}\mathscr{S}(T/\overline{U}^T) \cup \mathrm{M}\mathscr{S}(\overline{U}^T/U) \cup \mathrm{M}\mathscr{S}(U/R)$. By [37, Propositions 2.3 and 3.2], we have $\mathrm{M}\mathscr{S}(S/\overline{T}^S) \subseteq \mathscr{S}(\overline{T}^S/T) = \mathscr{S}(\overline{R}/\overline{R}^T) = \mathrm{M}\mathscr{S}(\overline{R}/R) = \mathrm{MSupp}(\overline{R}/R)$, $\mathrm{M}\mathscr{S}(T/\overline{U}^T) = \mathscr{S}(\overline{R}^T/R) \subseteq \mathscr{S}(\overline{R}/R) = \mathrm{Supp}(\overline{R}/R)$ and $\mathrm{M}\mathscr{S}(\overline{U}^T/U) = \mathscr{S}(\overline{R}^T/R) = \mathrm{Supp}(\overline{R}/R)$. To conclude, $\mathrm{MSupp}(S/R) = \mathrm{MSupp}(\widetilde{R}/R) \cup \mathrm{MSupp}(\overline{R}/R)$. □

Proposition 4.19 *Let $R \subset S$ be an FCP extension and $M \in \mathrm{MSupp}(S/R)$, then $\widetilde{R_M} = (\widetilde{R})_M$ if and only if $M \notin \mathrm{MSupp}(S/\widetilde{R}) \cap \mathrm{MSupp}(\widetilde{R}/R)$.*

Proof In fact, we are going to show that $\widetilde{R_M} \neq (\widetilde{R})_M$ if and only if $M \in \mathrm{MSupp}(S/\widetilde{R}) \cap \mathrm{MSupp}(\widetilde{R}/R)$.

Let $M \in \mathrm{MSupp}(S/\widetilde{R}) \cap \mathrm{MSupp}(\widetilde{R}/R)$. Then, $\widetilde{R_M} \neq R_M, S_M$ and $R_M \subset \widetilde{R_M} \subset S_M$. Since $R \subset \widetilde{R}$ is Prüfer, so is $R_M \subset \widetilde{R}_M$ by Proposition 1.2, giving $(\widetilde{R})_M \subseteq \widetilde{R_M}$ and $R_M \neq \widetilde{R}_M$. Therefore, $\widetilde{R_M} = S_M$ [37, Proposition 3.3] since R_M is local, and then $\widetilde{R_M} \neq (\widetilde{R})_M$.

Conversely, if $\widetilde{R_M} \neq (\widetilde{R})_M$, then, $R_M \neq \widetilde{R_M}$, so that $\widetilde{R_M} = S_M$, as we have just seen and then $R_M \subset S_M$ is integrally closed. It follows that $\overline{R_M} = \overline{R}_M = R_M$, so that $M \notin \mathrm{MSupp}(\overline{R}/R)$. Hence, $M \in \mathrm{MSupp}(\widetilde{R}/R)$ by Proposition 4.18(3). Moreover, $\widetilde{R}_M \neq S_M \Rightarrow M \in \mathrm{MSupp}(S/\widetilde{R})$. To conclude, $M \in \mathrm{MSupp}(S/\widetilde{R}) \cap \mathrm{MSupp}(\widetilde{R}/R)$. □

If $R \subseteq S$ is an extension, with $\dim(R) = 0$, $\widetilde{R_M} = (\widetilde{R})_M$ for any $M \in \mathrm{Max}(R)$. Indeed by Scholium A (2), the flat epimorphisms $R \to \widetilde{R}$ and $R_M \to (\widetilde{R})_M$ are bijective. This conclusion holds in another context.

Corollary 4.20 *Let $R \subset S$ be an FCP extension. Assume that one of the following conditions is satisfied:*

(1) $\mathrm{MSupp}(S/\widetilde{R}) \cap \mathrm{MSupp}(\widetilde{R}/R) = \emptyset$.
(2) $S = \overline{R}\widetilde{R}$, or equivalently, $R \subseteq S$ is almost-Prüfer.

Then, $\widetilde{R_M} = (\widetilde{R})_M$ for any $M \in \mathrm{Max}(R)$.

Proof (1) is Proposition 4.19. (2) is Proposition 4.12. □

Proposition 4.21 *Let $R \subset S$ be an almost-Prüfer FCP extension. Then, any $T \in [R, S]$ is the integral closure of $T \cap \widetilde{R}$ in $T\widetilde{R}$. Moreover, if $T \cap \overline{R} = R$, then $T = T\widetilde{R} \cap \overline{R}$; if $T\overline{R} = S$, then $T = (T \cap \overline{R})\widetilde{R}$; if $T\widetilde{R} = S$, then $T = (T \cap \widetilde{R})\overline{R}$.*

Proof Set $U := T \cap \widetilde{R}$ and $V := T\widetilde{R}$. Since $R \subset S$ is almost-Prüfer, $U \subseteq \widetilde{R}$ is Prüfer and $\widetilde{R} \subseteq V$ is integral and $\widetilde{R}$ is also the Prüfer hull of $U \subseteq V$. Because $R \subset S$ is almost-Prüfer, for each $M \in \mathrm{MSupp}_R(S/R)$, $R_M \subseteq S_M$ is either integral or Prüfer by Proposition 4.16, and so is $U_M \subseteq V_M$. But $\widetilde{R_M} = (\widetilde{R})_M$ by Corollary 4.20 is also the Prüfer hull of $U_M \subseteq V_M$. Let T' be the integral closure of U in V. Then, T'_M is the integral closure of U_M in V_M.

Assume that $U_M \subseteq V_M$ is integral. Then $V_M = T'_M$ and $U_M = (\widetilde{R})_M$, so that $V_M = T_M(\widetilde{R})_M = T_M$, giving $T_M = T'_M$.

Assume that $U_M \subseteq V_M$ is Prüfer. Then $U_M = T'_M$ and $V_M = (\widetilde{R})_M$, so that $U_M = T_M \cap (\widetilde{R})_M = T_M$, giving $T_M = T'_M$.

To conclude, $T_M = T'_M$ follows for each $M \in \mathrm{MSupp}_R(S/R)$. Since $R_M = S_M$, with $T_M = T'_M$ for each $M \in \mathrm{Max}(R) \setminus \mathrm{MSupp}_R(S/R)$, we get $T = T'$, whence T is the integral closure of $U \subseteq V$.

The last results are then obvious. □

We build an example of an FCP extension $R \subset S$ where $\widetilde{R_M} \neq (\widetilde{R})_M$ for some $M \in \mathrm{Max}(R)$. In particular, $R \subset S$ is not almost-Prüfer.

Example 4.22 Let R be an integral domain with quotient field S and $\mathrm{Spec}(R) := \{M_1, M_2, P, 0\}$, where $M_1 \neq M_2$ are two maximal ideals and P a prime ideal satisfying $P \subset M_1 \cap M_2$. Assume that there are R_1, R_2, and R_3 such that $R \subset R_1$ is Prüfer minimal, with $\mathcal{C}(R, R_1) = M_1$, $R \subset R_2$ is integral minimal, with $\mathcal{C}(R, R_2) = M_2$, and $R_2 \subset R_3$ is Prüfer minimal, with $\mathcal{C}(R_2, R_3) = M_3 \in \mathrm{Max}(R_2)$ such that $M_3 \cap R = M_2$ and $M_2R_3 = R_3$. This last condition is satisfied when $R \subset R_2$ is either ramified or inert. Indeed, in both cases, $M_3R_3 = R_3$; moreover, in the ramified case, we have $M_3^2 \subseteq M_2$ and in the inert case, $M_3 = M_2$ [36, Theorem 3.3]. We apply [14, Proposition 7.10] and [13, Lemma 2.4] several times. Set $R'_2 := R_1R_2$. Then, $R_1 \subset R'_2$ is integral minimal, with $\mathcal{C}(R_1, R'_2) =: M'_2 = M_2R_1$ and $R_2 \subset R'_2$ is Prüfer minimal, with $\mathcal{C}(R_2, R'_2) =: M'_1 = M_1R_2 \in \mathrm{Max}(R_2)$. Moreover, $M'_1 \neq M_3$, $\mathrm{Spec}(R_1) = \{M'_2, P_1, 0\}$, where P_1 is the only prime ideal of R_1 lying over P. But, $P = (R : R_1)$ by [17, Proposition 3.3], so that $P = P_1$. Set $R'_3 := R_3R'_2$. Then, $R'_2 \subset R'_3$ is Prüfer minimal, with $\mathcal{C}(R'_2, R'_3) =: M'_3 = M_3R'_2 \in \mathrm{Max}(R'_2)$ and $R_3 \subset R'_3$ is Prüfer minimal, with $\mathcal{C}(R_3, R'_3) = M''_1 = M_1R_3 \in \mathrm{Max}(R_3)$. We have therefore $\mathrm{Spec}(R'_3) = \{P', 0\}$ where P' is the only prime ideal of R'_3 lying over P. To end, assume that $R'_3 \subset S$ is Prüfer minimal, with $\mathcal{C}(R'_3, S) = P'$. Hence, R_2 is the integral closure of R in S. In particular, $R \subset S$ has FCP [13, Theorems 6.3 and 3.13] and is quasi-Prüfer. Since $R \subset R_1$ is integrally closed, we have $R_1 \subseteq \widetilde{R}$. Assume that $R_1 \neq \widetilde{R}$. Then, there exists $T \in [R_1, S]$ such that $R_1 \subset T$ is Prüfer minimal and $\mathcal{C}(R_1, T) = M'_2$, a contradiction by Proposition 4.16 since $M'_2 = \mathcal{C}(R_1, R'_2)$, with $R_1 \subset R'_2$ integral minimal. Then, $R_1 = \widetilde{R}$. It follows that $M_1 \in \mathrm{MSupp}(\widetilde{R}/R)$. But, $P = \mathcal{C}(R'_3, S) \cap R \in \mathrm{Supp}(S/\widetilde{R})$ and $P \subset M_1$ give $M_1 \in \mathrm{MSupp}(S/\widetilde{R})$, so that $\widetilde{R_{M_1}} \neq (\widetilde{R})_{M_1}$ by Proposition 4.19 giving that $R \subset S$ is not almost-Prüfer.

We now intend to refine Theorem 4.6, following the scheme used in [3, Proposition 4] for extensions of integral domains.

Proposition 4.23 *Let $R \subseteq S$ and $U, T \in [R, S]$ be such that $R \subseteq U$ is integral and $R \subseteq T$ is Prüfer. Then $U \subseteq UT$ is Prüfer in the following cases and $R \subseteq UT$ is almost-Prüfer.*

(1) $\mathrm{Supp}(\overline{R}/R) \cap \mathrm{Supp}(\widetilde{R}/R) = \emptyset$ *(for example, if $R \subseteq S$ has FCP).*
(2) $R \subseteq U$ preserves integral closure.

Proof (1) We have $\emptyset = \mathrm{MSupp}(U/R) \cap \mathrm{MSupp}(T/R)$, since $U \subseteq \overline{R}$ and $T \subseteq \widetilde{R}$. Let $M \in \mathrm{MSupp}((UT)/R)$. For $M \in \mathrm{MSupp}(U/R)$, we have $R_M = T_M$ and $(UT)_M = U_M$. If $M \notin \mathrm{MSupp}(U/R)$, then $U_M = R_M$ and $(UT)_M = T_M$, so that $U_M \subseteq (UT)_M$ identifies to $R_M \subseteq T_M$.

Let $N \in \mathrm{Max}(U)$ and set $M := N \cap R \in \mathrm{Max}(R)$ since $R \subseteq U$ is integral. If $M \notin \mathrm{Supp}(\overline{R}/R)$, then $R_M = \overline{R}_M = U_M$ and N is the only maximal ideal of U lying over M. It follows that $U_M = U_N$ and $(UT)_M = (UT)_N$ by [13, Lemma 2.4]. Then, $U_N \subseteq (UT)_N$ identifies to $R_M \subseteq T_M$ which is Prüfer. If $M \notin \mathrm{Supp}(\widetilde{R}/R)$, then $R_M = T_M$ gives $U_M = (UT)_M$, so that $U_N = (UT)_N$ by localizing the precedent equality and $U_N \subseteq (UT)_N$ is still Prüfer. Therefore, $U \subseteq UT$ is locally Prüfer, whence Prüfer by Proposition 1.1.

(2) The usual reasoning gives $U \otimes_R T \cong UT$, whence $U \subseteq UT$ is integrally closed. From $U \subseteq \overline{R}^{UT}$, we deduce $U = \overline{R}^{UT}$. Because $R \subseteq UT$ is almost-Prüfer, whence quasi-Prüfer, $U \subseteq UT$ is Prüfer. □

Next propositions generalize Ayache's results of [3, Proposition 11].

Proposition 4.24 *Let $R \subseteq S$ be a quasi-Prüfer extension, $T, T' \in [R, S]$ and $U := T \cap T'$. The following statements hold:*

(1) $\widetilde{T} = \widetilde{(T \cap \overline{R})}$ for each $T \in [R, S]$.
(2) $\widetilde{T} \cap \widetilde{T'} \subseteq \widetilde{T \cap T'}$.
(3) If $\mathrm{Supp}(\overline{T}/T) \cap \mathrm{Supp}(\widetilde{T}/T) = \emptyset$ (this assumption holds if $R \subseteq S$ has FCP), then, $T \subseteq T' \Rightarrow \widetilde{T} \subseteq \widetilde{T'}$.
(4) If $\mathrm{Supp}(\overline{U}/U) \cap \mathrm{Supp}(\widetilde{U}/U) = \emptyset$, then $\widetilde{T} \cap \widetilde{T'} = \widetilde{T \cap T'}$.

Proof (1) We observe that $R \subseteq T$ is quasi-Prüfer by Corollary 3.3. Since $T \cap \overline{R}$ is the integral closure of R in T, we get that $T \cap \overline{R} \subseteq T$ is Prüfer. It follows that $T \cap \overline{R} \subseteq \widetilde{T}$ is Prüfer. We thus have $\widetilde{T} \subseteq \widetilde{T \cap \overline{R}}$. To prove the reverse inclusion, we set $V := T \cap \overline{R}$ and $W := \widetilde{V} \cap \overline{T}$. We have $W \cap \overline{R} = \widetilde{V} \cap \overline{R} = V$, because $V \subseteq \widetilde{V} \cap \overline{R}$ is integral and Prüfer since we have a tower $V \subseteq \widetilde{V} \cap \overline{R} \subseteq \widetilde{V}$. Therefore, $V \subseteq W$ is Prüfer because $W \in [V, \widetilde{V}]$. Moreover, $T \subseteq \widetilde{T} \subseteq \widetilde{V}$, since $V \subseteq \widetilde{T}$ is Prüfer. Then, $T \subseteq W$ is integral because $W \in [T, \overline{T}]$, and we have $V \subseteq T \subseteq W$. This entails that $T = W = \widetilde{V} \cap \overline{T}$, so that $T \subseteq \widetilde{V}$ is Prüfer. It follows that $\widetilde{V} \subseteq \widetilde{T}$ since $T \in [V, \widetilde{V}]$.

(2) A quasi-Prüfer extension is Prüfer if and only if it is integrally closed. We observe that $T \cap T' \subseteq \widetilde{T} \cap \widetilde{T'}$ is integrally closed, whence Prüfer. It follows that $\widetilde{T} \cap \widetilde{T'} \subseteq \widetilde{T \cap T'}$.

(3) Set $U = T \cap \overline{R}$ and $U' = T' \cap \overline{R}$, so that $U, U' \in [R, \overline{R}]$ with $U \subseteq U'$. In view of (1), we thus can suppose that $T, T' \in [R, \overline{R}]$. It follows that $T \subseteq T'$ is integral and $T \subseteq \widetilde{T}$ is Prüfer. We deduce from Proposition 4.23(1) that $T' \subseteq T'\widetilde{T}$ is Prüfer, so that $\widetilde{T}T' \subseteq \widetilde{T'}$, because $\mathrm{Supp}(\overline{T}/T) \cap \mathrm{Supp}(\widetilde{T}/T) = \emptyset$ and $\overline{T} = \overline{R}$. Therefore, we have $\widetilde{T} \subseteq \widetilde{T'}$.

(4) Assume that $\mathrm{Supp}(\overline{U}/U) \cap \mathrm{Supp}(\widetilde{U}/U) = \emptyset$. Then, $T \cap T' \subset T, T'$ gives $\widetilde{T \cap T'} \subseteq \widetilde{T} \cap \widetilde{T'}$ in view of (3), so that $\widetilde{T \cap T'} = \widetilde{T} \cap \widetilde{T'}$ by (2). □

Proposition 4.25 *Let $R \subseteq S$ be a quasi-Prüfer extension and $T \subseteq T'$ a subextension of $R \subseteq S$. Set $U := T \cap \overline{R}$, $U' := T' \cap \overline{R}$, $V := T\overline{R}$ and $V' := T'\overline{R}$. The following statements hold:*

(1) $T \subseteq T'$ is integral if and only if $V = V'$.
(2) $T \subseteq T'$ is Prüfer if and only if $U = U'$.
(3) Assume that $U \subset U'$ is integral minimal and $V = V'$. Then, $T \subset T'$ is integral minimal, of the same type as $U \subset U'$.
(4) Assume that $V \subset V'$ is Prüfer minimal and $U = U'$. Then, $T \subset T'$ is Prüfer minimal.
(5) Assume that $T \subset T'$ is minimal and set $P := \mathcal{C}(T, T')$.

(a) If $T \subset T'$ is integral, then $U \subset U'$ is integral minimal if and only if $P \cap U \in \mathrm{Max}(U)$.
(b) If $T \subset T'$ is Prüfer, then $V \subset V'$ is Prüfer minimal if and only if there is exactly one prime ideal in V lying over P.

Proof In $[R, S]$, the extensions $U \subseteq U'$, $T \subseteq V$, $T' \subseteq V'$ are integral and $V \subseteq V'$, $U \subseteq T$, $U' \subseteq T'$ are Prüfer. Moreover, $\overline{R}$ is also the integral closure of $U \subseteq V'$.

(1) is gotten by considering the extension $T \subseteq V'$, which is both $T \subseteq V \subseteq V'$ and $T \subseteq T' \subseteq V'$.

(2) is gotten by considering the extension $U \subseteq T'$, which is both $U \subseteq T \subseteq T'$ and $U \subseteq U' \subseteq T'$.

(3) Assume that $U \subset U'$ is integral minimal and $V = V'$. Then, $T \subset T'$ is integral by (1) and $T \neq T'$ because of (2). Set $M := (U : U') \in \mathrm{Supp}_U(U'/U)$. For any $M' \in \mathrm{Max}(U)$ such that $M' \neq M$, we have $U_{M'} = U'_{M'}$, so that $T_{M'} = T'_{M'}$ because $U_{M'} \subseteq T'_{M'}$ is Prüfer. But, $U \subseteq T'$ is almost-Prüfer, giving $T' = TU'$. By Theorem 4.6, $(T : T') = (U : U')T = MT \neq T$ because $T \neq T'$. We get that $U \subseteq T$ Prüfer implies that $M \notin \mathrm{Supp}_U(T/U)$ and $U_M = T_M$. It follows that $T'_M = T_M U'_M = U'_M$. Therefore, $T_M \subseteq T'_M$ identifies to $U_M \subseteq U'_M$, which is minimal of the same type as $U \subset U'$ by [14, Proposition 4 6]. Then, $T \subset T'$ is integral minimal of the same type as $U \subset U'$.

(4) Assume that $V \subset V'$ is Prüfer minimal and $U = U'$. Then, $T \subset T'$ is Prüfer by (2) and $T \neq T'$ because of (1). Set $Q := \mathcal{C}(V, V')$ and $P := Q \cap T \in \mathrm{Max}(T)$ since $Q \in \mathrm{Max}(V)$. For any $P' \in \mathrm{Max}(T)$ such that $P' \neq P$, and $Q' \in \mathrm{Max}(V)$ lying over P', we have $V_{Q'} = V'_{Q'}$, so that $V_{P'} = V'_{P'}$. Therefore, $T'_{P'} \subseteq V'_{P'}$ is integral, so that $T_{P'} = T'_{P'}$ and $P' \notin \mathrm{Supp}_T(T'/T)$. Hence $T \subset T'$ is Prüfer minimal [13, Proposition 6.12].

(5) Assume that $T \subset T'$ is a minimal extension and set $P := \mathcal{C}(T, T')$.

(a) Assume that $T \subset T'$ is integral. Then, $V = V'$ and $U \neq U'$ by (1) and (2). We can use Proposition 4.5 getting that $P = (U : U')T \in \mathrm{Max}(T)$ and $Q := (U : U') = P \cap U \in \mathrm{Spec}(U)$. It follows that $Q \notin \mathrm{Supp}_U(T/U)$ so that $U_Q = T_Q$ and $U'_Q = T'_Q$. Then, $U_Q \subset U'_Q$ is integral minimal, with $Q \in \mathrm{Supp}_U(U'/U)$.

If $Q \notin \mathrm{Max}(U)$, then $U \subset U'$ is not minimal by the properties of the crucial maximal ideal.

Assume that $Q \in \mathrm{Max}(U)$ and let $M \in \mathrm{Max}(U)$, with $M \neq Q$. Then, $U_M = U'_M$ because $M + Q = U$, so that $U \subset U'$ is a minimal extension and (a) is gotten.

(b) Assume that $T \subset T'$ is Prüfer. Then, $V \neq V'$ and $U = U'$ by (1) and (2). Moreover, $PT' = T'$ gives $PV' = V'$. Let $Q \in \mathrm{Max}(V)$ lying over P. Then, $QV' = V'$ gives that $Q \in \mathrm{Supp}_V(V'/V)$. Moreover, we have $V' = VT'$. Let $P' \in \mathrm{Max}(T)$, $P' \neq P$. Then, $T_{P'} = T'_{P'}$ gives $V_{P'} = V'_{P'}$. It follows that $\mathrm{Supp}_T(V'/V) = \{P\}$ and $\mathrm{Supp}_V(V'/V) = \{Q \in \mathrm{Max}(V) \mid Q \cap T = P\}$. But, by [13, Proposition 6.12], $V \subset V'$ is Prüfer minimal if and only if $|\mathrm{Supp}_V(V'/V)| = 1$, and then if and only if there is exactly one prime ideal in V lying over P. □

This proposition has a simpler dual form in the FCP almost-Prüfer case.

Proposition 4.26 *Let $R \subseteq S$ be an FCP almost-Prüfer extension and $T \subseteq T'$ a subextension of $R \subseteq S$. Set $U := T \cap \widetilde{R}$, $U' := T' \cap \widetilde{R}$, $V := T\widetilde{R}$, and $V' := T'\widetilde{R}$. The following statements hold:*

(1) $T \subseteq T'$ is integral (and minimal) if and only if $U = U'$ (and $V \subset V'$ is minimal).
(2) $T \subseteq T'$ is Prüfer (and minimal) if and only if $V = V'$ (and $U \subset U'$ is minimal).

Proof In view of Proposition 4.21, T (resp. T') is the integral closure of U (resp. U') in V (resp. V'). The result is gotten by localizing at the elements of $\mathrm{MSupp}_U(V'/U)$ and using Proposition 4.16. □

Lemma 4.27 *Let $R \subseteq S$ be an FCP almost-Prüfer extension and $U \in [R, \overline{R}]$, $V \in [\overline{R}, S]$. Then $U \subseteq V$ has FCP and is almost-Prüfer. The same result holds when $U \in [R, \widetilde{R}]$ and $V \in [\widetilde{R}, S]$.*

Proof Assume first that $U \in [R, \overline{R}]$ and $V \in [\overline{R}, S]$. Obviously, $U \subseteq V$ has FCP and $\overline{R}$ is the integral closure of U in V. Proposition 4.16 entails that $\mathrm{Supp}_R(\overline{R}/R) \cap \mathrm{Supp}_R(S/\overline{R}) = \emptyset$. We claim that $\mathrm{Supp}_U(\overline{R}/U) \cap \mathrm{Supp}_U(V/\overline{R}) = \emptyset$. Deny and let $Q \in \mathrm{Supp}_U(\overline{R}/U) \cap \mathrm{Supp}_U(V/\overline{R})$. Then, $\overline{R}_Q \neq U_Q, V_Q$. If $P := Q \cap R$, we get that $\overline{R}_P \neq U_P, V_P$, giving $\overline{R}_P \neq R_P, S_P$, a contradiction. Another use of Proposition 4.16 shows that $U \subseteq V$ is almost-Prüfer. The second result is obvious. □

Theorem 4.28 *Let $R \subseteq S$ be an FCP almost-Prüfer extension and $T \subseteq T'$ a subextension of $R \subseteq S$. Set $U := T \cap \overline{R}$ and $V' := T'\overline{R}$. Let W be the Prüfer hull of $U \subseteq V'$. Then, W is also the Prüfer hull of $T \subseteq T'$ and $T \subseteq T'$ is an FCP almost-Prüfer extension.*

Proof By Lemma 4.27, we get that $U \subseteq V'$ is an FCP almost-Prüfer extension. Let $\widetilde{T}$ be the Prüfer hull of $T \subseteq T'$. Since $U \subseteq T$ and $T \subseteq \widetilde{T}$ are Prüfer, so is $U \subseteq \widetilde{T}$ and $\widetilde{T} \subseteq V'$ gives that $\widetilde{T} \subseteq W$. Then, $T \subseteq W$ is Prüfer as a subextension of $U \subseteq W$.

Moreover, in view of Proposition 4.17, W is the least U-subalgebra of V' over which V' is integral. Since $T' \subseteq V'$ is integral, we get that $W \subseteq T'$, so that $W \in [T, T']$, with $W \subseteq T'$ integral as a subextension of $W \subseteq V'$. It follows that W is also the Prüfer hull of $T \subseteq T'$ and $T \subseteq T'$ is an FCP almost-Prüfer extension. □

Remark 4.29 The result of this theorem may not hold if the FCP hypothesis is lacking. Take the example of [13, Remark 2.9(c)], where $R \subseteq S \subseteq T$ is almost-Prüfer, $R \subseteq S$ Prüfer, $S \subseteq T$ integral and $R \subseteq T$ has not FCP. Here, (R, M) is a one-dimensional valuation domain with quotient field S and $T = S[X]/(X^2) = S[x]$. Set $R' := R[x]$. Then, R' is local, with $\mathrm{Spec}(R') = \{P' := Rx,\ M' := M + Rx\}$. By the characterization of a Prüfer extension in Proposition 1.2 (3), $R' = \widetilde{R'}$, but $R' \subset T$ is not integral, so that $R' \subset T$ is not almost-Prüfer.

5 Fibers of Quasi-Prüfer Extensions

We intend to complete some results of Ayache-Dobbs [5]. We begin by recalling some features about quasi-finite ring morphisms. A ring morphism $R \to S$ is called quasi-finite by [39] if it is of finite type and $\kappa(P) \to \kappa(P) \otimes_R S$ is finite (as a $\kappa(P)$-vector space), for each $P \in \mathrm{Spec}(R)$ [39, Proposition 3, p. 40].

Proposition 5.1 *A ring morphism of finite type is incomparable if and only if it is quasi-finite and, if and only if its fibers are finite.*

Proof Use [40, Corollary 1.8] and the above definition. □

Theorem 5.2 *An extension $R \subseteq S$ is quasi-Prüfer if and only if $R \subseteq T$ is quasi-finite (equivalently, has finite fibers) for each $T \in [R, S]$ such that T is of finite type over R, if and only if $R \subseteq T$ has integral fiber morphisms for each $T \in [R, S]$.*

Proof Clearly, $R \subseteq S$ is an INC-pair implies the condition by Proposition 5.1. To prove the converse, write $T \in [R, S]$ as the union of its finite type R-subalgebras T_α. Now let $Q \subseteq Q'$ be prime ideals of T, lying over a prime ideal P of R and set $Q_\alpha := Q \cap T_\alpha$ and $Q'_\alpha := Q' \cap T_\alpha$. If $R \subseteq T_\alpha$ is quasi-finite, then $Q_\alpha = Q'_\alpha$, so that $Q = Q'$ and then $R \subseteq T$ is incomparable. The last statement is Proposition 3.8. □

Corollary 5.3 *An integrally closed extension is Prüfer if and only if each of its subextensions $R \subseteq T$ of finite type has finite fibers.*

Proof It is enough to observe that the fibers of a (flat) epimorphism have a cardinal ≤ 1, because an epimorphism is spectrally injective. □

An extension $R \subseteq S$ is called *strongly affine* if each of its subextensions $R \subseteq T$ is of finite type. The above considerations show that in this case $R \subseteq S$ is quasi-Prüfer if and only if each of its subextensions has finite fibers. For example, an FCP extension is strongly affine and quasi-Prüfer. We are also interested in extensions $R \subseteq S$ that are not necessarily strongly affine, whose subextensions have finite fibers.

Next lemma will be useful, its proof is obvious.

Lemma 5.4 *Let $R \subseteq S$ be an extension and $T \in [R, S]$.*

(1) *If $T \subseteq S$ is spectrally injective and $R \subseteq T$ has finite fibers, then $R \subseteq S$ has finite fibers.*
(2) *If $R \subseteq T$ is spectrally injective, then $T \subseteq S$ has finite fibers if and only if $R \subseteq S$ has finite fibers.*

Remark 5.5 Let $R \subseteq S$ be an almost-Prüfer extension, such that the integral extension $T := \widetilde{R} \subseteq S$ has finite fibers and let $P \in \mathrm{Spec}(R)$. The study of the finiteness of $\mathrm{Fib}_{R,S}(P)$ can be reduced as follows. As $\overline{R} \subseteq S$ is an epimorphism, because it is Prüfer, it is spectrally injective (see Scholium A). The hypotheses of Proposition 4.5 hold. We examine three cases. In case $(R : \overline{R}) \not\subseteq P$, it is well known that $R_P = (\overline{R})_P$ so that $|\mathrm{Fib}_{R,S}(P)| = 1$, because $\overline{R} \to S$ is spectrally injective. Suppose now that $(R : \overline{R}) = P$. From $(R : \overline{R}) = (T : S) \cap R$, we deduce that P is lain over by some $Q \in \mathrm{Spec}(T)$ and then $\mathrm{Fib}_{R,\overline{R}}(P) \cong \mathrm{Fib}_{T,S}(Q)$. The conclusion follows as above. Thus the remaining case is $(R : \overline{R}) \subset P$ and we can assume that $PT = T$ for if not $\mathrm{Fib}_{R,\overline{R}}(P) \cong \mathrm{Fib}_{T,S}(Q)$ for some $Q \in \mathrm{Spec}(T)$ by Scholium A (1).

Proposition 5.6 *Let $R \subseteq S$ be an almost-Prüfer extension. If $\widetilde{R} \subseteq S$ has finite fiber morphisms and $(\widetilde{R}_P : S_P)$ is a maximal ideal of $\widetilde{R}_P$ for each $P \in \mathrm{Supp}_R(S/\widetilde{R})$, then $R \subseteq \overline{R}$ and $R \subseteq S$ have finite fibers.*

Proof The Prüfer closure commutes with the localization at prime ideals by Proposition 4.12. We set $T := \widetilde{R}$. Let P be a prime ideal of R and $\varphi : R \to R_P$ the canonical morphism. We clearly have $\mathrm{Fib}_{R,.}(P) = {}^a\varphi(\mathrm{Fib}_{R_P,.P}(PR_P))$. Therefore, we can localize the data at P and we can assume that R is local.

In case $(T : S) = T$, we get a factorization $R \to \overline{R} \to T$. Since $R \to T$ is Prüfer so is $R \to \overline{R}$ and it follows that $R = \overline{R}$ because a Prüfer extension is integrally closed.

From Proposition 1.2 applied to $R \subseteq T$, we get that there is some $\mathfrak{P} \in \mathrm{Spec}(R)$ such that $T = R_{\mathfrak{P}}$, $R/\mathfrak{P}$ is a valuation ring with quotient field $T/\mathfrak{P}$ and $\mathfrak{P} = \mathfrak{P}T$. It follows that $(T : S) = \mathfrak{P}T = \mathfrak{P} \subseteq R$, and hence $(T : S) = (T : S) \cap R = (R : \overline{R})$. We have therefore a pushout diagram by Theorem 4.6:

$$\begin{array}{ccc} R' := R/\mathfrak{P} & \longrightarrow & \overline{R}/\mathfrak{P} := \overline{R'} \\ \downarrow & & \downarrow \\ T' := T/\mathfrak{P} & \longrightarrow & S/\mathfrak{P} := S' \end{array}$$

where $R/\mathfrak{P}$ is a valuation domain, $T/\mathfrak{P}$ is its quotient field, and $\overline{R}/\mathfrak{P} \to S/\mathfrak{P}$ is Prüfer by [25, Proposition 5.8, p. 52].

Because $\overline{R'} \to S'$ is injective and a flat epimorphism, there is a bijective map $\mathrm{Min}(S') \to \mathrm{Min}(\overline{R'})$. But $T' \to S'$ is the fiber at $\mathfrak{P}$ of $T \to S$ and is therefore finite. Therefore, $\mathrm{Min}(S')$ is a finite set $\{N_1, \dots, N_n\}$ of maximal ideals lying over the minimal prime ideals $\{M_1, \dots, M_n\}$ of $\overline{R'}$ lying over 0 in R'. We infer from Lemma 3.7 that $\overline{R'}/M_i \to S'/N_i$ is Prüfer, whence integrally closed. Therefore, $\overline{R'}/M_i$ is an integral domain and the integral closure of R' in S'/N_i. Any maximal ideal M of $\overline{R'}$ contains some M_i. To conclude it is enough to use a result of Gilmer [19, Corollary 20.3] because the number of maximal ideals in $\overline{R'}/M_i$ is less than the separable degree of the extension of fields $T' \subseteq S'/N_i$. □

Remark 5.7

(1) Suppose that $(\widetilde{R} : S)$ is a maximal ideal of $\widetilde{R}$. We clearly have $(\widetilde{R} : S)_P \subseteq (\widetilde{R}_P : S_P)$ and the hypotheses on $(\widetilde{R} : S)$ of the above proposition hold.
(2) In case $\widetilde{R} \subseteq S$ is a tower of finitely many integral minimal extensions $R_{i-1} \subseteq R_i$ with $M_i = (R_{i-1} : R_i)$, then $\mathrm{Supp}_{\widetilde{R}}(S/\widetilde{R}) = \{N_1, \dots, N_n\} \subseteq \mathrm{Max}(\widetilde{R})$ where $N_i = M_i \cap \widetilde{R}$. If the ideals N_i are different, each localization at N_i of $\widetilde{R} \subseteq S$ is integral minimal and the above result may apply. This generalizes the Ayache-Dobbs result [5, Lemma 3.6], where $\widetilde{R} \subseteq S$ is supposed to be integral minimal.

Theorem 5.8 *Let $R \subseteq S$ be a quasi-Prüfer ring extension. The following three conditions are equivalent:*

(1) $R \subseteq S$ has finite fibers.
(2) $R \subseteq \overline{R}$ has finite fibers.
(3) Each extension $R \subseteq T$, where $T \in [R, S]$ has finite fibers.

Proof (1) $\Leftrightarrow$ (2) Let $P \in \mathrm{Spec}(R)$ and the morphisms $\kappa(P) \to \kappa(P) \otimes_R \overline{R} \to \kappa(P) \otimes_R S$. The first (second) morphism is integral (a flat epimorphism) because deduced by base change from the integral morphism $R \to \overline{R}$ (the flat epimorphism $\overline{R} \to S$). Therefore, the ring $\kappa(P) \otimes_R \overline{R}$ is zero-dimensional, so that the second morphism is surjective by Scholium A (2). Set $A := \kappa(P) \otimes_R \overline{R}$ and $B := \kappa(P) \otimes_R S$, we thus have a module finite flat ring morphism $A \to B$. Hence, $A_Q \to B_Q$ is free for each $Q \in \mathrm{Spec}(A)$ [16, Proposition 9] and $B_Q \neq 0$ because it contains $\kappa(P) \neq 0$. Therefore, $A_Q \to B_Q$ is injective and it follows that $A \cong B$ giving (1) $\Leftrightarrow$ (2).

(2) $\Rightarrow$ (3) Suppose that $R \subseteq \overline{R}$ has finite fibers and let $T \in [R, S]$, then $\overline{R} \subseteq \overline{R}T$ is a flat epimorphism by Proposition 4.5(1) and so is $\kappa(P) \otimes_R \overline{R} \to \kappa(P) \otimes_R \overline{R}T$. Since $\mathrm{Spec}(\kappa(P) \otimes_R \overline{R}T) \to \mathrm{Spec}(\kappa(P) \otimes_R \overline{R})$ is injective, $R \subseteq \overline{R}T$ has finite fibers. Now $R \subseteq T$ has finite fibers because $T \subseteq \overline{R}T$ is integral and is therefore spectrally surjective.

(3) $\Rightarrow$ (1) is obvious. □

Remark 5.9 Actually, the statement (1) $\Leftrightarrow$ (2) is valid if we only suppose that $\overline{R} \subseteq S$ is a flat epimorphism. But this equivalence fails in case $\overline{R} \subseteq S$ is not a flat epimorphism as we can see in the following example. Let R be an integral domain with quotient field K and integral closure $\overline{R}$ such that $R \subset \overline{R}$ is a minimal extension. Then $R \subset \overline{R}$ has finite fibers. Consider the polynomial ring $S := K[X]$. It follows that $\overline{R}$ is also the integral closure of R in S. Moreover, $K \subset S$ and then $R \subset S$ have not finite fibers. Actually, $K \subset S$ and $\overline{R} \subset S$ are not flat epimorphisms.

Next result contains [5, Lemma 3.6], gotten after a long proof.

Corollary 5.10 *Let $R \subseteq S$ be an almost-Prüfer extension. Then $R \subseteq S$ has finite fibers if and only if $R \subseteq \overline{R}$ has finite fibers, and if and only if $\widetilde{R} \subseteq S$ has finite fibers.*

Proof By Theorem 5.8 the first equivalence is clear. The second is a consequence of Lemma 5.4(2). □

The following result is then clear and obviates any need to assume FCP or FMC.

Theorem 5.11 *Let $R \subseteq S$ be a quasi-Prüfer extension with finite fibers, then $R \subseteq T$ has finite fibers for each $T \in [R, S]$.*

Corollary 5.12 *If $R \subseteq S$ is quasi-finite and quasi-Prüfer, then $R \subseteq T$ has finite fibers for each $T \in [R, S]$ and $\widetilde{R} \subseteq S$ is module finite.*

Proof By the Zariski Main Theorem, there is a factorization $R \subseteq F \subseteq S$ where $R \subseteq F$ is module finite and $F \subseteq S$ is a flat epimorphism [39, Corollaire 2, p. 42]. To conclude, we use Scholium A in the rest of the proof. The map $\widetilde{R} \otimes_R F \to S$ is injective because $F \to \widetilde{R} \otimes_R F$ is a flat epimorphism and is surjective, since it is integral and a flat epimorphism because $\widetilde{R} \otimes_R F \to S$ is a flat epimorphism. □

Corollary 5.13 *An FMC extension $R \subseteq S$ is such that $R \subseteq T$ has finite fibers for each $T \in [R, S]$.*

Proof Such an extension is quasi-finite and quasi-Prüfer. Then use Corollary 5.12. □

[5, Example 4.7] exhibits some FMC extension $R \subseteq S$, such that $R \subseteq \overline{R}$ has not FCP. Actually, $[R, \overline{R}]$ is an infinite (maximal) chain.

Proposition 5.14 *Let $R \subseteq S$ be a quasi-Prüfer extension such that $R \subseteq \overline{R}$ has finite fibers and R is semi-local. Then T is semi-local for each $T \in [R, S]$.*

Proof Obviously $\overline{R}$ is semi-local. From the tower $\overline{R} \subseteq T\overline{R} \subseteq S$ we deduce that $\overline{R} \subseteq T\overline{R}$ is Prüfer. It follows that $T\overline{R}$ is semi-local [5, Lemma 2.5 (f)]. As $T \subseteq T\overline{R}$ is integral, we get that T is semi-local. □

The next proposition gives a kind of converse, but, before, we rewrite [4, Theorem 3.10] proved in the integral domains context, which holds in a more general context.

Theorem 5.15 *Let $R \subseteq S$ be an integrally closed extension with R semi-local. The following three conditions are equivalent:*

(1) $R \subseteq S$ is a Prüfer extension.
(2) $|\mathrm{Max}(T)| \leq |\mathrm{Max}(R)|$ for each $T \in [R, S]$.
(3) Each $T \in [R, S]$ is a semi-local ring.

Proof It is enough to mimic the proof of [4, Theorem 3.10] which is still valid for an arbitrary integrally closed extension of rings $R \subseteq S$. Indeed, $R \subseteq S$ is a Prüfer extension if and only if (R, S) is a residually algebraic pair such that $R \subseteq S$ is an integrally closed extension by Theorem 2.3 and Definition 2.1 . □

Proposition 5.16 *Let $R \subseteq S$ be an extension with $\overline{R}$ semi-local. Then $R \subseteq S$ is quasi-Prüfer if and only if T is semi-local for each $T \in [R, S]$.*

Proof If $R \subseteq S$ is quasi-Prüfer, $\overline{R} \subseteq S$ is Prüfer. Let $T \in [R, S]$ and set $T' := T\overline{R}$, so that $T \subseteq T'$ is integral, and $\overline{R} \subseteq T'$ is Prüfer (and then a normal pair). It follows from [5, Lemma 2.5 (f)] that T' is semi-local, and so is T.

If T is semi-local for each $T \in [R, S]$, so is any $T \in [\overline{R}, S]$. Then, $\overline{R} \subseteq S$ is Prüfer by Theorem 5.15 and $R \subseteq S$ is quasi-Prüfer. □

Acknowledgements The authors wish to thank the referee whose many suggestions improved the presentation of the paper.

References

1. Anderson, D.F., Dobbs, D.E.: Pairs of rings with the same prime ideals. Can. J. Math. **XXXII**, 362–384 (1980)
2. Ayache, A., Ben Nasr, M., Echi, O., Jarboui, N.: Universally catenarian and going-down pairs of rings. Math. Z. **238**, 695–731 (2001)
3. Ayache, A.: A constructive study about the set of intermediate rings. Commun. Algebra **41**, 4637–4661 (2013)
4. Ayache, A., Jaballah, A.: Residually algebraic pairs of rings. Math. Z. **225**, 49–65 (1997)
5. Ayache, A., Dobbs, D.E.: Finite maximal chains of commutative rings. J. Algebra Appl. **14**, 1450075-1–1450075-27 (2015)
6. Ben Nasr, M., Jarboui, N.: New results about normal pairs of rings with zero-divisors. Ricerche mat. **63**, 149–155 (2014)
7. Chang, G.W., Fontana, M.: Uppers to 0 in polynomial rings and Prüfer-like domains. Commun. Algebra **37**, 164–192 (2009)
8. Chatham, R.D.: Going-down pairs of commutative rings. Rendiconti del Circolo Matematico di Palermo, **Serie II, Tomo L**, 509–542 (2001)
9. Chatham, R.D., Dobbs, D.E.: On open-ring pairs of commutative rings. Houston J. Math. **31**, 65–74 (2005)
10. Dobbs, D.E.: On INC-extensions and polynomials with unit content. Can. Math. Bull. **23**, 37–42 (1980)
11. Dobbs, D.E.: On characterizations of integrality involving the lying-over and incomparability properties. J. Commun. Algebra **1**, 227–235 (2009)
12. Dobbs, D.E., Shapiro, J.: Pseudo-normal pairs of integral domains. Houston J. Math. **40**, 1–9 (2014)
13. Dobbs, D.E., Picavet, G., Picavet-L'Hermitte, M.: Characterizing the ring extensions that satisfy FIP or FCP. J. Algebra **371**, 391–429 (2012)
14. Dobbs, D.E., Picavet, G., Picavet-L'Hermitte, M., Shapiro, J.: On intersections and composites of minimal ring extensions. JP J. Algebra Number Theory Appl. **26**, 103–158 (2012)
15. Dobbs, D.E., Picavet, G., Picavet-L'Hermitte, M.: Transfer results for the FIP and FCP properties of ring extensions. Commun. Algebra **43**, 1279–1316 (2015)
16. Endo, S.: On semi-hereditary rings. J. Math. Soc. Jpn. **13**, 109–119 (1961)
17. Ferrand, D., Olivier, J.-P.: Homomorphismes minimaux d'anneaux. J. Algebra **16**, 461–471 (1970)
18. Fontana, M., Huckaba, J.A., Papick, I.J.: Prüfer Domains. Dekker, New York (1997)
19. Gilmer, R.: Multiplicative Ideal Theory. Dekker, New York (1972)
20. Grandet, M.: Une caractérisation des morphismes minimaux non entiers, C. R. Acad. Sci. Paris **271**, Série A 581–583 (1970)
21. Griffin, M.: Prüfer rings with zero divisors. J. Reine Angew. Math. **239**, 55–67 (1969)
22. Grothendieck, A., Dieudonné, J.: Eléments de Géométrie Algébrique. Springer, Berlin (1971)
23. Houston, E.: Uppers to zero in polynomial rings. In: Multiplicative Ideal Theory in Commutative Algebra, pp. 243–261. Springer, New York (2006)
24. Jarboui, N., Massaoud, E.: On finite saturated chains of overrings. Commun. Algebra **40**, 1563–1569 (2012)
25. Knebusch, M., Zhang, D.: Manis Valuations and Prüfer Extensions I. Springer, Berlin (2002)
26. Lazard, D.: Autour de la platitude. Bull. Soc. Math. Fr. **97**, 81–128 (1969)
27. Lazarus, M.: Fermeture intégrale et changement de base. Ann. Fac. Sci. Toulouse **6**, 103–120 (1984)

28. Lucas, T.G.: Some results on Prüfer rings. Pac. J. Math. **124**, 333–343 (1986)
29. Morita, K.: Flat modules, injective modules and quotient rings. Math. Z. **120**, 25–40 (1971)
30. Olivier, J.P.: Anneaux absolument plats universels et épimorphismes à buts réduits. Séminaire Samuel. Algèbre Commutative. Tome 2, exp. *n°* 6, p. 1–12 (1967–1968)
31. Olivier, J.P.: Montée des propriétés par morphismes absolument plats. J. Algebra Pure Appl. Université des Sciences et Technique du Languedoc, Montpellier, France (1971)
32. Olivier, J.P.: Going up along absolutely flat morphisms. J. Pure Appl. Algebra **30**, 47–59 (1983)
33. Picavet, G.: Submersion et descente. J. Algebra **103**, 527–591 (1986)
34. Picavet, G.: Seminormal or t-closed schemes and Rees rings. Algebra Represent. Theory **1**, 255–309 (1998)
35. Picavet, G.: Universally going-down rings, 1-split rings, and absolute integral closure. Commun. Algebra **31**, 4655–4685 (2003)
36. Picavet, G., Picavet-L'Hermitte, M.: About minimal morphisms. In: Multiplicative Ideal Theory in Commutative Algebra, pp. 369–386. Springer, New York (2006)
37. Picavet, G., Picavet-L'Hermitte, M.: Prüfer and Morita hulls of FCP extensions. Commun. Algebra **43**, 102–119 (2015)
38. Picavet, G., Picavet-L'Hermitte, M.: Some more combinatorics results on Nagata extensions. Palest. J. Math. **1, (Spec.1)**, 49–62 (2016)
39. Raynaud, M.: Anneaux locaux Henséliens. Lecture Notes in Mathematics, vol. 169. Springer, Berlin (1970)
40. Uda, H.: Incomparability in ring extensions. Hiroshima Math. J. **9**, 451–463 (1979)
41. Visweswaran, S.: Laskerian pairs. J. Pure Appl. Algebra **59**, 87–110 (1989)

A Note on Analytically Irreducible Domains

Roswitha Rissner

Abstract If D is a one-dimensional, Noetherian, local domain, it is well known that D is analytically irreducible if and only if D is unibranched and the integral closure D' of D is finitely generated as D-module. However, the proof of this result is split into pieces and spread over the literature. This paper collects the pieces and assembles them to a complete proof. Next to several results on integral extensions and completions of modules, we use Cohen's structure theorem for complete, Noetherian, local domains to prove the main result. The purpose of this survey is to make this characterization of analytically irreducible domains more accessible.

Keywords Unibranched • Analytically irreducible • Noetherian • Local • One-dimensional

MSC 2010: 13-02; 13B22, 13B35 13J05, 3J10, 13H05

1 Introduction

Let $(D, \mathbf{m})$ be a Noetherian, one-dimensional, local domain. It is well known that the question of whether $\widehat{D}$ has zero-divisors or nilpotents is strongly connected to certain properties of the integral closure D' of D.

Definition 1.1 Let $(D, \mathbf{m})$ be a Noetherian, local domain with integral closure D' and $\mathbf{m}$-adic completion $\widehat{D}$. We say D is

1. *unibranched*, if D' is local,
2. *analytically unramified*, if $\widehat{D}$ is a reduced ring and
3. *analytically irreducible*, if $\widehat{D}$ is a domain.

R. Rissner (✉)
Institute of Analysis and Number Theory, Kopernikusgasse 24, 8010 Graz, Austria
e-mail: rissner@math.tugraz.at

M. Fontana et al. (eds.), *Rings, Polynomials, and Modules*,
https://doi.org/10.1007/978-3-319-65874-2_17

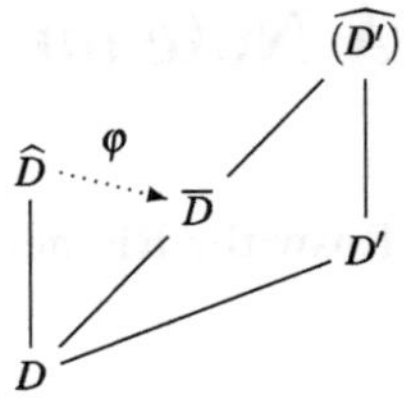

Fig. 1 Completions of D with respect to m and m' in case D' is local with maximal ideal m'

The aim of this paper is to give a complete proof of the following well-known theorem.

Theorem 1 *Let* $(D, \mathbf{m})$ *be a one-dimensional, Noetherian, local domain with integral closure* D'*. Then the following assertions are equivalent:*

1. *D is analytically irreducible.*
2. *D is unibranched and analytically unramified.*
3. *D is unibranched and D' is finitely generated as D-module.*
4. *D is unibranched and if $\mathbf{m}'$ denotes the maximal ideal of the integral closure, then the $\mathbf{m}$-adic topology on D coincides with subspace topology induced by $\mathbf{m}'$.*

Assume that D is unibranched and let $\mathbf{m}'$ be the unique maximal ideal of the integral closure D'. It follows from the Krull-Akizuki theorem that D' is a discrete valuation domain (see Corollary 2.5 below). In particular, D' is Noetherian and can be embedded into the $\mathbf{m}'$-adic completion $\widehat{(D')}$ of D'. Moreover, the valuation on D' can be extended to a valuation on $\widehat{(D')}$ which implies that $\widehat{(D')}$ is a domain (see Example 3.1). Since the completion $\overline{D}$ of D considered as a topological subspace of D' is the topological closure of D in $\widehat{(D')}$, it follows that $\overline{D} \subseteq \widehat{(D')}$ is a domain too.

On the other hand, D can be embedded into the $\mathbf{m}$-adic completion $\widehat{D}$ of D. Figure 1 above demonstrates the relationship between D, D' and the completions with respect to the different topologies. As usual, the solid lines represent inclusions. However, the dotted arrow deserves some additional explanation. Since $\mathbf{m}^n \subseteq \mathbf{m}'^n \cap D$ it follows that the $\mathbf{m}$-adic topology is finer than the $\mathbf{m}'$-adic subspace topology on D. This further implies that the inclusion $D \longrightarrow \overline{D}$ is a uniformly continuous homomorphism (where D is equipped with the $\mathbf{m}$-adic topology). Since $\overline{D}$ is complete, the inclusion can be uniquely extended to a uniformly continuous map $\varphi : \widehat{D} \longrightarrow \overline{D}$. Theorem 1 implies that D is analytically irreducible if and only if φ is an isomorphism. However, if $\widehat{D}$ is not a domain φ is not even injective.

Theorem 1 is well known but its proof is split into pieces and has to be assembled from several sources. This survey collects known results from different references in order to present a complete proof. We follow the approach of Nagata's textbook [7, (32.2)] for the implication (2) ⇒ (3). For the remaining implications, we pursue the suggestions of [3, Theorem III.5.2]. One can also refer to [8, Theorem 8] for the implication (4) ⇒ (1).

In order to make this survey more self-contained, we give a short introduction to integral ring extensions and completions in Sects. 2 and 3, respectively. Then, in Sect. 4 we discuss Cohen's structure theorem which allows us to prove that the integral closure of a complete, Noetherian, local, reduced ring R is a finitely generated R-module in Sect. 5. Finally, we give a proof of Theorem 1 in Sect. 6. It is worth mentioning that Sects. 4 and 5 are only needed for the implication (2) $\Rightarrow$ (3) whereas the remaining implications can be shown using the results in Sects. 2 and 3.

2 Integral Ring Extensions

In this section we recall some facts on integral ring extensions which we use throughout this paper.

Fact 2.1 (cf. [1, Proposition 5.1, Corollaries 5.3, 5.4]) *Let $R \subseteq S$ be a ring extension. We call $s \in S$* integral over R *if the following equivalent assertions are satisfied:*

1. *There exists a monic polynomial $f \in R[X]$ such that $f(s) = 0$.*
2. *$R[s]$ is finitely generated as R-module.*
3. *There exists a ring T containing $R[s]$ which is finitely generated as R-module.*

Let $R'_S = \{s \in S \mid s \text{ integral over } R\}$ denote the set of elements of S which are integral over R. Then $R \subseteq R'_S$ is a ring extension.

We call R'_S the integral closure of R in S *and if $R = R'_S$ we say R is* integrally closed in S. *If $S = R'_S$, we say $R \subseteq S$ is an* integral extension.

If $R \subseteq T \subseteq S$ is an intermediate ring such that both $R \subseteq T$ and $T \subseteq S$ are integral extensions, then $R \subseteq S$ is an integral extension. In particular, $R'_S = (R'_T)'_S$.

In case S is the total ring of quotients of R, we simplify and say $R' := R'_S$ is the integral closure of R *and R is* integrally closed *if $R = R'$.*

Fact 2.2 (Cohen-Seidenberg, cf. [1, Corollary 5.9, Theorem 5.11]) *Let $R \subseteq S$ be an integral extension. Then the following assertions hold:*

1. *If $Q_1 \subseteq Q_2$ are prime ideals of S such that $Q_1 \cap R = Q_2 \cap R$, then $Q_1 = Q_2$.*
2. *If $P_1, P_2 \in \mathsf{spec}(R)$ with $P_1 \subseteq P_2$ and $Q_1 \in \mathsf{spec}(S)$ with $Q_1 \cap R = P_1$, then there exists $Q_2 \in \mathsf{spec}(S)$ such that $Q_1 \subseteq Q_2$ and $Q_2 \cap R = P_2$.*
3. *$\mathsf{dim}(R) = \mathsf{dim}(S)$ and $\mathsf{max}(S) = \{P \in \mathsf{spec}(S) \mid P \cap R \in \mathsf{max}(R)\}$.*

As a first result we prove the so-called *Krull-Akizuki* theorem which is central to the remainder of this paper.

Proposition 2.3 (Krull-Akizuki, cf. [6, Theorem 11.7], [4, Theorem 4.9.2]) *Let D be a one-dimensional, Noetherian domain with quotient field K and L a finite field extension of K.*

Then the integral closure D'_L of D in L is a Dedekind domain. Moreover, if I is a nonzero ideal of D', then D'/I is a finitely generated D-module.

Proof We can reduce the proof to the case $L = K$ with the following argument. Let $b_1, \ldots, b_n$ form a K-basis of L. Without restriction we can assume that $b_i \in D'_L$.

Then the domain $R = D[b_1, \dots, b_n]$ is finitely generated as D-module and therefore $D \subseteq R$ is an integral extension according to Fact 2.1. Further, R is Noetherian and since $L = K[b_1, \dots, b_n]$ is the quotient field of R it follows that $R' = D'_L$ is the integral closure of R. Moreover, if I is a nonzero ideal of R' and R'/I is finitely generated as R-module, then $R'/I = D'_L/I$ is a finitely generated D-module.

Hence from this point on we assume that $L = K$ and $D = R$. Since $1 = \mathsf{dim}(D) = \mathsf{dim}(D')$ by Fact 2.2 and D' is integrally closed, we only need to prove that D' is Noetherian to conclude that D' is a Dedekind domain.

Let I be a nonzero ideal of D' and $s = \frac{a}{t} \in I$ be a nonzero element. Then $a = ts \in I \cap D$ is a nonzero element which implies that D/aD is a zero-dimensional, Noetherian ring and thus Artinian. Since $I_n = (a^n D') \cap D + aD$ for $n \in \mathbb{N}$ form a descending chain of ideals of D/aD, there exists an $m \in \mathbb{N}$ such that $I_m = I_n$ for all $n \geq m$.

If $a^m D' \subseteq a^{m+1} D' + D$, then

$$D'/aD' \simeq a^m D'/a^{m+1}D' \subseteq (a^{m+1}D' + D)/a^{m+1}D' \simeq D/(D \cap a^{m+1}D')$$

holds. This further implies that D'/aD' is a submodule of the Noetherian module $D/(D \cap a^{m+1}D')$ and hence a finitely generated D-module. Hence D'/aD' is Noetherian and the submodule I/aD' is finitely generated. Consequently I is a finitely generated ideal of D'. In addition, $D'/I \simeq (D'/aD')/(I/aD')$ is a quotient of the finitely generated D-module D'/aD' and therefore finitely generated.

It remains to prove that $a^m D' \subseteq a^{m+1}D' + D$. We can localize at each maximal ideal of D and prove the inclusion locally. So assume that D is local with maximal ideal $\mathbf{m}$.

If $a \notin \mathbf{m}$, then a is a unit in D and therefore $a^m D' = D' = a^{m+1}D' + D$. Now assume $a \in \mathbf{m}$ and let $x = \frac{b}{c} \in D' \setminus D$ where $b \in D$ and $c \in \mathbf{m}$. The radical of the nonzero ideal cD is then $\mathbf{m}$ and therefore there exists $n \geq m$ with $\mathbf{m}^{n+1} \subseteq cD$. It follows that

$$a^{n+1}x \in (a^{n+1}D') \cap D \subseteq I_{n+1} = I_{n+2} = (a^{n+2}D') \cap D + aD$$

and hence $a^n x \in a^{n+1}D' + D$. If $n > m$, then

$$a^n x \in (a^{n+1}D' + D) \cap a^n D' = a^{n+1}D' + \underbrace{D \cap a^n D'}_{\subseteq I_n = I_{n+1}} \subseteq a^{n+1}D' + aD$$

and therefore $a^{n-1}x \in a^n D' + D$. Repeating this argument completes the proof.

Corollary 2.4 *Let D be a one-dimensional, Noetherian, local domain with quotient field K and L a finite field extension of K.*

If $\mathbf{m}$ is a maximal ideal of D and $\mathbf{m}'$ a maximal ideal of the integral closure D'_L of D in L with $\mathbf{m}' \cap D = \mathbf{m}$, then the field extension $D/\mathbf{m} \subseteq D'_L/\mathbf{m}'$ is finite.

Proof It follows from Proposition 2.3, that $D'_L/\mathbf{m}'$ is a finitely generated D-module. Therefore $D'_L/\mathbf{m}'$ is finitely generated as $D/\mathbf{m}$-vector space as well.

Corollary 2.5 *Let D be a one-dimensional, Noetherian, local domain.*
If D is unibranched, then the integral closure D' is a discrete valuation domain.

Remark If D is a local, one-dimensional, Noetherian domain with maximal ideal $\mathbf{m}$, then Fact 2.2 implies that the maximal ideals of D' are the minimal primes of $\mathbf{m}D$ and therefore D' is always a semilocal Dedekind domain.

3 Completions

In this section we recall the necessary facts on topologies on rings and modules which are induced by ideals. Let R be a Noetherian ring, I an ideal of R and M an R-module. Then the submodules $(I^nM)_{n\in\mathbb{N}}$ form a filtration on M which induces a linear topology on M, that is, the sets $m + I^nM$ for $m \in M$ and $n \in \mathbb{N}$ form a basis of this topology. We call this the *I-adic topology* on M.

Addition, subtraction and scalar multiplication are continuous with respect to this topology. If M is a ring extension of R, then multiplication in M is continuous too.

Moreover, $M \setminus m + I^nM = \bigcup_y y + I^nM$ where the union runs over all $y \in M$ with $m - y \notin I^nM$ and hence each $m + I^nM$ is both open and closed.

The completion $\widehat{M}$ of M is the inverse limit of the inverse system M/I^nM together with the canonical projections $M/I^nM \longrightarrow M/I^mM$ for $n \geq m$, that is,

$$\widehat{M} = \varprojlim M/I^nM = \left\{ (a_n + I^nM)_n \in \prod_{n\in\mathbb{N}} M/I^nM \mid a_{n+1} \equiv a_n \pmod{I^nM} \right\}$$

A sequence $(x_k)_k$ in M is an I-adic Cauchy sequence, if for each n there exists k_n such that $x_{k_n} - x_{k_n+m} \in I^nM$ for all $m \in \mathbb{N}$. As usual, we say two Cauchy sequences $(x_k)_k$, $(y_k)_k$ are equivalent if $(x_k - y_k)_k$ converges to 0. In particular, (x_k) is equivalent to $(x_{k_n})_n$. Hence each equivalence class of Cauchy sequences in M contains the so-called *coherent* sequence $(a_n)_n$ which satisfies $a_{n+1} \equiv a_n \pmod{I^nM}$ for all n. Thus $\widehat{M}$ is isomorphic to the set of equivalence classes of Cauchy sequences.

If $\overline{\mathbf{0}} = \bigcap_{n\in\mathbb{N}} I^nM = \mathbf{0}$, then M is I-adically separated and we can embed M into $\widehat{M}$ via $m \longmapsto (m)_n$. We say that M is complete if $M \simeq \widehat{M}$.

Example 3.1 Let V be a discrete valuation domain with maximal ideal (t) and valuation v. By $\widehat{V}$ we denote the (t)-adic completion of V.

Moreover, let $(a_n)_n$ be a (t)-adic Cauchy sequence with limit $a \in \widehat{V}$. If $a = 0$, then for each $k \in \mathbb{N}$ there exists $n_0 \in \mathbb{N}$ such that $a_n \in (t^k)$ for all $n \geq n_0$ and hence $\lim v(a_n) = \infty$.

If $a \neq 0$, then there exist $k, m_0 \in \mathbb{N}$ such that $a_n \notin (t^k)$ for $n \geq m_0$. However, the sequence $(a_{n+1} - a_n)_n$ converges to 0 and hence there exists $m_1 \in \mathbb{N}$ such that

$a_{n+1} - a_n \in (t^k)$ for all $n \geq m_1$. If $m = \max\{m_0, m_1\}$, then for all $n \geq m$,

$$v(a_{n+1}) = v(a_{n+1} - a_n + a_n) \geq \min\{v(a_{n+1} - a_n), v(a_n)\} = v(a_n) < k$$

holds and the sequence $(v(a_n))_n$ stabilizes at $v(a_m)$.

For $a \in \widehat{V}$, we set $v(a) = \lim v(a_n) = v(a_m)$. This extends v to a discrete valuation on $\widehat{V}$.

Next, we present some basic results on the completion of finitely generated modules over a Noetherian ring R.

Fact 3.2 (Artin-Rees, cf. [6, Theorem 8.5]) *Let R be a Noetherian ring, I an ideal of R and M a finitely generated R-module.*

If N is an R-submodule of M, then there exists an integer r such that for all $k \geq 0$

$$I^{r+k}M \cap N = I^k(I^rM \cap N).$$

Corollary 3.3 (cf. [6, Theorem 8.9, Theorem 8.10]) *Let R be a Noetherian ring, I an ideal of R and M a finitely generated R-module.*

1. *If $N = \bigcap_{n\in\mathbb{N}} I^nM$, then there exists an $a \in R$ with $aN = \mathbf{0}$ and $1 - a \in I$.*
2. *If $I \subseteq \mathsf{Jac}(R)$, then M is I-adically separated and every submodule of M is I-adically closed.*

Proof (1): According to Fact 3.2, there exists $r \in \mathbb{N}$ such that $(I^{r+1}M) \cap N = I(I^rM \cap N) \subseteq IN$ and hence

$$IN \subseteq N = \bigcap_{n\in\mathbb{N}} I^nM \subseteq (I^{r+1}M) \cap N \subseteq IN.$$

Consequently, $N = IN$ and it follows from Nakayama's lemma that there exists an $a \in R$ with $1 - a \in I$ and $aN = \mathbf{0}$. (2): Since $I \subseteq \mathsf{Jac}(R)$, the element a from (1) is a unit of R. Therefore $\bigcap_{n\in\mathbb{N}} I^nM = \mathbf{0}$ and M is separated. Consequently, if P is a submodule of M, it follows that $\bigcap_{n\in\mathbb{N}} I^n(M/P) = \mathbf{0} = PM/P$ and therefore $\bigcap_{n\in\mathbb{N}}(P + I^nM) = P$.

Fact 3.4 (cf. [6, Theorem 8.7]) *Let R be a Noetherian ring, I an ideal and M a finitely generated module. Further, let $\widehat{R}$, $\widehat{M}$ be the I-adic completions of R and M, respectively.*

Then $\widehat{R} \otimes_R M \simeq \widehat{M}$ via $(\lim r_n, m) \mapsto \lim r_n m$. In particular, if R is I-adically complete, then M is I-adically complete.

Proposition 3.5 (cf. [6, Theorem 8.4]) *Let R be a complete ring with respect to an ideal I of R and M an I-adically separated R-module.*

If M/IM is a finitely generated R/I-module, then M is a finitely generated R-module.

Proof Let $m_1, \ldots, m_t \in M$ be elements such that their projections modulo IM generate M/IM as R/I-module. Then $M = \sum_{i=1}^{t} Rm_i + IM$ and for $x \in M$, there exist $r_{0,i} \in R$, $i_1 \in I$ and $x_1 \in M$ such that $x = \sum_{i=1}^{t} r_{0,i} m_i + i_1 x_1$. Then again, for $x_1 \in M$, there exist $r_{1,i} \in R$, $i_2 \in I$ and $x_2 \in M$ such that $x_1 = \sum_{i=1}^{t} r_{1,i} m_i + i_2 x_2$. For $j > 2$, we successively choose $r_{j-1,i} \in R$, $i_j \in I$ and $x_j \in M$ such that $x_{j-1} = \sum_{i=1}^{t} r_{j-1,i} m_i + i_j x_j$. Then $\left(\sum_{j=0}^{n} \left(\prod_{t=1}^{j} i_t \right) r_{j,i} \right)_{n \in \mathbb{N}}$ is a Cauchy sequence in R which has a limit $r_i \in R$. Moreover,

$$x - \sum_{i=1}^{t} r_i m_i \in \bigcap_{n \in \mathbb{N}} I^n M = \mathbf{0}$$

and therefore M is generated by $m_1, \ldots, m_t$.

Zero-Divisors in the Completion of R

Let V be a discrete valuation domain with valuation v and $\widehat{V}$ its completion. If $a, b \in \widehat{V}$ are nonzero elements, then $v(a), v(b) < \infty$ according to Example 3.1. Consequently, $v(ab) = v(a) + v(b) < \infty$ which implies that $\widehat{V}$ is a domain.

However, in general the completion of a domain may not be a domain.

Proposition 3.6 (cf. [3, Lemma III.3.4]) *Let R be a Noetherian domain and I an ideal of R.*

If there exist ideals J_1 and J_2 of R such that $I = J_1 \cap J_2$ and $R = J_1 + J_2$, then the I-adic completion $\widehat{R}$ of R is not a domain.

Proof Since J_1 and J_2 are coprime it follows that J_1^k and J_2^k are coprime as well. Hence there exist $b_k, c_k \in R$ such that

$$b_k \equiv 0 \bmod J_1^k, \qquad b_k \equiv 1 \bmod J_2^k$$
$$c_k \equiv 1 \bmod J_1^k, \qquad c_k \equiv 0 \bmod J_2^k$$

for all $k \in \mathbb{N}$. Since $b_{k+1} - b_k \equiv c_{k+1} - c_k \equiv 0 \bmod J_1^k \cap J_2^k = J_1^k J_2^k = I^k$, the sequences $(b_k)_k$ and $(c_k)_k$ converge I-adically. Let $b = \lim b_k \in \widehat{R}$ and $c = \lim c_k \in \widehat{R}$ their I-adic limits. By construction, $b \neq 0$ and $c \neq 0$. However, since $b_k c_k \equiv 0 \bmod J_1^k \cap J_2^k = I^k$ for all k, it follows that $bc = 0$. Hence b and c are nonzero zero-divisors.

It follows from Proposition 3.6 that the completion of a domain may contain zero-divisors (see also Proposition 3.8). However, constant sequences behave well as the next lemma states.

Lemma 3.7 *Let R be a Noetherian ring, I a proper ideal of R and $\widehat{R}$ the I-adic completion of R. If $d \in R$ is not a zero-divisor in R, then d is not a zero-divisor in $\widehat{R}$.*

Proof By Fact 3.2, there exists $r \in \mathbb{N}$ such that for all $n \in \mathbb{N}$

$$I^{n+r} \cap dR = I^n(I^r \cap dD) \subseteq dI^n \tag{1}$$

Let $x = \lim_n x_n \in \widehat{R}$ such that $dx = 0$. Then $(dx_n)_n$ is an I-adic Cauchy sequence with limit 0. Hence for each $n \in \mathbb{N}$ there exists $t_0 \in \mathbb{N}$ such that $dx_t \in I^{n+r}$ for all $t \geq t_0$. Then $dx_t \in dI^n$ by Equation (1) and therefore $x_t \in I^n$ which implies $0 = \lim_n x_n = x$.

Proposition 3.8 *Let $(D, \mathbf{m})$ be a Noetherian, local domain with quotient field K and $D \subset R \subset K$ be an intermediate ring such that R is finitely generated as D-module.*

Then the following assertions hold:

1. *R is semilocal,*
2. *the $\mathbf{m}$-adic topology on D coincides with subspace topology induced by the $\mathsf{Jac}(R)$-adic topology on R.*
3. *If R is not local, then the $\mathbf{m}$-adic completion $\widehat{D}$ of D is not a domain.*

Proof R is finitely generated as module over the Noetherian domain D and hence a Noetherian domain. Moreover, the ring extension $D \subseteq R$ is integral by Fact 2.1. Therefore, according to Fact 2.2, all prime ideals of R which lie over $\mathbf{m}$ are maximal and thus minimal prime ideals of $\mathbf{m}R$. Hence there are only finitely many maximal ideals $N_1, \ldots, N_n$ in R which proves (1).

(2): Since $\sqrt{\mathbf{m}R} = \bigcap_{i=1}^n N_i = \mathsf{Jac}(R)$, there exists $\ell \in \mathbb{N}$ such that $\mathsf{Jac}(R)^\ell \subseteq \mathbf{m}R$. Then

$$\mathsf{Jac}(R)^{\ell k} \subseteq (\mathbf{m}R)^k \subseteq \mathsf{Jac}(R)^k$$

and hence the $\mathsf{Jac}(R)$-adic and the $\mathbf{m}R$-adic topology coincide on R. Thus it suffices to prove that the $\mathbf{m}$-adic topology on D coincides with subspace topology induced by the $\mathbf{m}R$-adic topology on R. Clearly, $\mathbf{m}^k \subseteq \mathbf{m}^k R \cap D$ holds for all k. On the other hand, Fact 3.2 implies that there exists an integer r such that

$$\mathbf{m}^{k+r} R \cap D = \mathbf{m}^k(\mathbf{m}^r R \cap D) \subseteq \mathbf{m}^k$$

for all $k \geq 0$ and hence the topologies coincide.

(3): Let $M_1, \ldots, M_n$ be the maximal ideals of R with $n > 1$. We set $J_1 = M_1 \cdots M_{n-1} = M_1 \cap \cdots \cap M_{n-1}$ and $J_2 = M_n$. Then $J_1 \cap J_2 = \mathsf{Jac}(R)$ and $J_1 + J_2 = R$. By Proposition 3.6, the $\mathsf{Jac}(R)$-adic completion $\widehat{R}$ of R is not a domain.

By (2), $\widehat{D}$ is a topological subspace of $\widehat{R}$. Since $\widehat{R}$ is not a domain, there exist nonzero $b, c \in \widehat{R}$ with $bc = 0$. However, R is a finitely generated D-module and therefore there exists a nonzero element $d \in D$ with $d\widehat{R} \subseteq \widehat{D}$. Hence $(db)(dc) = 0$ with $db, dc \in \widehat{D}$. By Lemma 3.7, d is not a zero-divisor in $\widehat{D}$ and therefore $db \neq 0$ is a zero-divisor in $\widehat{D}$.

4 Structure Theorem for One-Dimensional Complete Local Domains

In this section we discuss the structure theorem for complete, Noetherian, local domains. We restrict our study to the one-dimensional case, since this is what we need later on. Nevertheless, it is worth mentioning that the results can be extended to higher dimensions but the proofs become more technical.

As Proposition 4.3 states, a complete, Noetherian, one-dimensional local domain $(D, \mathbf{m})$ contains a subring S such that D is finitely generated as S-module. Moreover, S is a certain complete discrete valuation domain whose residue field is isomorphic to $D/\mathbf{m}$. This result allows us in the next section to reduce the investigation to domains of this form.

Definition 4.1 Let $(D, \mathbf{m})$ be a complete, Noetherian, local domain. We say

1. D is of *equal characteristic*, if $\mathsf{char}(D) = \mathsf{char}(D/\mathbf{m})$ and
2. D is of *unequal characteristic*, if $\mathsf{char}(D) \neq \mathsf{char}(D/\mathbf{m})$.

If $\mathsf{char}(D/\mathbf{m}) = 0$, it follows that $\mathsf{char}(D) = 0$ and therefore $\mathbb{Z} \subseteq D$. However, $\mathbb{Z} \cap \mathbf{m} = \mathbf{0}$ which implies that every integer is invertible in D and thus $\mathbb{Q} \subseteq D$. Similarly, if $\mathsf{char}(D) = p > 0$, then $\mathsf{char}(D/\mathbf{m}) = p$ and $\mathbb{Z}/p\mathbb{Z}$ is contained in D. Hence, a domain of equal characteristic contains a field. On the other hand, if D contains a field k, then $\mathsf{char}(k) = \mathsf{char}(D)$. Let $\pi : D \longrightarrow D/\mathbf{m}$ denote the canonical projection. Then $\pi(k)$ is a subfield of $D/\mathbf{m}$ and since $\mathsf{char}(\pi(k)) = \mathsf{char}(k)$ it follows that D is of equal characteristic. Indeed, it is possible to show that a domain D of equal characteristic contains a field k with $\pi(k) = D/\mathbf{m}$, cf. Fact 4.2.

If D is a domain of unequal characteristic, then $\mathsf{char}(D) = 0$ and $\mathsf{char}(D/\mathbf{m}) = p > 0$. In this case it is possible to show that D contains a complete discrete valuation domain (R, pR) such that the residue fields of R and D are isomorphic. We summarize these results in Fact 4.2. However, the proof goes beyond the scope of this paper. We refer to Matsumura's textbook [6] for details.

Fact 4.2 (cf. [6, Theorem 28.3, Theorem 29.3]) *Let $(D, \mathbf{m})$ be a complete, Noetherian, local domain.*

1. *If D is of equal characteristic, then D contains a field k which is isomorphic to $D/\mathbf{m}$ via $d \longmapsto d + \mathbf{m}$. We say k is a* coefficient field of D.
2. *If D is of unequal characteristic and $\mathsf{char}(D/\mathbf{m}) = p$, then D contains a complete discrete valuation domain (R, pR) such that R/pR is isomorphic to $D/\mathbf{m}$ via $r + pR \longmapsto r + \mathbf{m}$. We say R is a* coefficient ring of D.

The existence of a coefficient field or coefficient ring, respectively, is crucial for the proof of the structure theorem which we state in the next proposition.

Proposition 4.3 (cf. [6, Theorem 29.4.(iii)]) *Let $(D, \mathbf{m})$ be a complete, Noetherian, one-dimensional, local domain.*

Then D contains a complete discrete valuation domain S such that D is finitely generated over S and

1. *in equal characteristic* $S \simeq k[\![X]\!]$ *where* k *is a coefficient field of* D.
2. *in unequal characteristic* S *is a coefficient ring of* D.

Proof Let $\mathbf{m}$ be the maximal ideal of D. First, we consider the case where D is of unequal characteristic and let $p = \mathsf{char}(D/\mathbf{m}) > 0$. According to Fact 4.2, D contains a coefficient ring S which is a complete discrete valuation domain with maximal ideal pS such that $S/pS \simeq D/\mathbf{m}$ via π.

Further, D/pD is a zero-dimensional, Noetherian ring and hence Artinian. Therefore D/pD has finite length as (D/pD)-module and hence as D-module. However, this is equivalent to the existence of a composition series $0 = N_0 \subsetneq N_1 \subsetneq \cdots \subsetneq N_r = D/pD$ of the D-module D/pD. Since N_{i+1}/N_i is simple for all $0 \leq i \leq r-1$ it follows that $N_{i+1}/N_i \simeq D/\mathbf{m} \simeq S/pS$ which implies that $(N_i)_{i=0}^r$ is a composition series of D/pD as S-module. Thus D/pD has finite length as S-module and is therefore a finite dimensional (S/pS)-vector space. Further, D is p-adically separated since D is Noetherian and we can conclude that D is a finitely generated S-module by Proposition 3.5.

If D is of equal characteristic, then D contains a coefficient field k which is a subfield of D such that $k \simeq D/\mathbf{m}$ via π according to Fact 4.2.

Let $T = k[\![X]\!]$ be the power series ring in the variable X and let $y \in D$ be a nonzero non-unit. We define the k-homomorphism $\varphi : T \longrightarrow D$ by $\varphi(X) = y$ and set $S = k[\![y]\!]$ to be the image of T under φ.

With the same argument as above we can conclude that D/yD is a finitely generated (S/yS)-module. Moreover, S is complete and D is separated with respect to the ideal yS. Hence D is finitely generated as S-module by Proposition 3.5. Furthermore, this implies $\dim(S) = \dim(D) = 1$ by Fact 2.2. However, since $S \simeq k[\![X]\!]/\ker(\varphi)$ and $\dim(k[\![X]\!]) = 1$ it follows that $\ker(\varphi) = \mathbf{0}$ and $S \simeq k[\![X]\!]$.

Remark The domain S is the so-called *regular local ring*, that is, a local, Noetherian domain S such that its maximal ideal is generated by $\mathsf{dim}(S)$ elements. Proposition 4.3 is a special case of Cohen's structure theorem which states that every complete Noetherian local domain D contains a regular local subring S such that D is finitely generated as S-module. Moreover, $S = R[\![X_1, \ldots, X_n]\!]$ is a power series ring where in equal characteristic R is a coefficient field and $n = \mathsf{dim}(D)$ and in unequal characteristic R is a coefficient ring and $n = \mathsf{dim}(D) - 1$. For details, we refer Matsumura's textbook [6, §28, §29].

5 Finiteness of the Integral Closure

Let D be a complete, one-dimensional, Noetherian, local domain with quotient field K and let $K \subseteq L$ be a finite field extension. The goal of this section is to prove that the integral closure D'_L of D in L is finitely generated as D-module (see Proposition 5.3). This allows us to conclude in Corollary 5.4 that the integral closure R' of a complete, one-dimensional, Noetherian, local, reduced ring R is finitely generated as R-module. The latter result is essential in the proof of Theorem 1 in the next section.

Following Nagata's textbook [7], we exploit the structure of complete, Noetherian, local domains. According to Proposition 4.3, D contains a subring S such that D is finitely generated as S-module (see Figure 2). If F is the quotient field of S, then the extension $F \subseteq L$ is finite. Moreover, by Fact 2.1, the extension $S \subseteq D$ is integral and hence $D'_L = S'_L$ is the integral closure of S in L.

If we show that S'_L is finitely generated as S-module, then it follows that D'_L is a finitely generated D-module. Therefore, Proposition 4.3 allows us to reduce the investigation to the case where S is a certain complete discrete valuation domain.

To prove that the integral closure of S in L is finitely generated, we distinguish between two cases, either the field extension $F \subset L$ is separable or it is inseparable.

Proposition 5.1 (cf. [2, Ch. V, 1.6, Corollary 1 of Proposition 18]) *Let S be an integrally closed, Noetherian domain with quotient field F and $F \subseteq L$ a finite field extension.*

If $F \subseteq L$ is separable, then the integral closure S'_L of S in L is finitely generated as S-module.

Proof Let $w_1, \ldots, w_n \in L$ be a K-basis of L. Without restriction we can assume that $w_j \in S'_L$ for $1 \leq j \leq n$. Further, let $L^\star = \mathsf{Hom}_K(L, K)$ be the dual space of L and $w'_i \in L^\star$ be the K-basis of $L^\star$ which is defined by $w'_i(w_j) = \delta_{ij}$ (Kronecker-delta) for $1 \leq i, j \leq n$.

Since L is a finite separable extension of K, L is isomorphic to its dual space $L^\star$ via the K-linear map

$$\begin{aligned} T : L &\longrightarrow L^\star \\ x &\longmapsto (y \mapsto \mathsf{tr}_{L/K}(xy)) \end{aligned}$$

where $\mathsf{tr}_{L/K} : L \longrightarrow K$ is the field trace with respect to the extension $K \subseteq L$ (cf. [5, Theorem 5.2]).

For $1 \leq i \leq n$, set $w^\star_i = T^{-1}(w'_i)$. Then $w^\star_1, \ldots, w^\star_n$ form a K-basis of L and $\mathsf{tr}_{L/K}(w^\star_i w_j) = \delta_{ij}$ holds for all $1 \leq i, j \leq n$. Hence, for $a \in S'_L$ there exist $a_i \in K$ such that $a = \sum_{i=1}^n a_i w^\star_i$. Moreover,

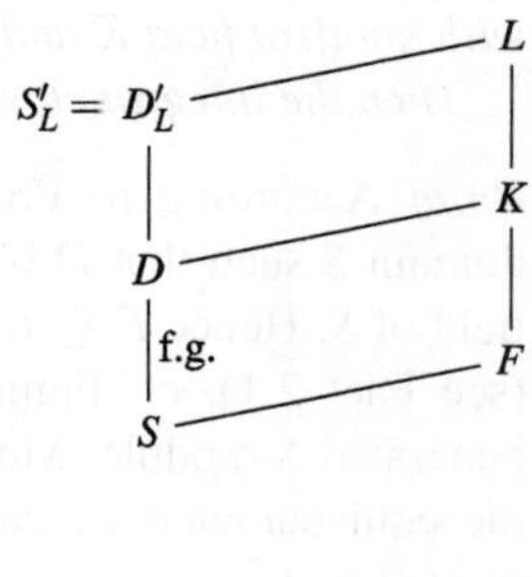

Fig. 2 D contains a subring S such that D is finitely generated as S-module which is either a complete discrete valuation ring or isomorphic to $k[\![X]\!]$ where k is a coefficient field of D

$$a_j = \sum_{i=1}^{n} a_i \operatorname{tr}_{L/K}(w_i^\star w_j) = \operatorname{tr}_{L/K}(aw_j)$$

holds for $1 \le j \le n$.

If $g_j \in K[X]$ is the minimal polynomial of aw_j and $z_1, \ldots, z_m$ are all roots of g_j in some field extension $\tilde{L}$ of K, then $\operatorname{tr}_{L/K}(aw_j) = \sum_{i=1}^{m} z_i \in K$ holds (cf. [5, p. 284]). Moreover, since aw_j is integral over S it follows that $g_j \in S[X]$ and $z_1, \ldots, z_m \in S'_{\tilde{L}}$ are integral over S as well. Therefore

$$a_j = \operatorname{tr}_{L/K}(aw_j) \in K \cap S'_{\tilde{L}} = S$$

where the last equality holds since S is integrally closed. It follows that S'_L is an S-submodule of the Noetherian module $\sum_{i=1}^{n} Sw_i^\star$ and therefore finitely generated.

Proposition 5.2 (cf. [6, p. 263]) *Let $S = k[\![X]\!]$ be a power series ring over a field k with quotient field F and L a finite purely inseparable field extension of F.*

Then the integral closure S'_L of S in L is finitely generated as S-module.

Proof Let $p > 0$ be the characteristic of the field F and $q = p^e = [L : F] < \infty$ be the degree of the field extension $F \subseteq L$. Since the extension is purely inseparable, every element $a \in L$ is a q-th root of an element in F.

Let $\overline{F}$ be an algebraically closed extension of F that contains L. Then $\overline{F}$ contains an element Y such that $X = Y^q$ and $\tilde{L} = L(Y)$ is a finite, purely inseparable field extension of K. Moreover, S'_L is an S-submodule of $S'_{\tilde{L}}$ and it therefore suffices to show that $S'_{\tilde{L}}$ is a finitely generated S-module. This allows us to assume that $L = \tilde{L}$ and $Y \in L$ from this point on.

If $a \in S'_L$ is an integral element, then $a^q \in S'_L \cap F$. However, S is a discrete valuation domain, so it is integrally closed and therefore $S'_L = \{a \in L \mid a^q \in S\}$.

If M is a maximal ideal of S'_L, then $M \cap S = XS$ by Fact 2.2. Hence $M = \{a \in L \mid a^q \in XS\}$ which implies that $M = YS'_L$ is the unique maximal ideal of S'_L. In addition, it follows from Corollary 2.4 that the field extension $k \simeq S/XS \subseteq S'_L/YS'_L$ is finite. Further, Corollary 3.3 implies that S'_L is X-adically separated. Finally, S is X-adically complete and we can conclude that S'_L is a finitely generated S-module by Proposition 3.5.

Proposition 5.3 *Let D be a complete, local, one-dimensional, Noetherian domain with quotient field K and $K \subseteq L$ a finite field extension.*

Then the integral closure D'_L of D in L is finitely generated as D-module.

Proof According to Proposition 4.3, D contains a complete discrete valuation domain S such that D is a finitely generated S-module. Let F denote the quotient field of S. Hence $F \subseteq L$ is a finite field extension, $S \subseteq D$ is integral and $S'_L = D'_L$ (see Fact 2.1), cf. Figure 2. Therefore, it suffices to show that S'_L is a finitely generated S-module. Moreover, S is a complete, discrete valuation domain and in the equicharacteristic case $S \simeq k[\![X]\!]$ where k is a field by Proposition 4.3 and $\dim(S) = 1$.

Fig. 3 Integral closures of S in the field extensions L, E and LE where E is a finite inseparable field extension of F such that LE is a finite separable extension of E

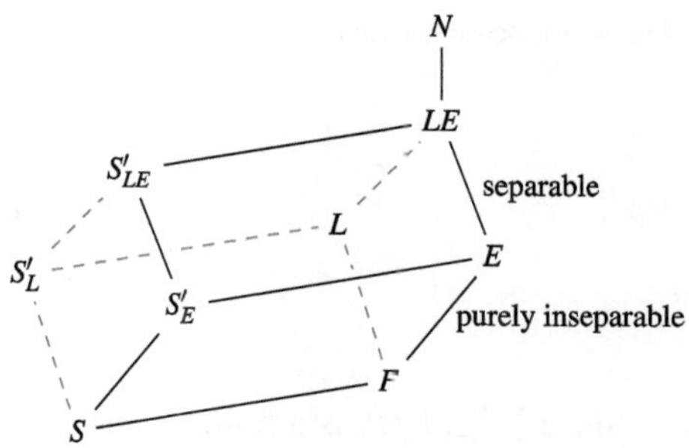

In particular, S is integrally closed. Consequently, if the field extension $F \subseteq L$ is separable, then the assertion follows from Proposition 5.1.

Let us assume that $F \subseteq L$ is inseparable. Then $\mathsf{char}(F) = p > 0$ and hence $\mathsf{char}(S) = \mathsf{char}(D) = p$ which implies that D is of equal characteristic. Therefore $S \simeq k[\![X]\!]$ where k is a field and $k \simeq D/\mathbf{m}$.

Let N be the normal hull of L and E be the fixed field of the automorphism group $\mathsf{Aut}_F(N)$ of $F \subseteq N$. Then $F \subseteq E$ is a purely inseparable extension and $E \subseteq N$ is a separable extension, cf. [5, Proposition V.6.11], see Figure 3.

Moreover, since $F \subseteq L$ is a finite extension, it follows that $F \subseteq N$ is finite which in turn implies that $F \subseteq E$ is a finite extension too.

Hence $E \subseteq LE$ is a finite separable extension and it follows from Proposition 5.1 that $S'_{LE} = (S'_E)'_{LE}$ is finitely generated as S'_E-module. In addition, S'_E is finitely generated as S-module according to Proposition 5.2.

Consequently, S'_{LE} is finitely generated as S-module and therefore a Noetherian S-module. However, S'_L is an S-submodule of S'_{LE} and thus finitely generated.

We conclude this section with the analogous assertion for complete, Noetherian, local, reduced rings.

Corollary 5.4 *Let R be a complete, Noetherian, one-dimensional, local ring.*

If R is reduced, then the integral closure R' of R is finitely generated as R-module.

Proof Let $P_1, \ldots, P_n$ be the minimal prime ideals of R. For $1 \leq i \leq n$, let Q_i be the quotient field of the Noetherian, one-dimensional, local domain R/P_i. Then $Q = Q_1 \times \cdots \times Q_n$ is the total ring of quotients of $R/P_1 \times \cdots \times R/P_n$. Moreover,

$$(R/P_1 \times \cdots \times R/P_n)'_Q = (R/P_1)'_{Q_1} \times \cdots \times (R/P_n)'_{Q_n}. \tag{2}$$

The Noetherian, local domain R/P_i is a finitely generated R-module and hence $\mathbf{m}$-adically complete by Fact 3.4. As the $\mathbf{m}$-adic topology ccincides with the $\mathbf{m}/P_i$-adic topology on R/P_i, it follows that $(R/P_i)'_{Q_i}$ is a finitely generated (R/P_i)-module according to Proposition 5.3. Together with Equation (2), it now follows that $(\prod_{i=1}^n R/P_i)'_Q$ is a finitely generated $\left(\prod_{i=1}^n R/P_i\right)$-module.

Fig. 4 Embeddings via ε

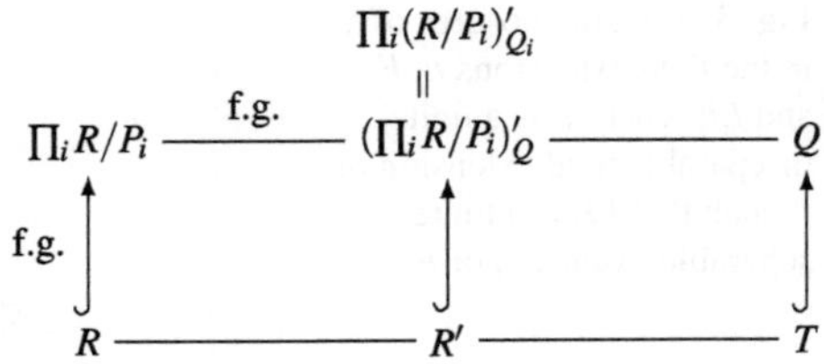

Since $\prod_{i=1}^n R/P_i$ is a finitely generated R-module, it follows that $(\prod_{i=1}^n R/P_i)'_Q$ is finitely generated as R-module too and therefore Noetherian. Let

$$\varepsilon : R \longrightarrow R/P_1 \times \cdots \times R/P_n$$
$$r \longmapsto (r + P_1, \ldots, r + P_n).$$

Then $\ker(\varepsilon) = \bigcap_{i=1}^n P_i = \mathsf{nil}(R) = \mathbf{0}$ since R is reduced by hypothesis. Hence we can embed R into $\prod_{i=1}^n R/P_i$ via ε. Similarly, we can embed R' into $(\prod_{i=1}^n R/P_i)'_Q$ since ε can be canonically extended to the total ring of quotients T of R, see Figure 4.

Thus R' is isomorphic to a submodule of the Noetherian R-module $(\prod_{i=1}^n R/P_i)'_Q$ and hence finitely generated.

6 Proof of the Theorem

Finally, we are ready to give a proof of Theorem 1. For the reader's convenience we restate it here. Recall that a Noetherian, local domain $(D, \mathbf{m})$ with $\mathbf{m}$-adic completion $\widehat{D}$ and integral closure D' is called

- *unibranched*, if D' is local,
- *analytically unramified*, if $\widehat{D}$ is a reduced ring and
- *analytically irreducible*, if $\widehat{D}$ is a domain

(cf. Definition 1.1).

Theorem 1 *Let $(D, \mathbf{m})$ be a one-dimensional, Noetherian, local domain with integral closure D'. Then the following assertions are equivalent:*

1. *D is analytically irreducible.*
2. *D is unibranched and analytically unramified.*
3. *D is unibranched and D' is finitely generated as D-module.*
4. *D is unibranched and if $\mathbf{m}'$ denotes the maximal ideal of the integral closure, then the $\mathbf{m}$-adic topology on D coincides with subspace topology induced by $\mathbf{m}'$.*

Proof (1) $\Rightarrow$ (2): By assumption, $\widehat{D}$ is a domain and therefore is a reduced ring.

Assume that D is not unibranched and let $M_1 \neq M_2$ be two different maximal ideals of the integral closure D' of D. Let $a_1 \in M_1 \setminus M_2$ and $a_2 \in M_2 \setminus M_1$. Then

$R = D[a_1, a_2] \subseteq D'$ is an integral extension of D and therefore finitely generated as D-module by Fact 2.1. Hence, according to Proposition 3.8, R is a semilocal domain. However, the extension $R \subseteq D'$ is integral and therefore $N_i = M_i \cap R$ are maximal ideals of R for $i = 1, 2$ according to Fact 2.2. Due to the choice of a_1 and a_2, $N_1 \neq N_2$ and R is semilocal but not local. It follows from Proposition 3.8 that $\widehat{D}$ is not a domain.

(2) $\Rightarrow$ (3): If D' is not finitely generated as D-module, then there exists an infinite strictly ascending chain of intermediate rings $D \subseteq D_i \subseteq D'$ which are finitely generated as D-modules.

Let K be the quotient field of D and D_i for all i and $\widehat{D_i}$ denote the $\mathbf{m}$-adic completion of D_i. If $\frac{a}{b} \in \widehat{D_i} \cap K$, then $a \in b\widehat{D_i} \cap K$. However, by Corollary 3.3, $b\widehat{D_i} \cap K = bD_i$ and hence $\frac{a}{b} \in D_i$. Hence $\widehat{D_i} \cap K = D_i \subsetneq D_{i+1} = \widehat{D_{i+1}} \cap K$ which implies that $\widehat{D_i} \subsetneq \widehat{D_{i+1}}$ for all i.

Moreover, according to Fact 3.4, $\widehat{D_i} \simeq D_i \otimes_D \widehat{D}$ is a finitely generated $\widehat{D}$-module. Hence $\widehat{D_i}$ is contained in the integral closure $(\widehat{D})'$ of $\widehat{D}$ in its total ring of quotients. Consequently, the extension $\widehat{D} \subseteq (\widehat{D})'$ contains the infinite strictly ascending chain of intermediate rings $\widehat{D_i}$. Thus $(\widehat{D})'$ is not finitely generated as $\widehat{D}$-module which implies that $\widehat{D}$ is not reduced by Corollary 5.4.

(3) $\Rightarrow$ (4): The assertion immediately follows from Proposition 3.8.

(4) $\Rightarrow$ (1): Let $\widehat{(D')}$ be the $\mathbf{m}'$-adic completion of D'. Since the $\mathbf{m}'$-adic topology induces the $\mathbf{m}$-adic topology on D, it follows that D is a topological subspace of $\widehat{(D')}$ and $\widehat{D}$ is the topological closure of D in $\widehat{(D')}$.

By assumption D' is local, so D' is a discrete valuation domain according to Corollary 2.5. As shown in Example 3.1, the $\mathbf{m}'$-adic completion $\widehat{(D')}$ of D' is also a discrete valuation domain.

Hence $\widehat{D}$ is a subring of the domain $\widehat{(D')}$ and thus it is a domain itself.

Remark There are examples of one-dimensional, Noetherian, local domains which are unibranched but not analytically irreducible, cf. [9].

Acknowledgements R. Rissner is supported by the Austrian Science Fund (FWF): P 27816-N26.

References

1. Atiyah, M.F., Macdonald, I.G.: Introduction to Commutative Algebra. Addison-Wesley, Reading, MA (1969)
2. Bourbaki, N.: Commutative Algebra, Chapters 1–7. Springer, Berlin (1989)
3. Cahen, P.-J., Chabert, J.-L.: Integer-Valued Polynomials. Mathematical Surveys and Monographs, vol. 48. American Mathematical Society, Providence (1997)
4. Huneke, C., Swanson, I.: Integral Closure of Ideals, Rings, and Modules. London Mathematical Society Lecture Note Series, vol. 336. Cambridge University Press, Cambridge (2006)
5. Lang, S.: Algebra. Graduate Texts in Mathematics, vol. 211. Springer, New York (2002)
6. Matsumura, H.: Commutative Ring Theory. Cambridge University Press, Cambridge (1986)

7. Nagata, M.: Local Rings. Interscience Tracts in Pure and Applied Mathematics, vol. 13. Wiley, New York (1962)
8. Northcott, D.G.: A general theory of one-dimensional local rings. Proc. Glasgow Math. Assoc. **2**, 159–169 (1956)
9. Olberding, B.: One-dimensional bad Noetherian domains. Trans. Am. Math. Soc. **366**(8), 4067–4095 (2014)

Integer-Valued Polynomials on Algebras: A Survey of Recent Results and Open Questions

Nicholas J. Werner

Abstract Given a commutative integral domain D with fraction field K, the ring of integer-valued polynomials on D is $\text{Int}(D) = \{f \in K[x] \mid f(D) \subseteq D\}$. In recent years, attention has turned to generalizations of $\text{Int}(D)$ where the polynomials act on D-algebras rather than on D itself. We survey the present activity on this topic and propose questions for further research.

Keywords Integer-valued polynomial • Algebra • P-ordering • Regular basis • Int-decomposable • Integral closure • Prüfer domain • Matrix • Quaternion • Octonion • Integer-valued rational function

MSC Primary: 13F20, 16S36. **Secondary**: 13F05, 13B22, 11R52, 11C99, 17D99

1 Introduction

Let D be a commutative integral domain with fraction field K. The ring of integer-valued polynomials on D is defined to be

$$\text{Int}(D) = \{f \in K[x] \mid f(D) \subseteq D\}.$$

The use of polynomials in $\text{Int}(\mathbb{Z})$ dates back at least to the seventeenth century [7, p. xiii]. The first systematic study of the algebraic properties of $\text{Int}(D)$ was done by Pólya [47] and Ostrowski [39] in 1919. Both Pólya and Ostrowski were primarily concerned with the module structure of $\text{Int}(D)$ when D is the ring of integers of a number field, and were interested in determining whether $\text{Int}(D)$ had a regular basis. Significant progress on understanding the ring structure of $\text{Int}(D)$ began in the 1970s with the work of Chabert [10], Cahen [5], and Brizolis [4], among others.

N.J. Werner (✉)
Department of Mathematics, Computer and Information Science, SUNY College at Old Westbury, Old Westbury, NY 11568, USA
e-mail: wernern@oldwestbury.edu

M. Fontana et al. (eds.), *Rings, Polynomials, and Modules*,
https://doi.org/10.1007/978-3-319-65874-2_18

The book [7] is the standard reference on this topic, and includes a comprehensive bibliography of articles published up to 1997.

More recently, attention has turned to generalization of Int(D) where the polynomials are evaluated at elements of a D-algebra A. For this construction—and unless noted otherwise—we assume that A is an associative torsion-free D-algebra such that $A \cap K = D$, and we let $B = K \otimes_D A$, which is the extension of A to a K-algebra. The maps $k \mapsto k \otimes 1$ and $a \mapsto 1 \otimes a$ allow us to identify K and A with their canonical images in B, and so we may evaluate polynomials in $B[x]$ or $K[x]$ at elements of A. The formality of tensor products is useful in some results, but for most purposes we may consider the elements of B to be fractions a/d with $a \in A$ and $d \in D, d \neq 0$.

Definition 1 We define $\text{Int}(A) = \{f \in B[x] \mid f(A) \subseteq A\}$ and $\text{Int}_K(A) = \text{Int}(A) \cap K[x] = \{f \in K[x] \mid f(A) \subseteq A\}$.

If B is noncommutative, then Int(A) contains polynomials with non-commuting coefficients. Following common conventions for dealing with polynomials over noncommutative rings (as in [32, Sec. 16]), we assume that the indeterminate x commutes with all elements of B and that polynomials in $B[x]$ satisfy right-evaluation, that is polynomials are evaluated with the indeterminate to the right of coefficients. With these conventions, any polynomial $f \in B[x]$ can be written as $f(x) = \sum_i b_i x^i$, and for any $a \in A$ we have $f(a) = \sum_i b_i a^i$. Although we shall not do so here, one may also consider integer-valued polynomials that satisfy left-evaluation, i.e., $f \in B[x]$ is written as $f(x) = \sum_i x^i b_i$ and $f(a) = \sum_i a^i b_i$. This approach is used in the work of Frisch [23].

Research articles devoted exclusively to Int(A) and $\text{Int}_K(A)$ began to appear around 2010, but the prospect of studying integer-valued polynomials on algebras was considered earlier, as can be seen in the 2006 survey [8] and the unpublished preprint of Gerboud [24] from 1998. Polynomials in $K[x]$ that act on the ring $M_n(D)$ of $n \times n$ matrices with entries in D were discussed in [19]. In fact, particular examples of polynomials in $\text{Int}_K(A)$ can be found much earlier. As pointed out in [30], a 1931 paper by Littlewood and Richardson [34] contains a construction for polynomials in $\mathbb{Q}[x]$ that are integer-valued on the ring of Hurwitz quaternions.

At the present time, numerous authors have contributed to the growing body of work surrounding $\text{Int}_K(A)$ and Int(A). Some articles in this area approach the subject broadly, and analyze $\text{Int}_K(A)$ and Int(A) for a general D-algebra A. These include [20–22, 42–45, 56], and [57]. Other papers concern results for specific algebras or classes of algebras A. For instance, Loper and the author [36] have studied $\text{Int}_{\mathbb{Q}}(A)$ when A is the ring of integers of a number field, as have Heidaryan, Longo, and Peruginelli [27, 41]. Related ideas were used by Chabert and Peruginelli in [12] to classify the overrings of Int($\mathbb{Z}$) in terms of polynomials that are integer-valued on subsets of the profinite completion of $\mathbb{Z}$.

Particular attention has been paid to the examples where A is the Lipschitz quaternions or the Hurwitz quaternions. Let **i**, **j**, and **k** be the imaginary quaternion units, which satisfy $\mathbf{i}^2 = \mathbf{j}^2 = -1$ and $\mathbf{ij} = \mathbf{k} = -\mathbf{ji}$. The Lipschitz quaternions $\mathbb{L}$ are defined to be

$$\mathbb{L} = \{a_0 + a_1\mathbf{i} + a_2\mathbf{j} + a_3\mathbf{k} \mid a_i \in \mathbb{Z} \text{ for all } i\}$$

and the Hurwitz quaternions $\mathbb{H}$ are defined to be

$$\mathbb{H} = \{a_0 + a_1\mathbf{i} + a_2\mathbf{j} + a_3\mathbf{k} \mid a_i \in \mathbb{Z} \text{ for all } i \text{ or } a_i \in \mathbb{Z} + \tfrac{1}{2} \text{ for all } i\}.$$

Integer-valued polynomials on $\mathbb{L}$, $\mathbb{H}$, and the split quaternions (a variation on $\mathbb{L}$ where $\mathbf{j}^2 = \mathbf{k}^2 = 1$ instead of $\mathbf{j}^2 = \mathbf{k}^2 = -1$) were examined in, respectively, [30, 53], and [13].

A good deal of research has been devoted to understanding $\text{Int}_K(A)$ and $\text{Int}(A)$ when A is a ring of matrices or triangular matrices. The author investigated the noncommutative ring $\text{Int}(M_n(D))$ in [55]. The corresponding commutative rings $\text{Int}_K(M_n(D))$ (and in particular $\text{Int}_{\mathbb{Q}}(M_n(\mathbb{Z}))$) are the subject of [17, 40], and [46]. Recently, Frisch [23] has examined the noncommutative ring $\text{Int}(A)$ where A is the ring of upper triangular matrices with entries in D; previously, Evrard, Fares, and Johnson [18] had considered the commutative ring $\text{Int}_K(A)$ where A is the ring of lower triangular matrices with entries in D.

While our focus in this survey will be on $\text{Int}(A)$ and $\text{Int}_K(A)$, we point out that Elliott [16] has begun studying integer-valued polynomials on general commutative rings (possibly with zero divisors) and modules. In this formulation, one takes R to be a commutative ring with unity and lets $T(R)$ denote the total quotient ring of R. Then, one may define $\text{Int}(R) = \{f \in T(R)[x] \mid f(R) \subseteq R\}$.

In Sect. 2, we will discuss some of the basic properties of $\text{Int}(A)$ and $\text{Int}_K(A)$, such as conditions under which $\text{Int}(A)$ has a noncommutative ring structure, or how $\text{Int}_K(A)$ compares to its subring $D[x]$ and its overring $\text{Int}(D)$. Section 3 examines ways to produce polynomials in $\text{Int}(A)$ by exploiting the relationships among integer-valued polynomials, P-orderings, and null ideals. In Sect. 4 we look at module decompositions of $\text{Int}(A)$, which provide a way to determine the extent to which the properties of $\text{Int}(A)$ follow from those of $\text{Int}_K(A)$. Section 5 focuses on $\text{Int}_K(A)$ and its integral closure; a key question here is whether or not the integral closure of $\text{Int}_K(A)$ is a Prüfer domain. Lastly, in Sect. 6 we describe some open problems that are largely untouched, but are good prospects for future research. We remark that while do not have space to study them here, common commutative objects of interest such as prime spectra and Krull dimension have been investigated for both $\text{Int}(A)$ and $\text{Int}_K(A)$. We refer the reader to the articles listed in this introduction for more information.

2 Basic Properties and Non-triviality Conditions

Under the definitions given in Sect. 1, one may easily verify that $\text{Int}_K(A)$ is subring of $K[x]$ containing $D[x]$; in fact, the condition $A \cap K = D$ is equivalent to having $\text{Int}_K(A) \subseteq \text{Int}(D)$. Moreover, if A (and hence B) is commutative, then $\text{Int}(A)$ is also a commutative ring. However, if A is noncommutative, then the evaluation of polynomials in $B[x]$ at $a \in A$ defines a multiplicative map $B[x] \to B$ if and only if a is central in B. Symbolically, let $f, g \in B[x]$ and let fg denote their product in $B[x]$. Then, if A is noncommutative it may be that $(fg)(a) \neq f(a)g(a)$. For a simple example let $a, b \in A$ be such that $ab \neq ba$, let $f(x) = x - a$ and let $g(x) = x - b$. Then, $f(a)g(a) = 0$, but $(fg)(x) = x^2 - (a+b)x + ab$, so $(fg)(a) = -ba + ab \neq 0$.

One may check that Int(A) is always a left $\text{Int}_K(A)$-module, but because of the above difficulty with evaluation, it is not clear at first glance whether Int(A) is closed under multiplication (and hence is a ring) when A is noncommutative. Nevertheless, there are conditions under which multiplicative closure can be guaranteed.

Theorem 2 ([55, Thm. 1.2]) *Assume that each $a \in A$ may be written as a finite sum $a = \sum_i c_i u_i$ for some $c_i, u_i \in A$ such that each u_i is a unit of A and each c_i is central in B. Then,* Int(A) *is closed under multiplication and hence is a ring.*

The condition in this theorem that each element of A be generated by units and central elements is sufficient for Int(A) to be a ring, but is not necessary. In [57, Ex. 3.8], examples of generalized quaternion algebras A over $\mathbb{Z}$ are given such that $A^\times = \{\pm 1\}$ and $Z(A) = \mathbb{Z}$, but Int(A) is still a ring. Additionally, it is shown in [23] that $\text{Int}(T_n(D))$ is a ring, where $T_n(D)$ is the ring of upper triangular matrices with entries in D. Since $T_n(D)$ is not generated by its units and central elements, this example also shows that the converse of Theorem 2 does not hold.

Question 3 What are necessary and sufficient conditions on A so that Int(A) is a ring? In particular, is Int(A) always a ring when A is finitely generated as a D-module?

To date, no examples have been found of an algebra A such that Int(A) is not a ring. However, as we shall see later in Sect. 6.3, if one considers the set $\text{Int}(S, A) = \{f \in B[x] \mid f(S) \subseteq A\}$ of integer-valued polynomials on a subset S of a noncommutative algebra A, then it is quite easy to produce examples where $\text{Int}(S, A)$ is not a ring.

We turn now to the commutative ring $\text{Int}_K(A)$. Because of the assumption that $A \cap K = D$, we always have the containments

$$D[x] \subseteq \text{Int}_K(A) \subseteq \text{Int}(D). \tag{1}$$

We say that $\text{Int}_K(A)$ is *trivial* if $\text{Int}_K(A) = D[x]$. Recall that for an element $q \in K$, an ideal of the form $(D :_D q) = \{d \in D \mid dq \in D\}$ is called a *conductor ideal*. When D is Noetherian, it is known [7, Thm. I.3.14] that Int(D) is nontrivial if and only if there exists a prime conductor ideal of D of finite index. It is shown in [20] that this same condition ensures the non-triviality of $\text{Int}_K(A)$.

Theorem 5 ([20, Thm. 4.3]) *Let D be a Noetherian domain. Then,* $\text{Int}_K(A)$ *is nontrivial if and only if there exists a prime conductor ideal of D of finite index.*

It is also possible to give non-triviality conditions that do not rely on the assumption that D is Noetherian. In [51, Cor. 1.7], Rush exhibited a double-boundedness condition that is necessary and sufficient for Int(D) to be nontrivial, and which holds for a general domain D. When A is finitely generated as a D-

module, Rush's result carries over to $\mathrm{Int}_K(A)$. In fact, we need only assume the weaker condition that A is an *integral algebra of bounded degree*, meaning that there exists a positive integer n such that each element of A satisfies a monic polynomial in $D[x]$ of degree at most n. Finally, when D is Dedekind we can further weaken the assumptions on A.

Theorem 6

(1) [45, Thm. 2.12] Let D be a domain and let A be an integral D-algebra of bounded degree. Then, $\mathrm{Int}_K(A)$ *is nontrivial if and only if* $\mathrm{Int}(D)$ *is nontrivial.*
(2) [45, Thm. 3.4] Let D be a Dedekind domain. Then, $\mathrm{Int}_K(A)$ *is nontrivial if and only if there exists a prime P of D such that* A/PA *is an integral* D/P*-algebra of bounded degree.*

Part (2) of Theorem 6 may apply to algebras that are not finitely generated. For instance, when $D = \mathbb{Z}$ we can take $A = \prod_{i \in \mathbb{N}} \mathbb{Z}$, and $\mathrm{Int}_{\mathbb{Q}}(A)$ is nontrivial (in fact, $\mathrm{Int}_{\mathbb{Q}}(A) = \mathrm{Int}(\mathbb{Z})$ [45, Ex. 3.1]). Similarly, with $D = \mathbb{Z}_{(p)}$, we can take $A = \mathbb{Z}_p$, the p-adic integers, and then $\mathrm{Int}_{\mathbb{Q}}(A) = \mathrm{Int}(\mathbb{Z}_{(p)})$ [45, Lem. 3.6].

The containment $\mathrm{Int}_K(A) \subseteq \mathrm{Int}(D)$ in (1) is also of interest. In the case of a Dedekind domain with finite residue rings, equality between $\mathrm{Int}_K(A)$ and $\mathrm{Int}(D)$ can be determined by examining the residue rings A/PA or the completions $\widehat{A}_P$.

Theorem 7 ([43, Thms. 2.11, 3.10]) *Let D be a Dedekind domain with finite residue rings. Let A be a D-algebra that is finitely generated as a D-module. Then, the following are equivalent:*

1. $\mathrm{Int}_K(A) = \mathrm{Int}(D)$.
2. *For each nonzero prime P of D, there exists a positive integer t such that* $A/PA \cong \bigoplus_{i=1}^{t} D/P$.
3. *For each nonzero prime P of D, there exists a positive integer t such that* $\widehat{A}_P \cong \bigoplus_{i=1}^{t} \widehat{D}_P$ *(here,* $\widehat{A}_P$ *and* $\widehat{D}_P$ *are the P-adic completions of A and D, respectively).*

When $D = \mathbb{Z}$, the conditions on A so that $\mathrm{Int}_{\mathbb{Q}}(A) = \mathrm{Int}(\mathbb{Z})$ become even more restrictive.

Theorem 8 ([43, Cor. 4.12]) *Let A be a* $\mathbb{Z}$*-algebra that is finitely generated as a* $\mathbb{Z}$*-module. Then,* $\mathrm{Int}_{\mathbb{Q}}(A) = \mathrm{Int}(\mathbb{Z})$ *if and only if there exists a positive integer t such that* $A \cong \bigoplus_{i=1}^{t} \mathbb{Z}$.

We close this section by mentioning some results on localizations of $\mathrm{Int}_K(A)$ and $\mathrm{Int}(A)$. In the case of traditional integer-valued polynomials, a useful and frequently used property of $\mathrm{Int}(D)$ is that it is often well-behaved with respect to localization at primes of D. If D is Noetherian, then [7, Thm. I.2.3] shows that $\mathrm{Int}(D)_P = \mathrm{Int}(D_P)$ for all primes P of D. In the case of algebras, we have the following analogous results.

Proposition 9 *Let D be a Noetherian domain and let P be a nonzero prime of D.*

1. $\mathrm{Int}_K(A)_P \subseteq \mathrm{Int}_K(A_P)$ *and* $\mathrm{Int}(A)_P \subseteq \mathrm{Int}(A_P)$.

2. *If A is finitely generated as a D-module, then* $\mathrm{Int}_K(A)_P = \mathrm{Int}_K(A_P)$ *and* $\mathrm{Int}(A)_P = \mathrm{Int}(A_P)$.
3. *If D is Dedekind, then* $\mathrm{Int}_K(A)_P = \mathrm{Int}_K(A_P)$.

Proof (1) can be proved by using a method of Rush [51, Prop. 1.4] involving induction on the degrees of the polynomials. The reverse containments in (2) and (3) are demonstrated in [56, Prop. 3.2] and [45, Lem. 3.2].

Note that in part (3) of Proposition 9 there is no assumption that A is finitely generated as a D-module.

3 *P*-Orderings, Regular Bases, and Null Ideals

A basis for $\mathrm{Int}(\mathbb{Z})$ as a $\mathbb{Z}$-module is given by the set of binomial polynomials

$$\binom{x}{n} = \frac{x(x-1)\cdots(x-(n-1))}{n!}.$$

In this basis, there is one polynomial of degree n for each $n \geq 0$; such a basis is called a *regular basis*. The focus of the research done by Pólya and Ostrowski [39, 47] was to determine when $\mathrm{Int}(D)$ had a regular basis in the case where D is the ring of integers of a number field. Such a characterization can be made for any domain R such that $D[x] \subseteq R \subseteq K[x]$ via the use of characteristic ideals [7, Sec. II.1].

When D is a Dedekind domain and $S \subseteq D$, a regular basis for $\mathrm{Int}(S, D) = \{f \in K[x] \mid f(S) \subseteq D\}$ can be found (if it exists) by using P-orderings and P-sequences, which were introduced by Bhargava in [2]. Given a nonzero prime ideal P of D, let v_P be the corresponding valuation. A *P-ordering* of $S \subseteq D$ is a sequence $\{a_0, a_1, \ldots\} \subseteq S$ such that for each $k > 0$, a_k minimizes $v_P(\prod_{i=0}^{k-1}(a - a_i))$ as a ranges over all elements of S. A P-ordering gives rise to a *P-sequence*, which is a sequence of ideals $\{\nu_0(S,P), \nu_1(S,P), \ldots\}$ defined by taking $\nu_k(S,P)$ to be the highest power of P containing $\prod_{i=0}^{k-1}(a - a_i)$. Bhargava has shown [2, Thm. 1] that the P-sequence for S is independent of the P-ordering chosen. The relation to regular bases of $\mathrm{Int}(S, D)$ is given by the next theorem.

Theorem 10 ([2, Thm. 14]) *Let* $\nu_k(S) = \prod_{P\ prime} \nu_k(S,P)$. *Then,* $\mathrm{Int}(S, D)$ *has a regular basis if and only if* $\nu_K(S)$ *is a nonzero principal ideal for all* $k \geq 0$. *In particular, in the case where D is a discrete valuation ring, then a regular basis for* $\mathrm{Int}(S, D)$ *is given by* $\prod_{i=0}^{k-1}(x - a_i)/(a_k - a_i)$, *where* $k = 0, 1, \ldots$.

Johnson [29] has extended the notion of P-orderings to certain noncommutative rings.

Definition 11 ([29, Def. 1.1]) Let K be a local field with valuation v, $\mathscr{D}$ a division algebra over K to which the valuation v extends, R the maximal order in $\mathscr{D}$, and S a subset of R. Then, a *v-ordering* of S is a sequence $\{a_0, a_1, \ldots\} \subseteq S$ with the property that for each $k > 0$, a_k minimizes the quantity $v(f_k(a_0, \ldots, a_{k-1})(a))$ over $a \in S$, where $f_0 = 1$ and, for $k > 0$, $f_k(a_0, \ldots, a_{k-1})(x)$ is the minimal polynomial (in the sense of [33]) of the set $\{a_0, a_1, \ldots, a_{k-1}\}$. The sequence of valuations $\{v(f_k(a_0, \ldots, a_{k-1})(a_k)) \mid k = 0, 1, \ldots\}$ is called the *v-sequence* of S.

Theorem 12 ([29, Prop. 1.2]) *With notation as in Definition 11, let $\pi \in R$ be a uniformizing element. Then, the v-sequence $\{\alpha_S(k) = v(f_k(a_0, \ldots, a_{k-1})(a_k)) \mid k = 0, 1, \ldots\}$ depends only on the set S and not on the choice of v-ordering. Moreover, the sequence of polynomials*

$$\{\pi^{-\alpha_S(k)} f_k(a_0, \ldots, a_{k-1})(x) \mid k = 0, 1, \ldots\}$$

forms a regular R-basis for Int(S, R).

This approach has been used to good effect in [29] and [17]. In [29], a recursive formula is given for the v-sequence of the Hurwitz quaternions $\mathbb{H}$ localized at the maximal ideal generated by $1 + \mathbf{i}$ (since $\mathbb{H}$ is a noncommutative ring, "localization" here means Ore localization, as discussed in [31]). Similar formulas and algorithms are given in [17] for the maximal order in a division algebra $\mathscr{D}$ of degree 4 over the field of p-adic numbers $\mathbb{Q}_p$. These formulas can be used to produce basis elements for Int($\mathbb{H}$) and the integral closure of $\text{Int}_{\mathbb{Q}}(M_2(\mathbb{Z}))$, respectively.

Even in cases where a regular basis for Int(A) or $\text{Int}_K(A)$ does not exist or is computationally expensive to compute, it is possible to produce integer-valued polynomials by exploiting the relationship between integer-valued polynomials and elements of null ideals.

Definition 13 Let R be a ring. The *null ideal* $N(R)$ of R is defined to be $N(R) = \{f \in R[x] \mid f(R) = 0\}$.

Under our convention that polynomials satisfy right-evaluation, one may easily check that $N(R)$ is always a left ideal of $R[x]$. When R is commutative, $N(R)$ is clearly a two-sided ideal of $R[x]$. When R is noncommutative, it is not known whether $N(R)$ is always a two-sided ideal of $R[x]$. Indeed, this question is closely related to the problem of determining if Int(A) is a ring when A is noncommutative. More details on this topic (along with a proof that $N(R)$ is a two-sided ideal for many classes of finite rings) can be found in [57].

The connection between Int(A) and $N(R)$ is encapsulated in the following easily verified correspondence lemma. Versions of this lemma are often used (sometimes implicitly) when null ideals are employed to study integer-valued polynomials, e.g., in [19, Lem. 3.4] or [57, Sec. 2].

Lemma 14 *With our standard notation, let $f(x) = g(x)/d \in B[x]$, where $g(x) \in A[x]$ and $d \in D$. Then, $f \in$ Int(A) if and only if the residue of g mod A/dA is in $N(A/dA)$.*

Thus, to verify that $(1 + \mathbf{i} + \mathbf{j} + \mathbf{k})(x^2 - x)/2 \in \text{Int}(\mathbb{L})$, one need only check that $(1 + \mathbf{i} + \mathbf{j} + \mathbf{k})(x^2 - x)$ is in $N(\mathbb{L}/2\mathbb{L})$. Similarly, $(1 + \mathbf{i})(x^4 - x)/2 \in \text{Int}(\mathbb{H})$ because $(1 + \mathbf{i})(x^4 - x)$ sends each element of the finite ring $\mathbb{H}/2\mathbb{H}$ to 0. As noted by [30], this correspondence between integer-valued polynomials and null ideals goes back at least to [34], where it is used (under different terminology) to show that for each prime p the polynomial $x^2(x^{p^3-p}-1)/p \in \mathrm{Int}_{\mathbb{Q}}(\mathbb{H})$.

In the case of matrix rings, it is known [3, Thm. 3] that for each positive integer n and prime power q, the polynomial

$$\Phi_{q,n}(x) = (x^{q^n}-x)(x^{q^{n-1}}-x)\cdots(x^q-x)$$

is an element of (in fact, it is the generator of) the null ideal $N(M_n(\mathbb{F}_q))$, where $\mathbb{F}_q$ is the field with q elements. Consequently, if $\pi \in D$ is such that $D/\pi D \cong \mathbb{F}_q$, then $\Phi_{q,n}(x)/\pi \in \mathrm{Int}_K(M_n(D))$. Moreover, for each odd prime p we have $\mathbb{L}/p\mathbb{L} \cong \mathbb{H}/p\mathbb{H} \cong M_2(\mathbb{F}_p)$ (see [15, Sec. 2.5] or [26, Ex. 3A]), so the polynomial $(x^{p^2}-x)(x^p-x)/p$ is in both $\mathrm{Int}_{\mathbb{Q}}(\mathbb{L})$ and $\mathrm{Int}_{\mathbb{Q}}(\mathbb{H})$.

The study of null ideals is an active area of research in its own right. We direct the reader toward the recent papers [28, 49], and [50] on this topic for more information and further references.

4 Module Decomposition

Frisch was the first to notice the following property of $\mathrm{Int}(M_n(D))$.

Theorem 15 ([21, Thm. 7.2]) *Let D be a domain. Then,* $\mathrm{Int}(M_n(D)) \cong M_n(\mathrm{Int}_K(M_n(D)))$.

That is, $\mathrm{Int}(M_n(D))$ is itself a matrix ring, where the entries of the matrix are polynomials in $\mathrm{Int}_K(M_n(D))$. The isomorphism in the theorem is obtained by associating a polynomial with matrix coefficients to a matrix with polynomial entries. For example, with $M_2(\mathbb{Z})$,

$$\begin{pmatrix}1&0\\0&0\end{pmatrix}\frac{(x^4-x)(x^2-x)}{2}+\begin{pmatrix}0&1\\-1&0\end{pmatrix}x^2+3x \in \mathrm{Int}(M_2(\mathbb{Z}))$$

corresponds to

$$\begin{pmatrix}\frac{(x^4-x)(x^2-x)}{2}+3x & x^2\\ -x^2 & 3x\end{pmatrix} \in M_2(\mathrm{Int}_{\mathbb{Q}}(M_2(\mathbb{Z})))$$

This example led the author to search for other algebras with a similar property. The key property of the matrix example can be generalized by considering how a D-module basis of A corresponds to an $\mathrm{Int}_K(A)$-module basis for $\mathrm{Int}(A)$.

Definition 16 ([56, Def. 1.2]) Let D be domain with fraction field K. Let A be a free D-algebra of finite rank, and let $\{\alpha_1,\ldots,\alpha_t\}$ be a D-module basis for A, so that $A = \bigoplus_{i=1}^t D\alpha_i$. If $\mathrm{Int}(A) = \bigoplus_{i=1}^t \mathrm{Int}_K(A)\alpha_i$, then A is said to be Int_K-*decomposable*.

So, an Int_K-decomposable algebra A is a free D-algebra whose D-basis extends to an Int_K-basis for $\mathrm{Int}(A)$. There are equivalent ways to view the notion of Int_K-decomposability. For instance, given $A = \bigoplus_{i=1}^t D\alpha_i$, for any $f \in B[x]$ we may write $f = \sum_{i=1}^t f_i\alpha_i$ for some $f_i \in K[x]$. The algebra A is Int_K-decomposable if having $f \in \mathrm{Int}(A)$ implies that each $f_i \in \mathrm{Int}_K(A)$. We can also think of an Int_K-decomposable algebra A as one for which $\mathrm{Int}(A)$ is generated (as a subring of $B[x]$) by $\mathrm{Int}_K(A)$ and A.

Aside from matrix rings, most of the common choices for A are not Int_K-decomposable. For instance, if $A = \mathbb{Z}[\mathbf{i}]$ is the Gaussian integers, then $(1+\mathbf{i})(x^2 - x)/2 \in \mathrm{Int}(A)$, but $(x^2 - x)/2 \notin \mathrm{Int}_{\mathbb{Q}}(A)$ because $(\mathbf{i}^2 - \mathbf{i})/2 \notin \mathbb{Z}[\mathbf{i}]$; hence, $\mathbb{Z}[\mathbf{i}]$ is not Int_K-decomposable. Similar noncommutative examples arise with the Lipschitz and Hurwitz quaternions.

Example 17 The polynomial $(1 + \mathbf{i} + \mathbf{j} + \mathbf{k})(x^2 - x)/2 \in \mathrm{Int}(\mathbb{L})$ (this follows from Lemma 14), but $(x^2 - x)/2 \notin \mathrm{Int}_{\mathbb{Q}}(\mathbb{L})$ because $(\mathbf{i}^2 - \mathbf{i})/2 \notin \mathbb{L}$. Similarly, $(1 + \mathbf{i})(x^4 - x)/2 \in \mathrm{Int}(\mathbb{H})$ (this can be proved by Lemma 14 and is also shown in [24]), but $(x^4 - x)/2 \notin \mathrm{Int}_{\mathbb{Q}}(\mathbb{H})$ because $(\mathbf{i}^4 - \mathbf{i})/2 \notin \mathbb{H}$. Thus, neither $\mathbb{L}$ nor $\mathbb{H}$ is $\mathrm{Int}_{\mathbb{Q}}$-decomposable.

There do exist Int_K-decomposable algebras other than matrix rings. The unifying property turns out to be that for each prime P of D, the residue ring A/PA is a direct sum of copies of a matrix ring.

Theorem 18 ([56, Thm. 6.1]) *Let D be a Dedekind domain with finite residue rings. Let A be a free D-algebra of finite rank. Then, A is Int_K-decomposable if and only if for each nonzero prime P of D, there exist positive integers n and t and a finite field $\mathbb{F}_q$ such that $A/PA \cong \bigoplus_{i=1}^t M_n(\mathbb{F}_q)$. In particular, if A is commutative, then A is Int_K-decomposable if and only if for each P there exists a finite field $\mathbb{F}_q$ such that $A/PA \cong \bigoplus_{i=1}^t \mathbb{F}_q$ for some t.*

Using this theorem, examples of Int_K-decomposable algebras can be produced that are not direct sums $\bigoplus_{i=1}^t M_n(D)$. However, the work in [56] relied on the assumption that A is free. Subsequent work by the author and Peruginelli resulted in [43], where a more general definition of Int_K-decomposability was established, and alternate characterizations of such algebras were given.

Definition 19 ([43, Def. 2.3]) Let D be a domain and A a torsion-free D-algebra. We say that A is Int_K*-decomposable* if the tensor product $\mathrm{Int}_K(A) \otimes_D A$ is isomorphic (as a D-algebra) to $\mathrm{Int}(A)$ via the map $\mathrm{Int}_K(A) \otimes_D A \to \mathrm{Int}(A)$ sending $f(x) \otimes a \mapsto f(x) \cdot a$.

Even without the assumption that A is free, Definition 19 formalizes the idea that $\mathrm{Int}(A)$ is generated by $\mathrm{Int}_K(A)$ and A. As shown in [43, Prop. 2.5], Definition 19 reduces to Definition 16 when A is a free D-algebra of finite rank. Theorem 18 carries over to the case where A is torsion-free and finitely generated as a D-module, but the flexibility of the tensor product definition of Int_K-decomposability allows for other characterizations of these algebras. First, instead of focusing on the residue rings of A, one can examine the completions $\widehat{A}_P$ of A at primes P of D.

Theorem 20 ([43, Thm. 2.10, Thm. 3.6]) *Let D be a Dedekind domain with finite residue rings. Let A be a torsion-free D-algebra that is finitely generated as D-module. Then, the following are equivalent:*

1. *A is* Int_K*-decomposable.*
2. *For each nonzero prime P of D, there exist positive integers n and t and a finite field $\mathbb{F}_q$ such that $A/PA \cong \bigoplus_{i=1}^{t} M_n(\mathbb{F}_q)$.*
3. *For each nonzero prime P of D, there exist positive integers n and t such that the P-adic completion $\widehat{A}_P$ of A satisfies $\widehat{A}_P \cong \bigoplus_{i=1}^{t} M_n(\widehat{T}_P)$, where $\widehat{T}_P$ is a complete discrete valuation ring with finite residue field and fraction field that is a finite unramified extension of $\widehat{K}_P$.*

Second, there is also a global variant [43, Thm. 4.10] of this theorem, which characterizes Int_K-decomposable algebras in terms of the extended K-algebra $B = K \otimes_D A$. The statement and proof of [43, Thm. 4.10] make extensive use of the theory of maximal orders (as presented in [48]). The statement of the theorem is quite technical and we omit it for the sake of space, but it does lead to some very clean corollaries when either $D = \mathbb{Z}$ or A is the ring of integers of a number field.

Corollary 21 ([43, Cor. 4.12]) *Let A be a torsion-free $\mathbb{Z}$-algebra that is finitely generated as a $\mathbb{Z}$-module. Then, A is* $\mathrm{Int}_{\mathbb{Q}}$*-decomposable if and only if there exist positive integers n and t such that $A \cong \bigoplus_{i=1}^{t} M_n(\mathbb{Z})$.*

Corollary 22 ([43, Cor. 4.11]) *Let $K \subseteq L$ be number fields with rings of integers O_K and O_L, respectively. Consider O_L as an O_K-algebra. Then*

1. *O_L is* Int_K*-decomposable if and only if L/K is an unramified Galois extension.*
2. $\mathrm{Int}_K(O_L) = \mathrm{Int}(O_K)$ *if and only if $L = K$.*

In particular, Corollary 22 shows that rings of integers of number fields can provide examples of Int_K-decomposable algebras that are not direct sums of matrix rings. For a noncommutative example, let p be an odd prime, $D = \mathbb{Z}_{(p)}$, and $A = D \oplus D\mathbf{i} \oplus D\mathbf{j} \oplus D\mathbf{k}$ (which is the localization of $\mathbb{L}$ at p). Then, $A/pA \cong \mathbb{L}/p\mathbb{L} \cong M_2(\mathbb{F}_p)$ (see [15, Sec. 2.5] or [26, Ex. 3A]), so A is $\mathrm{Int}_{\mathbb{Q}}$-decomposable. But, A cannot be isomorphic to a direct sum of matrix rings because it is contained in the division algebra $\mathbb{Q} \oplus \mathbb{Q}\mathbf{i} \oplus \mathbb{Q}\mathbf{j} \oplus \mathbb{Q}\mathbf{k}$.

We close this section by remarking that Int_K-decomposability is not the only form of decomposition possible with $\mathrm{Int}(A)$. For each $n > 0$, let $T_n(D)$ denote the ring of $n \times n$ upper triangular matrices with entries in D. Frisch studied $\mathrm{Int}(T_n(D))$ in [23] and proved the following theorem.

Theorem 23 ([23, Cor. 5.3]) *Let D be a domain. Let $T_n(D)$ be the ring of upper triangular matrices with entries in D. Then,*

$$\mathrm{Int}(T_n(D)) \cong \begin{pmatrix} \mathrm{Int}_K(T_n(D)) & \mathrm{Int}_K(T_{n-1}(D)) & \cdots & \mathrm{Int}_K(T_2(D)) & \mathrm{Int}_K(T_1(D)) \\ 0 & \mathrm{Int}_K(T_{n-1}(D)) & \cdots & \mathrm{Int}_K(T_2(D)) & \mathrm{Int}_K(T_1(D)) \\ & & \ddots & & \\ 0 & 0 & \cdots & \mathrm{Int}_K(T_2(D)) & \mathrm{Int}_K(T_1(D)) \\ 0 & 0 & \cdots & 0 & \mathrm{Int}_K(T_1(D)) \end{pmatrix}$$

Question 24 Will other algebras admit decompositions similar to that of Theorem 23?

5 Prüfer Conditions and Integral Closure

One of the long-standing questions regarding $\mathrm{Int}(D)$ was to determine necessary and sufficient conditions on D so that $\mathrm{Int}(D)$ is a Prüfer domain. For Noetherian D, this is the case if and only if D is a Dedekind domain with finite residue fields [11, 37]. For general D, a necessary and sufficient double-boundedness condition was given in [35].

It is natural to consider whether $\mathrm{Int}_K(A)$ can be a Prüfer domain. To date, both examples and non-examples of this phenomenon have been found. When A is the ring of integers of a number field, [36, Thm. 3.7] shows that $\mathrm{Int}_{\mathbb{Q}}(A)$ is a Prüfer domain. On the other hand, $\mathrm{Int}_K(M_n(D))$ is never Prüfer.

Lemma 25 *Let D be a domain. For all* $n \geq 2$, $\mathrm{Int}_K(M_n(D))$ *is not Prüfer.*

Proof This is an adaptation of an example given in [36, p. 2488]. Let $d \in D$ be a nonzero non-unit. Let $N \in M_n(D)$ be the nilpotent matrix with 1 in the $(1, n)$-entry and 0 elsewhere. Then, $N^2 = 0$ and $N/d \notin M_n(D)$. Now, it is well known [25, Chap. IV] that any overring of a Prüfer domain is again a Prüfer domain, and that Prüfer domains are integrally closed. Consider the ring $R = \mathrm{Int}_K(\{N\}, M_n(D)) = \{f \in K[x] \mid f(N) \in M_n(D)\}$. This is an overring of $\mathrm{Int}_K(M_n(D))$ in $K(x)$, so if $\mathrm{Int}_K(M_n(D))$ were Prüfer, then R would be Prüfer, and hence integrally closed. However, the polynomial $x^2/d^2 \in R$ but $x/d \notin R$. Thus, R is not integrally closed, and therefore $\mathrm{Int}_K(M_n(D))$ is not a Prüfer domain.

If D is Dedekind, then it is known [46, Cor. 3.4] that $\mathrm{Int}_K(M_n(D))$ is not even integrally closed when $n \geq 2$. However, [36, Thm. 4.6] shows that the integral closure of $\mathrm{Int}_{\mathbb{Q}}(M_n(\mathbb{Z}))$ is a Prüfer domain. Determining when this holds for $\mathrm{Int}_K(A)$ in general is an open question.

For simplicity, let us assume that D is integrally closed, so that $\mathrm{Int}(D)$ is also integrally closed [7, Prop. IV.4.1]. Since one of our assumptions on A is that $A \cap K = D$, we have the following containments:

$$D[x] \subseteq \mathrm{Int}_K(A) \subseteq \mathrm{Int}(D).$$

Moreover, if A can be finitely generated as a D-module by n elements, then by [45, Lem. 2.7] we have

$$D[x] \subseteq \mathrm{Int}_K(M_n(D)) \subseteq \mathrm{Int}_K(A) \subseteq \mathrm{Int}(D).$$

Thus, when D is integrally closed, a necessary condition for the integral closure of $\mathrm{Int}_K(A)$ to be Prüfer is that $\mathrm{Int}(D)$ be Prüfer. Furthermore, if D is such that the integral closure of $\mathrm{Int}_K(M_n(D))$ is Prüfer for all n, then the integral closure of $\mathrm{Int}_K(A)$ is Prüfer whenever A is finitely generated. This leads us to the following version of the Prüfer question for $\mathrm{Int}_K(A)$.

Question 26 Let D be an integrally closed domain and let A be a finitely generated D-algebra. When is the integral closure of $\mathrm{Int}_K(A)$ a Prüfer domain? In particular, is the integral closure of $\mathrm{Int}_K(M_n(D))$ a Prüfer domain?

The study of the integral closure of $\mathrm{Int}_K(A)$ is interesting in its own right, even without the connection to Prüfer domains. Different descriptions of the integral closure of $\mathrm{Int}_K(A)$ have been given, particularly for $\mathrm{Int}_{\mathbb{Q}}(M_n(\mathbb{Z}))$:

- [36, Thm. 4.6] and [41] Let $n \geq 2$ and let $\mathcal{O}_n$ be the set of algebraic integers of degree n. Then, the integral closure of $\mathrm{Int}_{\mathbb{Q}}(M_n(\mathbb{Z}))$ is equal to $\mathrm{Int}_{\mathbb{Q}}(\mathcal{O}_n) = \{f \in \mathbb{Q}[x] \mid f(\mathcal{O}_n) \subseteq \mathcal{O}_n\}$.
- [17, Prop. 2.1] Let p be a prime and let R_n be the maximal order in a division algebra of degree n^2 over the field of p-adic numbers. Then, the integral closure of $\mathrm{Int}_{\mathbb{Q}}(M_n(\mathbb{Z})_{(p)})$ is $\mathrm{Int}_{\mathbb{Q}}(R_n) = \{f \in \mathbb{Q}[x] \mid f(R_n) \subseteq R_n\}$.
- [44, Thm. 13] Let D be an integrally closed domain with finite residue rings. Let $A' \subseteq B$ be the set of elements of B that solve a monic polynomial in $D[x]$. Then, the integral closure of $\mathrm{Int}_K(A)$ is equal to $\mathrm{Int}_K(A, A') = \{f \in K[x] \mid f(A) \subseteq A'\}$.

It is also possible to give constructions for polynomials that lie in the integral closure of $\mathrm{Int}_K(A)$ but not in the ring itself. In [17], Evrard and Johnson derive explicit formulas for the p-sequences and p-orderings of the integral closure of$\mathrm{Int}_{\mathbb{Q}}(M_2(\mathbb{Z}))$. These are then used [17, Cor. 3.6] to prove that the degree 10 polynomial

$$x(x^2 + 2x + 2)(x - 1)(x^2 + 1)(x^2 - x + 1)(x^2 + x + 1)/4$$

is integral over $\mathrm{Int}_{\mathbb{Q}}(M_2(\mathbb{Z}))$ but is not in the ring itself. Furthermore, this is a polynomial of minimal degree with that property. More generally, for any discrete valuation ring V with fraction field K, [46, Construction 2.1] defines an explicit polynomial that is integral over $\mathrm{Int}_K(M_n(V))$ but not in the ring itself. In some instances, the same construction can be applied to other D-algebras such as the Lipschitz quaternions or the Hurwitz quaternions [46, Cor. 3.5, Ex. 3.6].

The questions considered in this section can also be asked of the ring $\mathrm{Int}_K(S, A) = \{f \in K[x] \mid f(S) \subseteq A\}$, where $S \subseteq A$. For a finite subset $S \subseteq D$, McQuillan proved [38] that $\mathrm{Int}(S, D)$ is Prüfer if and only if D is Prüfer (the classification of all such subsets $S \subseteq D$ is an open problem worthy of a survey of its own). Peruginelli has shown that the analogous theorem holds for the integral closure of $\mathrm{Int}_K(S, A)$.

Theorem 27 ([42, Co. 1.1]) *Let D be integrally closed and let $S \subseteq A$ be finite. Then, the integral closure of* $\mathrm{Int}_K(S, A)$ *is Prüfer if and only if D is Prüfer.*

6 Further Questions

In this final section, we introduce three topics for further study: integer-valued polynomials on nonassociative algebras, integer-valued rational functions on algebras, and integer-valued polynomials on subsets of algebras. Some work has been done on the last topic (as mentioned at the end of Sect. 5), but to date the first two areas are completely untouched. The work in this section should be considered "proof of concept" and will hopefully serve as motivation for further research.

6.1 Nonassociative Algebras

We know that it is possible to define and work with integer-valued polynomials on noncommutative algebras. What if we relax our assumptions further and consider polynomials that act on nonassociative algebras?

The following forms of "weak" associativity are discussed in standard references on nonassociative algebras such as [52]. A D-algebra A is called an *alternative algebra* if the relations $a(ab) = (aa)b$ and $(ba)a = b(aa)$ hold for all $a, b \in A$. It is a theorem of Artin [52, p. 18] that if A is an alternative algebra then $K[a, b]$ is associative for all $a, b \in K$. The algebra A is *power associative* if the usual addition rules for exponents hold for powers of an element $a \in A$; that is, if $a^{n+m} = a^n a^m$ for all $a \in A$. This is equivalent to $K[a]$ being associative for each $a \in A$. Every alternative algebra is power associative, but the converse does not hold in general.

Power associativity is sufficient for the definition of $\mathrm{Int}_K(A)$ to make sense.

Lemma 28 *Let A be a power associative D-algebra. Then,* $\mathrm{Int}_K(A)$ *is a well-defined (associative) subring of* $K[x]$.

Proof The elements of K are unaffected by the lack of associativity in A and B. Hence, for each $b \in B$, the algebra $K[b]$ is associative, and we can define evaluation of $f \in K[x]$ at $b \in B$ in the usual way. Consequently, for all $a \in A$, $\mathrm{Int}_K(\{a\}, A) = \{f \in K[x] \mid f(a) \in A\}$ is a well-defined (associative) subring of $K[x]$. Then, $\mathrm{Int}_K(A) = \bigcap_{a \in A} \mathrm{Int}_K(\{a\}, A)$ is also an associative subring of $K[x]$.

Question 29 Let A be a D-algebra that is power associative or alternative but not associative. To what extent (if at all) does the lack of associativity in A affect the algebraic properties of $\mathrm{Int}_K(A)$?

So, if A is power associative, then $\text{Int}_K(A)$ can be defined as usual. What about the analogue of $\text{Int}(A)$? To preserve some semblance of sanity, we will not attempt to work in any generality, but will instead confine ourselves to a particular alternative algebra over $\mathbb{Z}$: the integral octonions.

The octonions are a nonassociative extension of the quaternions.[1] References for this material include the book [14] and the survey article [1] by Baez. When defined over the real numbers, the octonions comprise an 8-dimensional (nonassociative) normed division algebra. We denote the basis for this algebra by $\{1, \mathbf{e}_1, \mathbf{e}_2, \mathbf{e}_3, \mathbf{e}_4, \mathbf{e}_5, \mathbf{e}_6, \mathbf{e}_7\}$. Each of the $\mathbf{e}_i$ satisfies $\mathbf{e}_i^2 = -1$, and $\mathbf{e}_i\mathbf{e}_j = -\mathbf{e}_j\mathbf{e}_i$ for all distinct i and j. Other multiplicative relations among the $\mathbf{e}_i$ can be expressed with a table, but it is more concise to express them via the Fano plane, as in [1, p. 152]:

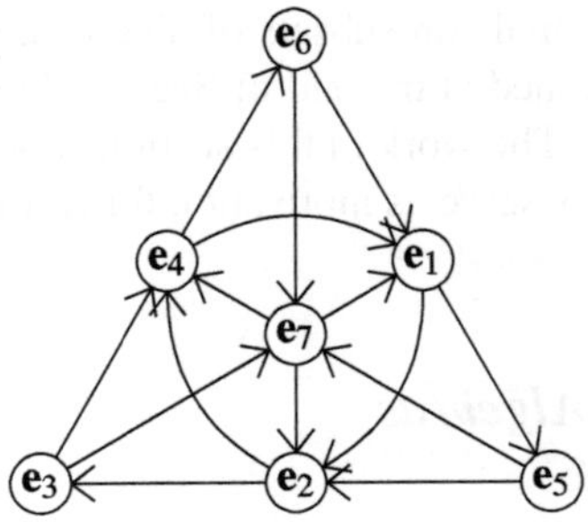

Straight line paths "wrap around," so that we imagine directed edges joining $\mathbf{e}_5$ to $\mathbf{e}_6$, $\mathbf{e}_1$ to $\mathbf{e}_3$, etc. Each pair of units $\{\mathbf{e}_i, \mathbf{e}_j\}$ appears as part of a straight line or circular cycle $\{\mathbf{e}_i, \mathbf{e}_j, \mathbf{e}_k\}$ for some k. Traversing a cycle in the direction of the arrows corresponds to the multiplication $\mathbf{e}_i\mathbf{e}_j = \mathbf{e}_k$; traversing a cycle in the opposite direction gives $\mathbf{e}_j\mathbf{e}_i = -\mathbf{e}_k$. Thus, for example, $\mathbf{e}_1\mathbf{e}_2 = \mathbf{e}_4$; $\mathbf{e}_6\mathbf{e}_1 = \mathbf{e}_5$; $\mathbf{e}_4\mathbf{e}_6 = \mathbf{e}_3$; and $\mathbf{e}_2\mathbf{e}_5 = -\mathbf{e}_3$.

With these rules, the nonassociativity of the octonions becomes evident, since for distinct i, j, and k such that $\mathbf{e}_i\mathbf{e}_j \neq \pm\mathbf{e}_k$ we have $\mathbf{e}_i(\mathbf{e}_j\mathbf{e}_k) = -(\mathbf{e}_i\mathbf{e}_j)\mathbf{e}_k$. Nevertheless, the octonions are an alternative algebra and exhibit a multiplicative norm. Given $\alpha = c_0 + \sum_{i=1}^{7} c_i\mathbf{e}_i$ with $c_i \in \mathbb{R}$, the conjugate of α is $\overline{\alpha} = c_0 - \sum_{i=1}^{7} c_i\mathbf{e}_i$, and the norm of α is $||\alpha|| = \alpha\overline{\alpha} = \sum_{i=0}^{7} c_i^2$. Then, for all α, β, we have $||\alpha\beta|| = ||\alpha|| \cdot ||\beta||$.

For our purposes, we define

$$\mathbb{O}_{\mathbb{Z}} = \{c_0 + c_1\mathbf{e}_1 + \cdots + c_7\mathbf{e}_7 \mid c_i \in \mathbb{Z}\}$$

and

$$\mathbb{O}_{\mathbb{Q}} = \{c_0 + c_1\mathbf{e}_1 + \cdots + c_7\mathbf{e}_7 \mid c_i \in \mathbb{Q}\}.$$

[1] To quote Baez [1]: "The quaternions, being noncommutative, are the eccentric cousin who is shunned at important family gatherings. But the octonions are the crazy old uncle nobody lets out of the attic."

We are interested in polynomials with coefficients in $\mathbb{O}_\mathbb{Q}$ that map elements of $\mathbb{O}_\mathbb{Z}$ back to $\mathbb{O}_\mathbb{Z}$, and we define

$$\text{Int}(\mathbb{O}_\mathbb{Z}) = \{f \in \mathbb{O}_\mathbb{Q}[x] \mid f(\mathbb{O}_\mathbb{Z}) \subseteq \mathbb{O}_\mathbb{Z}\}.$$

Since we are allowing polynomials with coefficients from a nonassociative ring, care must be taken when evaluating polynomials. We still insist that the indeterminate x commutes with all elements and that polynomials satisfy right-evaluation. To deal with the lack of associativity we adopt the following: given $f(x) = \sum_i a_i x^i \in \mathbb{O}_\mathbb{Q}[x]$ and $b \in \mathbb{O}_\mathbb{Q}$, we define $f(b) = \sum_i (a_i)(b^i)$. Thus, the powers b^i are evaluated first, then multiplied with the coefficients a_i, and finally the resulting monomials are added. In particular, if $f(x) = \sum_i a_i x^i \in \mathbb{O}_\mathbb{Q}[x]$, $g \in \mathbb{Q}[x]$, and $b \in \mathbb{O}_\mathbb{Q}$, then for each i all three of a_i, $g(b)$, and b^i lie in $\mathbb{Q}[a_i, b]$, which is associative because $\mathbb{O}_\mathbb{Q}$ is alternative. Hence, evaluation of fg is defined without ambiguity as $(fg)(b) = \sum_i a_i g(b) b^i$. With these conventions, we can derive information about the algebraic structure of $\text{Int}(\mathbb{O}_\mathbb{Z})$.

Lemma 30 $\text{Int}(\mathbb{O}_\mathbb{Z})$ *is a left* $\text{Int}_\mathbb{Q}(\mathbb{O}_\mathbb{Z})$*-module.*

Proof It is clear that $\text{Int}(\mathbb{O}_\mathbb{Z})$ is an Abelian group under addition, and $\text{Int}_\mathbb{Q}(\mathbb{O}_\mathbb{Z})$ is a commutative ring by Lemma 28. Let $f(x) = \sum_i a_i x^i \in \text{Int}(\mathbb{O}_\mathbb{Z})$, $g \in \text{Int}_\mathbb{Q}(\mathbb{O}_\mathbb{Z})$, and $a \in \mathbb{O}_\mathbb{Z}$. Then, keeping in mind the conventions of the last paragraph, we have

$$(gf)(a) = (fg)(a) = \sum_i a_i g(a) a^i = \sum_i a_i a^i g(a) = f(a)g(a).$$

Hence, $gf \in \text{Int}(\mathbb{O}_\mathbb{Z})$ and so $\text{Int}(\mathbb{O}_\mathbb{Z})$ is a left $\text{Int}_\mathbb{Q}(\mathbb{O}_\mathbb{Z})$-module.

Thus, $\text{Int}(\mathbb{O}_\mathbb{Z})$ has—at the minimum—a module structure, and contains both $\mathbb{O}_\mathbb{Z}[x]$ and $\text{Int}_\mathbb{Q}(\mathbb{O}_\mathbb{Z})$ as subrings. However, there exist polynomials in $\text{Int}(\mathbb{O}_\mathbb{Z})$ that are in neither $\mathbb{O}_\mathbb{Z}[x]$ nor $\text{Int}_\mathbb{Q}(\mathbb{O}_\mathbb{Z})$.

Lemma 31 *Let* $\mu = 1 + \mathbf{e}_1 + \cdots + \mathbf{e}_7$*. Then,* $\mu(x^2 - x)/2 \in \text{Int}(\mathbb{O}_\mathbb{Z})$.

Proof Let $R = \mathbb{O}_\mathbb{Z}/2\mathbb{O}_\mathbb{Z}$. It suffices to show that for each $\alpha \in R$ we have $\mu(\alpha^2 - \alpha) = 0$ (i.e., that $\mu(x^2 - x)$ is in the null ideal of R). Observe that since R has characteristic 2, R is a commutative and associative ring with unity: for all i, j, and k we have $\mathbf{e}_i\mathbf{e}_j = -\mathbf{e}_j\mathbf{e}_i = \mathbf{e}_j\mathbf{e}_i$ and $\mathbf{e}_i(\mathbf{e}_j\mathbf{e}_k) = -(\mathbf{e}_i\mathbf{e}_j)\mathbf{e}_k = (\mathbf{e}_i\mathbf{e}_j)\mathbf{e}_k$.

Now, for each $\alpha = c_0 + \sum_{i=1}^7 c_i\mathbf{e}_i \in R$, $||\alpha|| = 0$ or 1, depending on whether the number of nonzero c_i (for $0 \leq i \leq 7$) is even or odd. Moreover, the set of non-units of R forms a maximal ideal M, and $\alpha \in M$ if and only if $||\alpha|| = 0$. This ideal M is generated by $\{1 + \mathbf{e}_i \mid 1 \leq i \leq 7\}$, because for each $i \neq j$ there exists k such that $\mathbf{e}_i + \mathbf{e}_j = \mathbf{e}_i(1 + \mathbf{e}_k)$. Then, we have $\mu M = 0$, because $\mu(1 + \mathbf{e}_i) = \mu + \mu = 0$ for each i.

Next, given any $\alpha \in R$, either α or $\alpha - 1$ is in M. Thus, the polynomial $x^2 - x$ moves R into M. Since $\mu M = 0$, we conclude that $\mu(x^2 - x)$ sends all of R to 0, and therefore $\mu(x^2 - x)/2 \in \text{Int}(\mathbb{O}_\mathbb{Z})$.

So, despite the inconvenience of losing associativity, $\mathrm{Int}(\mathbb{O}_{\mathbb{Z}})$ contains nontrivial elements and has some manner of algebraic structure.

Question 32 What algebraic structure does $\mathrm{Int}(\mathbb{O}_{\mathbb{Z}})$ have? Is $\mathrm{Int}(\mathbb{O}_{\mathbb{Z}})$ closed under multiplication? Is it a nonassociative ring? Besides the polynomial of Lemma 31, what are some other elements of $\mathrm{Int}(\mathbb{O}_{\mathbb{Z}})$ that are not in $\mathbb{O}_{\mathbb{Z}}[x]$ or $\mathrm{Int}_{\mathbb{Q}}(\mathbb{O}_{\mathbb{Z}})$?

6.2 Integer-Valued Rational Functions[2]

A natural extension of the idea of integer-valued polynomials is that of integer-valued rational functions. Instead of considering $\mathrm{Int}(D) = \{f \in K[x] \mid f(D) \subseteq D\}$, one may study $\mathrm{Int}^R(D) = \{\varphi \in K(x) \mid \varphi(D) \subseteq D\}$, which is called the *ring of integer-valued rational functions on D*. The rings $\mathrm{Int}^R(D)$ are not as well-studied as $\mathrm{Int}(D)$, but some research has been conducted [6, 9] and $\mathrm{Int}^R(D)$ is discussed in [7, Chap. X].

Here, we consider what occurs with the analogous construction on a D-algebra. Hence, we make the following definitions.

Definition 33 A polynomial $g(x) \in K[x]$ is said to be *unit-valued on B* if $g(b) \in B^\times$ for all $b \in B$. The set of *integer-valued rational functions* $\mathrm{Int}_K^{\mathrm{R}}(A)$ is defined to be

$$\mathrm{Int}_K^{\mathrm{R}}(A) = \{f(x)/g(x) \in K(x) \mid g \text{ is unit-valued on } B \text{ and } f(a)/g(a) \in A \text{ for all } a \in A\}.$$

We denote the set of polynomials in $K[x]$ that are unit-valued on B by $\mathscr{U}$, and the set of polynomials in $D[x]$ that are unit-valued on A by $\mathscr{U}_D$.

The stipulation that g be unit-valued on B is made so that the evaluation of rational functions in $K(x)$ is well-defined at elements of B. While A and B may be noncommutative, for all $b \in B$ the elements $f(b)$ and $g(b)$ lie in the commutative algebra $K[b]$. Hence, if g is unit-valued on B, the fraction $f(b)/g(b)$ is equal to both $g(b)^{-1}f(b)$ and $f(b)g(b)^{-1}$, and so is well-defined. However, this requirement means that $\mathrm{Int}_K^{\mathrm{R}}(A)$ is simply a localization of $K[x]$: $\mathrm{Int}_K^{\mathrm{R}}(A) = \mathscr{U}^{-1}K[x]$.

Question 34 When A is noncommutative, can evaluation of a rational function $f(x)/g(x) \in K(x)$ and the definition of $\mathrm{Int}_K^{\mathrm{R}}(A)$ be well-defined without the assumption that g is unit-valued on B?

Even with the restriction that $\mathrm{Int}_K^{\mathrm{R}}(A) = \mathscr{U}^{-1}K[x]$, there are two interesting questions we can ask about $\mathrm{Int}_K^{\mathrm{R}}(A)$. First, it is clear that $\mathrm{Int}_K^{\mathrm{R}}(A)$ is a subring of $K(x)$ containing $\mathrm{Int}_K(A)$. Is it possible to have a strict containment $\mathrm{Int}_K(A) \subsetneqq \mathrm{Int}_K^{\mathrm{R}}(A)$? Second, we will always have $\mathscr{U}_D^{-1}D[x] \subseteq \mathrm{Int}_K^{\mathrm{R}}(A)$. Is it possible to have a strict containment $\mathscr{U}_D^{-1}D[x] \subsetneqq \mathrm{Int}_K^{\mathrm{R}}(A)$? The answer to both questions is yes, and can be demonstrated with examples involving matrix algebras.

[2]The results of this subsection are joint work with Alan Loper.

Lemma 35 *Let $n > 1$. Then, $g \in K[x]$ is unit-valued on the matrix ring $M_n(K)$ if and only if each irreducible factor of g has degree greater than n.*

Proof ($\Leftarrow$) Let $g \in K[x]$. Since the set of polynomials in $K[x]$ that is unit-valued on $M_n(K)$ is closed under multiplication, it suffices to consider the case where g itself is irreducible and $\deg g > n$.

Let $b \in M_n(K)$. The matrix $g(b)$ is invertible if and only if $g(b)$ does not have 0 as an eigenvalue. It is well known that the eigenvalues of $g(b)$ are precisely $g(\lambda)$, where λ is an eigenvalue of b. If $g(\lambda) = 0$ for some eigenvalue λ of b, then λ is algebraic over K and the minimal polynomial of λ divides g. Since g is irreducible, this minimal polynomial must have degree equal to $\deg g$. However, the eigenvalues of b are the roots of the minimal polynomial of b, which has degree at most n. Thus, $g(\lambda) \neq 0$ for each λ, and hence $g(b)$ is invertible. Since this holds for each b, g is unit-valued on $M_n(K)$.

($\Rightarrow$) Suppose that g has an irreducible factor h of degree less than or equal to n. Let $d = \deg h$, and let $C \in M_d(K)$ be the companion matrix for h. Let b be the block diagonal matrix $b = \left(\begin{smallmatrix} C & 0 \\ 0 & 0 \end{smallmatrix}\right) \in M_n(K)$. Then, $g(b) = \left(\begin{smallmatrix} 0 & 0 \\ 0 & g(0) \end{smallmatrix}\right)$, which has determinant 0 and hence is not invertible. Thus, g is not unit-valued on $M_n(K)$.

Now, we give an example of a domain D, an algebra A, and non-constant polynomials $f, g \in D[x]$ such that g is not unit-valued on A, but $\varphi = f/g \in \text{Int}_K^{\text{R}}(A)$. The rational function φ provides a positive answer to both questions posed prior to Lemma 35, since $\varphi \notin \text{Int}_K(A)$ because g is non-constant, and $\varphi \notin \mathscr{U}_D^{-1}D[x]$ because $g \notin \mathscr{U}_D$.

Example 36 Let F be a field, and let D be the valuation domain $F[t]_{(t)}$, which has fraction field $K = F(t)$. Let $A = M_2(D)$, and let $g(x) = x^4 + t \in D[x]$. Since the matrix $\alpha = \left(\begin{smallmatrix} 0 & t \\ 1 & 0 \end{smallmatrix}\right)$ squares to t, we have $g(\alpha) = t^2 + t$, which has determinant in tD. Since $t \notin D^\times$, $g(\alpha)$ is not a unit in A and so g is not unit-valued on A.

However, we claim that the rational function $\varphi(x) = t/(x^4 + t) \in K(x)$ is integer-valued on A. To see this, let $a \in A$. The polynomial $x^4 + t$ is irreducible over K by Eisenstein's Criterion, so $a^4 + t$ is a unit of $M_n(F)$ by Lemma 35. Let $a^4 + t = \left(\begin{smallmatrix} a_{11} & a_{12} \\ a_{21} & a_{22} \end{smallmatrix}\right)$. Then,

$$\varphi(a) = t(a^4 + t)^{-1} = \frac{t}{\det(a^4 + t)} \begin{pmatrix} a_{22} & -a_{12} \\ -a_{21} & a_{11} \end{pmatrix}.$$

Now, if $\det(a^4 + t) \notin tD$, then $\det(a^4 + t)$ is a unit of D and $\varphi(a) \in A$. So, assume that $\det(a^4 + t) \in tD$. We compute that $\det(a^4 + t) = \det(a^4) + t(\text{Tr}(a^4) + t)$, so this means that $\det(a^4) \in tD$. If $\text{Tr}(a^4) \notin tD$, then $t/(\det(a^4) + t(\text{Tr}(a^4) + t)) \in D$ and once again we have $\varphi(a) \in A$. So, assume that both $\det(a^4)$ and $\text{Tr}(a^4)$ are in tD.

The characteristic polynomial of a^4 is $x^2 - \text{Tr}(a^4)x + \det(a^4)$. It follows that $a^8 = \text{Tr}(a^4)a^4 + \det(a^4) \in tA$. Since $A/tA \cong M_2(D/tD) \cong M_2(F)$, we have (up to isomorphism) that $a^8 \equiv 0$ in $M_2(F)$. The maximum nilpotency of a matrix in a 2×2 matrix ring over a field is 2, so we must have $a^2 \equiv 0$ in $M_2(F)$. Thus, $a^2 \in tA$ and $a^2 = t\beta$ for some $\beta \in A$. Hence,

$$\varphi(a) = t/(a^4 + t) = t/(t^2\beta^2 + t) = 1/(t\beta + 1).$$

Since $\det(t\beta + 1) \notin tD$, the matrix $t\beta + 1$ is invertible in A. Hence, $\varphi(a) \in A$.

We have considered all the possible cases, so $\varphi(a) \in A$ for all $a \in A$. Therefore, the rational function $t/(x^4 + t)$ is integer-valued on A even though the denominator $x^4 + t$ is not unit-valued on A.

Question 37 For which domains D and D-algebras A do we have $\text{Int}_K^{\text{R}}(A) \neq \mathcal{U}_D^{-1}D[x]$?

6.3 Integer-Valued Polynomials on Subsets of Algebras

We close this survey by considering integer-valued polynomials on subsets of noncommutative algebras. Given a noncommutative D-algebra A and a subset $S \subseteq A$, we define $\text{Int}(S, A) = \{f \in B[x] \mid f(S) \subseteq A\}$. For any $S \subseteq A$, the corresponding set $\text{Int}_K(S, A) = \{f \in K[x] \mid f(S) \subseteq A\}$ is a commutative ring, and as with $\text{Int}(A)$ one may easily verify that $\text{Int}(S, A)$ always has the structure of a left $\text{Int}_K(S, A)$-module. Our main question is whether or not $\text{Int}(S, A)$ is closed under multiplication, and hence is a ring.

Definition 38 A subset $S \subseteq A$ is called a *ringset* if $\text{Int}(S, A)$ is a ring.

If S consists of central elements, then one may easily check that S is a ringset, but when S contains non-central elements it is nontrivial to determine whether or not $\text{Int}(S, A)$ is a ring. In Sect. 2, we conjectured that $\text{Int}(A) = \text{Int}(A, A)$ is always a ring when A is finitely generated as a D-module. However, it is easy to construct examples where $S \neq A$ and $\text{Int}(S, A)$ is not a ring.

Example 39 Let D be a Noetherian domain and assume that there exist $a, b \in A$ such that $ab \neq ba$. If $ab - ba \in dA$ for all $d \in D$, then $ab - ba = 0$ contrary to our assumption, so there exists a nonzero $d \in D$ such that $ab - ba \notin dA$. Let $f(x) = (x - a)/d$ and $g(x) = x - b$. Then, both f and g are elements of $\text{Int}(\{a\}, A)$, but their product is not, because $(fg)(x) = (x^2 - (a + b)x + ab)/d$ and $(fg)(a) = (-ba + ab)/d \notin A$. Thus, $\text{Int}(\{a\}, A)$ is not a ring.

This example shows that a singleton set $S = \{a\}$ is a ringset if and only if $a \in Z(A)$. As we shall see below, ringsets consisting of non-central elements do exist, and they can have as few as two elements. Before giving examples of such sets, we prove some general properties of ringsets.

Proposition 40 *Let $S, T \subseteq A$.*

1. *If S and T are ringsets, then $S \cup T$ is a ringset.*
2. *S is a ringset if and only if $fa \in \text{Int}(S, A)$ for all $f \in \text{Int}(S, A)$ and all $a \in A$.*
3. *If S is a ringset, then $f(usu^{-1}) \in A$ for all $f \in \text{Int}(S, A)$, $s \in S$, and $u \in A^{\times}$.*

4. *Assume that there exists a finite set $U = \{u_1, \ldots, u_n\}$ of units of A such that each element of A can be written as a sum $\sum_{j=1}^{n} c_j u_j$, where each $c_j \in Z(A)$. If $uSu^{-1} \subseteq S$ for all $u \in U$, then S is a ringset.*

Proof (1) is true because (as sets) we have $\text{Int}(S \cup T, A) = \text{Int}(S, A) \cap \text{Int}(T, A)$.

For (2), if S is a ringset, then $\text{Int}(S, A)$ is closed under multiplication, so $fa \in \text{Int}(S, A)$ because both f and a are in $\text{Int}(S, A)$. Conversely, assume that $\text{Int}(S, A)$ is closed under right multiplication by constants in A. Let $f, g \in \text{Int}(S, A)$ and let $s \in S$. Write $f(x) = \sum_i b_i x^i$ for some $b_i \in B$ and let $a = g(s) \in A$. Then, $(fg)(x) = \sum_i b_i g(x) x^i$, so $(fg)(s) = \sum_i b_i g(s) s^i = (fa)(s) \in A$ because $fa \in \text{Int}(S, A)$. Thus, $\text{Int}(S, A)$ is closed under multiplication, and hence is a ring.

For (3), when S is a ringset we have $fu \in \text{Int}(S, A)$ for all $f \in \text{Int}(S, A)$ and all $u \in A^\times$. Let $f(x) = \sum_i b_i x^i$. Then, for all $s \in S$, we have

$$(fu)(s) = \sum_i b_i u s^i = \sum_i b_i u s^i u^{-1} u = \sum_i b_i (usu^{-1})^i u = f(usu^{-1})u.$$

Since $(fu)(s) \in A$, so are $f(usu^{-1})u$ and $f(usu^{-1})$.

Finally, for (4), assume that $uSu^{-1} \subseteq S$ for all $u \in U$. Let $f(x) = \sum_i b_i x^i \in \text{Int}(S, A)$ and let $a = c_1 u_1 + \cdots + c_n u_n \in A$. Then, for all $s \in S$, we have

$$\begin{aligned}(fa)(s) &= \sum_i b_i (c_1 u_1 + \cdots + c_n u_n) s^i \\ &= c_1 \sum_i b_i u_1 s^i + \cdots + c_n \sum_i b_i u_n s^i \\ &= c_1 f(u_1 s u_1^{-1}) u_1 + \cdots + c_n f(u_n s u_n^{-1}) u_n.\end{aligned}$$

By part (3), each $c_i f(u_i s u_i^{-1}) u_i \in A$, so $(fa)(s) \in A$. Thus, S is a ringset by part (2).

Proposition 40 implies the following useful corollary.

Corollary 41 *Assume that A can be generated by central elements and a finite set of units. If S is a union of conjugacy classes, then S is a ringset.*

This corollary can be applied to many common choices of A such as matrix algebras, group rings, or certain quaternion algebras. We will give several examples involving subsets of the Lipschitz quaternions $\mathbb{L}$. These examples come from an unpublished portion of the author's doctoral dissertation [54], which examined the ringsets of $\mathbb{L}$ in greater detail.

Example 42 The unit group of the Lipschitz quaternions $\mathbb{L}$ is $\mathbb{L}^\times = \{\pm 1, \pm \mathbf{i}, \pm \mathbf{j}, \pm \mathbf{k}\}$. It is easily verified that $u\mathbf{i}u^{-1} = \pm \mathbf{i}$ for all $u \in \mathbb{L}^\times$, and similarly for $\mathbf{j}$ and $\mathbf{k}$. By Corollary 41, $S = \{\mathbf{i}, -\mathbf{i}\}$ is a ringset of $\mathbb{L}$.

The converse of Corollary 41 is not true, as we demonstrate with another example involving $\mathbb{L}$.

Example 43 Let $S = \{\mathbf{i}, \mathbf{j}\}$ and $T = \{\pm\mathbf{i}, \pm\mathbf{j}\}$. Then, T is a ringset by Corollary 41. We show that $\mathrm{Int}(S, A) = \mathrm{Int}(T, A)$, which implies that S is also a ringset.

Since $S \subseteq T$, we certainly have $\mathrm{Int}(T, A) \subseteq \mathrm{Int}(S, A)$. For the other inclusion, let $f \in \mathrm{Int}(S, A)$. Each element of T satisfies the polynomial $x^2 + 1$. Working over $B[x]$ (where $B = \mathbb{Q} \oplus \mathbb{Q}\mathbf{i} \oplus \mathbb{Q}\mathbf{j} \oplus \mathbb{Q}\mathbf{k}$), we may divide f by $x^2 + 1$ to get $f(x) = q(x)(x^2+1)+\alpha x+\beta$ for some $q(x), \alpha x+\beta \in B[x]$. By assumption, $f(\mathbf{i}) = \alpha\mathbf{i}+\beta \in A$ and $f(\mathbf{j}) = \alpha\mathbf{j} + \beta \in A$. So, A also contains $f(\mathbf{i}) - f(\mathbf{j}) = \alpha(\mathbf{i} - \mathbf{j})$ and

$$(f(\mathbf{i}) - f(\mathbf{j}))(-\mathbf{i} + \mathbf{j}) = \alpha(\mathbf{i} - \mathbf{j})(-\mathbf{i} + \mathbf{j}) = 2\alpha.$$

So, $2\alpha \in A$. This is relevant because

$$f(\mathbf{i}) - f(-\mathbf{i}) = (\alpha\mathbf{i} + \beta) - (\alpha(-\mathbf{i}) + \beta) = 2\alpha\mathbf{i}$$

so $f(-\mathbf{i}) \in A$. Similarly, $f(-\mathbf{j}) \in A$. It follows that $\mathrm{Int}(S, A) = \mathrm{Int}(T, A)$, and so S is a ringset.

Part (1) of Proposition 40 shows that unions of ringsets are ringsets. Unfortunately, the intersection of two ringsets need not be a ringset.

Example 44 By Examples 42 and 43, both $\{\mathbf{i}, -\mathbf{i}\}$ and $\{\mathbf{i}, \mathbf{j}\}$ are ringsets of $\mathbb{L}$. But, $\{\mathbf{i}, -\mathbf{i}\} \cap \{\mathbf{i}, \mathbf{j}\} = \{\mathbf{i}\}$ is not a ringset by Example 39.

The technique of Example 43 can be generalized to other subsets $S \subseteq \mathbb{L}$, but first we need to establish some basic properties of elements of $\mathbb{L}$. Given $a = a_0 + a_1\mathbf{i} + a_2\mathbf{j} + a_3\mathbf{k} \in \mathbb{L}$, the conjugate of a is $\overline{a} = a_0 - a_1\mathbf{i} - a_2\mathbf{j} - a_3\mathbf{k}$ and the norm of a is $||a|| = a\overline{a} = a_0^2 + a_1^2 + a_2^2 + a_3^2$. If $a \notin \mathbb{Z}$, then the minimal polynomial of a is $x^2 - 2a_0x + ||a||$, which has coefficients in $\mathbb{Z}$. Finally, note that for each $u \in \mathbb{L}^\times$, conjugating a by u merely changes some of the signs on a_1, a_2, and a_3. That is, $uau^{-1} = a_0 \pm a_1\mathbf{i} \pm a_2\mathbf{j} \pm a_3\mathbf{k}$. This means that $a - uau^{-1} \in 2\mathbb{L}$ for all $a \in \mathbb{L}$ and all $u \in \mathbb{L}^\times$.

Proposition 45 *Let $S \subseteq \mathbb{L}$ be such that $S \cap \mathbb{Z} = \varnothing$, each element of S has the same minimal polynomial, and* $\gcd(\{||a - b|| \mid a, b \in S\}) = 2$. *Then, S is a ringset.*

Proof Let $S^* = \{usu^{-1} \mid s \in S, u \in \mathbb{L}^\times\}$. Then, S^* is a ringset and $\mathrm{Int}(S^*, \mathbb{L}) \subseteq \mathrm{Int}(S, \mathbb{L})$. As in Example 43, we will show that $\mathrm{Int}(S, \mathbb{L}) = \mathrm{Int}(S^*, \mathbb{L})$.

Let $f \in \mathrm{Int}(S, \mathbb{L})$ and let $m(x) \in \mathbb{Z}[x]$ be the common minimal polynomial of the elements of S. Divide f by m to get $f(x) = q(x)m(x) + \alpha x + \beta$ for some $q(x), \alpha x + \beta \in B[x]$. Then, for all $a \in S^*$, we have $f(a) = \alpha a + \beta \in \mathbb{L}$, and if $a, b \in S$, then

$$f(a) - f(b) = \alpha(a - b) \in \mathbb{L}. \tag{2}$$

Now, the condition $\gcd(\{||a - b|| \mid a, b \in S\}) = 2$ means that there exist $a_1, \ldots, a_t, b_1, \ldots, b_t \in S$ such that

$$\gcd(||a_1 - b_1||, \ldots, ||a_t - b_t||) = 2.$$

Hence, there exist $n_1, \ldots, n_t \in \mathbb{Z}$ such that

$$2 = n_1||a_1 - b_1|| + \cdots + n_t||a_t - b_t||.$$

Thus,

$$\begin{aligned} 2\alpha &= n_1\alpha||a_1 - b_1|| + \cdots + n_t\alpha||a_t - b_t|| \\ &= n_1\alpha(a_1 - b_1)\overline{(a_1 - b_1)} + \cdots + n_t\alpha(a_t - b_t)\overline{(a_t - b_t)}. \end{aligned}$$

By (2), each $\alpha(a_i - b_i) \in \mathbb{L}$, so $2\alpha \in \mathbb{L}$.

Finally, given $uau^{-1} \in S^*$, we have $a - uau^{-1} \in 2\mathbb{L}$ and hence $f(a) - f(uau^{-1}) = \alpha(a - uau^{-1}) \in \mathbb{L}$. Since $f(a) \in \mathbb{L}$, we get $f(uau^{-1}) \in \mathbb{L}$. It follows that $\mathrm{Int}(S, \mathbb{L}) = \mathrm{Int}(S^*, \mathbb{L})$, and thus S is a ringset.

Clearly, the determination of ringsets in noncommutative algebras is a nontrivial problem. More theorems regarding finite ringsets of $\mathbb{L}$ can be found in [54], but for other algebras this question has not been explored.

Question 47 When A is noncommutative, which subsets of A are ringsets? In particular, what are the finite ringsets of the Hurwitz quaternions $\mathbb{H}$? What are the finite ringsets of the matrix algebra $M_n(D)$?

References

1. Baez, J.: The octonions. Bull. Am. Math. Soc. (N.S.) **39**(2), 145–205 (2002)
2. Bhargava, M.: P-orderings and polynomial functions on arbitrary subsets of Dedekind rings. J. Reine Angew. Math. **490**, 101–127 (1997)
3. Brawley, J., Carlitz, L., Levine, J.: Scalar polynomial functions on the $n \times n$ matrices over a finite field. Linear Algebra Appl. **10**, 199–217 (1975)
4. Brizolis, D.: On the ratios of integer-valued polynomials over any algebraic number field. Am. Math. Mon. **81**, 997–999 (1974)
5. Cahen, J.-P.: Polynomes à valeurs entières. Can. J. Math. **24**, 747–754 (1972)
6. Cahen, J.-P.: Fractions rationnelles à valeurs entières. Ann. Sci. Univ. Clermont Math. **16**, 85–100 (1978)
7. Cahen, J.-P., Chabert, J.-L.: Integer-Valued Polynomials. American Mathematical Society Surveys and Monographs, vol. 48. American Mathematical Society, Providence (1997)
8. Cahen, J.-P., Chabert, J.-L.: Old Problems and New Questions Around Integer-Valued Polynomials and Factorial Sequences. Multiplicative Ideal Theory in Commutative Algebra, vol. 89–108. Springer, New York (2006)
9. Cahen, P.-J., Loper, K.A.: Rings of integer-valued rational functions. J. Pure Appl. Algebra **131**(2), 179–193 (1998)
10. Chabert, J.-L.: Anneaux de "polynômes à valeurs entières" et anneaux de Fatou. Bull. Soc. Math. Fr. **99**, 273–283 (1971)
11. Chabert, J.-L.: Un anneau de Prüfer. J. Algebra **107**(1), 1–16 (1987)
12. Chabert, J.-L., Peruginelli, G.: Polynomial overrings of Int($\mathbb{Z}$). J. Commut. Algebra **8**(1), 1–28 (2016)
13. Cigliola, A., Loper, K.A., Werner, N.J.: Split quaternions and integer-valued polynomials. Commun. Algebra **43**(1), 182–196 (2015)

14. Conway, J.H., Smith, D.: On Quaternions and Octonions: Their Geometry, Arithmetic, and Symmetry. A K Peters, Natick, MA (2003)
15. Davidoff, G., Sarnak, P., Valette, A.: Elementary Number Theory, Group Theory, and Ramanujan Graphs. London Mathematical Society Student Texts, vol. 55. Cambridge University Press, Cambridge (2003)
16. Elliott, J.: Integer-valued polynomials on commutative rings and modules. Preprint. https://arxiv.org/abs/1608.00171
17. Evrard, S., Johnson, K.: The ring of integer valued polynomials on 2 × 2 matrices and its integral closure. J. Algebra **441**, 660–677 (2015)
18. Evrard, S., Fares, Y., Johnson, K.: Integer valued polynomials on lower triangular integer matrices. Monatsh. Math. **170**(2), 147–160 (2013)
19. Frisch, S.: Polynomial separation of points in algebras. In: Arithmetical Properties of Commutative Rings and Monoids. Lect. Notes Pure Appl. Math., vol 241, pp. 253–259. Chapman & Hall/CRC, Boca Raton, FL (2005)
20. Frisch, S.: Integer-valued polynomials on algebras: a survey. Actes du CIRM **2**(2), 27–32 (2010)
21. Frisch, S.: Integer-valued polynomials on algebras. J. Algebra **373**, 414–425 (2013)
22. Frisch, S.: Corrigendum to "Integer-valued polynomials on algebras" [J. Algebra 373 (2013) 414–425]. J. Algebra **412**, 282 (2014)
23. Frisch, S.: Polynomial functions on upper triangular matrix algebras. To appear in Monatsh. Math. https://arxiv.org/abs/1605.06027 (Dec. 2016)
24. Gerboud, G.: Polynômes à Valuers Entières sur L'anneu des Quaternions de Hurwitz. Preprint (1998)
25. Gilmer, R.: Multiplicative Ideal Theory. Queen's Papers in Pure and Applied Mathematics, vol. 90. Kingston, Ontario (1992)
26. Goodearl, K., Warfield Jr., R.: An Introduction to Noncommutative Noetherian Rings. Lond. Math. Soc. Stud. Texts, vol. 16. Cambridge University Press, Cambridge (2004)
27. Heidaryan, B., Longo, M., Peruginelli, G.: Galois structure on integral valued polynomials. J. Number Theory **171**, 198–212 (2017)
28. Heuberger, C., Rissner, R.: Computing J-ideals of a matrix over a principal ideal domain. Preprint (2016)
29. Johnson, K.: P-orderings of noncommutative rings. Proc. Am. Math. Soc. **143**(8), 3265–3279 (2015)
30. Johnson, K., Pavlovski, M.: Integer-valued polynomials on the Hurwitz ring of integral quaternions. Commun. Algebra **40**(11), 4171–4176 (2012)
31. Lam, T.Y.: Lectures on Modules and Rings. Graduate Texts in Mathematics, vol. 189. Springer, New York (1999)
32. Lam, T.Y.: A First Course in Noncommutative Rings. Graduate Texts in Mathematics, vol. 131, 2nd edn. Springer, New York (2001)
33. Lam, T.Y., Leroy, A.: Wedderburn polynomials over division rings. I. J. Pure Appl. Algebra **186**(1), 43–76 (2004)
34. Littlewood, D.E., Richardson, A.R.: Fermat's equation in real quaternions. Proc. Lond. Math. Soc. **S2–32**(1), 235–240 (1931)
35. Loper, K.A.: A classification of all D such that Int(D) is a Prüfer domain. Proc. Am. Math. Soc. **126**(3), 657–660 (1998)
36. Loper, K.A., Werner, N.J.: Generalized rings of integer-valued polynomials. J. Number Theory **132**(11), 2481–2490 (2012)
37. McQuillan, D.L.: On Prüfer domains of polynomials. J. Reine Angew. Math. **358**, 162–178 (1985)
38. McQuillan, D.L.: Rings of integer-valued polynomials determined by finite sets. Proc. Roy. Irish Acad. Sect. A **85**(2), 177–184 (1985)
39. Ostrowski, A.: Über ganzwertige Polynome in algebraischen Zahlkörpen. J. Reine Angew. Math. **149**, 117–124 (1919)

40. Peruginelli, G.: Integer-valued polynomials over matrices and divided differences. Monatsh. Math. **173**(4), 559–571 (2014)
41. Peruginelli, G.: Integral-valued polynomials over sets of algebraic integers of bounded degree. J. Number Theory **137**, 241–255 (2014)
42. Peruginelli, G.: The ring of polynomials integral-valued over a finite set of integral elements. J. Commut. Algebra **8**(1), 113–141 (2016)
43. Peruginelli, G., Werner, N.J.: Decomposition of integer-valued polynomial algebras. https://arxiv.org/abs/1604.08337 (under review)
44. Peruginelli, G., Werner, N.J.: Integral closure of rings of integer-valued polynomials on algebras. In: Commutative Algebra, pp. 293–305, Springer, New York (2014)
45. Peruginelli, G., Werner, N.J.: Non-triviality conditions for integer-valued polynomials on algebras. Monatsh. Math. 13pp. (2016). http://dx.doi.org/10.1007/s00605-016-0951-8
46. Peruginelli, G., Werner, N.J.: Properly integral polynomials over the ring of integer-valued polynomials on a matrix ring. J. Algebra **460**, 320–339 (2016)
47. Pólya, G.: Über ganzwertige Polynome in algebraischen Zahlkörpen. J. Reine Angew. Math. **149**, 97–116 (1919)
48. Reiner, I.: Maximal Orders. London Mathematical Society Monographs. New Series, vol. 28. The Clarendon Press, Oxford University Press, Oxford (2003). Corrected reprint of the 1975 original.
49. Rissner, R.: Null ideals of matrices over residue class rings of principal ideal domains. Linear Algebra Appl. **494**, 44–69 (2016)
50. Rogers, M., Wickham, C.: Polynomials inducing the zero function on local rings. Int. Electron. J. Algebra, 22 (2017), To appear.
51. Rush, D.E.: The conditions $\mathrm{Int}(R) \subseteq R_S[X]$ and $\mathrm{Int}(R_S) = \mathrm{Int}(R)_S$ for integer-valued polynomials. J. Pure Appl. Algebra **125**(1–3), 287–303 (1998)
52. Schafter, R.: An Introduction to Nonassociative Algebras. Dover, New York (1995). Corrected reprint of the 1966 original.
53. Werner, N.J.: Integer-valued polynomials over quaternion rings. J. Algebra **324**(7), 1754–1769 (2010)
54. Werner, N.J.: Integer-valued polynomials over quaternion rings, 156 pp Ph.D. Thesis The Ohio State University. (2010)
55. Werner, N.J.: Integer-valued polynomials over matrix rings. Commun. Algebra **40**(12), 4717–4726 (2012)
56. Werner, N.J.: Int-decomposable algebras. J. Pure Appl. Algebra **218**(10), 1806–1819 (2014)
57. Werner, N.J.: Polynomials that kill each element of a finite ring. J. Algebra Appl. **13**(3), 1350111, 12 pp. (2014)

MIX
Papier aus verantwortungsvollen Quellen
Paper from responsible sources
FSC® C105338

If you have any concerns about our products,
you can contact us on
ProductSafety@springernature.com

In case Publisher is established outside the EU,
the EU authorized representative is:
Springer Nature Customer Service Center GmbH
Europaplatz 3, 69115 Heidelberg, Germany

Printed by Libri Plureos GmbH
in Hamburg, Germany